Applied Agroclimatology

Applied Agroclimatology

O.P. Bishnoi

Oxford Book Company
Jaipur ▪ New Delhi

Published By

Oxford Book Company

Regd. Off. "Bony Residency", Gate No. 2
Opp. Tilak Public School, Vishweswariya Nagar,
Gopalpura Road, Jaipur - 302018, India
Ph: 0141 - 2761280, Telefax: 2761381
email - oxfordbookjpr@gmail.com
website - oxfordbookcompany.com

Delhi Office: 102, 1st Floor, Satyam House,
4327/3, Ansari Road, Darya Ganj,
New Delhi - 110002, India
Ph: 011-45652440

Distributed by
KEYA INTERNATIONAL PRESS LLP
E-mail: keyainternationalpressllp@gmail.com

Published in India, 2023

ISBN : 9789380179117

Typeset by Shivangi Computers, Jaipur

Printed and bound in India by Replika Press Pvt. Ltd.

Contents

Preface

This book on "Applied Agroclimatology" has a comprehensive account of agroclimatological related problems and phenomenon in agriculture. This book has been developed to benefit the postgraduate students in Meteorology/Agricultural Meteorology covering the syllabus of University Grant Commission/Agricultural Scientist Recruitment Board, and other agrorelated agencies confronting with problems as how to use and interpret the vast continuously recorded climatological and agrometeorological data in relation to their specialized aspects, i.e., to exploit the climate as a resource in agriculture, hydrological problems, agroclimatic classification for specific crops/vegetations, comfortness to human and animals to generate maximum output, growth and productivity of crops and animals, development and adaptation of new crops with their genotypes, characteristics of monsoon rainfall and its variability due to ENSO phenomena, to avert weather vagaries agroadvisory services on national perspective, and impact of climate change and climatic variability on agriculture.

This book will provide a sound base on how to use and interpret the vast continuously recorded agrometeorological data in order to express the biological behaviour interactions in a prediction format/nomograms to benefit the user agencies. Agricultural productions is deeply influenced by the agroclimatic variables throughout the crop/vegetation growing season, and therefore agroadvisory services forms an important component to derive the maximum benefit for exploiting climate as a non-cash resource.

With advancements in local, national, regional and global climate models for weather forecast at short, medium and long range levels, consistent efforts have been made during the last two decades to tailor forecasts prescriptions to meet the requirements in agriculture to benefit the farmers and agrorelated agencies so that they are in a position to take quick right decisions with confidence. Advancements in agricultural technology should counteract the increasing climate change and climatic variability issues threatening to meet the basic human and animal needs in terms of adequate food, fodder, water, energy, safe shelter, and a healthy environment. Through well-defined interactions and appropriate choice of interacting agricultural technologies will certainly reduce the long term agricultural vulnerability to climate change, there is a strong need to train national user agencies in agroclimatological applications in agriculture so that they can share knowledge with extension agencies toiling to improve the economic status of farming systems existing with farmers.

This book has been divided into two volumes. Volume I, after a brief introduction (Chapter 1) describes the use of agroclimatological data for different problems surrounding the farmers and other agencies. On how to use and interpret data in a scientific acceptable format (Chapter 2), hydrological problems influencing the water availability for utilization (Chapter 3), agroclimatic classification for better use of natural gift of soil and climate natural resources to exploit the resource for reproduction of better adapted species of crops and animals (Chapter 4), use of available thermal energy to develop efficient thermal genotypes (Chapter 5), adaptation of human and animals to environments to generate maximum output (Chapter 6), vagaries of monsoon features influencing agricultural production and its variability due to droughts and floods linked with ENSO phenomena (Chapter 7), and different types of microclimates in different agricultural systems (Chapter 8). Volume II has been devoted to use the qualitative climate prediction prescription scenarios in agriculture for the benefits of the farmers in his local format as advisory services (Chapter 9). The burning issue of global climate change with its impact on agriculture/vegetations is very crucial to sustain agricultural production (Chapter 10).

The author wishes to express his gratitude to all the reviewers for their constructive criticisms. Thanks are also to all who helped in bringing out this book with full support and cooperation particularly my wife Anila Bishnoi and the full team of M/s Oxford Book Company, Jaipur.

O.P. Bishnoi

Volume-I

Agroclimatological Problems

1

Introduction

Applied Agroclimatology/Agrometeorology is an intimately related science of Agricultural Meteorology dealing with agroclimatological/agrometeorological applications directly related to economic development of farmers. How to integrate the impact of weather features over a period of time on the growth, development and performance of different living organisms for economic benefit of farmer's community? Agroclimate is an element of physical environment of a vegetative field, whereas agroclimatology explores the relationships of climatology to other biosciences and considers its potential for adapting and improving the productivity of living organisms confronting the complexities of climate change and modification of agroclimates in different cropping systems. The proliferation of specialized studies have led to a wider scope with variety of climate related problems and phenomena termed as agromicroclimatology, building design climatology, human comfort climatology, statistical climatology, hydroclimatology, animal comfort climatology, urban climatology, air pollution climatology, bioclimatology—an unending chain in our surroundings. Therefore in the words of Austin Bourke (1968) the scope of agroclimatology "the tasks of agroclimatologist as to apply every relevant climatological skill to helping the farmers to make most efficient use of his environment, with the prime aim of improving agricultural production, both in quantity and quality." The statement reflects the role of an agroclimatologist in diverse facets of agricultural activities from sowing to harvest in variable monsoonal weather.

In agricultural production there is largely small scale microclimate modification, therefore how the atmosphere is affected by the production of food (Monteith, 1985) is our approach for agrometeorological forecast advisories. The meteorological information can be used for weather forecasting for agriculture, agroclimatology and agrometeorological forecasting (WMO, 1981). How to interpret the vast continuously recorded climatological data? Database is enriched with data on climatological, phenological events in plants and animals; genetic resources of plants and animals; crop pest, diseases and weeds; crop yield and their variations. Therefore operational agricultural aspects are gaining momentum to help the agro-related user agencies. With advancement of scientific technology in weather forecasting and fast growing age of information technology, it as essential that weather based advisories for daily/weekly periods will provide advance planning for day to day field operational and seasonal management activities. Short, medium and long term weather forecasting will help in developing the applied agroclimatological aspects to provide a wide varieties of strategies for operational applications in different agroclimatic zones and farming systems.

India is a vast country enriched with a wide range of 20 different agroecological zones (Sehgal and Mandal, 1996) with different agricultural potentials and constraints. Being an agricultural developing sustainable production country, it is essential to make efficient use of natural resources, maintaining soil health without deteriorating the environment to meet the demands of burgeoning population. National Centre for Medium Range Weather Forecasting (NCMRWF) is helping the farmers and agro-related user agencies through agroadvisory service in 127 agrometeorological zones (Fig. 1.1)—a smaller areal homogeneous unit having difference in farming systems; crop production potential; variation in growing season and crop varieties; performance of crops and animals; pest, disease and weed problems; differences of soil lypes, irrigation potentials, etc. Therefore due to the variability in different zones it would demand for a different types of agrometeorological format service by developing the productivity of farming systems to approach the potential productivity levels, and even would further benefit if it could be executed at homogeneous clustering of villages.

Use of crop simulation and statistical models for strategic and tactical applications to evaluate alternate strategies or any tactical decision during crop growing season would help the farmers for better crop management and higher production. An advance yield forecasting would help in food policy issues and food security system at state, regional and national level. Simulation models also help in projecting the impact of climatic variability and climate change with different expected scenarios. The increase of global air temperature by 0.6°C over past 100 years is due to mismanagement of natural resources to meet the demands of rising population. The studies carried out by Inter-Governmental Panel on Climate Change (IPCC) of United Nations has confirmed the global warming trend of 1.4 to 5.8°C in different parts of the world particularly in post-monsoon and winter seasons with increasing variations as we go away from equator towards the poles in both hemispheres. Such global climate changes will affect agriculture considerately through its direct and indirect affect on crops, cropping systems, area under different crops, livestock population, pest and disease occurrence and their population ultimately affecting the food security. The impact of climate change will strongly influence the crop management efficiency in terms of fertilizer uptake, carbon sequestration, availability and management of soil water use, changes in homogenous

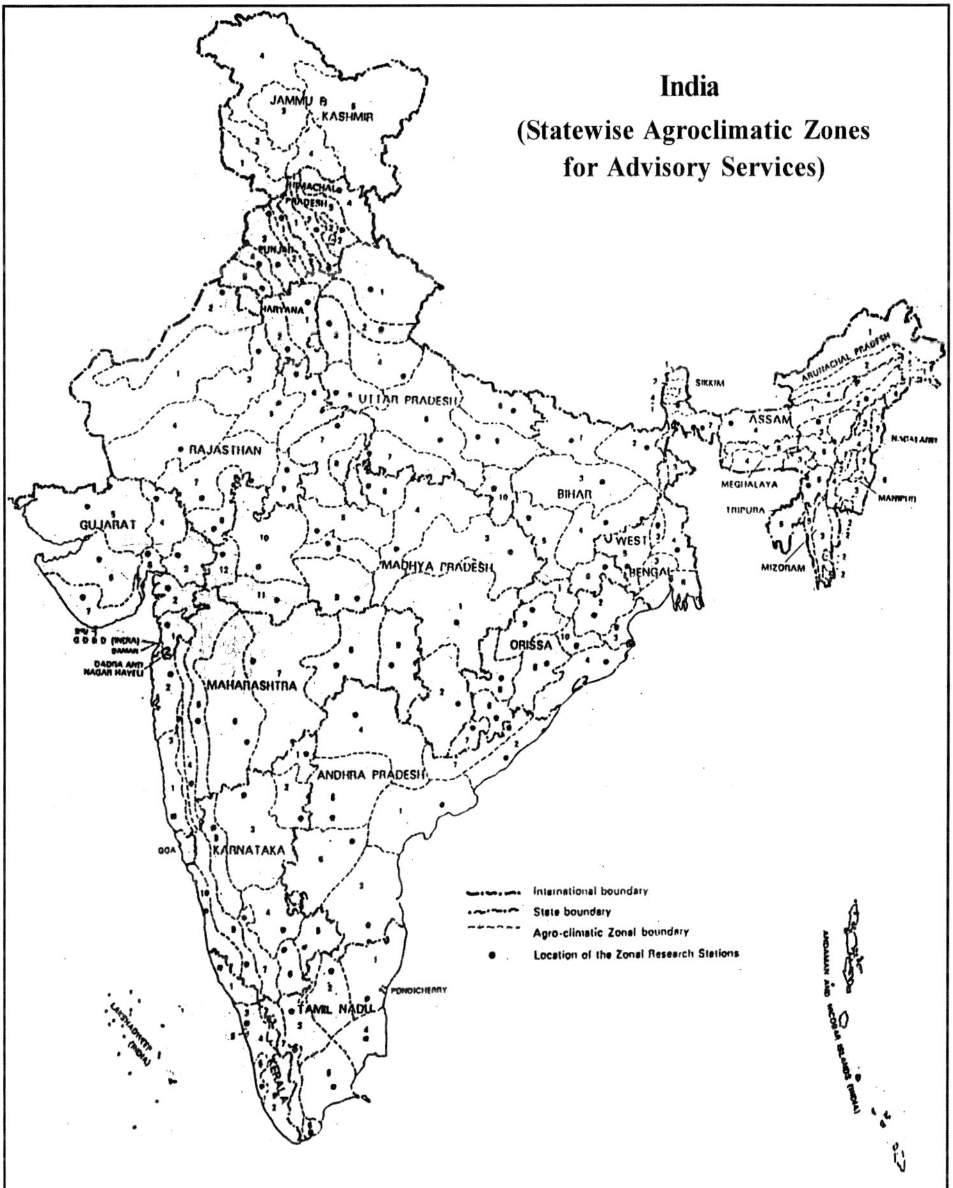

Fig. 1.1: Agroclimatic zones for weather advisories. DST (1986)

soil-climatic or agroecological zones for land use, improving rainwater management through watershed development etc. Therefore agroclimatology has to play a significant role to play in these continuous monitoring nature of different applications.

Animal and human comfort zones require an optimal environmental range for their better growth and maximum output. Domestic animals depending on species and productivity level have an optimal adaptive zone for optimum growth, lactation and reproductive functions. Heat or cold stress influence adversely on thermal regulation or heat balance, energy balance, water balance, and endocrine changes influencing the reduced performance. The thermal aspects of climate will influence the body's temperature, heat balance, harmonal and water balance with resultant effects on growth, reproduction, milk and egg production. Several associated factors are also influenced such as feed intake, feed utilization, milk quality production and reproductivily with the microclimatic environment of the animal. Comfort zones for livestock breeds are highly desired for effective utilization of milk yield estimation and for developing an early warning for heat or cold stress for livestock and poultry in the prevailing environment. Therefore weather based strategies will improve the advisory services.

Climatic water budget applications play fundamental role in many aspects of irrigation scheduling, climatic classification, to quantify run off conditions with influence on land use patterns, drought and wet spells quantifications, and soil moisture retention and storage capabilities. Climate and comfort for increasing human efficiency to generate more output, climate and crop scheduling with use of thermal time concept, and climate and building orientation are some of the applications involving climatic variability influences.

Monsoon climatology have remained a challenge in the Asian region as agriculture is strongly influenced by its hidden features. With the continuous records of experimental data under several National and International Programmes, now it has been possible to understand the wave nature of this global phenomenon and the impact of El Neno-Southern Oscillation. A clear understanding of the different monsoonal features would benefit agrometeorologist to bring out its impact on farmer's field. A transcription of monsoonal features in farmer's format and agricultural landuse in monsoon Asia needs no emphasis (Yoshino, 1984).

Microclimatic management and manipulation in traditional farming have been a continuous effort during the last 50 years for better use the crop microclimate in landuse planning, agroecological zoning, and improving the crop production with operational results for farmers. Agrometeorological forecasting for crop development, frost protection, crops and animal's pest and disease development and their multiplication, and crop yield assessments will improve the farmer's economy. Quantitative long-term onfarm advice on such management and agronomic manipulation practices termed as weather advisories (Stigter, 1984) will increase quantity, quality and protection of crops and animals. (Stigter and Weiss, 1986). Such operational weather advisories for small, medium and large holding farmers, adapted to their physical and social environment, would certainly improve socio-economic status and sustained yield agroecosystem. Pest-disease control based on a through understanding of weather elements—pest/disease relationships in relation to growth cycles of the organisms through control conditions experimentations.

During the last 30 years, Agricultural Meteorology has advanced from a geographical description to a quantitative science based on physical and biophysiological principles and hypothesis (Monteith, 1975; Bishnoi, 2007). However, agrometeorological applications for farmers have gained momentum during the last 15 years with the cooperation between two services-agrometeorological service (Under Ministry of Earth Sciences) and agricultural services (under Ministry of Agriculture) confronting with third component of unconvinced farming community? The development of an effective agrometeorological education and applied training programmes remained a challenge for farming community and needs to be developed for specific farming systems in different agroecological zones due to substantial difference in agricultural potentials and constraints in farming productivity. The agrometeorological components will differ due to different requirements at national, regional and international levels, but the basic principles of adoption remains the same.

Our ultimate aim is to assist the agricultural industry to produce economally qualitative commodities with reduction of climatic risks and sustainable system environmentally compatible with the requirement of society. Therefore it is necessary to rectify the lack of agrometeorological/agroclimatological knowledge at all levels in agricultural and horticultural industry. It has been demonstrated that 35 percent of the interannual variability in crop yields is due to climatic variations, and another 10 to 15 per cent to the interaction between climate and planting dates, fertilizer applications, and pest and diseases control. Therefore it as anticipated that agrometeorological/agroclimatological applications be given greater emphasis to minimize risk and maximize economic benefits by paying adequate attention on natural and managed agroecosystems. Development of risk management strategies based on climatic data is desired to interpret the crop-climate interactions in order to provide crop insurance to the farmers and raising the livelihood of farmers (Stigter, 2008). A close coordination between agrometeorological forecast services and agricultural extension and research workers would be richly rewarding for farming community. It is hoped that services of proven use to the agricultural community will succeed in the 21st century as free market economy gaining ground with adoption of WTO regulations.

References

Auston Bourke, P.M. 1968. The Aims of Agrometeorology. Agroclimatelogical Methods—Proc. Reading Symposium, UNESCO, p. 11-15.

Bishnoi, O.P. 2007. Principles of Agricultural Meteorology, Oxford Book Company, Jaipur, India.

Department of Science & Technology, 1986. Science and Technology Project: National Centre for Medium Range Weather Forecasting and Development of Agrometeorological Services, DST, New Delhi, p. 1-43.

Monteith, J.L. 1975. Vegetation and Atmosphere, Vol. 1, Principles; Vol. 2 Case Studies, Academic Press, London, U.K.

Monteith, J.L. 1985. Weather and Agriculture, In: Recent Advances in Meteorology and Physical Oceanography, Royal Meteorological Soc., Barksh., England.

Sehgal, J.L. and D.K. Mandal 1996. Agro-ecological Regions of India and Climate Change. In: Climate Variability and Agriculture. (eds.), V.P. Abrol. S. Gadgil and G.B. Pant. Narora Publishing House, New Delhi, p. 204-22.

Stigter, C.J. 1984. Forecasting and Weather Services for Agriculture in Africa. In: Hawskworth (ed.).

Stigter C.J. and A. Weiss 1986. In Quest of Tropical Micrometeorology for on Farm Weather Advisories, Guest Editorial. In: Agril. For. Meteorol. 36, p. 289-96.

Stigter, C.J. 1986. Microclimate Management and Manipulation in Traditional Farming. Report of Working Group on Microclimate Management and Manipulation in Traditional Farming. WCP–1762, WMO, Geneva.

Stiger, Kees 2008. Coping with climate risks in agriculture needs farmer oriented research and extension policies. Scientia Agricola (Brazil).

WMO, 1981. Guide to Agricultural Meteorological Practices. WMO No. 134, Geneva, Switzerland.

Yoshino, M.M. (ed.) 1984. Climate and Agricultural Landuse in Monsoon Asia, University of Tokyo Press, Tokyo, Japan.

2

Agroclimatic Statistics

2.1 Introduction

During the past 50 years agrometeorological data is being continuously recorded, whereas rainfall data is available for more than 100 years. This voluminous data would require some technique for its summarization and interpretation in the user's language. Statistical techniques have been developed to quantify correlation, regression, proability analysis, periodicity and risk analysis of any climatological data series. Complex statistical procedures are available for different types of agroclimatological problems of crop yield forecasting, extreme value determination for hydrological problems, incomplete gamma distribution for rainfall probability etc. Since analysis of any climatological series of radiation, temperature, rainfall, wind speed etc., involves the use of the statistical terms, therefore it is worthwhile to present these basic techniques of agroclimatic statistics with use of agrometeorological data for clear interpretation and their physical significance in order to solve and have the physical concept of agricultural problems.

2.2 Statistical Treatment of Data Series

(i) **Mean**: Mean is the average value of a given climatological data series. For example let us use daily temperature data as given in Appendix I where daily maximum and minimum temperature data is presented. The daily mean temperature is average of maximum and minimum temperature values. But the mean monthly temperature is the sum of individual daily means divided by the number of observations. Therefore using the January mean temperature data for the past 40 years of record (Appendix I) at Hisar (Latitude 29°10'N, Longitude 75°40'E) we can arrive at following mean temperature values with evaluation of

$$\text{Arithmetic Mean (A.M.)} = \sum_{i=1}^{n} \frac{x_i}{n} \quad \ldots (2.1)$$

Mean maximum January temperature = 19.9°C
Mean minimum January temperature = 4.7°C
Mean January temperature = 12.3°C

Sometimes geometric mean is also used in climatological data which is determined as

$$\text{Geometric Mean (GM)} = (x_1 . x_2 . x_n)^{\frac{1}{n}} \quad \ldots (2.2)$$

(ii) **Frequency**: The temperature data can be grouped into a number of classes conveniently determined by the term 5 log N, where N is the total number of observations. For ex. When N is 40, the number of classes are around 8. Since mean January temperature range from 10.1 to 14.2°C during 40 years of record, a difference of 4.1°C results in a class interval of 0.5°C. Using Appendix I data of mean January temperature the frequency distribution table is given below:

Class Interval °C	Class Mak	Class Frequency	Cumulative Frequency	Percent Frequency	Cumulative Percent Frequency
10.1-10.5	10.25	2	2	5.0	5.0
10.6-11.0	11.75	2	4	5.0	10.0
11.1-11.5	11.25	2	6	5.0	15.0
11.6-12.0	11.75	14	20	35.0	50.0
12.1-12.5	12.25	3	23	7.5	57.5
12.6-13.0	12.75	7	30	17.5	75.0
13.1-13.5	13.25	4	34	10.0	85.0
13.6-14.0	13.75	5	39	12.5	97.5
14.1-14.5	14.25	1	40	2.5	100.0

A graphical representation of the frequency distribution is called histogram or bargraph. Each observation is represented by a unit of area. The area under the bar for each class interval is equal to the number of observations in that class interval. Percentage frequency is determined with percentage of all observations which fall into each class interval. By accumulating the frequencies in each class interval from the lowest to highest interval we can obtain the cumulative frequency distribution as well as the cumulative percent frequency. The graphical representation of the cumulative frequency distribution is called ogive.

(iii) **Normals**: World Meteorological Organization (WMO) adopted a figure of atleast 30 years latest period as the proper time interval for computation of normal of the meteorological variables (Thom, 1971). Therefore normal refers to the airthmatic mean of a meteorological variable during the latest 30 years period. Due to variable nature of climatic record, with the existence of trends, the length of record with minimum variance is taken as the normal. Therefore the idea of normal needs to be developed in the light of climatic trends and variations.

(iv) **Median and mode**: These are measures of central value of a data series. Median is the middle value in a series of observations when ranked in ascending or descending order. It is taken as the 50th percentile value as obtained from an ogive. There may be slight differences in the results because median is a particular observation and involves discrete numbers, whereas an ogive will provide the plotted observations and are in the form of a continuous variable. The median value of the January maximum, minimum and mean temperature (Appendix 1) is 19.9, 4.7 and 12.0 °C respectively.

The mode is that value (of class interval if data is grouped) which occur most frequently. For ex. mode value of January maximum, minimum and mean temperatures is 19.8, 3.8, 4.5 and 6.3°C (each three times) respectively.

(v) **Measure of variation**: It is a measure of spread or dispersion of the data around the central value. Temperature, being a conservative quantity, is evenly distributed around the mean value, whereas rainfall vary widely around the mean value and the distribution is skewed due to more lower values as compared to few higher values. The most commonly used measure of variation is the standard deviation (SD), a measure of dispersion of individual observed values (x_i) about their mean ($\bar{x}$). S.D. is expressed as

$$S.D. = \sqrt{\sum_{i=1}^{n} \frac{(x_1 - \bar{x})^2}{N}}, \qquad \ldots (2.3)$$

where N is number of observations. S.D. normally refers to standard deviation of a given sample, whereas σ referred to the standard deviation of the universe from which the sample has been taken. σ is estimated as

$$\sigma = S.D. \cdot \sqrt{\frac{N}{N-1}}$$

The SD and σ for January maximum, minimum and mean temperature are 1.24, 1.26, 1.48, and 1.50, 0.99, 1.00 °C respectively.

Variance of the distribution is the square of the standard deviation expressed as

$$(S.D.)^2 = \frac{1}{N}\left[\sum_{i=1}^{n}(x_1 - \bar{x})^2\right] = \frac{1}{N-1}\left[\Sigma x_1^{\ 2} - \frac{1}{N}(\Sigma x_i)^2\right] \qquad \ldots(2.4)$$

The universe variance is

$$\sigma^2 = \frac{1}{N-1}\left[\sum_{i=1}^{n}(x_1 - \bar{x})^2\right] \qquad \ldots(2.5)$$

(vi) **Coefficient of variation (CV):** In data series, the deviation of parameters from mean can be expressed in absolute terms, however it is not clear that magnitude of deviation depends in part on the magnitude of the mean itself. When comparison between different location values are made this influence of actual magnitude of the mean is not convenient, therefore this influence is eliminated by expressing the deviation value as a percentage of the mean expressed as $\frac{S.D.}{mean} \times 100$. Coefficient of variations of January maximum, minimum and mean temperature is 6.2, 26.3 and 8.1 per cent respectively.

(vii) **Skewness:** It is a measure of symmetry of distribution around mean. When mean, mode, median values are equal there is perfect symmetry otherwise there is lack of symmetry. There is negative skewness if mean < medium < mode and positive skewness when mean > medium > mode. For January temperatures of Appendix I data series the skewness values are assessed as S_k = 3 (Mean – Mode)/S.D. and their values are

Skewness for maximum temperature in January = +0.24

Skewness for minimum temperature in January = –0.40

Skewness for mean temperature in January = –1.82

In case of rainfall data a large number of lower values will provide negative deviations from the mean which are balanced by just few large positive deviations. The lowest limit of rainfall data can be zero when there no rainfall, but there is no limit for upper value. There will always be more number of rainfall observations below the mean than above it and will show a positive skewness.

(viii) **Homogeneity of data:** Many statistical tests require the homogeneity of the data series. A simple well known method for the assessment of homogeneity by considering the sign of deviation of each individual value from the mean of the series. If one deviation is followed by another deviation of the same sign, a sequence (S) of deviation is created otherwise with a different sign, a change of deviation (C) is recorded. If the series is homogeneous or without trend, the difference between the number of sequences (S) in the time series and the number of changes (C) should be zero within the limits of the probable error depending on the length of series expressed as

$$S - C = 0 \pm \sqrt{N-1} \qquad \dots (2.6)$$

If in a given data series this equation is satisfied, then the data series is homogeneous. If the umber of sequences is significantly greater than the number of changes, then some trend in the data is indicated. If the number of changes is significantly greater than the number of sequences, then some form of oscillation of the mean is indicated.

Example 1:

Test the homogeneity of January temperature data of Appendix 1.

In case of maximum temperature $S - C = 0 \pm \sqrt{N-1}$ becomes $6 = \pm 6.24$

In case of minimum temperature $-1 = \pm 6.24$

In case of mean temperature $+1 = \pm 6.24$

Therefore temperature data is homogeneous series.

(ix) **Correlation Coefficients**: The coefficient of correlation between two variables is a mathematical expression of their degree of relationship truly representing the cause and effect. Good correlation coefficients may occur even when there is no real relationship between the variables or when both are related to a third variable but not directly to each other. For ex. Solar radiation is the source of energy at the earth's surface and its intensity should be related to the net radiation available at the earth's surface. Similarly adveted energy directly influence the air temperature throughout day and night and are strongly correlated. To quantify correlation coefficient (r) and linear regression equation it is necessary to compute Σx, Σx^2, Σy, Σy^2, Σxy of the two variables (x and y) involved. A linear expression $y = m\,x + c$ and correlation coefficient are given by

$$r = \frac{N\Sigma xy - \Sigma x.\Sigma y}{\sqrt{N\Sigma x^2 - (\Sigma x)^2}.\sqrt{N\Sigma y^2 - (\Sigma y)^2}} \qquad \dots (2.7)$$

$$m = \frac{N\Sigma xy - \Sigma x.\Sigma y}{N\Sigma x^2 - (\Sigma x)^2} \qquad \dots (2.8)$$

$$c = \frac{\Sigma y.\Sigma x^2 - \Sigma x.\Sigma xy}{N\Sigma x^2 - (\Sigma x)^2} \qquad \dots (2.9)$$

Example 2:

Determine mean, mode, median, standard deviation, variance, coefficient of variation and skewness for the monsoon seasonal rainfall data at a sub-tropical semiarid location.

Year	Rainfall (mm) (x)	Year	Rainfall (mm) (x)	Year	Rainfall (mm) (x)
1916	539	1931	363	1946	301
1917	860	1932	184	1947	582
1918	171	1933	618	1948	372
1919	249	1934	357	1949	285
1920	178	1935	432	1950	349
1921	302	1936	222	1951	172
1922	413	1937	513	1952	368
1923	536	1938	117	1953	329
1924	363	1939	207	1954	369
1925	335	1940	273	1955	390
1926	446	1941	150	1956	389
1927	379	1942	627	1957	536
1928	309	1943	216	1958	604
1929	564	1944	383	1959	502
1930	429	1945	749	1960	656

Solution:

$N = 45$
$\sum x^2 = 8375020$ mm
$\sum x = 17888$ mm
$\bar{x} = 397.5$ mm
σ_N (SD) = 167.6 mm
$\sigma_{N-1} = 169.5$ mm

Variance = 28096 mm^2
C.V. (%) = 42
Mode = 363, 563 mm
Median = 363 mm
Skewness = –0.93

Example 3:

Test the homogeneity of rainfall data series given in example 2 and exercise 3.

Solution:

Mean of total series rainfall data = 384.4 mm
Total number of observations = (N) = 90
Total number of sequences = (S) = 41
Total number of changes = (C) = 48

Therefore $S - C = 0 \pm\sqrt{N-1}$

$$41 - 48 = \pm\sqrt{90-1}$$

$$-7 = \pm 9.43$$

Thus –7 lies within the limits of –9.43 to 9.43, therefore homogenous.

Example 4:

Use the following data to develop a linear regression expression between open pan evaporation rate and daily temperature data:

(a) Plot the data alongwith regression best fit line.

(b) Express what does square of the correlation coefficient explain?

(c) Determine standard error of estimate and draw their plotted lines in the graph.

Open Pan Evaporation (mm/day) (y)	Temperature °C (x)
1.0	13.5
1.7	14.3
3.0	14.0
2.8	15.5
4.1	18.0
3.7	18.4
5.6	18.5
5.6	20.8
4.0	22.7
5.4	22.4
6.5	18.5
6.1	19.0
7.6	21.0
6.5	22.5
10.1	34.4
10.4	31.4
12.8	32.0
11.7	30.5
10.3	31.5
7.9	32.6
11.7	31.7
14.0	36.0
13.1	32.5
15.2	35.1
19.4	35.3
7.4	32.9
10.0	30.5
2.0	14.1
3.6	12.0
1.0	12.1

Solution:

$N = 30$

$\Sigma y^2 = 2302.3$ $\Sigma x^2 = 19356.89$

$\Sigma y = 224.2$ $\Sigma x = 721.7$

$\Sigma xy = 6404.31$ $\bar{x} = 24.056$

$\bar{y} = 7.47$ $\sigma_{N,x} = 8.155$

$\sigma_{N,y} = 4.57$ $\sigma_{N-1,x} = 8.294$

$\sigma_{N-1}, y = 4.65$

Regression coefficient A = –4.714

Regression coefficient B = 0.5066

Correlation coefficient r = 0.90

$\hat{x} = 11.1$°C

$\hat{y} = 0.9$ (mm/day)

Linear regression equation

Open pan evaporation rate (mm/day) = –4.714 + 0.5066 Temperature (in °C), $R^2 = 0.81$

(a) Plotting of data is shown in Fig. 2.1

(b) 81 percent of evaporation rate is explained by the temperature variable

(c) $$SE = \pm\, Sy, x = \sqrt{\frac{\Sigma y^2 - A\ \Sigma y - B\Sigma xy}{N}}$$

$$= \sqrt{\frac{2302.6 + 4.714 \times 224.2 - 0.5066 \times 6404.31}{30}}$$

$$= \sqrt{\frac{2302.6 + 1056.88 - 3244.42}{30}}$$

$$S.E. = \sqrt{\frac{115.06}{30}} = \pm 1.96$$

Draw two dashed lines parallel to the regression line at a distance of ± Sy, x from the line in the y direction. Sixty percent of data points should fall within those two lines.

(x) **Moving averages**: Climatic data may exhibit some trends or persistent long term movement of the series due to climatic changes and global warming. The rainfall or temperature observations are fluctuating strongly due to vast changes in global man made activities and there may be some trend. A moving average technique is used to reduce the some of the fluctuations to provide a smooth curve to indicate any possible trend.

Example 5:

Compute the 2 to 5 year moving average of monsoon season temperature values and determine trend if any with use of following data.

Year	Temperature	°C Moving average temperature °C			
		5yr	4yr	3yr	2yr
1971	30.8	-	-	-	-
1972	31.0	-	-	31.2	30.9
1973	31.9	30.8	32.0	32.4	31.4
1974	34.4	31.5	30.9	30.8	33.2
1975	26.1	32.9	31.7	31.6	30.3
1976	34.2	31.5	33.1	32.7	30.2
1977	37.8	32.3	30.8	32.4	36.0
1978	25.1	35.7	34.0	34.0	31.5
1979	39.0	36.0	36.1	35.5	32.0
1980	42.5	35.6	35.4	38.8	40.8
1981	35.0	38.1	38.3	38.0	38.8
1982	36.5	36.5	37.8	36.2	35.8
1983	37.2	34.2	35.0	34.9	36.8
1984	31.1	33.4	34.0	33.1	34.2
1985	31.0	31.9	32.6	31.1	31.1
1986	31.1	30.7	30.6	30.4	31.0
1987	29.1	30.6	30.6	30.4	30.1
1988	31.0	31.4	30.5	30.3	30.1
1989	30.9	32.2	31.5	32.3	30.9
1990	34.9	32.1	33.0	33.6	32.5
1991	35.0	32.0	33.0	33.7	35.0
1992	31.1	32.5	32.3	31.4	33.0
1993	28.1	31.3	32.0	30.9	29.6
1994	33.6	30.0	30.4	30.2	30.8
1995	28.8	31.1	29.8	30.3	31.2
1996	28.6	31.7	31.8	31.2	28.7
1997	36.3	31.4	31.3	32.1	32.5
1998	31.3	31.8	32.1	33.3	33.8
1999	32.2	31.9	32.6	31.3	31.8
2000	30.4	31.5	30.8	30.7	31.3
2001	29.4	31.2	31.5	31.3	29.9
2002	34.1	31.4	30.9	31.1	31.8
2003	29.7	31.4	31.6	32.4	31.9
2004	33.3	-	31.9	31.1	31.5
2005	30.3	-	-	-	31.8

The temperature of S.W. monsoon season varied from 25.1 to 42.5°C, however 2 to 5 years moving averages varied from 28.7 to 40.8°C, 30.3 to 38.8°C, 30.4 to 38.3°C and 30.0

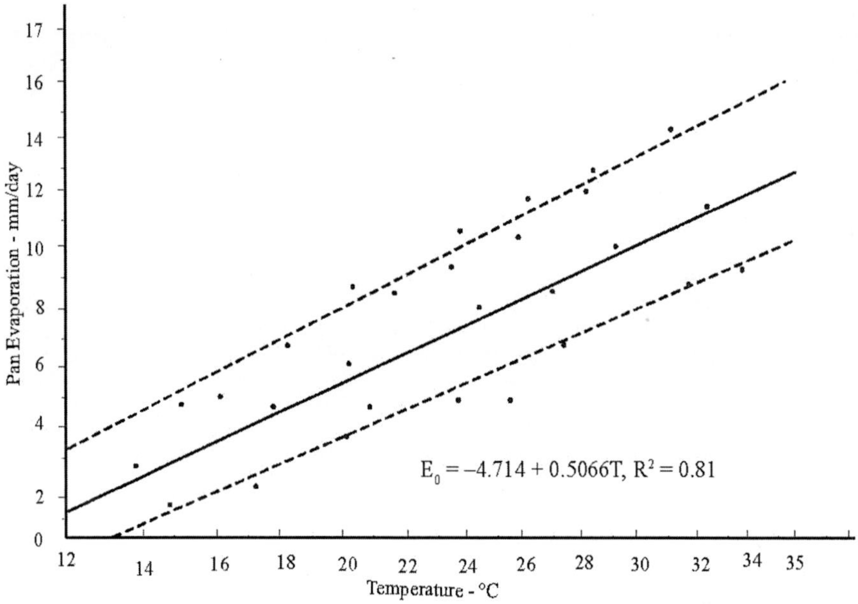

Fig. 2.1: Relationship beween open pan evaporation and air temperature.

to 38.1°C respectively. The range continuously reduced from 17.4 to 12.1, 8.5, 7.9 and 8.1°C respectively and becomes stable around 4/5th year moving average. This moving average reveals the maxima around 1980/81 with lower values before and after this period. Plotting of these moving averages Fig. 2.2 shows that a variable trend or oscillation exists in temperature data. 1978 to 1983 years have been excessively warm with slight warming around 1989 to 1992, and still less warming 1996 to 1999. However a cooling trend prevails around 1975-76, 1986-88, 1994-95, and 2000-01 years.

2.3 Normal Distribution

The meteorological variables, except rainfall, follow the statistically normal distribution. The normal distribution is a symmetrical, bell-shaped frequency distribution curve extending to infinity in both positive and negative directions. The normal curve which fits for a particular data series is a curve having same area, same mean and same standard deviation as the sample itself. The idealized expression for frequency or probability density function is

$$F(x) = \frac{Ni}{\sigma\sqrt{2\pi}} \exp\left\{\frac{-(x_i - \bar{x})^2}{2\sigma^2}\right\} \quad \ldots (2.10)$$

where F(x) is probability function for a given x_i, N is total number observations, i is range of class interval x as real value between – ∝ to +∝. Two constant parameters – mean ($\bar{x}$) and standard deviation (σ) are important in normal distribution when $x_i = \bar{x}$, $F(x)_{max} = \frac{N_i}{\sigma\sqrt{2\pi}}$.

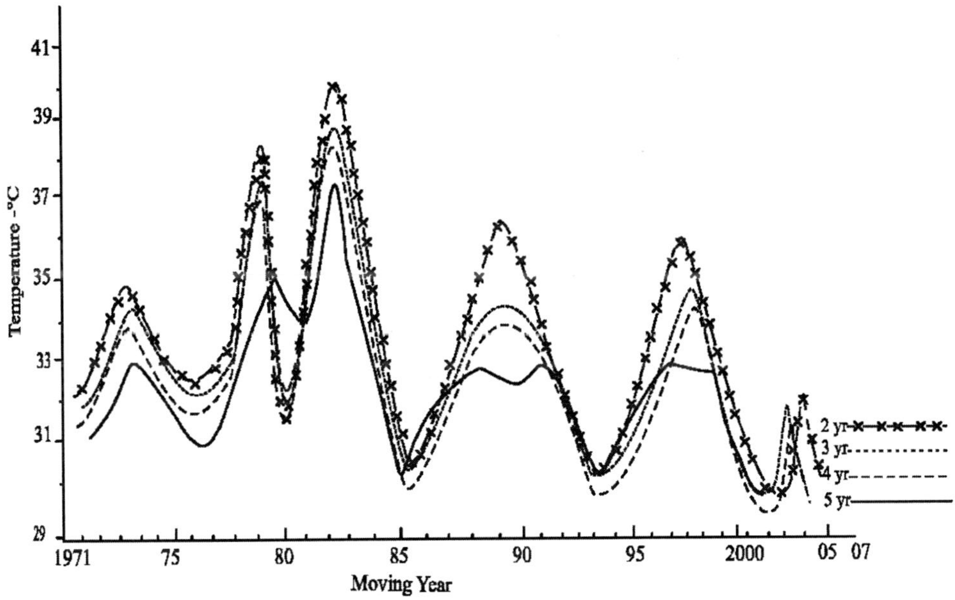

Fig. 2.2: Variation of temperature with moving averages over 2 to 5 years.

The normal curve corresponds to two parameters $\bar{x}$ and σ can have different normal curves. Standard tables are provided for distribution with $\bar{x} = 0$ and $\sigma = 1$. Therefore to use this table with $\bar{x}$ and σ, there is need to rescale x so that $\bar{x} = 0$ and $\sigma = 1$. The rescaled measurement will be expressed in terms a parameter z expressed as $z = (x - \bar{x})/\sigma$ where z is called standard normal variate. z can be transformed to x scale by the expression $x = \bar{x} + \sigma z$. Table is provided to give area under the normal curve for different values of standard normal variate to determine probabilities under the normal distribution. Therefore using this variable z, the distribution function becomes

$$F(z) = \frac{1}{\sqrt{2\pi}} \int_{-\alpha}^{\alpha} \exp\left(\frac{-z^2}{2}\right) dz = 1 \qquad \ldots (2.11)$$

The cumulative distribution tables are more frequently used so as to provide for any value of z the area under the curve from origin to z point. If A is area under the curve (Appendix II) then the probabilities are given by

Probability of a Value	Expression
(i) between 0 and z	A
(ii) between –z and z	2A
(iii) outside the interval (-z, z)	1 – 2A
(iv) less than +z	0.5 + A

(v) less than –z 0.5 – A
(vi) greater than + z 0.5 – A
(vii) greater than –z 0.5 + A

Appendix II provides the area under standard normal curve between 0 to z for computing probability. Appendix III provides the relative height of the ordinate.

Example 6:

Use the following monsoon seasonal temperature data of a dry subhumed, subtropical location to compute frequency to fit the normal curve and determine the frequency distribution.

Year	Temperature °C
1968	25.6
1969	30.9
1970	27.7
1971	30.8
1972	31.0
1973	31.9
1974	34.4
1975	26.1
1976	34.2
1977	37.8
1978	26.1
1979	39.0
1980	39.5
1981	35.0
1982	36.5
1983	37.2
1984	31.1
1985	31.0
1986	31.1
1987	29.1
1988	30.0
1989	29.1
1990	34.9
1991	31.1
1992	29.1
1993	33.6
1994	28.8
1995	28.6
1996	36.3
1997	32.5

1998	32.4
1999	33.6
2000	26.5
2001	34.5
2002	36.0
2003	32.2
2004	31.6
2005	28.5

Solution:

Temperature data varies from 25.6 to 39.5°C having a range of 13.9°C with 38 observations. Therefore we will have a class interval of 2°C in order to have adequate graphical points.

Class Interval °C	Class Mark (X_i) °C	x_i Values	$\left\lvert\frac{x_i - \bar{x}}{\sigma}\right\rvert$	Relative height of Ordinate Y_z*	Computed Frequency Distribution	Actual Frequency Distribution
23.1 to 25.0	24	-	–2.22	0.0339	0.7 ~ 1	-
25.1 to 27.0	26	4	–1.66	0.1006	2.13 ~ 2	4
27.1 to 29.0	28	4	–1.1086	0.2155	4.56 ~ 5	4
29.1 to 31.0	30	8	–0.0055	0.3429	7.26 ~ 7	8
31.0 to 33.0	32	8	+0.0055	0.3989	8.44 ~ 8	8
33.1 to 35.0	34	7	0.5626	0.3410	7.22 ~ 7	7
35.1 to 37.0	36	3	1.1198	0.2131	4.51 ~ 5	3
37.1 to 39.0	38	3	1.6769	0.0973	2.06 ~ 2	3
39.1 to 41.0	40	1	2.2334	0.0332	0.70 ~ 1	1
41.1 to 43.0	42	-	2.7911	0.0081	0.17 ~ 0	-
					38	38

$N = 38$

$\sum x^2 = 39357.55$

$\sum x = 1215.3$

$\bar{x} = 31.98$

$\sigma_N = 3.59$

$\sigma_{N-1} = \sim 3.64$

$$\frac{Ni}{\sigma} = 21.1699$$

$$\text{Computed frequency} = \frac{Ni}{\sigma} \times \text{Relative height of ordinate.}$$

The simplest method of testing a given distribution for normality is to plot the distribution on a probability graph. The cumulative normal distribution will appear as a straight line.

Probability for any value of the meteorological variable can be determined from the graph so drawn.

Since extreme temperatures are very important during summer as well as in winter influencing the agroclimatic conditions, Baier and Russelo (1968) determined the risks from 99 to 1 percent for summer maximum and winter minimum temperatures from the long records of meteorological data. Here the term risk is synomuous with term probability that the temperature will exceed the value specified value in the Table 2.1 (Bishnoi, 1985). Therefore impact of low temperature for frost conditions can be assessed for taking suitable measures to avoid detrimental effects on winter crops. Occurrence of low air temperature is very harmful for plant leaves, buds, flowers and fruits for many ornamental and cultivated native plants when the tender parts attain 4 to 5°C lower temperature than the air (Shaw, 1964).

Table 2.1

Weekly minimum temperature for a given risk at Hisar (°C)

Standard	Lowest	Percentage Risk							Highest
Weeks	Observed	99	90	60	50	10	5	1	Observed
44	9.1	8.3	9.8	11.4	11.8	13.6	14.2	15.7	15.8
45	8.6	4.3	6.9	9.5	10.1	13.3	14.3	16.0	14.4
46	5.4	4.7	6.4	8.2	8.6	10.8	11.3	12.8	13.7
47	4.0	3.0	4.8	6.8	7.2	9.5	10.3	11.6	12.6
48	3.2	2.4	3.7	5.1	5.4	7.1	7.7	8.6	9.3
49	4.2	2.0	3.2	4.5	4.8	6.3	6.7	7.6	8.2
50	2.2	1.7	2.9	4.1	4.4	5.8	6.3	7.1	8.1
51	2.5	1.4	2.6	3.8	4.1	5.6	6.1	7.0	7.5
52	-0.5	1.1	2.2	3.2	3.4	4.7	5.1	5.8	6.1
1	1.1	0.8	1.7	2.6	2.8	3.9	4.2	4.8	5.5
2	1.9	0.8	2.1	3.4	3.7	5.3	5.8	6.7	7.8
3	1.7	1.1	2.5	3.9	4.7	6.1	6.6	7.6	8.8
4	1.8	0.2	2.5	4.0	5.3	6.2	6.8	7.8	9.1
5	2.4	0.5	2.0	3.5	3.9	5.7	6.3	7.3	9.3
6	-0.7	0.8	1.7	3.1	3.5	5.2	5.8	6.8	7.5
7	3.2	1.2	3.3	5.4	6.0	8.5	9.4	10.8	12.7
8	4.4	4.3	5.6	6.9	7.2	8.9	9.4	10.3	11.1
9	3.9	4.0	5.6	7.3	7.7	9.7	10.4	11.5	12.2
10	6.0	5.9	7.2	8.5	8.7	10.5	11.2	11.9	12.1
11	5.8	5.4	6.9	8.4	8.8	10.7	11.2	12.3	12.5
12	8.8	8.3	9.5	10.6	10.9	12.3	12.8	13.5	15.0
27	34.3	33.8	35.4	37.1	37.5	39.5	40.2	41.3	41.3

28	32.9	31.6	33.5	35.4	35.9	38.2	38.9	40.2	39.7
29	31.3	28.3	30.8	33.3	33.9	37.0	38.0	39.7	38.4
30	30.2	39.9	32.1	34.2	34.8	37.4	38.3	39.8	39.8
31	31.3	28.9	31.3	33.8	34.4	37.3	38.3	39.9	41.0
32	31.7	27.4	29.8	32.1	32.7	35.6	36.5	38.2	38.7

Ventskevich (1958) reported damaging effects on tender plant portions for 1 to 2°C at flowering, 1 to 4°C at fruiting, and 2 to 8°C at germination in food grain, vegetables and fruit crops.

2.4 Weibull Distribution

x is said be follow Weibull distribution with parameters α and β if the probability distribution function of x is

$$F(x,\ \alpha,\ \beta) = \frac{\alpha}{\beta^{\alpha}} x^{\alpha-1} e^{-\left(\frac{x}{\beta}\right)^{\alpha}} \quad \text{for } \alpha, \beta, x > 0 \qquad \ldots (2.12)$$

The mean is $\beta\,\Gamma\left(1+\frac{1}{\alpha}\right)$ and variance is $\beta^2\left[\Gamma\left(1+\frac{2}{\alpha}\right)-\left\{\Gamma\left(1+\frac{1}{\alpha}\right)\right\}^2\right]$

This type of distribution is applied for temperature independent (constant temperature data) distribution of normalized flowering times of seedling Johnson grass. (Bridges and Chandler, 1989). $\Gamma(\alpha)$ is an ordinary gamma function of (α). The cumulative Weibull distribution F(x) was expressed as (Baxendale et al., 1984a; 1984b).

$$F(x) = 1 - \text{Exp}\ \{-[(x - \gamma)/\eta]^{\beta}\} \qquad \ldots (2.13)$$

where F(x) is the probability of complete development at normalized time x, and gamma (γ), beta (β) and eta. η are parameters estimated by the nonlinear regression. For Johnson grass flowering Bridges and chandler (1989) developed the Weibull distribution function as

$$F(x) = 1 - \text{Exp}\ \{-[(x - 0.626)/0.42]^{3.611}\},\ R^2 = 0.99 \qquad \ldots (2.14)$$

Similarly Baxendale et al. (1984) with constant temperature data between 20 to 36°C, developed nondiapasing sorghum midges, Contarinia sorghicola (Coquillett), from egg to adult emergence. Development time decreased with increase of temperature. A Poikilotherm rate equation along with cumulative Weibull distribution provided accurate description of the development. Weibull expression was of the form

$$F(x) = 1 - \text{Exp}\ \{[(x - 0.\ 8085)/0.2391]^{1.7269}\},\ R^2 = 0.99 \qquad \ldots (2.15)$$

2.5 Incomplete Gamma Distribution

The gamma distribution is applied for rainfall probabilities as it has zero lower bound values and fits well to rainfall series. The incomplete gamma distribution is expressed as (Thom, 1971)

$$F(x) = \int_{r}^{x} \frac{e^{-x/\beta} x^{\alpha-1} dx}{\beta^{\alpha} \overline{|(\alpha)}} \quad \text{where x, } \alpha, \beta \text{ are} > 0 \qquad \ldots (2.16)$$

$$= 0 \text{ for } x \leq 0.$$

β is a scale parameter, α is shape parameter. $\overline{|(\alpha)}$ is an ordinary gamma function of α. These values are estimated as

$$\hat{\alpha} = \frac{1}{4A}\left(1 + \sqrt{1 + \frac{4A}{3}}\right) \qquad \ldots (2.17)$$

$$\hat{\beta} = \frac{\overline{x}}{\hat{\alpha}} \qquad \ldots (2.18)$$

$$A = \log \overline{x} - \frac{1}{n}\sum_{i=1}^{n} \log x_i \qquad \ldots (2.19)$$

knowing $\hat{\alpha}$, $\hat{\beta}$, and $\overline{|(\alpha)}$, tables of incomplete gamma functions, the f(x)—the rainfall probability for x amount of rainfall is determined. Pearson's tables of incomplete gamma function (Pearson, 1951 or Mooley, 1974) tables (Appendix IV) gives the probability values for different values of x in terms of z values as $x/\hat{\beta}$. In case rainfall data contains zeroes of mixed distribution function of zeroes and continuous precipitation amounts be expressed as

$$H(x) = q + p\, f(x) \qquad \ldots (2.20)$$

where q is the probability of zero rainfall and p = 1 – q. Therefore when x = 0, H(0) = q. If m is the number of zeroes in a rainfall series data, then q = m/N.

The maximum likelihood estimates ($\hat{\alpha}$, $\hat{\beta}$) of parameters of gamma distribution are given by

$$\Psi(\hat{\alpha}) - \text{Ln}(\hat{\alpha}) - \text{Ln}(R) = 0$$

and $$\hat{\beta} = \overline{x}/\hat{\alpha} \qquad \ldots (2.21)$$

where $\overline{x}$ is airthmatic mean and R is the ratio of geometric mean to airthmatic mean of x the non-zero values of variable x_i and $\Psi(\hat{\alpha})$ is digamma function given by $\frac{\partial}{\partial \hat{\alpha}}\left[\text{Ln}(|\overline{\alpha})\right]$. The values of $\hat{\alpha}$ is uniquely determined by R.

The probability density function is given by

$$f(x) = \int \frac{z^{\hat{\alpha}-1} e^{-z}}{|(\hat{\alpha})} \quad \text{where} \quad z = \frac{x}{\hat{\beta}} \qquad \ldots (2.22)$$

and the probability function P_z (between 0 and 1) is

$$P_z = \int_{0}^{z} \frac{z^{\hat{\alpha}-1} e^{-z}}{|(\hat{\alpha})} dz \qquad \ldots (2.23)$$

Example 7:

Fit the incomplete gamma distribution function for following monthly rainfall data and determine probability of getting 100 mm rainfall.

Solution:

Monthly rainfall data:

Year	Rainfall (mm) (x)	Ln x
1972	315.1	5.7529
1973	157.3	5.0581
1974	194.3	5.2694
1975	192.5	5.2601
1976	193.0	5.2627
1977	20.0	2.9957
1978	185.0	5.2203
1979	346.6	5.8482
1980	487.8	6.1900
1981	479.2	6.1721
1982	85.9	4.4532
1983	253.2	5.5342
1984	122.0	4.8040
1985	247.9	5.5130
1986	17.0	2.8382
1987	52.2	3.9551
1988	509.9	6.2342
1989	113.8	4.7344
1990	364.2	5.8979
1991	29.2	3.3742
1992	395.0	5.9789
1993	205.4	5.3249
1994	180.6	5.1963
1995	300.0	5.7038
1996	286.5	5.6577
1997	175.8	5.1693
1998	211.0	5.3518
1999	304.4	5.7183
2000	115.0	4.7449
2001	205.6	5.3259
2002	154.0	5.0369
2003	89.7	4.4965
2004	151.7	5.0219
2005	129.6	4.8644

$N = 34$ $\bar{x} = 213.835$

$\sum x^2 = 2113777.38$ $\sum Ln\, x_i = 173.9544$

$\sum x = 7270.4$ $Ln\, \bar{x} = 5.3652$

$\sigma_N = 128.23$

$\sigma_{N-1} = 130.16$

$$A = Ln\,\bar{x} - \frac{\sum Ln\, x_i}{N}$$

$$A = 5.3652 - \frac{173.9544}{34} = 0.24889$$

$$\hat{\alpha} = \frac{1}{4A}\left(1+\sqrt{1+\frac{4A}{3}}\right) = \frac{1}{4\times 0.24889}\left(1+\sqrt{1+4\times\frac{0.24889}{3}}\right)$$

$$\hat{\alpha} = \frac{1}{0.99596}\left(1+\sqrt{1+0.33198}\right) = 2.1628$$

$$\hat{\beta} = \frac{\bar{x}}{\hat{\alpha}} = \frac{213.835}{2.1628} = 98.8695$$

From tables of gamma distribution (Appendix III) to determine the probability of getting 100 mm rainfall (x) corresponding to

$$\hat{\alpha} = 2.16 \text{ and } z = \frac{x}{\hat{\beta}} = \frac{100}{98.8695} = 1.0114$$

Probability of monthly zero rains = 0.0

Probability of getting rain less than 100 mm (Table) is 28 percent or 0.28 and probability of getting rain more than 100 mm will be 0.72, 50 percent probability exists for getting rainfall $z = 1.827$, therefore $x = 1.827 \times 98.8695 = 180.6$ mm of rainfall.

Example 8:

Determine the weekly rainfall probability of getting 25 mm rain for sowing of rainy season crop in the first week of July month using the past records given below.

Solution:

Number of non-rainy weeks in the past records of 52 years = 28

24 rainy events recorded weekly rainfalls of 47.3, 1.0, 16.0, 10.2, 9.0, 25.0, 6.6, 1.0, 0.5, 12.5, 13.8, 50.5, 3.8, 44.3, 46.3, 15.6, 3.4, 32.7, 3.5, 17.9, 41.4, 1.8, 85.9 and 3.0 mm.

$N = 52$

Zero rainfall frequency (m) = 28

Probability of zero rainfall = 28/52 = 0.5385

Probability of non-zero rainfall (p) = 1 – 0.5885 = 0.4615

Using rainfall events data.

$\Sigma x = 493$

$\bar{x} = 20.5417$

$\Sigma \text{Ln } x_i = \frac{55.0544}{24} = 2.2939$

Therefore A = 20.5417 – 2.2939 = 0.7285

$$\hat{\alpha} = \frac{1}{4A}\left(1 + \sqrt{\frac{4A}{3}}\right)$$

$$= \frac{1}{4 \times 0.7285}\left(1 + \frac{\sqrt{1 + 4 \times 0.7285}}{3}\right)$$

$$\hat{\alpha} = \frac{2.4040}{2.914} = 0.825$$

$$\hat{\beta} = \frac{\bar{x}}{\hat{\alpha}} = \frac{20.5417}{0.825} = 24.899$$

Probability of getting 25 mm of rainfall = $\frac{x}{\hat{\beta}} = \frac{25}{24.899}$, $z = 1.0040 \simeq 1, 0$

Using tables probability of getting rain less than 25 mm = 0.72

Therefore probability of getting less than 25 mm rain = 0.5385 + 0.72 × 0.4615 = 0.87

Probability of getting more than 25 mm rain = 0.13.

2.6 Extreme Value Distribution

In designing of river dams, the climatological variables of annual/seasonal extremes of rainfall are of specific interest so that it can accommodate upper limit extreme in the region. The distribution of upper annual extreme values will provide the prediction of rainfall. Gumbel (1958) or Fisher – Tippett Type I distribution has been widely used in such cases expressed as

$$F(x) = \exp\left\{-\exp^{\pm}\frac{x-\alpha}{\beta}\right\}, \qquad \ldots (2.24)$$

where ± is indicates the minimum and maximum values. F(x) is cumulative probability that any extreme value will be less than x. Type II distribution which is exponential transformation of Type I is also used for extreme winds in climatological studies.

The return period (T) of an extreme value ≥ x is given by the expression

$$T = \frac{1}{1 - F(x)} = \frac{1}{1 - e^{-e^{-y}}}, \qquad \ldots (2.25)$$

where $y = \mp\frac{x-\alpha}{\beta}$, i.e., $\mp x = \alpha + \beta y$

It can also be expressed as

$$y = -\text{Ln}\left\{\text{Ln}\left(\frac{T}{T-1}\right)\right\} \simeq \text{Ln } T \text{ leaving higher terms.} \quad \ldots (2.26)$$

According to Gumbel (1958) the return period T is also expressed as

$$T = \frac{N+1}{m}, \quad \ldots (2.27)$$

where N is total number of years of record and m is the rank of the ordered series

Expression $x = \alpha + \beta y$ can also be reduced in terms of T as $x = \bar{x} - 0.45\sigma + 0.78\sigma \text{ Ln } T$. This is simplest procedure for T.

As with most skewed distributions the moments give foor estimates of the parameters, therefore Lieblein (1954) provided a simple method for fitting type I distribution which give estimates of probabilities or quantiles with minimum variance. It involves, maintaining original order of the climatological series, dividing into suitable subgroups for computations. The following table of weights is needed for computations.

Suppose a climatological series of n values are divided into m subgroups which can take values from 2 to 6 (Table 2.2). Each column of x is first summed to obtain S_j. These are multiplied by a_j and summed to obtain the second row sum. Exponent $\frac{(x-\alpha)}{\beta}$ is a standarised variable located at α and scaled in β.

Table 2.2

Statistical weights in order form

m		x.1	x.2	x.3	x.4	x.5	x.6
1							
2	a.j.	0.91637	0.08363				
	b.j.	-0.72135	0.72135				
3	a.j.	0.65632	0.25571	0.08797			
	b.j.	-0.63054	0.25582	0.37473			
4.	a.j.	0.51100	0.26394	0.15368	0.07138		
	b.j.	-0.55862	0.08590	0.22392	0.24880		
5.	a.j.	0.41893	0.24628	0.16761	0.10882	0.05835	
	b.j.	-0.50313	0.00653	0.13045	0.18166	0.18448	
6.	a.j.	0.35545	0.22549	0.16562	0.12105	0.08352	0.04887
	b.j.	-0.45928	-0.03599	0.07319	0.12673	0.14953	0.14581

After: Lieblein, T (1954)

The minimum variance estimates for α and β are

$$\hat{\alpha} = \sum_{j=1}^{m} a.j\,S.j/k \qquad \ldots (2.28)$$

$$\hat{\beta} = \sum_{j=1}^{m} b.j\,S.j/k \qquad \ldots (2.29)$$

Type II distribution, useful for extreme winds, is expressed as

$$F(x) = \exp\left\{-\left(\frac{x}{\beta_2}\right)^{\gamma}\right\}$$

where F(x) is probability of an extreme value being $\leq x$, β_2 is a scale parameter, γ is shape factor. It can also be expressed as

$$x = \exp\left\{\left(\mathrm{Ln}(\beta_2) - \frac{1}{\gamma}\mathrm{Ln\,Ln}\left(\frac{1}{F(x)}\right)\right)\right\} \qquad \ldots (2.30)$$

where $F(x) = 1 - \frac{1}{T}$, T is return period, Lieblein (1954) procedure can be adopted to fit the data series.

Example 9:

Fit the extreme value distribution functions Type I and Type II for the following monsoon season rainfall data.

Solution:

Year	**Seasonal Rainfall (mm) (R)**	**X = LnR**
2005	425	6.0521
2004	555	6.3190
2003	415	6.0283
2002	554	6.3172
2001	426	6.0544
2000	360	5.8861
1999	507	6.2285
1998	189	5.2417
1997	587	6.3750
1996	295	5.6869
1995	346	5.8464

1994	488	6.1903
1993	350	5.8579
1992	565	6.3368
1991	465	6.1420
1990	516	6.2461
1989	288	5.6630
1988	469	6.1506
1987	633	6.4505
1986	548	6.3063
1985	797	6.6809
1984	460	6.1312
1983	262	5.5683
1982	342	5.8348
1981	326	5.7969
1980	443	6.0935
1979	500	6.2146
1978	580	6.3630
1977	200	5.2983
1976	625	6.4377

$\overline{R} = 450.53$

$\sigma_N = 135.616$

Using statistical weight table.

j	1	2	3	4	5	6
a.j.	0.35545	0.22549	0.16562	0.12105	0.08352	0.04887
b.j.	-0.45925	-0.03599	0.07319	0.12673	0.14953	0.14581
i	X_{i1}	X_{i2}	X_{i3}	X_{i4}	X_{i5}	X_{i6}
1	5.2983	5.7869	6.0935	6.2146	6.3630	6.4377
2	5.5683	5.8348	6.1312	6.3063	6.4505	6.6809
3	5.6630	5.8579	6.1420	6.1506	6.2461	6.3368
4	5.2417	5.6869	5.8464	6.1903	6.2285	6.3750
5	5.8861	6.0283	6.0521	6.0544	6.3172	6.3190
S.j.	27.6574	29.1948	30.2652	30.9162	31.6043	32.1794

$$\sum_{j=1}^{6} \frac{a.j\,S.j}{5} = \frac{29.3796}{5} = 5.87592$$

$$\sum_{j=1}^{6} \frac{a.j\,S.j}{5} = \frac{1.79423}{5} = 0.35885$$

Therefore parameters for Type I distribution of Ln (rainfall)

$\hat{\alpha} = 5.87592$

$\hat{\beta} = 0.35885$

The parameters for Type II distribution will be given by

$\hat{\beta}_2 = \exp\ \hat{\alpha} = \exp(5.87592) = 356.35$

$$\hat{\gamma} - \frac{1}{\hat{\beta}} = \frac{1}{0.35885} = 2.78668 = 2.79$$

Quantiles are obtained by taking logarithms twice as

$$\text{Ln}\ x = \text{Ln}\ \hat{\beta}_2 - \frac{\text{Ln.Ln}\left(\frac{1}{F}\right)}{\hat{\gamma}}$$

For F = 0.98 (a design value)

$$x = \exp\left\{\text{Ln}\ \hat{\beta}_2 - \frac{\text{Ln.Ln}\left(\frac{1}{F}\right)}{\hat{\gamma}}\right\}$$

$= \exp\{5.87591 + 1.39854\}$

$x = \exp(7.27445) = 1442.96$ mm

$x \simeq 1443$ mm of upper limit of rainfall

for F value of 0.5

$x(0.5) = \exp(6.00601) = 405.8$ mm

Type I distribution $F(x) = \exp\left[-\exp\left(\pm\frac{x-\alpha}{\beta}\right)\right]$

$$\text{Ln.Ln}\left(\frac{1}{F}\right) = \pm\frac{x-\alpha}{\beta}$$

$x = \alpha + \beta\ \{\text{Ln. Ln}(1/F)\}$ for positive sign

$x = \alpha + \beta\ \text{LnT} = 5.8759 + 0.35885\ \text{Ln T}$

$x = \alpha - \beta\{\text{Ln. Ln}(1/F)\}$ for negative sign

$x = \alpha - \beta\ \text{LnT} = 5.8759 - 0.35885\ \text{Ln T}$

Therefore for different values of return period (T)

T	+x	–x	+ Rainfall (mm)	– Rainfall (mm)
2	6.12	5.63	454.9	278.7
5	6.45	5.30	632.7	200.3

10	6.70	5.05	812.4	156.0
20	6.95	4.80	1043.1	121.5
50	7.28	4.47	1451.0	87.3
100	7.53	4.22	1863.1	68.0

This can also be examined by the simplest expression for upper limits as a first approximation.

$x = \bar{x} - 0.45\sigma + 0.78\sigma \text{ Ln } T$

with $\bar{x}$ = 450.5 mm and σ = 135.62 mm

T (years)	x (years)
2	462.8
5	559.7
10	633.0
20	706.4
50	803.3
100	876.6

2.7 Bionomial Distribution

When we are concerned with problems of occurrence or non-occurrence of an event (may be rain, thunderstorm or hails), the relative frequency or probability of occurrence (x) in a sample of data series

$$P(x) = N_{C_x}\ p^x\ q(N - x) \text{ for } x = 1, 2, \ldots N \qquad \ldots (2.31)$$

where p is probability of occurrence of event and q = 1 – p. N_{C_x} denotes the number of combination of N things taken x at a time. The statistical parameters are

mean = N.p.

Standard deviation = $\sqrt{pqN}$

Skewness = $\dfrac{(q-p)}{\sqrt{pqN}}$

when p = q it becomes symmetrical commonly used form of discrete distribution.

Simple Markov – Chain model fits the observed frequency better than the random models as confirmed by the chi-square tests. Markov - Chain model describes the persistence of dry or wet spells. The model assumes that the probability of rain occurring on any day depends on whether it occurred or not on the previous day. The probability model of Markov – Chain depends on the two conditional probabilities denoted as p_0 and $(1 - p_1)$, where p_0 is probability of a wet day given the previous day dry, and $(1 - p_1)$ is the probability of a dry day given the previous day wet.

Therefore p_1 = Prob (w/w) ... (2.32)

p_0 = Prob (w/d) ... (2.33)

$(1 - p_1)$ = Prob (d/w) ... (2.34)

$(1 - p_0)$ = Prob (d/d) ... (2.35)

In continuity probability of a dry spell of length n days = $p_0(1 - p_0)^{n-1}$ ... (2.36)

Probability of a wet spell of length n days = $(1 - p_1)p_1^{n-1}$ ... (2.37)

Bishnoi and Saxena (1979) found that third order Markov Chain Model represents better the daily precipitation occurrence during the S.W. monsoon season over the first order Markov Chain Model commonly used. When the length of record is less than 2000 days for a dry subhumid location and 4000 days over an arid location, there is tendency for low order chain to be misrepresented as a proper model (Table 2.3)

Table 2.3

Observed distribution of 4 day sequences compared with first, second and third order Markov Chain Models for a dry subhumid location.

Sequence	Observed	M_1	M_2	M_3	M_1-Obs	M_2-Obs	M_3-Obs
d d d d	1714	1543	1692	1712	-171	-22	-2
d d d w	244	326	266	243	82	22	-1
d d w d	163	208	177	164	45	14	1
d d w w	145	188	136	145	43	-9	0
d w d d	208	208	209	209	0	1	1
d w d w	68	44	66	68	-24	-2	0
d w w d	105	120	100	105	15	-5	0
d w w w	99	108	108	99	9	9	0
w d d d	307	327	315	306	20	8	-1
w d d w	57	69	49	57	12	-8	0
w d w d	74	44	66	75	-30	-8	1
w d w w	42	40	49	42	-2	7	0
w w d d	163	188	147	164	2	-16	1
w w d w	41	40	49	41	-1	8	0
w w w d	102	108	106	103	6	6	1
w w w w	128	99	123	127	-29	-5	-1
x^2 value					110.90	11.94	0.06

(After: Bishnoi and Saxena, 1979).

The Poisson distribution is derived from Biomomial distribution when N is infinitely large and p probability of success is small so that N p = λ a positive real number. Then the probability of success is given by

$$P(x) = \frac{e^{-\lambda}\lambda^x}{x|}, \text{ where } x| \text{ is a factorial} \quad \ldots (2.38)$$

Mean and variance of poisson distribution is λ. Distribution is useful for several biological events.

If we consider the length of interval as d instead of unit length, then average number of occurrences of events is λd in d length. The probability density function is

$$P(x) = e^{-\lambda}(\lambda d)^x/x! \text{ for } x = 0, 1, 2, \qquad \ldots (2.39)$$

Example 10:

Based on following weekly rainfall data (mm) determine the length of dry spells of 2 to 4 weeks and wet spells of 2 to 3 weeks in S.W. Monsoon season.

Solution:

	Standard Weeks													
Year	**25**	**26**	**27**	**28**	**29**	**30**	**31**	**32**	**33**	**34**	**35**	**36**	**37**	**38**
1985	3.8	4.3	8.6	-	3.3	3.0	40.4	-	-	-	-	-	39.1	-
1986	6.1	-	19.5	16.0	20.3	10.2	45.7	221.2	-	-	182	-	57.2	17.8
1987	16.0	-	33.8	-	14.2	-	8.4	14.0	32.0	32.2	26.2	485	29.0	355.0
1988	-	3.6	-	-	-	-	-	-	35.5	48.0	-	19.5	-	-
1989	-	-	-	8.1	33.0	35.5	8.1	6.4	-	-	78.5	33.0	-	6.0
1990	-	-	2.5	-	10.9	70.6	26.4	-	-	-	-	-	-	-
1991	2.5	-	-	39.4	-	47.0	13.7	52.8	-	-	-	10.7	-	78.7
1992	-	26.7	-	-	25.9	33.0	77.2	-	-	5.0	-	130.8	27.4	43.4
1993	2.1	-	10.7	21.1	113.8	50.5	30.7	62.2	57.1	64.0	45.2	-	-	-
1994	6.8	-	13.0	89.4	-	-	-	5.6	-	-	2.8	63.0	11.2	-
1995	12.2	17.8	31.2	-	54.9	15.0	79.2	-	-	-	-	-	-	-
1996	63.0	-	-	18.0	48.3	16.0	20.0	17.0	49.5	85.6	62.5	-	-	-
1997	40.1	-	120	57.6	68.1	92.4	-	30.2	13.2	-	40.1	2.5	-	-
1998	-	33.0	43.2	5.0	-	10.2	-	-	-	114.3	6.0	-	-	-
1999	-	2.5	94.0	2.5	30.5	7.6	38.1	15.2	-	5.0	33.0	7.5	-	-
2000	12.7	3.0	6.1	61.2	58.9	-	86.4	66.0	2.5	-	-	-	-	-
2001	3.8	-	3.3	-	-	-	-	4.6	41.1	104.1	147.8	-	2.5	115.0
2002	31.2	-	11.2	43.6	41.6	106.1	15.7	-	114.3	19.1	-	-	-	-
2003	-	-	22.1	-	24.4	7.1	5.0	10.0	-	5.0	5.0	7.5	-	48.3
2004	-	18.3	-	20.5	-	10.0	-	18.6	25.4	-	-	1.0	-	20.0
2005	93.5	-	-	-	5.0	34.3	21.6	52.3	3.3	-	4.0	18.0	-	23.0
Prob. (w/w) = P_1		0.14	0.24	0.38	0.38	0.62	0.62	0.48	0.38	0.28	0.33	0.33	0.14	0.19
Prob. (w/d) = P_0		0.24	0.43	0.14	0.13	0.14	0.10	0.19	0.10	0.19	0.24	0.19	0.14	0.24
Prob. For dry spell of length														
2 weeks		0.18	0.24	0.12	0.22	0.12	0.09	0.15	0.09	0.15	0.18	0.15	0.12	0.18
3 weeks		0.14	0.14	0.10	0.15	0.10	0.08	0.12	0.08	0.12	0.14	0.12	0.10	0.14
4 weeks		0.10	0.08	0.09	0.10	0.09	0.07	0.10	0.07	0.10	0.10	0.10	0.09	0.10
Prob. of wet spell of length														
2 weeks		0.12	0.18	0.23	0.23	0.23	0.23	0.25	0.23	0.20	0.22	0.22	0.12	0.15
3 weeks		0.02	0.04	0.09	0.09	0.15	0.15	0.12	0.09	0.06	0.07	0.07	0.02	0.03

2.8 Beta Distribution

The incomplete β-function is defined as

$$\beta_x(p,q) = \int_0^x t^{p-1}(1-t)^{q-1}\,dt \qquad \ldots (2.40)$$

For x = 1, the complete β function is

$$\beta(p,q) = \int_0^1 t^{p-1}(1-t)^{q-1}\,dt = \frac{\Gamma(p)\Gamma(q)}{\Gamma(p+q)} \qquad \ldots (2.41)$$

In order that the function $\beta_x(p, q)$ may represent a cumulative distribution function of variable x varied between 0 and 1, $\beta_x(p, q)$ is divided by $\beta(p, q)$ to obtain the standardized incomplete beta function ratio as

$$I_x(p,q) = \frac{\beta_x(p,q)}{\beta(p,q)} = \frac{\Gamma(p+q)}{\Gamma(p)\Gamma(q)} \cdot \int_0^x t^{p-1}(1-t)^{q-1}\,dt \qquad \ldots (2.42)$$

where p and q are two positive parameters. This function represents a two parameter family of distribution and tables are provided (Pearson, 1934; 1951) for various values of p and q.

The estimates of parameters p and q are derived the first two moment coefficients given by Peasson (1934).

The characteristics of beta distribution variable x is similar to any agroclimatic variable varying between 0 and 1 such as the relative leaf area index, relative crop yield, actual to potential evapotranspiration, water availability in the root zone, and therefore these analogies suggest the use of beta distribution for probability studies of these agroclimatic variables. Yao (1969) and Bishnoi (1980) used this concept for probability distribution of moisture adequacy index (AET/PET) to assess irrigation requirement, optimum evapotranspiration needs, water management, landuse pattern and crop planning aspects for optimum utilization of available natural resources in the region. Fig. 2.3 shows the seasonal variation of moisture adequacy index and the corresponding probability levels of beta distribution. Table 2.4 is given to indicate the probability of getting moisture adequacy index equal to or greater than 0.6 for summer and winter crops at different location of arid to moist-subhumid climates. The level of 0.6 is required for maintaining optimum evapotranspiration to maintain crop yields without shortage of moisture availability to the plants. Whenever the level is getting below 75 percent (an assured level) the moisture adequacy index needs to be maintained with external irrigation resources.

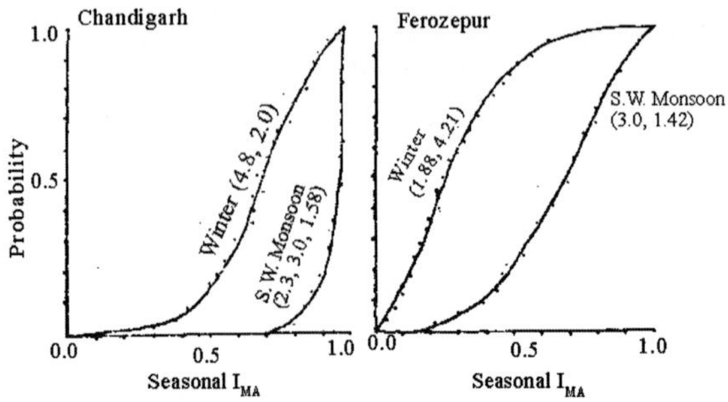

Fig. 2.3

Table 2.4

Probabilities of getting AET/PET > 0.6 during rainy and winter seasons at different locations

Location	Climate type	Rainy Season				Winter Season				
		Jul.	Aug.	Sep.	Oct.	Nov.	Dec.	Jan.	Feb.	Mar.
Sri Ganganagar	Arid	17	44	8	6	2	9	17	19	9
Ferozpur	Arid	78	84	44	11	25	26	69	41	37
Hisar	Arid	48	71	64	26	11	14	42	37	18
Alwar	Semi arid	83	89	75	63	36	26	37	32	4
New Delhi	Semi arid	60	91	85	65	36	36	68	44	17
Amritsar	Semi arid	63	82	46	57	37	56	74	89	52
Patiala	Semi arid	93	90	78	57	44	51	79	81	43
Karnal	Semi arid	66	96	79	70	60	48	79	69	36
Ambala	Semi arid	92	95	86	70	47	47	77	83	52
Ludhiana	Semi arid	82	91	69	57	30	42	77	75	59
Chandigarh	Dry subhumid	100	99	98	84	61	71	86	82	68
Pathankot	Dry subhumid	100	100	97	94	87	76	96	100	94

(After: Bishnoi, 1980).

2.9 Goodness of Fit

Two commonly used methods are the multiple correlation coefficients and chi-square test. Appendix V gives the critical value of rank correlation coefficient at 5 and 1 percent level of acceptance.

In chi-square test to test the goodness of fit, the difference between observed and expected data is expressed as a ratio to the expected frequencies in the form known as chi-square expressed as

$$\chi^2 = \sum_{i=1}^{n} \frac{(o_i - e_i)}{e_i} \qquad \ldots (2.43)$$

Larger the χ^2, smaller is the probability that the observed and expected frequencies are from the same population. Values of χ^2 for a given number of degrees of freedom should be less than the tabulated values for the null hypothesis to be accepted

Degree of Freedom	P = 0.05	P = 0.01
1	3.84	6.64
2	5.99	9.21
3	7.82	11.34
4	9.49	13.28
5	11.07	15.09
10	18.31	23.21
20	31.41	37.57
30	43.77	50.89

Example 11:

Apply χ^2 test to the temperature data of example 6 for fitting normal distribution curve for temperature data.

Solution:

Temperature Range °C	Observed (O_i) Frequency	Expected (e_i) Frequency	$(o_i - e_i)^2$	$(o_i - e_i)^2/e_i$
23-25	-	1	1	1.00
25-27	4	2	4	2.00
27-29	4	5	1	0.20
29-31	8	7	1	0.14
31-33	8	8	0	0.00
33-35	7	7	0	0.00
37-37	3	5	4	0.80
37-39	3	2	1	0.50
39-41	1	1	0	0.00
41-43	-	0	0	0.00

Degree of freedom = k – 1 – m, where k is number of classes in grouped data and m is number of population statistics estimated from the sample.

Therefore hypothesis is accepted = 10 – 1 – 2 = 7

2.10 Student's t Test

It is used to determine trends and to evaluate the statistical significance of the difference between the groups of the data. It involves the computation of an index, t, which is the ratio of the difference between the means to the standard error of the difference. It quantifies how many times greater than the standard error the observed differences actually compares with the table values. It is computed for two variables x and y as

$$t = \frac{|\bar{x} - \bar{y}|}{\sqrt{\frac{\sigma x^2}{n_x} + \frac{\sigma y^2}{n_y}}} \quad \ldots (2.44)$$

The significance of a particular value of t decreases as the number of occurrences involved also decreases. Student's 't' test value is also used with the concept of degree of freedom i.e., the number of values that can be assigned arbitrarily assuming that the sample total and mean remain unchanged which is normally one less than the total number of occurrences.

Example 12:

Apply 't' test to the following maximum January temperature data series.

Year	**January Temperature (°C)**
Data Series X_1	
1976	20.8
1977	20.4
1978	20.8
1979	20.4
1980	19.8
1981	21.3
1982	20.2
1983	20.9
1984	18.5
1985	21.2
1986	19.9
1987	20.5
1988	20.6
1989	20.2
1990	19.9
Data Series X_2	
1991	20.0
1992	19.5
1993	19.6
1994	19.8

1995	19.8
1996	20.6
1997	22.2
1998	19.2
1999	22.2
2000	19.7
2001	20.2
2002	19.7
2003	20.6
2004	16.6
2005	20.1

$\overline{X}_1 = 20.36°C$ $\overline{X}_2 = 20.17°C$

$\sigma_1 = 0.6630°C$ $\sigma_2 = 0.9686°C$

t value is

$$t = \frac{\overline{X}_1 - \overline{X}_2}{\sqrt{\frac{\sigma_1^{\ 2}}{N_1 - 1} + \frac{\sigma_2^{\ 2}}{N_2 - 1}}} = \frac{20.36 - 20.17}{\sqrt{\frac{(0.663)^2}{14} + \frac{(0.9686)^2}{14}}}$$

From table 't' value for 0.05 level of significance for 28 degree of freedom, the null hypothesis that the two sample means (X_1, X_2) come from the same large sample is acceptable as 0.6 is less than the table value corresponding to 28 degrees of freedom being 2.048 standard errors.

Example 13:

If the average number of plants infected per week by a disease is 1.5. Determine the probability that there will be 5 or more plants infected in a month.

Solution: d = 1 month i.e. 4 weeks

$\lambda = 1.5$

$\lambda d = 4 \times 1.5 = 6$

Probability that there will be 5 or more plants infected in a month

is $\text{Prob.}\ (x \geq 5) = \sum_{x=5}^{\infty} \frac{e^{-6}6^x}{x|}$

It can be also expressed as

$\text{Prob.}\ (x \geq 5) = 1 - P(x \leq 4)$

$$= 1 - \sum_{x=0}^{4} \frac{e^{-6}6^x}{x|}$$

$$= 1 - e^{-6}\left\{\frac{6^0}{0|} + \frac{6^1}{1|} + \frac{6^2}{2|} + \frac{6^3}{3|} + \frac{6^4}{4|}\right\}$$

$$= 1 - 0.002\ \{1 + 6 + 18 + 36 + 54\}$$

$$= 1 - 0.23 = 0.77, \qquad 77 \text{ percent probability.}$$

2.11 Weather and Crop Yield Relation

Weather is an essential component of crop yield responses. Crop yield is the resultant parameter of the interacting variables of crop, soil, weather and management inputs. McQuigg (1975) reported that there are three major sources of variability in grain yield over a period of years as technological change, meteorological variability and random noise. Technological change is most important due to impact of improved seeds, fertilizer use, control of pest and diseases and improved adoption of package of practices in different crops. Meteorological variability is most important in order to identify specific variables of temperature, soil moisture availability, radiation influencing crop yield variability at different phenophases. McQuigg (1975) analysis concluded that technology contributed 70-80 per cent of total variance, weather about 12-18 percent and random noise about 5-10 percent for a sample of yield from a region over a long series of years.

It is worth to explain these points with the use of following pearl millet yield data plotted over a period of 1966 – 2004 in the Haryana State. Fig 2.4 shows that yield data over the

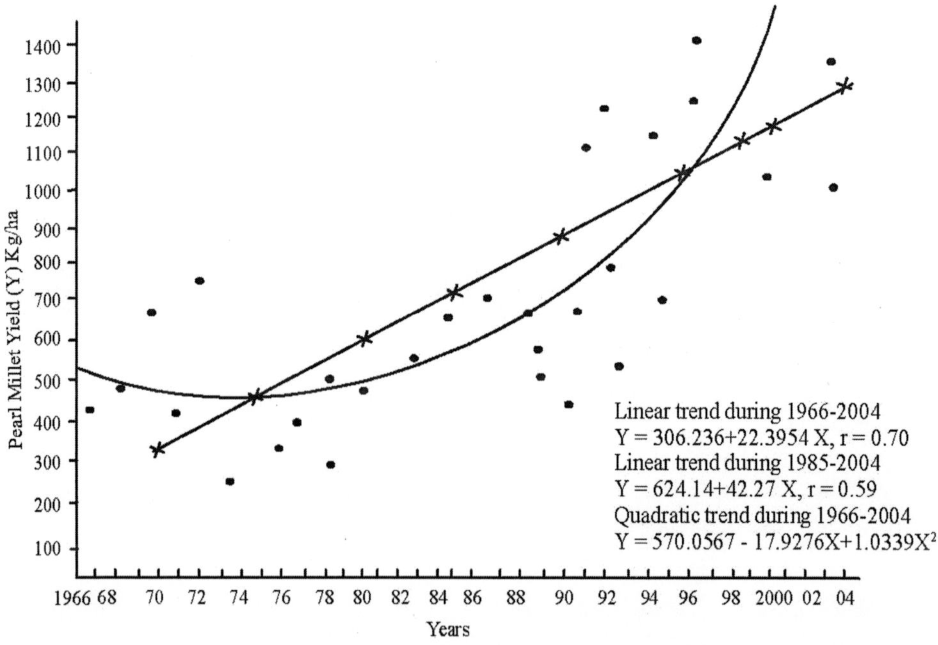

Fig. 2.4: Technological trend in Pearl Millet

years follow a linear response for technological trend with 22.39 kg/ha increase in yield per year. However in the beginning period from 1966 to 1973 a stable yield level exists without any trend and there is slight decrease from 1974 to 1986 leading to a stagnant yield level. After 1987 there is sharp rise in technology trend with 31.41 kg/ha increase in yield per year. A quadratic form of technology trend appeared better in this situation to explain the yield variations with time. In order to quantify yield variability with meteorological variables the contribution of technology trend is subtracted and then impact of weather variables on yield is determined.

A vast literature is available on empirical statistical relationships between crop yield of different crops and weather variables in different regions all over the world and these relations can be used for regional forecasts of crop yields based on weather conditions every year in order to monitor the food availability. The linear correlation between crop yield and weather variable over growth phases will yield significant weather variables. Bishnoi (1987) developed pearl millet yield weather relationships after removing the technology trend as

$$\begin{aligned} y(\text{kg/ha}) = 390.32 \; &+ \underset{(0.67)}{0.021\, X_1} - \underset{(1.92)^*}{5.96\, X_2} + \underset{(2.08)^*}{5.944\, X_5} \qquad \ldots (2.45) \\ &- \underset{(-1.34)}{1.505\, X_6} + \underset{(1.23)}{3.54\, X_3} + \underset{(2.02)}{1.045\, X_4} - \underset{(-0.58)}{0.138\, X_7} \qquad (R^2 = 0.91,\; F = 7.43^*) \end{aligned}$$

where X_1 (rainfall in mm), X_2 (maximum temperature °C), X_3 (pan evaporation in mm/day), X_4 (relative humidity in percent) during vegetative phase; X_5 (minimum temperature in °C) during reproductive phase; X_6 (maximum temperature in °C) and X_7 (relative humidity in percent) during physiological maturity.

A stepwise yield forecast can be made using weather variables during the different phenophases. With use of X_1 and X_4 variables during vegetative phase the pearl yield first forecast explaining 82 percent yield variability was expressed as

$$y(\text{kg/ha}) = +191.48 + \underset{(3.34)^*}{0.062\, X_1} + \underset{(3.14)^*}{0.757\, X_4} \qquad (R^2 = 0.82,\; F = 22.05) \qquad \ldots (2.46)$$

A second forecast of yield variability upto 87 per cent was explained using the weather variables of vegetative and reproductive phenophases expressed as:

$$y(\text{kg/ha}) = 245.59 + \underset{(2.88)^*}{0.05\, X_1} + \underset{(1.98)^*}{2.177\, X_5} + \underset{(3.42)^*}{0.726\, X_4} \qquad (R^2 = 0.87,\; F = 20.29) \qquad \ldots (2.47)$$

An important contribution in this direction was made by Baier (1973)—the crop weather analysis model with analysis of daily contribution of upto three selected agrimeteorological variables for crop yield with consideration of responses at different phenophases. It was assumed that basically crop yields depend on three agrometeorological variables particularly solar energy, temperature, and soil moisture or evapotranspiration. These variables modify each other on any particular day during crop life cycle producing a positive or negative contribution on the final yield expressed as

$$y = \sum_{t=0}^{m} V_1 . V_2 . V_3, \qquad \ldots (2.48)$$

where y is dependent variable, $\sum_{t=0}^{m}$ is summation of daily V-values from biometeorological time t = 0 to t = m according to Robertson (1968) time scale. V_1, V_2 and V_3 are functions of selected independent agrometeorological variables. Each V-function represents the functional relationship linear or quadratic in its contribution to final yield. It is complex approach and requires specific computer programmes to run the model.

Similarly factorial yield weather model (Robertson, 1974) involved the summation of products of several quadratic functions for different weather variables expressed as

$$\hat{y}_t = V_1(\hat{y}_{t-1}, R_1) . V_2(T_{min}) . V_3(R_s) . V_4(ET) \qquad \ldots (2.49)$$

where $\hat{y}_t$ is estimated crop yield, $\hat{y}_{t-1}$ is estimated crop yield at the end of previous stage, R_1 is rainfall between the stages; V_1, V_2, V_3 V_4 are functions based on significant weather variables in quadratic forms. Crop yield are assessed at different crop growth stages with interactions of contributions by different variables towards crop yield.

Gangopadhyaya and Sarkar (1964) used curvilinear technique to predict sugarcane yield with crop characteristics namely height, girth, number of canes per clump explaining 80-90 percent of yield variations, but linear regression only accounted for 35-70 percent of variations. Therefore in addition to weather variables, crop characteristics can also be used alongwith significant weather variables to improve the crop yield predictions. Subbaramayya and Rupa Kumar (1980) developed second degree multiple regression equation involving maximum and minimum temperature, relative humidity in third month after sowing, coinciding with tillering phase of sugarcane, expressed as

$$\begin{aligned} y(\text{tones/ha}) = &-1253.3 + 281.7\, T_{Max} - 292.5\, T_{Min} - 4.36\, (T_{Max})^2 \qquad \ldots (2.50) \\ &+ 1.49(T_{Min} - RH) - 1.11(T_{Max}\, RH) + 4.84(T_{Max} . T_{Min}). \quad (R^2 = 0.79) \end{aligned}$$

where T_{Max} °C, T_{Min} °C, RH (per cent) during tillering phase (3rd month after sowing).

Bishnoi (1998) assessed pearl millet crop yield with the progress of apical development based plant biological characters. Emergence of plant characters is the resultant effect of exploitation of plant's genetic potential under the interaction of plant – weather – soil water and nutrients. Grain yield per plant (gm) was correlated with the emergence of total number of leaves (N_l) on effective tillers producing grain, total number of nodes (N_n), and total tillers (N_t) per plant in Hybrid pearl millet varieties HHB-50 and MHB-110.

$Y = 1.23 + 1.351\ N_n,\ \ R^2 = 0.89$

$Y = 9.558 + 5.443\ N_t,\ \ R^2 = 0.78$

$Y = -4.82 + 1.14,\ \ N_l,\ R^2 = 0.91$ . . . (2.51)

The next stage of yield prediction at fully developed canopy with maximum leaf area index and ear head emergence and filling is initiated. Pearl millet crop yield variations upto 91 percent was explained by the maximum leaf area index and dry matter per plant at anthesis expressed as

Yield (q/ha) for MHB 110 = $9.2698 + 0.1291\ DM_a - 0.4331\ LAI_{Max}$ . . . (2.52)
($F = 10.6,\ R^2 = 0.91$)

Yield (q/ha) for HHB 50 = $-2.274 - 0.0245\ DM_a + 1.99$ LAIMax
($F = 9.87,\ R^2 = 0.91$)

Therefore dry matter at anthesis and maximum LAI are important plant characters explaining 91 per cent yield variability in pearl millet. These characters can be estimated with the use of remote sensing techniques over large area scale yield prediction of pearl millet.

The next stage of yield prediction is the utilization of plant characters in reproductive phenophase for final yield prediction. These are sink development parameters in the ear head.

This stage determines the total kernal number and weight of seeds characterized by harvest index which in these genotype is of the order of 0.24 to 0.3. The kernal number (KNO) is also a useful parameter influenced by radiation and temperature during this period. KNO being the final outcome of plant during pre-anthesis period from the available tillers, spikelets and florets that it will decide the number of seeds to be set provided environmental parameter and availability of soil moisture are not limitations. In these cultivars of pearl millet KNO showed a significant relation with accumulated dry matter at anthesis expressed as

KNO = $653.72 + 55.54\ DM_a\ \ (R^2 = 0.62)$ . . . (2.53)

Test weight of 1000 grains were also strongly related to accumulated dry matter at anthesis. In the same genotype, the relationship was expressed as

Test weight of 1000 grains (gm) = $4.01 + 0.05\ DM_a,\ (R^2 = 85)$ . . . (2.54)

Therefore grain yield can be predicted at different phenophases with the growth and development of the plant to help in developing agroadvisories during the pearl millet growing season at different phenophases. Similarly such prescriptions can be developed for different crops in each region for quick assessment of crop yield attributes in predicting crop yield of field crops.

1.12 Fitting of Curves with Functions

In agroclimatology/agrometeorology the biological functional relationship are developed between the sets of data arbitrarily independent variables (radiation, temperature) and experimentally determined dependent variable (crop yield). For example in a field experiment to study dry matter production with intercepted PAR radiation, the intercepted PAR data is independent variable and dry matter is dependent variable. The mathematical/statistical equations to represent the functional relationship in agricultural climatology experiments are of the types given below:

Mathematical Function	Functional Relationship	Example
(i) $Y = A + BX$	Linear	Relation between dry matter and intercepted PAR, grain yield and transpiration, crop phonological development and total thermal time.
(ii) $Y = A + B_1X_1 + B_2X_2 \ldots B_nX_n$	Multiple Linear	Crop yield variation with weather variables; pest/disease multiplication with weather variables.
(iii) $Y = A + B_1X + B_2X^2 \ldots B_nX^n$	Polynomical of n degree	Production functions, crop yield with one independent variable available water or AET/PET ratio.
(iv) $Y = AX^B + C$	Logrithmic or power	Variation of wind speed with height.
(v) $Y = \exp(A+BX) + C$	Exponential or semilogarithmic	Relation between leaf area index and evapotranspiration, radiation pentration through crop foliage,
(vi) $Y = \frac{X}{BX - A}$	Hyperbola	Crop yield variations with environmental variables
(vii) $Y = AB^{C^X}$	Gompertz equation	Relation between crop root weight and growth period.
$Y = AB^{C^X} + D$	Modified Gompertz equation	Variation of sea surface water temperature with time due to water heating.
(viii) $Y = \frac{A}{1 + B\exp(-ex)}$	Logistic growth curve	Relationship between the growth of organism or tissue with time.
(ix) $Z = \frac{X - X_1}{A + BX}$, where $Z = \log\left\{\frac{20Y}{\log(200 - Y)}\right\}$	Sigmoid Curve	Relationship between accumulated dry weight with tune.
(x) $Y = A + B\,\mathrm{Sin}\,(\omega t + \theta)$ $A + B\cos(\omega t + \theta)$	Sine Curve Cosine curve	Variation of air or soil temperatures, net radiation, temperature and humidity fluxes with time.
(xi) $Y = AX^B \exp(CX)$	Horl's function	Rainfall-runoff relations.
(xii) $Y = AX_1^B X_2^C$	Cobb-Douglas function	Variation of response of input variables on grain yield.

In order to develop the desired function and to have a graphical presentation of the response between the datasets, method of least squares is commonly adopted. In order to fit these curves let us use the data set with evaluation of different parameters involved.

(i) ***Linear equation***: The method of least square is based on the principle that the difference of observed and estimated values of 'Y' should be minimum for any number of observations. We have 6 observations on grain yield (Y) and AET/PET ratio's (X) and these should follow the least square minima expressions in the following example.

Example 14:

Develop linear relation of variations in pearl millet grain yield and moisture adequacy index (AET/PET) during the growing season.

Grain yield (q/ha)	AET/PET $-X_1$	Thermal Time (day °C)$-(X_2)$
5	0.5	1050
7	0.6	1100
11	0.7	1150
17	0.8	1400
20	0.9	1550
22	1.0	1600

$\sum_1^6 Y = 82$ $\quad \sum_1^6 X_1 = 4.5$ $\quad \sum_1^6 X_2 = 7850$ $\quad \sum_1^6 X_1X_2 = 6105$

$\sum_1^6 X_1Y = 68$ $\quad \sum_1^6 X_1^2 = 3.55$ $\quad \sum_1^6 YX_2 = 115600$ $\quad \sum_1^6 X_2^2 = 10557500$

$\sum_1^6 X_1^2Y = 58.24$ $\quad \sum_1^6 X_1^3 = 2.925$ $\quad \sum_1^6 X_1^4 = 2.4979$

Therefore
$$\sum_1^6 Y = B\sum_1^6 X + 6A = 4.5B + 6A$$

or
$$82 = 4.5 \times B + 6A \qquad \ldots (2.55)$$

$$\sum_1^6 XY = B\sum_1^6 X^2 + A\sum_1^6 X$$

or
$$68 = B \times 3.55 + 4.5A \qquad \ldots (2.56)$$

Solving these equations we get B = 4.53, A = 10.269

Linear representative equation is $Y = -8.476 + 29.5238X$, $r = 0.98$ and graphically represented in Fig. 2.5(a)

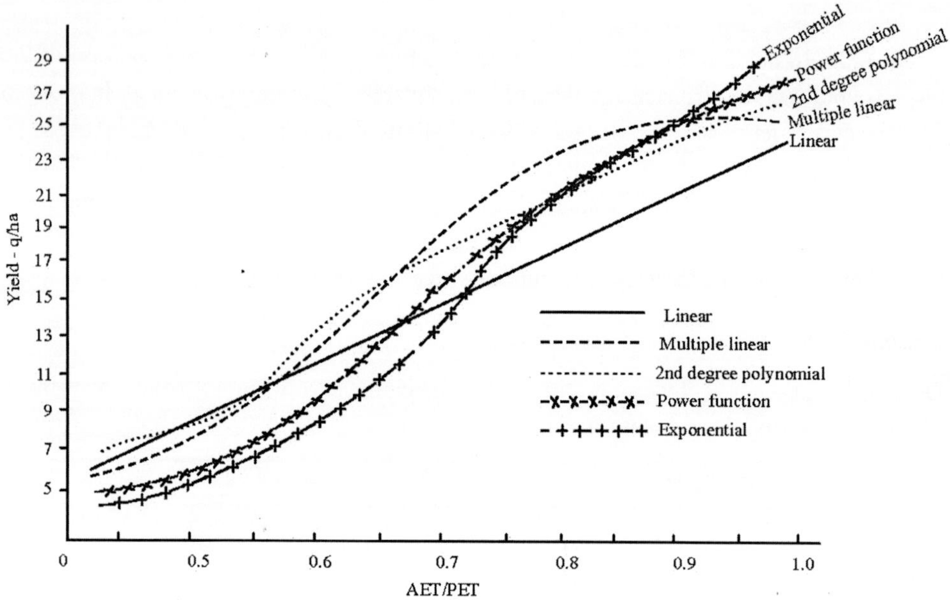

Fig. 2.5: (a) Linear (b) Multilinear, (c) Polynomial of 2nd degree, (d) Power function (e) Exponential form relationship between grain yield and moisture adequacy index (AET/PET).

(ii) ***Multiple linear equation***: We have two independent variables AET/PET ratio (X_1) and total thermal time (X_2), using the method of least squares to estimate constants A_0, A_1, and A_2, we need to solve following three equations (using data of example 14)

$$82 = 6A_0 + 4.5A_1 + 7850A_2$$

$$68 = 4.5A_0 + 3.55A_1 + 6105A_2$$

$$115600 = 7850A_0 + 6105A_1 + 10557500A_2,$$

The multiple linear equation after solving for A_0, A_1 and A_2, we get

It is graphically represented in Fig. 2.5b.

$$y = -21.784 + 12.413X_1 + 0.01998X_2 \qquad \ldots (2.57)$$

(iii) ***Polynomial of n degree of freedom:***

$$y = A + B_1X + B_2X^2 + \ldots B_nX^n \qquad \ldots (2.58)$$

Let us consider second degree polynomial of AET/PET ratio alone. The least square equations will be of the form (using data of example 14)

$$82 = 6A + B_1\, 4.5 + 3.5B_2$$

$$68 = 4.5A + 3.5B_1 + 2.925B_2 \qquad \ldots (2.59)$$

$$58.26 = 3.55A + 2.925B_1 + 2.4979B_2$$

Solving these equations for A, B_1 and B_2 wet get

$$y = -1.913 + 6.01\frac{AET}{PET} + 18.981\frac{AET}{PET}$$

and it is graphically represented in Fig. 2.5(c)

(iv) ***Logarithmic function***:

$$y = AX^{B} + C \qquad \ldots (2.60)$$

$$\text{Log}\,(y - C) = \text{Log}\,A + B \log X$$

C is estimated from the expression

$$C = \frac{Y_1 Y_2 - Y_3}{Y_1 + Y_2 - 2Y_3} \qquad \ldots (2.61)$$

where Y_1 and Y_2 are ordinates for two points (X_1, Y_1) and (X_2, Y_2) selected near extremities of a smooth curve drawn by plotting the data series. Y_3 is ordinate of the point (X_3, Y_3) such that $X_3 = \sqrt{X_1 X_2}$. Using data of example 14 let us take two points (0.5, 6.2), (1, 22.2) and (0.70, 11.6) we get C value of 0.59. A straight line curve is fitted between (Y – C) and log X providing constant log A as 1.38515 and B as 2.4483. Therefore power function fitted is of the form.

$$Y = 24.2745\, X^{2.4483} + 0.59,\ r = 0.988 \qquad \ldots (2.62)$$

It is graphically presented in Fig. 2.5(d)

(v) ***Exponential form***: The general form of exponential equations is

$$Y = \exp\{A + BX\} + C \qquad \ldots (2.63)$$

$$\text{Ln}\,(Y - C) = A + BX$$

A straight line function is fitted between Ln (Y – C) and X to obtain constant values of A as –0.0839 and B as 3.3368

Therefore exponential function to the data series fitted is of the form given below and graphically represented in Fig. 2.5(e)

$$Y = \exp\{-0.0839 + 3.3368\} + 0.59,\ r = 0.97 \qquad \ldots (2.64)$$

Another exponential form useful in agroclimatology is of the form

$$Y = A\,\exp^{BX} + C \qquad \ldots (2.65)$$

After evaluating the constants it will appear as

$$Y = 0.9195\,\exp\{3.3368X\} + 0.59,\ r = 0.97 \qquad \ldots (2.66)$$

(vi) ***Hyperbola function***:

$$Y = \frac{X}{BX \pm A} \qquad \ldots (2.67)$$

It can be expressed as

$$\frac{1}{Y} = B \pm A.\frac{1}{X}$$

A linear regression is filled between 1/Y and 1/X to evaluate the constants B and A. Since the data series is not of hyperbola type. Therefore it is not plotted here. Among all the above fitted curves in Fig. 2.5, the multilinear curve is better fitted with observed data on moisture adequacy index and thermal time data being additional factor.

(vii) ***Gompertz function***: The expression of the function are

$$Y = A B^{C^X} \quad \ldots (2.68)$$

$$Y = A B^{C^X} + D \quad \ldots (2.69)$$

These curves are fitted with data having small initial slope followed by a period of rapid increasing slope, then followed by a period of small decreasing slope. Accumulation of dry matter, pest or disease occurrence followed by rapid multiplication will suit to this type of curve. Table 2.5 provides a data of dry matter accumulation in pearl millet for 24 days after tillering. The data is divided into three equal segments for equal entries. The columns of Y, log Y, log(Y – 68) are prepared. Then sum of log Y is each group are denoted as S_1, S_2, S_3. Then constants A, B and C are assessed as

Table 2.5

Dry matter accumulation of pearl millet (gm/plant) from tillering to anthesis on different days after tillering is recorded as

Time in Days after Tillering	Accumulated Dry matter (Y)	log Y	Y – 68	log(Y – 68)	δY	$\frac{\Delta Y}{\Delta X}$
Tillering (X)	68					
2	70	1.8451	2	0.3010	2	1.0
4	73	1.8633	5	0.6989	3	1.5
6	80	1.9031	12	1.0792	7	3.5
8	95	1.9777	17	1.2304	15	7.5
10	106	2.0253	38	1.5798	11	5.5
12	120	3.0792	52	1.7160	14	7.0
14	135	2.1303	67	1.8865	15	7.5
16	145	2.1613	77	1.8865	10	5.0
18	160	2.2041	92	1.9638	15	7.5
20	173	2.2380	105	2.0211	13	6.5
22	182	2.2601	114	2.0569	9	4.5
24	200	2.2787	132	2.1206	8	4.0

$S_1 = 7.5892$ $S_1 = 3.3095$

$S_2 = 8.3961$ $S_2 = 7.0083$

$S_3 = 8.9809$ $S_3 = 8.1624$

$n = 4$ $n = 4$

$$C = \left[\frac{S_2 - S_3}{S_1 - S_2}\right]^{\frac{1}{n}} \quad \ldots (2.70)$$

$$\log A = \frac{1}{n}\left[S_1 - \frac{S_1 - S_2}{1 - C^n}\right] \quad \ldots (2.71)$$

$$\log B = \frac{(S_1 - S_2)(1 - C)}{(1 - C^n)^2} \quad \ldots (2.72)$$

when X is zero Y value is 68 is taken as D and value of log(Y – D) is determined instead of log Y to fit the curve. Using table 2.5 data these expressions takes the form.

$$Y = 426(0.1503)^{0.9226^x} \quad \ldots (2.73)$$

$$Y = 148(0.0106)^{0.7474^x} + 68.0 \quad \ldots (2.74)$$

Modified Gompertz equation appeared closer to observed data but not better curve fit (Fig. 2.6)

(viii) ***Logistic Growth Curve:***

The most common form of the growth curve is

$$Y = \frac{A}{\left(1 + Be^{-CX}\right)}, \quad \ldots (2.75)$$

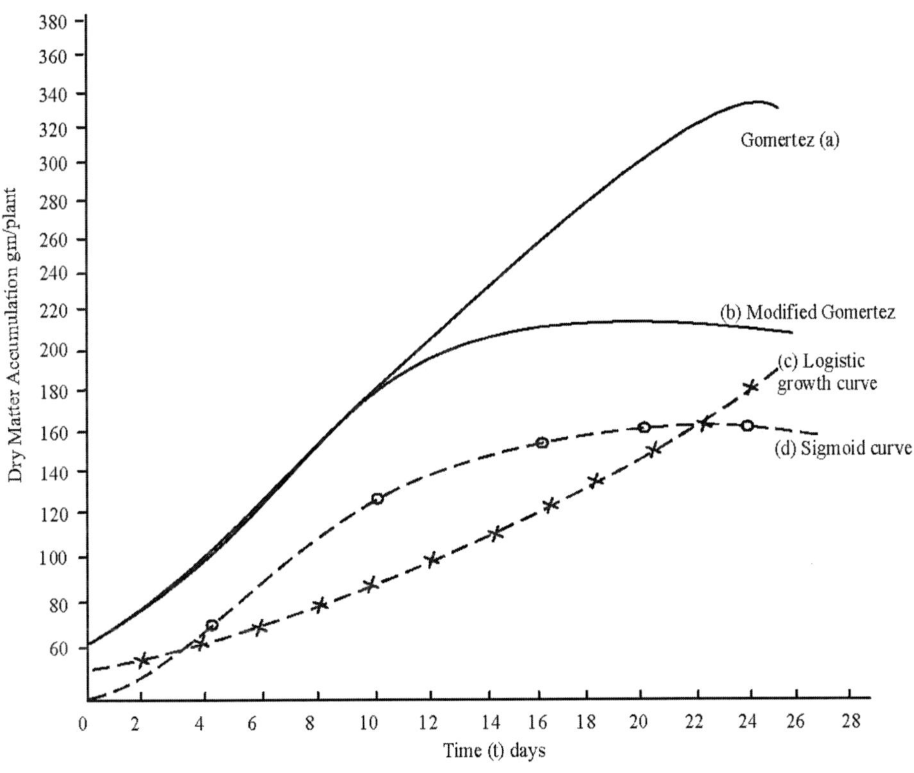

Fig. 2.6: (a) Gomertez (b) Modified Gomertez (c) Logistic growth curves (d) Sigmoid curve.

where X as time variable and Y is growth size. Constants A, B, C are characteristics of the organism dry matter, rate of growth, rate of multiplication of organism population. In this growth curve the time (X) at which the growth rate is maximum is

$$X = \frac{1}{C}\text{Ln B},$$

At this time of maximum growth, Y will become A/2 and growth rate $\frac{dY}{dX}$ will be AC/4. Therefore from the existing data we can determine the value of X when growth rate is maximum.

Using data of dry matter in Table 2.5, the growth rate is maximum after 8, 14 and 18 days of tillering. Let us take, the observation as maxima growth rate around 18 days after tillering as X, when growth is 160 gm/plant. The growth rate $\frac{dY}{dX}$ is 7.5 gm/plant/day. Using the above expression we get A of 320, B as 5.4059 and C as 0.09375 and the logistic curve takes the form

$$Y = \frac{320}{1+5.4059\{\exp(-0.09375X)\}} \quad \ldots (2.76)$$

This curve fits much closer to the observed data (Fig. 2.6c). It can be also fitted as a straight line best fit between $\text{Ln}\left(\frac{1}{Y}-\frac{1}{A}\right)$ and X by using A as 320. Constant B and C are given by intercept on Y-axis as $\text{Ln}\left(\frac{B}{A}\right)$ and by slope as –C.

(ix) ***Sigmoid curve***: The general form of sigmoid curve is

$$Z = \frac{X - X_1}{A + BX}, \text{ where} \quad \ldots (2.77)$$

$$Z = \log\left\{\frac{20Y}{\log(200 - Y)}\right\} \text{ and } X_1 \text{ is value of X at } Y = 0$$

This expression can be expressed so as to fit a straight line

$$\frac{X - X_1}{z} = A + BX \quad \ldots (2.78)$$

After evaluating the constants for the biomass data in Table 2.5, the sigmoid curve takes the form as (neglecting the data at X = 0 and 24)

$$\frac{X}{z} = 0.5414 + 0.4048X,\ r = 0.99 \quad \ldots (2.79)$$

The shape of curve is shown in Fig. 2.6d.

(x) ***Sine Function***: A symmetric sine function is very important in agroclimatology to generate weather data from observed data series and for simulating variety of applications in crop models to generate data during the period for which it has not been recorded or filling the gap of missing data. The general form of the function is

$$Y = \overline{Y} + A \sin\left(\frac{2\pi}{k}t - \theta\right) \quad \ldots (2.80)$$

$$Y - \overline{Y} + A \cos\left(\frac{2\pi}{k}t - \theta\right)$$

where $\overline{Y}$ is mean response of function over the complete time period, A is half the range of maximum and minimum value of Y, k is number units of time in a single cycle to angular measure in radiation or temperature and θ is phase angle.

$$Y = \overline{Y} + A \cos\frac{2\pi}{k}t \cos\theta + A \sin\frac{2\pi}{k}t \sin\theta$$

$$Y = \overline{Y} + A_1 \cos\frac{2\pi}{k}t + A_2 \sin\frac{2\pi}{k}t, \text{ where, } A_1 = A\cos\theta,\ A_2 = A\sin\theta \quad \ldots (2.81)$$

$$A = \sqrt{A_1^2 + A_2^2} \text{ and } \tan\theta = \frac{A_2}{A_1}$$

Therefore equation is linear in A_1 and A_2. From the signs of coefficient A_1 and A_2, phase angle θ is determined. To illustrate computations let us use mean monthly temperature data in table 2.6 for fitting of sine curve.

Let U_1 is $\cos\frac{2\pi}{k}t$ and U_2 is $\sin\frac{2\pi}{k}t$, then the expression in a linear form can be written as

$$Y = \overline{Y} + A_1U_1 + A_2U_2 \quad \ldots (2.82)$$

Using the principle of least square, this equation is solved for constants for n replicated values of Y at t, the regression coefficients are estimated as

$$\overline{Y} = \frac{\Sigma Y}{n.k} = \frac{\Sigma T_t}{n.k} \quad \ldots (2.83)$$

$$A_1 = \frac{\Sigma YU_1}{n.\Sigma U_1^2} = \frac{\Sigma T_tU_1}{n.\Sigma U_1^2} \quad \ldots (2.84)$$

$$A_2 = \frac{\Sigma YU_2}{n.\Sigma U_2^2} = \frac{\Sigma T_tU_2}{n.\Sigma U_2^2} \quad \ldots (2.85)$$

Assuming that all evenly spaced series of the same length k it may hold

$$\Sigma U_1^2 = \Sigma U_2^2 = \frac{k}{2}.$$

Therefore $A_1 = \dfrac{\Sigma T_t U_1}{\frac{1}{2} k f}$ and $A_2 = \dfrac{\Sigma T_t U_2}{\frac{1}{2} k f}$

where T_t is total of values of Y at each t for n replications.

Example 15:

Use table 2.6 data of monthly mean temperature of 21 years at an semiarid tropical location for fitting sine curve.

Solution:

Replication of data series (n) = 20
Intervals per cycle of temperature data (k) = 12

Estimated value of $\frac{2\pi}{k} = \frac{360}{12} = 30°$ or 0.52359 radian

Assessed values of U_4 and U_2 for different t values are expressed in tabular data.
Estimated values of $\Sigma T_t U_1 = 937.58$ and $\Sigma T_t U_2 = 467.18$
Therefore values of constants of sine curve are

$$\overline{Y} = \frac{5757.6}{20 \times 12} = 23.99 \simeq 24°C$$

$$A_1 = \frac{937.58}{20 \times 5.9996} = 7.8137$$

$$A_2 = \frac{467.18}{20 \times 5.9996} = 3.8934$$

Therefore the representative relationship can be expressed as
$Y = 23.99 + 7.8137\, U_1 + 3.8934\, U_2$

or

$$Y = 23.99 + 7.8137 \cos\frac{2\pi}{k}t + 3.8934 \sin\frac{2\pi}{k}t$$

The seasonal range of temperature or amplitude in the mean temperature is

$$A = \sqrt{A_1^2 + A_2^2} = \sqrt{61.0539 + 15.1585} = 8.73$$

$$A = 8.73$$

Phase angle $\tan\theta = \frac{A_2}{A_1} = 0.4983$

Since A_1, A_2 are both positive, therefore θ will fall in first quadrant. Therefore θ = 26.4870° or 0.4622 radian (360° = 6.283185 radians) using trigonometric tables.

Value of months from t_0 in the cycle at which the maximum temperature is reached = $\frac{12\theta}{2\pi}$ = 0.8828 months. Since t_0 corresponds to mid June, this places the maximum temperature around 26th June and minimum 6 months after around December 26th.

$$Y = 24 + 8.73 \cos(0.52359\,t - 0.4622)$$

$$Y - 24 + 8.73 \cos\left(\frac{2\pi}{k}t - 26.5°\right),$$

where t is the number of the month and this curve is graphically represented in Fig. 2.7.

Shape of the response curve will appear as

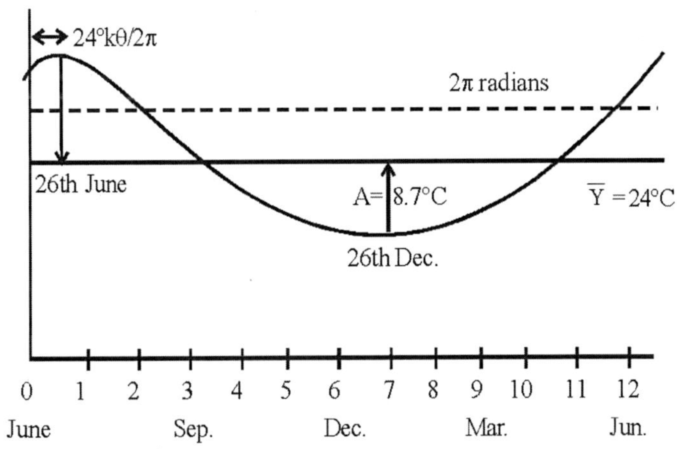

Fig. 2.7

(xii) ***Horl's functions***: These functions are useful for hydrometeorological studies between two variables such as relationship between rainfall and uneffective rainfall which is not usable towards meeting evapotranspiration needs, rainfall and runoff relations where the dependent variable increases with independent variable to a certain maximum and then decreases with further increase of independent variables. Such variations are expressed as Horl's function of the form

$$Y = A X^B \exp(CX) \qquad \ldots (2.86)$$

It can be expressed as linear form of the shape

$$\text{Ln } Y = \text{Ln } A + B \text{ Ln } X + CX$$

Let us use following data of unutilized rainfall based on 100 mm available soil moisture

Table 2.6

Monthly maximum, minimum and mean temperature (°C) for 20 years at Hisar station (Lat 29°10'N and Long 75°46'E)

Year	Jan.	Feb.	Mar.	Apr.	May	Jun.	Jul.	Aug.	Sep.	Oct.	Nov.	Dec.	Temperature
1	19.8	25.5	31.7	36.3	41.9	39.9	35.1	34.9	34.4	31.6	28.4	22.4	Max.
	5.5	5.3	11.8	18.2	23.6	27.1	26.1	25.9	23.0	15.7	10.0	7.9	Min.
	12.6	15.4	21.8	27.2	32.8	33.5	30.5	30.4	29.2	23.6	19.2	15.2	Mean
2	20.5	23.0	30.2	36.5	40.3	40.9	36.5	33.9	35.2	33.8	28.5	29.0	
	5.3	7.4	12.7	18.7	23.2	27.1	26.1	25.6	23.2	17.0	10.2	5.9	
	12.9	15.2	21.5	27.6	31.7	34.0	31.3	29.2	29.2	25.4	19.4	17.5	
3	20.6	27.7	29.0	36.8	35.9	41.7	40.3	38.4	37.7	35.2	30.2	23.1	
	5.6	8.8	13.0	16.0	19.3	25.8	28.4	28.0	23.4	17.6	8.9	5.2	
	13.1	16.8	21.0	26.4	27.6	33.8	34.3	33.2	30.5	26.4	19.5	14.2	
4	22.2	24.0	28.9	38.0	43.1	40.2	34.6	33.5	34.1	32.4	28.1	22.7	
	5.2	7.8	11.3	18.2	25.3	27.0	27.0	25.3	23.5	16.4	10.2	6.7	
	13.7	15.9	20.1	28.1	34.2	33.6	30.8	29.4	28.8	24.4	19.2	14.7	
5	18.9	22.6	28.0	34.6	41.5	39.0	37.2	36.7	35.7	34.5	28.1	19.4	
	4.1	5.2	11.2	14.3	21.4	26.0	25.9	26.1	22.2	15.4	10.4	5.6	
	11.5	13.9	19.6	24.5	31.5	32.5	31.5	31.4	29.0	25.0	19.2	12.5	
6	22.2	21.8	26.7	35.1	40.2	40.8	35.2	35.1	33.9	32.0	27.7	22.8	
	6.1	9.1	11.3	17.6	24.9	26.5	26.5	25.4	23.7	15.6	9.9	6.0	
	14.2	15.5	19.0	26.3	32.5	33.6	30.8	30.2	28.8	23.8	18.8	14.4	
7	19.7	21.7	27.5	33.6	39.7	39.2	39.3	34.9	35.3	33.2	28.0	22.6	
	4.0	7.5	11.2	16.3	22.3	25.9	27.6	25.6	22.2	13.9	9.9	7.6	
	13.8	14.6	19.4	25.0	31.0	32.6	33.5	30.2	28.7	23.5	18.9	15.1	
8	20.8	21.1	27.8	34.4	38.7	40.8	35.0	33.7	33.6	32.0	27.7	24.0	
	6.3	6.6	10.4	15.5	21.3	25.7	24.7	24.9	20.8	14.7	9.6	6.3	
	13.5	13.8	19.1	25.0	30.0	33.2	29.8	29.3	27.2	23.4	18.7	15.2	

Contd.

Year	Jan.	Feb.	Mar.	Apr.	May	Jun.	Jul.	Aug.	Sep.	Oct.	Nov.	Dec.	Temperature
9	19.7	25.1	26.5	35.4	41.4	40.7	35.3	37.5	33.7	34.1	29.6	23.9	
	6.0	8.8	10.6	16.2	23.2	26.1	25.4	25.7	23.4	14.9	10.3	4.4	
	12.8	16.9	18.6	25.8	32.3	33.4	30.4	31.6	28.5	24.5	20.0	14.2	
10	20.7	22.8	30.7	35.0	41.5	41.5	35.2	33.2	33.7	33.4	28.5	23.8	
	6.3	6.1	10.5	15.6	23.4	26.4	26.2	25.2	21.9	14.9	10.3	4.4	
	13.5	14.5	20.6	25.3	32.5	34.0	30.7	29.2	27.8	24.2	19.3	15.2	
11	18.6	22.3	27.4	34.4	40.9	42.1	37.4	32.4	34.3	34.1	29.4	22.9	
	5.4	8.2	10.5	16.6	22.7	25.9	27.3	25.8	23.5	17.1	29.4	22.9	
	12.0	15.3	19.0	25.5	31.8	34.0	32.4	29.1	28.9	25.6	19.4	14.4	
12	20.1	23.6	28.8	36.7	39.8	37.5	36.7	34.0	34.4	33.2	28.7	23.1	
	4.3	6.8	11.7	16.0	22.3	24.8	25.8	24.6	22.3	15.7	8.1	2.3	
	12.2	15.2	20.2	26.4	31.1	31.2	31.2	29.3	28.4	24.5	18.4	12.7	
13	20.1	23.2	27.4	33.2	37.6	37.5	37.2	33.5	35.3	28.0	25.2	6.6	
	1.9	4.8	11.1	16.3	20.4	23.8	25.4	24.1	23.0	16.2	10.3	6.6	
	11.0	14.0	19.2	24.8	29.0	30.6	31.3	29.3	29.2	22.1	17.8	11.4	
14	19.0	22.5	26.2	36.3	41.9	40.0	36.1	36.3	33.5	31.4	27.9	21.6	
	3.8	7.5	13.2	17.9	23.2	25.7	26.5	26.8	24.6	19.6	11.4	6.4	
	11.4	15.0	19.7	27.1	32.5	32.	31.3	31.5	29.0	25.5	19.6	14.0	
15	17.2	22.7	30.3	39.7	40.9	39.6	37.5	36.4	36.6	34.6	30.4	23.6	
	7.7	8.6	11.1	16.2	23.0	25.1	26.9	24.6	23.7	15.8	9.0	3.7	
	12.5	15.6	20.7	28.0	32.0	32.3	32.2	30.5	30.2	25.2	19.7	13.7	
16	18.4	20.3	29.0	38.8	41.5	39.5	35.6	36.7	36.5	36.5	29.5	24.4	
	5.4	6.0	9.5	17.5	25.5	25.8	25.2	24.9	21.3	14.8	8.9	3.1	
	11.9	13.2	19.3	28.2	33.5	32.7	30.4	30.8	28.9	25.6	19.2	13.7	
17	18.2	24.5	30.0	35.6	40.2	36.3	34.8	34.4	36.4	34.6	29.7	22.8	
	3.0	5.6	9.5	16.4	22.9	23.3	24.5	23.5	21.6	17.6	9.4	6.3	

Contd.

Year	Jan.	Feb.	Mar.	Apr.	May	Jun.	Jul.	Aug.	Sep.	Oct.	Nov.	Dec.	Temperature
	10.6	15.0	19.7	26.0	31.5	29.8	29.6	29.0	29.0	26.1	19.5	14.5	
18	19.8	22.5	29.4	38.2	42.0	40.0	39.8	37.0	34.4	33.9	26.7	23.8	
	4.5	6.4	11.5	18.3	26.8	27.6	28.5	27.1	21.6	16.9	9.8	6.4	
	12.2	14.5	20.5	28.2	34.4	33.8	34.2	32.0	28.0	25.4	18.2	15.1	
19	17.4	23.8	31.7	38.3	38.8	38.6	39.2	33.9	36.4	31.1	28.8	22.4	
	6.4	6.7	11.7	19.6	23.3	24.6	27.4	25.0	22.5	16.4	9.3	6.2	
	11.9	15.3	21.7	29.0	31.1	31.1	33.3	29.5	29.5	23.8	19.1	14.3	
20	18.3	20.5	27.3	35.3	39.0	41.0	34.6	36.0	33.4	33.2	28.4	21.4	
	4.0	7.4	13.1	15.2	21.3	25.6	26.1	26.8	23.2	15.7	9.5	2.8	
	11.2	14.0	20.2	25.3	30.2	33.3	30.4	31.4	28.3	24.5	19.0	12.1	
ΣMean (T_t) Temp-erature	248.5	304.6	400.9	529.7	643.2	655.8	629.9	606.5	578	493.3	382.1	285.1	ΣT_t =5757.6
Value of t	7	8	9	10	11	0	1	2	3	4	5	6	
$u_1 = \cos\frac{2\pi t}{k}$	-0866	-0.5	0	0.5	0.866	1	0.866	0.5	0	-0.5	-0.866	-1	
$u_2 = \sin\frac{2\pi t}{k}$	-0.5	-0.866	-1	-0.866	-0.5	0	0.5	0.866	1	0.866	0.5	0	
u_1T_t	-215.20	-152.30	0	106.18	557.01	655.8	545.49	303.25	0	-246.65	330.90	-285.1	$\Sigma u_1T_t = 937.58$
u_2T_t	-124.25	-263.78	-400.9	-458.72	-321.6	0	314.95	525.23	578.00	427.20	191.05	0.0	$\Sigma u_2T_t = 467.18$

capacity during the month of July at a semiarid subtropical location with loamy sand soil type.

July rainfall (mm)	30	50	75	100	150	200	250	300	350	400	450	500	700	1000
Uneffective rainfall (mm)	5	12	30	50	50	115	140	160	190	225	250	280	390	800

Uneffective rainfall is the prepitation falling during the growing season of crop that is not available towards meeting the consumptive use of the crops.

This equation is of the multilinear equation form when constant Ln A, B and C are determined using least square equations procedure already given. On solving these expressions for three constants Ln A= -3.1102, B = 1.4699 and C = -0.0004134. Therefore Horl's equation takes the form as

$$Y = 0.04459\, X^{1.4699} \exp\{-0.00004134\} \qquad \ldots (2.87)$$

(xii) ***Cobb-Douglas function***: This function represents the response of input parameters in productivity such as fertilizer uptake (X_1), soil moisture utilization (X_2). The function is of the form expressed as

$$Y = A X_1^{B} \times X_2^{C} \qquad \ldots (5.88)$$

The values of the constants may be determined using least square method by transforming the above expression as

$$\text{Log } Y = \log A + B \log X_1 + C \log X_2$$

This a linear expression among log Y and log X, and log X_2.

Example 16:

Use the following field experiment data of pearl millet yield data (kg/ha) in relation to N nutrient uptake (kg/ha) and soil moisture utilized from sowing to maturity (in cm) in sandy loam soil.

Pearl Millet Yield (kg/ha) (Y)	**N-Nutrient Uptake (kg/ha) (X_1)**	**Soil Moisture Used (cm) (X_2)**	**Biomass Straw Yield (kg/ha)**
884	65	14	3158
1188	92	20	3610
1288	114	21	4113
1248	95	23	3722
984	74	21	3515
1558	90	23	4945
1674	102	23	5165
1645	100	23	5047
625	32	37	4523

762	41	39	5511
901	50	40	6137
900	49	40	6112
1335	111	32	3705
1700	132	32	4190
1920	170	33	4385
1708	142	38	4225
1344	101	33	4628
1964	117	35	5301
2105	126	36	5490
1992	122	35	5240
702	36	39	4790
783	44	40	5530
903	53	41	6150
895	51	41	6120

N=24

Let $\alpha = \log y$, $\beta = \log X_1$ and $\gamma = \log X_2$

$\Sigma\alpha = 73.92$ $\Sigma\beta = 45.59$ $\Sigma\gamma = 35.68$

$\Sigma\alpha^2 = 228.34$ $\Sigma\beta^2 = 87.60$ $\Sigma\gamma^2 = 53.37$

$\Sigma\alpha\beta = 141.3$ $\Sigma\alpha\gamma = 109.92$ $\Sigma\beta\gamma = 67.57$

Using these values, he three least square method expressions are

$$73.92 = 24\,A_0 + 45.59\,B + 35.68\,C$$

$$141.30 = 45.59\,A_0 + 87.6\,B + 67.57\,C$$

$$109.92 = 35.68\,A_0 + 67.57\,B + 53.37\,C$$

Solving these expression for A, B and C, we get

$A_0 = 0.10045$, $B = 1.023$ and $C = 0.6974$

$\text{Log}\,Y = 0.10045 + 1.023\ \text{lox}\,X_1 + 0.6974 \log X_2$

Therefore $Y = 1.2602 \times X_1^{1.023} \times X_2^{0.6974}$

Substituting X_1 and X_2 values, estimated yield is compared with observed grain yield.

Exercise 1

Ex. 1: Using Appendix I data for July month, determine mean, mode, median, and skewness values of maximum, minimum and mean temperature data.

Ans. July maximum temperature 36.7, 34.2, 34.6°C; 36.3°C; 3.32

July minimum temperature 26.3, 26.9, 26.1°C; 26.3°C, -0.54

July mean temperature 31.5; 32.2, 31.3, 30.7, 30.3; 31.2°C; 0.81°C

Ex. 2: Determine standard deviation, variance and coefficient of variation of July temperature values of Appendix 1 data.

	July		
	Max.	Min.	Mean
S.D.	2.08	1.10	1.48
σ	2.10	1.12	1.50
Variance	4.33	1.21	2.19
CV(%)	5.67	4.17	4.70

Ex. 3: Determine mean, mode, median, standard deviation, variance, coefficient of variation and skewness for the monsoon season rainfall data (mm) at a subtropical semiarid location.

Year	**Rainfall**
1961	375
1962	490
1963	418
1964	358
1965	176
1966	444
1967	602
1968	171
1969	634
1970	506
1971	322
1972	246
1973	194
1974	253
1975	664
1976	714
1977	439
1978	538
1979	348
1980	194

1981	352
1982	255
1983	391
1984	330
1985	425
1986	300
1987	173
1988	561
1989	156
1990	453
1991	300
1992	355
1993	492
1994	366
1995	471
1996	360
1997	453
1998	414
1999	184
2000	122
2001	538
2002	92
2003	432
2004	189
2005	458

Ans: 371.3; 194, 300, 453, 538 mm; 366 mm; 149.8 mm; 22437 mm²; 40%; yes small

Ex. 4. Test the homogeneity of July temperature data given in appendix I.

Ex. 5. Test the homogeneity of rainfall data given in exercise 3 above.

Ex. 6. Compute correlation coefficient and plot the regression line with use of following data.

Minimum Temperature (°C) (Y)	Sunshine Hours (X)
9.5	4.5
13.4	1.8
7.6	4.6
12.0	1.2
2.4	7.2
1.7	8.5

1.2	8.8
4.8	2.2
11.1	2.6
13.2	5.9
7.5	7.0
2.5	8.2
-1.5	9.1
1.1	8.8
7.8	1.7

(Ans: Y = 13.259 – 1.275, r = –0.77, SE = ±3)

Ex. 7. Using S.W monsoon seasonal rainfall data given in example 2 and exercise 3 in continuity, determine the 2 to 5 years moving averages and examine for any trend or periodicity in rainfall data. (Fig. 2.8)

Ex. 8. Use the following relative humidity data to fit the normal distribution curve recorded during the months of March and April, 1997 at semiarid subtropical location. Also (i) apply the chi-square test for goodness of frequency distribution, and (ii) determine the probability of relative humidity greater than 80 per cent and less than 40 per cent values.

March, 1997	RelativeHumidity (%)	April, 1997	Relative Humidity (%)
1	61	1	41
2	79	2	33
3	70	3	31
4	75	4	24
5	61	5	31
6	62	6	51
7	59	7	25
8	71	8	27
9	67	9	53
10	66	10	44
11	68	11	40
12	61	12	52
13	89	13	29
14	75	14	26
15	72	15	46
16	65	16	47
17	57	17	38
18	59	18	47
19	55	19	37
20	49	20	69

21	59	21	62
22	49	22	66
23	47	23	63
24	59	24	73
25	58	25	79
26	39	26	85
27	50	27	84
28	47	28	95
29	56	29	84
30	36	30	78
31	20		

Ans.

R.H. Range(%)	Normal Distribution Frequency	Observed Frequency
0-10	-	-
10.1-20	1	1
20.1-30	3	5
30.1-40	7	8
40.1-50	11	10
50.1-60	14	11
60.1-70	12	13
70.1-80	8	8
80.1-90	4	4
90.1-100	1	1
	61	61

Prob. for RH >80% = 82 per cent

Prob. for RH <20% = 1.6 per cent.

Ex. 9. Determine the extreme values of wind speed at F value of 0.98 and 0.50 using following wind speed data at the standard height of 10m recorded on a coastal location frequently receiving cyclones from the Bay of Bengal. Use type II distribution function.

Year	Wind Speed in mph
2005	80
2004	56
2003	43
2002	76
2001	110
2000	90
1999	75
1998	120

1997	150
1996	100
1995	78
1994	158
1993	60
1992	180
1991	104
1990	86
1989	175
1988	70

(**Ans:** $\beta_2 = 90.94$, $\gamma = 23.18$, wind speed (0.98) = 107.6 mph., wind speed (0.50) = 92.4 mph.)

Ex. 10. Using the wind speed data given in exercise above use type I distribution to quantify wind speed values at return period 2, 5 and 10 years.

(Ans. $\hat{\alpha} = 4.5102$, $\hat{\beta} = 0.04313$)

Return period	Wind Speed Limits in mph	
2 yr	93.7	88.3
5 yr	97.5	84.8
10 yr	100.8	82.3

Ex. 11. Use the following daily rainfall data (mm) in the first fortnight of July to determine the probability of a wet spell of 3 days and a dry spell of 5 days for a fruitful germination of rainy season crops.

JULY

Year	1	2	3	4	5	6	7	8	9	10	11	12	13	14	15
1991	8.1	-	11.0	10.0	1.2	11.0	-	-	-	-	-	-	32.5	2.5	-
1992	-	11.0	-	-	-	-	-	-	-	-	10.0	-	-	50.0	7.6
1993	39.4	-	-	-	-	-	-	-	19.8	19.0	-	7.6	-	-	-
1994	-	22.9	23.0	2.5	5.0	2.5	12.0	-	-	5.0	2.5	50.4	7.6	-	-
1995	-	-	21.3	-	4.0	-	-	-	-	-	-	32.0	-	-	-
1996	-	-	9.6	-	100.5	-	-	-	1.0	2.5	7.5	2.0	33.0	-	-
1997	4.3	-	1.3	6.3	6.0	12.0	-	31.0	4.0	-	0.8	5.4	2.0	-	-
1998	-	-	-	-	38.0	7.6	-	-	-	-	-	-	-	15.0	-
1999	-	-	-	5.0	-	-	63.0	35.0	30.0	7.6	5.0	10.0	-	-	-
2000	35.5	56.0	-	-	-	-	-	-	2.5	-	12.0	45.0	-	-	2.5
2001	-	-	-	10.0	-	-	22.5	22.6	22.5	7.6	25.4	-	-	-	-
2002	-	7.6	33.0	-	-	2.5	7.6	17.5	-	30.0	-	5.0	15.0	-	-
2003	6.4	82.5	-	63.0	11.4	40.5	-	-	-	-	12.0	-	-	15.0	10.0
2004	-	-	20.0	35.6	-	-	20.0	17.5	22.5	40.0	45.0	37.0	-	20.0	5.0
2005	2.5	7.6	31.0	11.9	2.6	-	-	-	-	1.5	-	70.0	16.0	37.0	-

Ex. 12. Use the following data of rainy month (August) rainfall data (mm) to represent the

(a) frequency distribution with 20 mm of class interval

(b) determine mean, mode, median, standard deviation, variance, coefficient of variation, skewness.

(c) fit the normal distribution curve and apply χ^2 test to test the null hypothesis.

Year	Rainfall
1976	250
1977	144
1978	83
1979	108
1980	171
1981	62
1982	67
1983	119
1984	157
1985	23
1986	78
1987	79
1988	85
1989	18
1990	105
1991	48
1992	41
1993	44
1994	133
1995	158
1996	54
1997	72
1998	49
1999	110
2000	100
2001	125
2002	57
2003	206
2004	107
2005	144

Ex. 13. Develop linear regression between the following two variables alongwith correlation coefficient and standard error of estimate.

Year	Pearl Millet Grain Yield (kg/ha)	Seasonal Rainfall (mm)
1966	418	165
1967	519	280
1968	265	158
1969	552	320
1970	939	450
1971	708	212
1972	502	145
1973	725	144
1974	233	205
1975	605	299
1976	574	531
1977	321	324
1978	440	437
1979	337	332
1980	549	190
1981	582	275
1982	648	269
1983	656	283
1984	639	348
1985	488	371
1986	454	238
1987	232	20
1988	1219	701
1989	696	76
1990	863	293
1991	565	193
1992	1168	267
1993	646	416
1994	1260	535
1995	711	5
1996	1134	320

Ex. 14. Determine statistical variables for long series of rainfall and temperature data of your location.

Ex. 15. Determine the assured probability levels (80 per cent) and critical probability level (50 per cent) values for weekly rainfall data at your location using at least 30 years of past record?

Ex. 16. Conduct risk analysis of weekly minimum temperature during December and January, and weekly maximum temperature during the May and June at your location? And determine the probability of extreme temperatures i.e., above 45°C and below 3°C?

Ex. 17. Using record of crop yield data (25 to 30 years) of main crops at your region determine the

(i) Technology trend of different main crops at your location

(ii) Develop multiple regressions for assessing crop yield of important crops of your region with the use of corresponding years of agrometeorological data of your agromet. observatory.

Ex. 18. Test the crop yield multiple regressions developed under the above exercises for the current year prediction of crop yield of main crops at your location using the observatory agromet data, and develop a format for food grain policy for your region?

Ex. 19. Fit the modified Gompertz equation to the following data of variation of pest populations in cotton plant with time in minutes at favourable temperature and relative humidity.

Time (t) in Minutes	Pest Populations Average Number/Plant
0 (starting)	66.7
2	68.8
4	72.5
6	79.5
8	90.5
10	105.0
12	123.0
14	140.0
16	156.0
18	171.0
20	185.0
22	196.5
24	206.0
26	215.4

Ans. $[Y = 66.7 + 173\,(0.116)^{0.76^{t}}]$

Ex. 20. Fit the logistic growth curve to the following data on dry matter accumulation of 5 plants of rice crop as a function of time.

Time in days (t)	Accumulated root dry matter (gm)
20	2.00
30	3.84
40	6.20
50	9.90
60	14.14
70	27.50
80	47.50
90	54.46
120	58.33
140	58.65

Ans: $Y = \left[\dfrac{95}{1 + 706.3\exp(-0.082t)}\right]$

Ex. 21. Fit the sine curve to the maximum and minimum temperatures data of Table 2.6 and determine the variations of maxima and minima temperatures w.r. to the mean temperature variations of Fig. 2.7.

Ex. 22. Fit the sine curve to monthly soil temperature at 5 cm soil depth recorded at your location. Determine the lag between soil and air temperature variations (exercise 21) at your locations.

Ex 23. Develop Cobb-Douglas functions of the form grain yield = $A x_1^B x_2^C$, where x_1 and x_2 are two important input variables influencing crop yield variations at your locations (fertilizer, soil moisture, PAR interception, plant protection controls) for main crops of your location assuming that package of practices for the crops are fully adopted.

Ex. 24. Develop the curve fitting for pest and disease multiplication data of main crops using agromet data of the observatory of your location and develop a format for issuing weather advisory to the farmers of the region.

Ex. 25. Emergence of sorghum midges, Contarinia Sorghicola (Coquillett), were observed to be a function of soil temperature and rainfall. Adults midges initiated spring emergence from April 1 above a threshold 10 cm soil temperature of 14.8°C as (Baxendale and Teetes, 1983)

Initiation of emergence = 431°C days
50% emergence = 679°C days
95% emergence = 977°C days

Develop the heat unit development model for emergence of midges in the spring.

Ans. $\left[\text{Cumlulative \% Emergence} = \dfrac{100}{1 + \exp(8.8452 - 0.0127\sum HU)}\right]$

References

Baier, W. 1973. Crop weather analysis model: Review and model development. Jour. of Applied Meteorology, 12, 6, p. 937-47.

Baier, W. and Russela, D.A. 1968. A computer programme for estimating risks of irrigation requirements from climatic data. Tech. Bull. 2059, p. 1-48.

Baxendale, F.P. and G.L. Teetes 1983. Thermal requirements for emergence of over wintered sorghum midge. Environ. Entomol., 12, 4, p. 1078-82.

Baxendale, F.P., G.L. Teetes, P.J.H. Sharpe 1984a. Temperature – dependent model for sorghum midge spring emergence. Environ. Entomol., 13, 6,, 1566-71.

Baxendale, F.P., G.L. Teetes, P.J.H. Sharpe and H.Wu. 1984b. Temperature – dependent model for development of nondiapausing sorghum midges. Environ. Entomol., 13, 6, p. 1572-76.

Bishnoi, O.P. 1998. Predicting crop yields through phonological and plant's biophysiological parameters. Jour. of Applied Hydrology, XI, 283, p. 1-10.

Bishnoi, O.P. 1980. The behaviour of moisture adequacy index and its utilization for exploiting the agricultural potential in Punjab and Haryana. Mausam, 31, 1, p. 157-64.

Bishnoi, O.P. 1985 Risk analysis of temperatures in the northwestern plains, p. 1-5. Proceeding of 2nd India Agrometeorological society Meetings held at GAU, Anand.

Bishnoi, O.P. 1987. Annual progress report of DST project on a study of growth and water use simulation submodels in pearl millet. DST|ES|92|011|87, p. 1-35.

Bishnoi, O.P. and Saxena, K.K. 1979. Precipitation simulation process with markov chain modeling, Proc. First International Conf. on Stat. Climatology, held in Tokyo, Nov. 29 – Dec. 1, 1979.

Bridges, D.C. and J.M. Chandler. 1989. A population level temperature dependent model of seedling johnsongrass flowering. Weed Scien., 37, p. 471-77.

Gangopadhyaya, M. and R.P. Sarkar 1964. Curvilinear study on the effect of weather on the growth of sugarcane. Indian Jour. Met. Geophys., 15, p. 215-26.

Gumbel, E.J. 1954. Applied Math Series No. 33, National Bureau Standards, U.S. Deptt. of Commerce, NOAA.

Gumbel, E.J. 1958. Statistics of extremes. Columbia University Press, New York.

Lieblein, J. 1954. A new method of analyzing extreme value data. U.S. Nat. Adv. Comm. For Aero., Tech. Note No. 3053.

McQuigg, J.D. 1975. Economic impacts of weather variability. University of Missouri - Columbia., Deptt. of Atmospheric Science, p. 256.

Mooley, D.A. 1973. Gamma distribution probability model for Asian summer nonsoon monthly rainfall. Monthly Weather Review, US Deptt. of Commerce, NOAA, Vol. 101, 2, p. 160.76.

Mooley, D.A. 1974. Suitable tables for application of gamma probability model to rainfall. Research report - RR018, Indian Institute of Tropical Meteorology, Poona, p. 1-19.

Pearson, K. 1934. Tables of incomplete Beta function. Biometrika Office, University College, London, pp. 1-494.

Pearson, K., et al., 1951. Tables of incomplete gamma T-function. Cambridge Univ. Press, London.

Robertson, G.W. 1968. A biometrological time scale for a cereal crop involving day and night temperature and photoperiod. Int. J. Biometeorol., 12, 3, p. 191-213.

Robertson, G.W. 1974 Wheat yield for 50 years at swift current, Saskatchewan, in relation to weather Can. J. Plant Sci., 54, p. 625-50.

Shaw, R.H. 1954. Leaf and air temperatures under freezing conditions. Plant Physiology, 29, p. 102-04.

Subbaramayya, I and Rupa Kumar, K. 1980. Crop weather relationships of sugarcane and yield prediction in northeast Andhra Pradesh, India. Agricultural Meteorology, 21, p. 265-79.

Thom, H.C.S. 1971. Some methods of climatological analysis. WMO Tech. Note 81, WMO, Geneva, p. 1-52.

Ventskevich, G.Z. 1958. Agricultural Meteorology, Gidrometeorizdat, Leningrad, p. 376.

Yao, Y.M. 1969. Agricultural Meteorology, 6, p. 259-73.

3

Applied Climatic Water Budgeting

3.1 Introduction

Climatic water budgeting is a simple book keeping procedure of daily/weekly/monthly/yearly comparison of precipitation with the climatic demand of water at a place. The hydrological balance equation is a useful balance fo parameters of water availability and water needs at a place expressed as

$$P + I = AET/PET \pm \Delta SM + RO + D \qquad \ldots (3.1)$$

where P is prepitation, I is amount of irrigation applied, AET or PET is actual or potential evapotranspiration, $\pm\Delta SM$ is the change in soil moisture storage which is either depleted or replenished, RO is surface runoff and D is deep percolation or subsurface runoff below the root zone. All terms are expressed in the same units. In any water budget procedure, we are encountered with problems of either soil moisture deficit or surplus with respect to the climatic demand. If still water surplus exists after meeting the climatic demand, it may accumulate to produce runoff at the surface if intensity of rainfall is much higher than the percolation rate or at the subsurface level below root zone where plants are unable to extract it. In order to assess quantitatively these parameters it is desired to collect rainfall data and other meteorological variables to quantify potential evapotranspiration with the help of expressions given by several research workers. The most popular among these are Penman (1948) and its modified forms of the combination approaches depending on the availability of data. In arid to moist subhumid and tropical to sub-tropical locations, the climatic water budgeting procedure provides several useful parameters of interest to explain soil moisture variability with time. Therefore the availability of soil moisture in arid to moist subhumid and tropical to subtropical regions is most crucial and different applications of climatic water balance procedure will answer the ways and means to tackle such problems.

3.2 Climatic Water Budgeting

Thornthwaite (1948), Thornthwaite and Mather (1955) brought out this concept to quantify the moisture status at a place. The procedure is illustrated with the following example of computing monthly climatic water balance at Chandigarh (a subhumid location) assuming the sandy loam soils have a soil moisture detention value of 150 mm in the root zone of vegetations. Table 3.1 shows the monthly variations of precipitation (P) and potential evapotranspiration (PET). Subtraction of PET from P shows either positive or negative values of (P – PET) indicating when water supply is greater or less than the need in that month. Cumulative positive value of (P – PET) will build up the soil water storage. Storage is recorded corresponding to Σ(P – PET) value (Appendix IV). When storage is above the upper limit of soil moisture retention of soil, it will percolate as subsurface runoff. Actual evapotranspiration (AET) equal PET when soil moisture is at water holding capacity of the soil, otherwise equals the precipitation plus change in soil moisture storage i.e., utilization of reserved soil water. Water surplus is the amount above the water holding capacity of the soil root zone, whereas water deficit is the difference between PET and AET.

$$\text{AET} = \text{PET when P} > \text{PET otherwise AET} = \text{P} + \Delta\text{ST}$$

$$\text{Storage} = \text{Field capacity}\left[\exp\{-\Sigma(\text{P} - \text{PET})/\text{field capacity}\}\right] \quad \ldots(3.2)$$

There are two check equations to verify the correctness of water budgeting.

$$\text{Total AET} + \text{Total WD} = \text{Total PET} \quad \ldots(3.3)$$

$$\text{Total AET} + \text{Total WS} = \text{Total P} \quad \ldots(3.4)$$

Here depth of water stored in the root zone of soil at field capacity is an important soil moisture regulating parameter. Depth of the rooting of the plants varies within different species. Sandy soils have deeper roots as compared to less deeply rooted in clays. The total amount of water that can be held in the root zone of the soil should be either measured or estimated from soil and vegetation types. The different parameters can also be plotted on a graph called climatic water budget diagram (Fig. 3.1). Figure clearly represents the periods

Table 3.1
Monthly climatic water balance at Chandigarh

(All values in mm)

Parameter	Jan.	Feb.	Mar.	Apr.	May	Jun	Jul.	Aug.	Sep.	Oct.	Nov.	Dec.	Year
PET	20	38	105	150	180	200	165	150	120	88	52	36	1304
P	50	30	25	20	10	100	250	300	180	58	20	30	1073
P-PET	30	-8	-80	-130	-170	-100	85	150	60	-30	-32	-6	-231
∑(P-PET)	-28	-36	-116	-246	-416	-516				-30	-62	-68	
Storage (ST)	124	117	68	28	9	5	90	150	150	122	98	94	
Chang in storage (ΔST)	+30	-7	-49	-40	-19	-4	+85	+65	0	-28	-24	-4	
AET	20	37	74	60	29	104	165	150	120	86	44	34	923
Water surplus (WS)	0	0	0	0	0	0	0	90	60	0	0	0	150
Water deficit (WD)	0	1	31	90	150	96	0	0	0	2	8	2	381
Water surplus carried over from previous month	6	3	1	0	0	0	0	0	45	52	23	13	
Total available surplus	6	3	1	0	0	0	0	90	105	52	26	13	
Estimated Runoff (½WS) (RO)	3	2	1	0	0	0	0	45	53	26	13	7	150

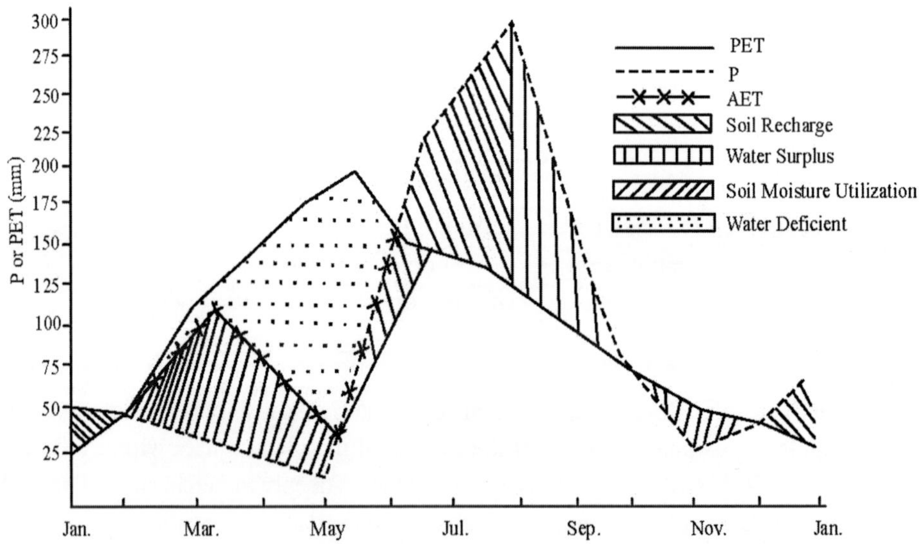

Fig. 3.1: Monthly climatic water balance.

of utilization and recharge of soil moisture. AET equals PET until precipitation drops below PET. The amount and period of moisture surplus, moisture deficit, soil moisture utilization and soil moisture recharge can be clearly identified. All there parameters are very useful for different applications in hydrometeorology and agroclimatology particularly irrigation scheduling, agroclimatic classification, runoff estimation, drought and wet spells quantification and crop yield estimations.

Example 1:

Compute the monthly climatic water balance of a location with following rainfall and potential evapotranspiration values having soil moisture retention value of 200 mm.

Solution:

Climatic water balance (all values in mm)

Parameter	Jan.	Feb.	Mar.	Apr.	May	Jun	Jul.	Aug.	Sep.	Oct.	Nov.	Dec.	Year
PET	53	58	65	86	99	129	87	84	88	66	63	55	943
P	62	73	87	149	106	71	66	127	90	257	283	142	1513
P–PET	9	15	22	63	07	-58	-31	43	2	191	220	87	570
Σ(P–PET)					0	-58	-89						
Storage	200	200	200	200	200	150	128	171	173	200	200	200	
ΔST	0	0	0	0	0	-50	-22	+43	+2	+27	0	0	
AET	53	58	65	86	99	121	88	84	88	66	63	55	926
WD	0	0	0	0	0	8	9	0	0	0	0	0	17
WS	9	15	22	63	7	0	0	0	0	164	220	87	587

3.3 Climatic Water Balance with Two Soil Layers Model

Baier and Robertson (1966) considered the soil profile to be made up of a large number of layers to quantify recharge and water loss of moisture from soil. Palmer (1965) adopted a two layer model of varying field capacity one upper surface layer of 25 mm of field capacity and second underlying layer of a certain quantity of available moisture depending on the water holding capacity of soil of the location. With arrival of rains, after evaporation upper layer is first recharged to its full capacity and then lower layer is recharged. Runoff takes place after both the soil layers have attained field capacity. However when rainfall is less than potential evapotranspiration or there is no rain, then loss of moisture takes place from upper layer at potential rate and then moisture is extracted from lower layer. It is slightly complex method, but provides better results for moisture status in the root zone and can be adopted after comparison with actual measured soil moisture in situ.

Procedure is illustrated with the following example.

Example 2:

Using Palmer (1965) approach of two soil layers – upper surface layer of 25 mm and lower layer of 125 mm of available water holding capacity of the soil in the root zone determine moisture status with following data.

(All values in mm)

Month	P	PET	P-PET	ΔS_s	ΔS_l	Ss	S_l	S	PR	R	PL	L	AET	RO	WD
Jan.	12	53	-39	0	-13	0	37	37	100	0	18	13	25	0	28
Feb.	23	58	-35	0	-9	0	28	28	113	0	14	9	32	0	26
Mar.	17	65	-48	0	-9	0	19	19	122	0	12	9	26	0	39
Apr.	9	86	-77	0	-10	0	9	9	137	0	11	10	19	0	67
May	6	99	-93	0	-6	0	3	3	141	0	6	6	12	0	87
Jun	71	129	58	0	-1	0	2	2	147	0	3	1	72	0	127
Jul.	166	97	69	25	4	25	46	71	148	69	1	0	97	0	0
Aug.	127	84	43	0	3	25	89	114	79	43	43	0	84	0	0
Sep.	90	88	2	0	2	25	91	116	36	2	62	0	88	0	0
Oct.	25	66	-41	-25	-10	0	81	81	34	0	50	35	60	0	6
Nov.	3	63	-60	0	-32	0	49	49	59	0	34	32	35	0	28
Dec.	14	55	-41	0	-13	0	36	36	101	0	18	13	27	0	28

ΔS_s – loss of moisture from surface layer = S_s' or (PET – P) whichever is smaller

ΔS_l – loss of moisture from lower layer = (PET – P - ΔSs) S_l'/AWC ...(3.5)

S'_s – available moisture in surface layer in beginning of the month

S_l' – available moisture in lower layer in beginning of the month

S', S – total available moisture in root zone at start and at end of the month

AWC – combined available water capacity of both layers

PR – potential recharge to bring both layers to field capacity i.e., (AWC – S') where S' is total available moisture in the beginning of month. To initiate the calculations we have assumed a figure of 50 mm at the beginning of January month. R is actual recharge of the month

PL – potential loss is the maximum loss of moisture from both the layers assuming there is no rainfall i.e., PL = $PL_s + PL_l$

PL_s = PET or S'_s whichever is smaller

PL_l = (PET – PL_s) S_l'/AWC ...(3.6)

PRO = potential runoff is taken as total available moisture at beginning of month

RO = Actual runoff after recharge of both layers to field capacity.

3.4 Applications of Climate Water Budgeting

3.4.1 Water Budget over Different Periods

Climatic water budget can be done over different periods of interest from daily to weekly, monthly, seasonal or over annual periods in order to tackle the problem over a location, region or global spatial scale. Day to day soil moisture variations are of critical importance to crop growers and planning of off road operations. There are several factors which influence the soil water balance, such topography, soil type, infiltration rate, soil moisture retention capacity of the soils, root zone depth etc. Computation of daily water balance is illustrated with following example for a location having 150 mm of water stored in the beginning of the month.

Example 2:

Daily climatic water balance of a location with rainfall and potential evapotranspiration values (mm) given below:

Date	P	PET	P-PET	Σ(P-PET)	Storage (S) (150 mm)	±ΔS	AET	WD	WS
1	-	8	-8	-8	142	-8	8	0	0
2	12	9	3	-5	145	3	9	0	0
3	-	9	-9	-14	136	-9	9	0	0
4	30	8	22	0	150	14	8	0	8
5	-	9	-9	-9	141	-9	9	0	0
6	-	9	-9	-18	132	-9	9	0	0
7	-	9	-9	-27	125	-7	7	2	0
8	10	8	2	-25	127	2	8	0	0
9	-	+9	-9	-34	119	-8	8	1	0
10	3	8	-5	-39	114	-5	8	0	0
11	-	8	-8	-47	109	-5	5	3	0
12	-	8	-8	-55	103	-6	6	2	0
13	2	8	-6	-61	99	-4	6	2	0
14	1	9	-8	-69	96	-6	7	2	0
15	-	8	-8	-77	89	-4	4	4	0
16	-	8	-8	-85	84	-5	5	3	0
17	15	8	7	-73	91	7	8	0	0
18	-	6	-6	-79	87	-4	4	2	0
19	-	6	-6	-85	84	-3	3	3	0
20	20	8	12	-66	96	12	8	0	0
21		8	-8	-74	90	-6	6	2	0
22	1	8	-7	-81	86	-4	5	3	0
23	-	9	-9	-90	82	-4	4	5	0
24	-	10	-10	-100	76	-6	6	4	0
25	-	10	-10	-110	71	-5	5	5	0
26	-	10	-10	-120	66	-5	5	5	0
27	-	10	-10	-130	62	-4	4	6	0
28	-	10	-10	-140	58	-4	4	6	0
29	05	10	-5	-145	56	-2	7	3	0
30	4	10	-6	-151	53	-3	7	3	0
Month	103	258					192	66	8

Therefore computing water balance over shorter periods provide more refined picture of different variables of climatic water balance parameter as compared to larger periods of monthly or seasonal basis.

3.4.2 Scheduling of Irrigation

Irrigation scheduling is based on the availability of soil moisture conditions in the root zone to meet the climatic demand. With the daily water budgeting, the soil moisture depletion will indicate how much to irrigate and at what time? Shallow rooted crops will have to be irrigated more frequently but with less amount of irrigation. If irrigation is regulated according to the track of the soil moisture conditions, no significant moisture deficiency can develop in the soil to limit growth and development with much loss of available water. To illustrate these features let us solve the following example by computing water balance using two different irrigation schedules.

Example 3:

Using the example 2 daily data of precipitation and potential evapotranspiration examine which one of following irrigation schedule criteria is better (a) an irrgation of 25 mm soil water when storage in the root zone is less than 125 mm, (b) an irrigation of 50 mm soil water where storage in the root zone is less than 100 mm. Assuming field capacity of root zone being 150 mm.

Solution:

Computing the water balance parameter in both situations will answer it (a) 25 mm of irrigation when storage becomes less than 125 mm.

(All values in mm)

Date	P	PET	P-PET	Σ(P-PET) (150 mm)	Storage (S)	±ΔS	AET	WD	WS
1	-	8	-8	-8	142	-8	8	0	0
2	12	9	+3	-5	145	3	9	0	0
3	-	9	-9	-14	136	-9	9	0	0
4	30	8	22	0	150	14	8	0	8
5	-	9	-9	-9	141	-9	9	0	0
6	-	9	-9	-18	132	-9	9	0	0
7	-	9	-9	-27	125	-7	7	2	0
8	10	8	2	-25	127	2	8	0	0
9	-	9	-9	-34	119	-8	8	1	0
10	3+(25)	8	-23	-8	142	23	8	0	0
11	-	8	-8	-16	134	-8	8	0	0
12	-	8	-8	-24	127	-7	7	1	0
13	2	8	-6	-30	122	-5	7	1	0

14	1+(25)	9	16	-12	138	16	9	0	0
15	0	8	-8	-20	131	-7	7	1	0
16	0	8	-8	-28	124	-7	7	1	0
17	15+(25)	8	32	0	150	26	8	0	5
18	0	6	-6	-6	144	-6	6	0	0
19	0	6	-6	-12	138	-6	6	0	0
20	20	8	12	0	150	12	8	0	0
21	0	8	-8	-8	142	-8	8	0	0
22	1	8	-7	-15	135	-7	8	0	0
23	0	9	-9	-24	127	-8	8	1	0
24	0	10	-10	-34	119	-8	8	2	0
25	0(+25)	10	15	-16	134	15	10	0	0
26	0	10	-10	-26	126	-8	8	2	0
27	0	10	-10	-36	117	-9	9	1	0
28	0+(25)	10	15	-18	132	15	10	0	0
29	5	10	-5	-23	128	-4	9	1	0
30	4	10	-6	-29	123	9	1	0	0
Month	103+(125) = 228	258					243	15	13

(b) Similarly computing water balance with second criteria:

Date	**P**	**PET**	**P-PET**	**Σ(P-PET)**	**Storage (S) (150 mm)**	**±ΔS**	**AET**	**WD**	**WS**
1	0	8	-8	-8	142	-8	8	0	0
2	12	9	3	-5	145	3	9	0	0
3	0	9	-9	-14	136	-9	9	0	0
4	30	8	22	0	150	14	8	0	8
5	0	9	-9	-9	141	-9	9	0	0
6	0	9	-9	-18	-132	-9	9	0	0
7	0	9	-9	-27	125	-7	7	2	0
8	10	8	2	-25	127	2	8	0	0
9	0	9	-9	-34	119	-8	8	1	0
10	3	8	-5	-39	114	-5	8	0	0
11	0	8	-8	-47	109	-5	5	3	0
12	0	8	-8	-55	103	-6	6	2	0
13	2	8	-6	-61	99	-4	6	2	0
14	1+(50)	9	+42	-9	141	42	8	0	0
15	0	8	-8	-17	133	-8	8	0	0
16	0	8	-8	-25	127	-6	6	2	0

17	15	8	7	-16	134	+7	8	0	0
18	0	6	-6	-22	129	-5	5	1	0
19	0	6	-6	-28	124	-5	5	1	0
20	20	8	12	-14	136	12	8	0	0
21	0	8	-8	-22	129	-7	7	1	0
22	1	8	-8	-30	122	-7	8	0	0
23	0	9	-9	-39	114	-8	8	1	0
24	0	10	-10	-49	107	-7	7	3	0
25	0	+10	-10	-59	100	-7	7	3	0
26	0+(50)	10	+40	-10	140	40	10	0	0
27	0	10	-10	-20	131	-9	9	1	0
28	0	10	-10	-30	122	-9	9	1	0
29	5	10	-5	-35	118	-4	9	1	0
30	4	10	-6	-41	113	-5	9	1	0
Month	103+(100)	258					232	26	8

Therefore the water balance situation in all cases are as:

Criteria	PET	P	Irrigation (I)	P+I	AET	WD	WS
No irrigation	258	103	0	103	194	64	0
(a) 25 mm irrigation every time when storage is 125 mm or less	258	103	125	228	243	15	13
(b) 50 mm irrigation every time when storage is 100 mm or less	258	103	100	203	232	26	8

Therefore comparing the results it is better to provide 5 irrigations of 25 mm each time when storage capacity falls below 125 mm where the AET is higher and water deficiency is less.

3.4.3 Climatic Classification

Thornthwaite and Mather (1955) Carter and Mather (1966) outlined the water balance procedure to provide parameters for climatic classification where moisture and thermal indices are basic characteristics for biological environment. Four basic parameters defined were

(i) **Moisture Index (M.I.)**: It is expressed as ratio of availability of water w.r. to the water need at a place expressed as (P - PET)×100/PET demarcating the land area into different zones as for following criteria:

Moisture Index	Moisture Zone	Representation
<-66.7	Arid	E
-33.3 to -66.7	Semiarid	D
0 to -33.3	Dry subhumid	C_1
0 to +20	Moist subhumid	C_2
20 to 40	Humid	B_1
40 to 60	Humid	B_2
60 to 80	Humid	B_3
80 to 100	Humid	B_4
>100	Per humid	A

(ii) **Seasonal variation of effective moisture**: In moist climates aridity index, whereas in dry climates humidity index are quantified to examine seasonal variations of moisture index according to following criteria:

In Moist Climates	Aridity Index (WD/PET × 100)	Representation
Little or no water deficiency	0-10	r
Moderate summer water deficiency	10-20	s
Moderate winter water deficiency	10-20	w
Large summer water deficiency	20-40	s_2
Large winter water deficiency	20-40	w_2
In Dry Climates	**Humidity Index (WS/PET × 100)**	**Representation**
Little or no water surplus	0-16.7	d
Moderate winter water surplus	16.7-33.3	s
Moderate summer water surplus	16.7-33.3	w
Large winter water surplus	>33.3	s_2
Large summer water surplus	>33.3	w_2

(iii) **Index of thermal efficiency and (d) summer concentration type**: Thermal energy of the region is assessed with PET value at a location i.e. available energy will be total utilized for its coversion into water vapours. These are categorized as

Thermal Efficiency PET (cm)	Climate Type	Presentation	Summer Concentration (%)	Summer Concentration Type
<14.2	Frost	E'		
14.2-28.5	Tundra	D'	>88	d'
28.5-42.7	Microthermal	C_1'	76.3-88.0	c_1'
42.7-57.0	Microthermal	C_2'	68.0-76.3	c_2'
57.0-71.2	Mesothermal	B_1'	61.6-68.0	b_1'
71.2-85.5	Mesothermal	B_2'	56.3-61.6	b_2'
85.5-99.7	Mesothermal	B_3'	51.9-56.3	b_3'
99.7-114.0	Mesothermal	B_4'	48.0-51.9	b_4'
114.0-128.2	Megathermal	A_1'	43.0-48.0	a_1'
128.2-142.4	Megathermal	A_2'	38.0-43.0	a_2'
142.4-156.6	Megathermal	A_3'	33.0-38.0	a_3'
156.6-171.0	Megathermal	A_4'	<33.0	a_4'

Table 3.2

Climatic shifts

Location	Climate Type	Percentage number of years having the climate types of					Percentage shift from normal climate towards	
		Arid	Semiarid	Dry Subhumid	Moist Subhumid	Humid	Drier Climate	Wetter Climate
Hisar	Arid	55	45				-	45
Ambala	Semiarid	4	62	27	4	3	4	34
Chandigarh	Dry subhumid	-	48	44	4	4	48	8
New Delhi	Semiarid	13	77	7	3	-	13	10
Pathankot	Dry subhumid	-	8	56	24	12	8	36

Precipitation is highly variable component as compared to other climatic variables in nature and therefore moisture regime also varies temporarily due to changes of climatic conditions from their normal behaviour almost in all regions, but more prominent in arid to semiarid and tropical to subtropical conditions. Therefore when the climatic shift takes place towards positive moisture regime, the excess moisture status are useful in subsequent dry periods. However when the climatic shift is towards negative moisture regime, the moisture scarcity deteriorates the normal conditions. Normally the arid locations have more likely shifts towards water scarcity, but semiarid locations have shifts towards both sides. Subhumid locations have higher shifts towards the higher moisture regimes. Table 3.2 shows the trends of few locations as an example (Bishnoi 1980). Such temperaray climatic shifts are exploited for better moisture conservation and shifting for scheduling improved crop planning of the region.

Example 4:

Represent the climatic classification of following locations with the use of corresponding climatic water balance parameters.

Solutions:

	Locations (All values in mm)				
Parameter	**A**	**B**	**C**	**D**	**E**
Annual precipitation	430	700	800	1100	2000
Annual PET	1625	1450	1420	1300	800
Water deficit	1250	810	675	400	100
Water surplus	55	60	55	200	1300
AET	375	640	745	900	700
Moisture Index %	-73.5	-51.7	-43.7	-15.4	150
Aridity Index %	76.9	55.9	47.5	30.8	12.5
Humidity Index %	3.4	4.1	3.9	15.4	162.5
Summer Concentration %	25	35	39	45	65
Climatic classification Representation	$EdA_4'a_4'$	$DdA_3'a_3'$	$DdA_2'a_2'$	$C_1dA_2'a_1'$	$AsB_2'b_1'$

3.4.4 Drought and Wet Spells

(i) *Use of Aridity and Humidity Indices*

Aridity and humidity indices are used for quantifying the drought and wet spell conditions. Drought or wet spell's inception, withdrawal, its intensity and duration are determined using aridity index or humidity index over weekly or monthly periods of interest using water balance procedure for at lest 30 years of data of the location. The aridity or humidity index values are arranged in ascending or descending order and its mean, median and standard deviation values are determined. The drought/wet spells are categorized as per the following scale.

Mean + S.D./2	Moderate drought/slightly wet
Mean + (S.D./2 to S.D.)	Large drought/wet
Mean + (S.D. to 2 S.D.)	Severe drought/very wet
Mean + (>2 S.D.)	Disastrous drought/extremely wet.

These aspects are easy to compute and are illustrated with the following example.

Example 5:

Determine the years of different categories of drought and wet spells with use of following data on aridity index for an arid location.

Solution:

Year	Aridity Index (%)	Humidity Index (%)
1915	83.4	4.5
1916	65.8	12.6
1917	36.8	30.4
1918	89.5	4.0
1919	87.2	3.5
1920	86.2	3.5
1921	83.4	5.2
1922	76.2	6.7
1923	66.3	7.0
1924	72.4	6.5
1925	81.7	4.0
1926	68.7	8.1
1927	75.6	7.0
1928	70.6	6.2
1929	66.4	7.8
1930	75.6	6.8
1931	77.0	7.0
1932	85.6	4.0
1933	62.7	8.2
1934	73.1	6.0
1935	70.3	7.0
1936	85.3	4.0
1937	67.3	12.6
1938	92.0	2.5
1939	84.7	4.2
1940	81.9	5.0
1941	85.1	5.0
1942	64.5	7.0
1943	87.5	5.0
1944	69.0	6.2
1945	56.7	18.2

1946	79.7	7.2
1947	67	10.8
1948	75.7	7.2
1949	83.8	4.0
1950	79.3	5.2
1951	84.9	4.3
1952	79.3	6.5
1953	78.1	7.0
1954	73.3	6.8
1955	67.2	9.7
1956	65.4	15.7
1957	63.2	20.3
1958	65.0	22.7
1959	62.0	25.6
1960	60.1	26.7
1961	68.8	20.2
1962	71.4	10.7
1963	75.6	7.2
1964	78.4	6.4
1965	86.9	4.2
1966	73.7	5.0
1967	60.7	14.8
1968	87.6	14.8
1969	63.2	20.8
1970	68.6	18.7
1971	72.7	7.2
1972	79.9	8.0
1973	78.9	7.6
1974	83.9	5.2
1975	71.6	6.7
1976	51.0	25.4
1977	66.3	20.1
1978	61.0	20.9
1979	71.1	6.7
1980	82.2	7.2
1981	68.2	10.2
1982	71.4	8.0
1983	65.2	9.0
1984	78.5	7.0
1985	70.0	6.0
1986	78.7	7.2
1987	81.9	5.2
1988	63.9	12.7
1989	89.4	5.0
1990	68.8	7.9
1991	77.9	5.0
1992	73.8	4.2
1993	65.9	10.6
1994	74.4	8.0
1995	65.8	12.7

Solution:

Steps for drought and wet spells determination:

(i) Arrange the data of aridity/humidity index values in ascending or descending order.
(ii) Determine mean, median and standard deviation of the series.

	Aridity Series	Humidity Series
Mean	73.6	9.3
Median	73.7	7.0
Standard Deviation	9.6	6.2
N	81	81

(iii) Verify the years with impact points of agricultural and animal husbandry statistical data of different years as

Mild drought: AI (73.6 to 78.4%)	1922, 1927, 1930, 1931, 1948, 1953, 1963, 1964, 1966, 1991, 1992, 1994
Moderate drought: AI (78.4 to 83.2%)	1919, 1925, 1928, 1939, 1940, 1946, 1950, 1952, 1972, 1973, 1980, 1984, 1986, 1987
Severe drought: AI (83.2 to 92.6%)	1915, 1918, 1920, 1921, 1932, 1936, 1938, 1941, 1943, 1949, 1951, 1965, 1968, 1974, 1989
Extreme drought: AI (>92.8%)	Nil

Categories of different wet spells are

Moderately wet: HI (9.3 to 12.4%)	1947, 1955, 1962, 1981, 1993
Largely wet: HI (12.4 to 15.5%)	1916, 1937, 1967, 1988, 1995
Severely wet: HI (15.5 to 21.7%)	1945, 1956, 1957, 1961, 1969, 1970, 1977, 1978
Extremely wet: HI (>21.7%)	1917, 1958, 1959, 1960, 1976

(ii) *Palmer's concept of drought and wet spells*

(a) **Intensity of drought and wet spells**: Palmer (1965) described a better technique for computing the severity of drought and wet spells by computing he rainfall anomaly which is the difference between actual rainfall occurred at a place and the 'Climatically Appropriate for Existing Conditions' known as CAFEC precipitation given by

$$\hat{P} = \widehat{AET} + \hat{R} + \widehat{RO} - \hat{L} \quad \ldots (3.7)$$

All water balance parameters in terms of CAFEC for a location are expressed as

$$\hat{p} = \alpha \,.\, PE + \beta PR + \gamma\, PRO - \delta \,.\, PL \quad \ldots (3.8)$$

where α, β, γ and δ coefficients of evapotranspiration, recharge, runoff and loss determined with the long-term average values of these parameters computed in the two layer model of water balance procedure.

Therefore rainfall anomaly is given by $P - \hat{P}$ for the period. In order to make this anomaly comparable in space and time it is multiplied by a suitable weighing factor 'K' given by te expression (Palmer, 1965) as

$$K = 1.5 \log\left\{ \frac{\dfrac{\overline{PET} + \overline{R} + \overline{RO}}{\overline{P} + \overline{L}} + 2.80}{\overline{D}} \right\} + 0.50 \quad \ldots (3.9)$$

where $\overline{D}$ is mean of absolute values of anomaly of a allocation. Therefore a weighted anomaly (z) is computed as $(P - \hat{P})K/25.4$ which is comparable in time and space for computation of drought and wet spells in Palmer's Scale.

Palmer observed that on plotting the highest accumulated values of negative anomalies (z) over two climatically different regions against the corresponding period followed a linear relation representing extreme drought and dividing the ordinate of this line into 4 equal lengths to describe severe, moderate and mild droughts. The drought severity (X) was expressed as

$$X = \frac{\sum z}{0.309t + 2.691}, \quad \ldots (3.10)$$

where t is first month of drought.

The intensity of drought severity of month (X_i) accumulating with previous months X_{i-1} values and Z_i values of the current month expressed as

$$X_i = 0.897\, X_{i-1} + \frac{Z_i}{3} \quad \ldots (3.11)$$

Based on X_i values of index, Palmer described following qualitative description of drought and wet spells.

X_i Index Values	Description
≥ 4.00	Extremely wet
3.00 to 3.99	Very wet
2.00 to 2.99	Moderately wet
1.00 to 1.99	Slightly wet
0.50 to 0.99	Incipient wet spell
+0.49 to -0.49	Near normal
-0.50 to -0.99	Incipient drought
-1.00 to -1.99	Mild drought
-2.00 to -2.99	Moderate drought
-3.00 to 3.99	Severe drought
≤ - 4.00	Extreme drought

A drought is taken as established when index attained the value of -1.0 or less, and a wet spell is established when index acquires the value of 1.0 or greater. Drought or wet spell will cease as soon as the index reaches normal category between -0.5 to 0.5. Therefore intensity, beginning or end of a drought/wet spell, duration of drought or wet spell is better described in this procedure so as to quantify the real impact of drought/wet spells.

Example6:

The long-term average of the coefficient of evapotranspiration, recharge, runoff and moisture loss at a place are given as below for different months. Compute the CAFEC precipitation, and anomaly in rainfall of the example 2 data given above.

Month	α	β	γ	δ	K
Jan.	0.39	0.02	0.00	0.51	1.978
Feb.	0.33	0.00	0.00	0.92	2.153
Mar.	0.20	0.00	0.00	0.82	2.264
Apr.	0.17	0.00	0.00	0.88	3.275
May	0.07	0.00	0.00	0.78	3.003
Jun.	0.39	0.08	0.00	0.98	1.817
Jul.	0.84	0.28	0.00	0.00	0.820
Aug.	0.81	0.41	0.04	0.00	0.877
Sep.	0.82	0.30	0.01	0.04	0.850
Oct.	0.66	0.00	0.00	0.49	2.01
Nov.	0.43	0.00	0.00	0.28	3.05
Dec.	0.26	0.00	0.00	0.29	2.896

Solution:

Using example 2 data, PET, PR, PRO and PL monthly data the CAFEC precipitation are computed as

	Jan.	Feb.	Mar.	Apr.	May	Jun.	Jul.	Aug.	Sep.	Oct.	Nov.	Dec.
$\hat{P}$	14	6	3	5	2	59	122	103	82	19	17	9
P	12	23	17	9	6	71	166	127	90	25	3	14
$(P-\hat{P})$	-2	17	14	4	4	12	44	24	8	6	-14	5
Z_i	-0.15	1.44	1.25	0.51	0.47	0.86	1.42	0.83	0.27	0.47	1.68	0.57
X_i	0.05	0.43	0.81	0.90	0.96	1.15	1.51	1.63	1.55	1.55	0.83	0.94

Therefore slightly wet conditions remained for June to October months.

(b) **End and beginning of drought wet spell**: The anomaly Z_e required to end the prevailing drought or wet spell in one month can be determined by expression by substituting $X_i = \pm 0.5$ in case of drought and wet spell respectively. Therefore $Z_e = -2.691\ X_{i-1} \pm 1.5$

for end of drought and wet spell respectively. Also the index value can be near normal between -0.5 to +0.5 and with no change in index value $(X_i - X_{i-1}) = 0$. The Z value is $\pm$ 0.15 giving effective value of dryness or wetness of the month will be Z – 0.15 or Z + 0.15. Incipient wet spell will occur after the establishment of the drought, for the first time Z becomes < -0.15. Effective wetness of month will be Z + 0.15 and a monthly anomaly value required to end the prevailing drought at the start of month is $-2.691\, X_{i-1} - 1.5$. Therefore the ratio of $(Z + 0.15)/(-2.691\, X_{i-1} - 1.5)$ will provide the probability of end of the drought. If probability level is not significant (<1), it will be evaluated for the next month. The drought severity will decrease slowly and ultimately when probability level approaches 1.0.

Similarly incipient drought spell's probability is determined after the establishment of wet spell so that for the first time Z becomes ≤ 0.15. The effective dryness of the month is Z – 0.15, whereas Z value required to end in one month the prevailing wet spell at the start of the month is $-2.691\, X_{i-1} + 1.5$. Therefore the probability of enduing of wet spell is the ratio $(z - 0.15)/(-2.691\, X_{i-1} + 1.5)$. Once the index is > 1.0 for wet spell or <-1.0 for dry spell the conditions are established for wet or dry spell. It may be added that drought spells are establishing with accumulating effects of dry weather sequences of several months, but a wet spell can occur instantaneously due to heavy rains. The spatial and time aspects of these drought or wet spells can be depicted over a successive maps of continuous months. Therefore floods and inundations may occur several times a years, without warning while a long period rainless period give rise to a drought situation. Drought may be prolonged or short lived depending upon the nature and duration of water deficiency.

(iii) **Crop moisture index**: Palmer's meteorological drought index clearly demonstrates the intensity, severity, beginning and end of drought or wet spells and their durations in dry or wet weather conditions. However to monitor the effect of dry or wet weather conditions on crops, Palmer (1968) evolved a crop moisture index (PCMI) for "Weekly Weather and Crop Bulletin Reports" to represent the macroscale crop moisture conditions. Palmer's crop moisture index is a function of evapotranspiration deficit term of water balance procedure expressed as

$$\mathrm{PCMI} = \gamma + G, \qquad \ldots (3.12)$$

where γ is he index of evapotranspiration deficit expressed as

$$\gamma_i = 0.67\, \gamma_{i-1} + 1.8\, \mathrm{DE}, \qquad \ldots (3.13)$$

where DE is relative evapotranspiration anomaly for the week period and G is index of excessive moisture but never less than zero.

$$\mathrm{DE} = \frac{\mathrm{AET} - \overline{\mathrm{AET}}}{\sqrt{\alpha}} = \frac{\mathrm{AET} - \alpha\overline{\mathrm{PET}}}{\sqrt{\alpha}} \qquad \ldots (3.14)$$

$G_i = G_{i-1} + H + MR + RO$, where H is a "return to normal" factor, a function of G_{i-1}. H is 0, <0.5, 0.5 and 0.5 G_{i-1} for G_{i-1} values of 0, <0.5, 0.5 to 1.0 and >1.0 respectively. R is moisture recharge in the root zone, RO is runoff and M is average field capacity during the week expressed as

$$M = \frac{S'+S_s+S_L}{2(S'+PR)}, \qquad \ldots (3.15)$$

where S_s and S_L are soil water at the end of week in both layers of root zone, S' + PR is water storage capability of root depth zone. Regular weekly mps of crop moisture index are issued by U.S. Department of Agriculture and Commerce jointly (Palmer, 1968). Bishnoi (1980, 1984) developed crop moisture index (CMI) using moisture adequacy index (AET/PET) to represent the crop moisture conditions. When moisture adequacy index is comparable with optimum evapotranspiration rate crop will not suffer, but when it is lower than evapotranspiration needs crop will reflect stress conditions. Therefore the ratio of actual evapotranspiration to the CAFEC evapotranspiration for a location is greater than 1.0 reflects that the moisture supply is adequate to meet evapotranspirational needs above the CAFEC value. However when this ratio is less than 1.0 the crop conditions starts deteriorating with time. Therefore crop moisture index was expressed as

$$CMI_i = 0.3\ CMI_{i-1} + X_i, \text{ where} \qquad \ldots (3.16)$$

$$X_i = M_i\left(\frac{AET_i}{\alpha \times \overline{\overline{PET_i}}}\right), \qquad \ldots (3.17)$$

$$M_i = \frac{S'_{i-1}+S_i+RO_i}{2AWC}, \qquad \ldots (3.18)$$

where S_{i-1}', S_i, RO_i are soil moisture storage in the beginning and end of the week, runoff for the week respectively. α is optimum evapotranspiration rate to meet the crop water need without stress conditions. CMI index explained weekly crop conditions satisfactorily under rainfed conditions.

3.4.5 Natural Vegetation Distribution

Vegetative growth is in full bloom when conditions of moisture and heat are most favourable. The distribution of thermal efficiency and its seasonal variation are very essential for meeting the thermal time requirement and phenological development of vegetations. Moisture effectivity and its seasonal variation meet the evapotranspiration requirement of vegetations. The duration of dryness and wetness conditions governs the growth and development of vegetation sub-types. Therefore vegetation groups were compared with relative magnitude of aridity and humidity indices. Table 3.3 Shows the march of natural vegetations. Aridity indices are responsible for tropical thorn and dry forests, but with decrease of aridity index and increase of humidity index the vegetation types of temperature and wet evergreen forests increased. With a more wider spread of aridity and humidity indices, Carter and

Mather (1966) produced several classes of natural vegetations in United State closely linked (Fig. 3.2).

Table 3.3

Maturation vegetation types in relation to aridity and humidity indices

Location	Aridity Index (%)	Humidity Index (%)	Natural Vegetation
Barmer	80.7	0.0	Tropical thorn forest
Jodhpur	77.2	0.0	Tropical thorn forest
Ajmer	62.6	0.0	Tropical thorn forest
Ambala	40.6	0.0	Tropical dry deciduous
Buxar	29.3	0.0	Tropical dry deciduous
Bahraich	22.0	1.0	Tropical dry deciduous
Jamshedpur	20.7	9.2	Tropical dry deciduous
Bhopal	33.0	12.8	Moist temperate
Balasore	12.8	17.7	Moist temperate
Trivandrum	12.5	12.3	Wet temperate
Tejpur	2.7	42.9	Wet temperate
Shellong	0.6	155.5	Subtropical wet forest
Lumding	3.3	2.3	Tropical wet evergreen
Darjeeling	0.4	364.4	Wet evergreen

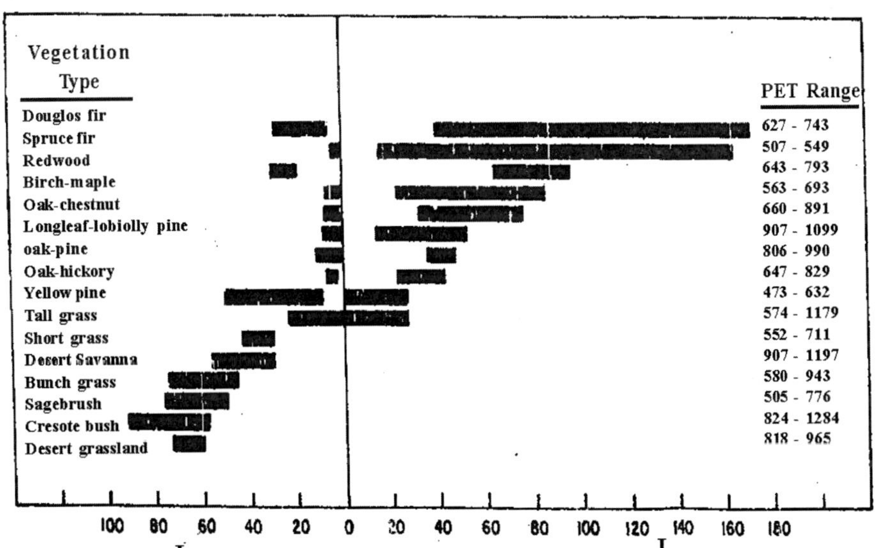

(*After*: Carter and Mather, 1966)

Fig. 3.2 Humidity (I_h) and Aridity (I_a) Indices and Annual Potential Evapotranspiration for representative vegetation types

The moisture index's seasonal variation also reflect the crop and vegetation distribution as it meets the water requirement of the vegetation provided thermal regime is favourable. Therefore for crop distribution seasonal march of moisture index, its duration, moisture surplus and moisture deficiency during the crop growing season are taken into account, the ratio of AET to PET i.e., moisture adequacy index provides the moisture status of the location in relation to water need. Lower values reflect poor moisture availability. Seasonal distribution of moisture adequacy index matches with the distribution of crops. For example for rice, maize, sorghum, and millets the ratio of 1.0, 0.9, 0.8 and 0.6 are ideal for their growth and development. The isoline of 0.4 index value limits the boundary for crop cultivation, and below which only pastures and rangeland is possible (Subrahmayam et al., 1963).

3.4.6 Hydrometeorological Applications

Since water balance procedure determines the various hydrologic parameters of soil moisture storage, water surplus, water deficit, runoff components, these parameters can be used for quantitative answers to stream flow, flood forecasting, reservoir construction, sharing of runoff water of a basin, surface and ground water availability. Hydrometeorologist is concerned with these quantities for monitoring of river basins and release of water for as per needs of agricultural crops in the region.

(i) **Precipitation intensity**: Intensity of precipitation is very useful for hydrological studies as surface runoff occurs when precipitation intensity is greater than infiltration capacity of the soil. Indian Ocean data on measurements of intense, short periods of rainfall for different time provided the following data (Paulhus, 1965).

Duration (minutes)	Depth (mm)	Intensity (mm/min)
1	32	32.00
5	62	12.40
8	120	15.00
15	200	13.33
20	208	18.40
40	300	7.50
125	480	3.84
160	550	3.44
275	788	2.86
550	1100	2.00
700	1340	1.90
1110	1680	1.52
1450	1870	1.30
2880	2500	0.87
4320	3250	0.75
10080	4120	0.41
12500	5000	0.40

Occasionally due to cloud brusts such intense heavy downpour also occur over the land surface leading to heavy flood situations beyond over control. Climatic changes induces also intensify such short period of heavy rainfall particularly during S.W. monsoon activity.

***Example 7*:**

(a) Plot the extreme value point rainfall (given above) with time scale and depict the behaviour of rainfall intensity?

(b) How such short period intense rainfall are measured?

Solution:

It is an exponential behaviour. Either plot the data on a log-log paper or plot a graph and then fit the linear curve of rainfall depth with time. Ordinary rain gauges with large bucket size are useful, whereas automatic rain gauges with sensitive recording and large bucket are useful.

Log (rainfall) (mm) = 1.575 + 0.5306 Log (duration time in minutes), (R^2 = 0.99, N = 17)

Intensity of rainfall (mm/minute) = 37.585 exp{-1.0809 log t}, r = -0.99. Intensity of rainfall exponentialy decreases with time, but rainfall amount increases with time (Fig. 3.3).

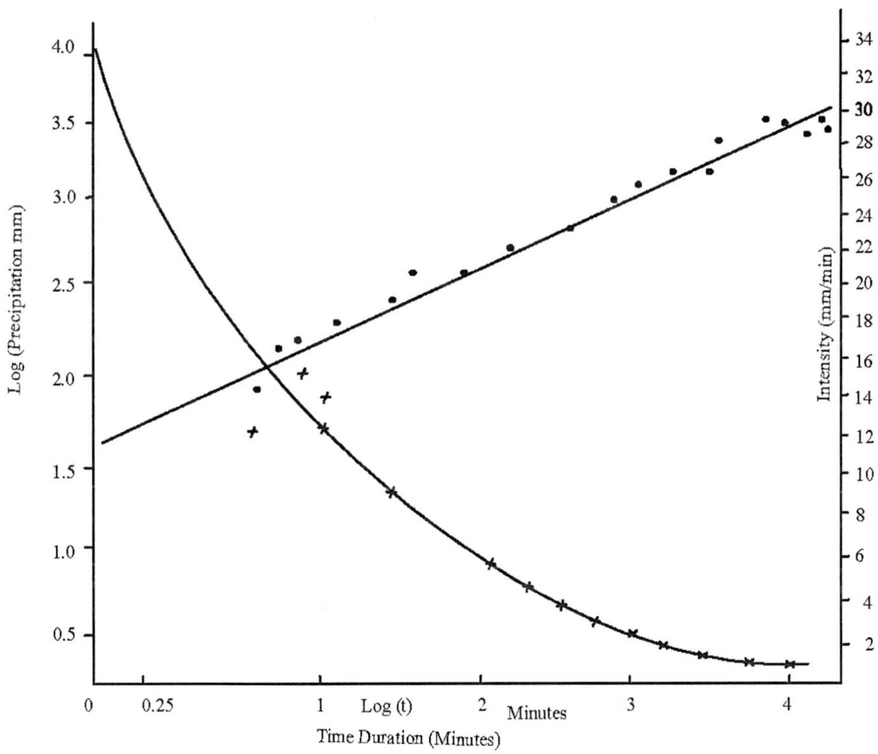

Fig. 3.3: Relation between precipitation amount, its intensity with time.

(ii) **Hydrological cycle**: The hydrological balance equation is a conservative balance equation over space and time where precipitation and evapotranspiration are main balancing parameters. The hydrological cycle between land-ocean-atmosphere systems is a unending flow cycle of water vapours to maintain the ecological balance. Budyko (1956) and Benton et al. (1950) reported that about 10 to 15 percent of total annual precipitation can from evapotranspiration/evaporation over land surface and remaining is contributed by evaporation from ocean surface. Over global basis precipitation and evaporation/evapotranspiration are main parameters of hydrological equal as other parameters of changes in storages and detention are very small. Water being a conservative quantity whatever evaporation/evapotranspiration takes place must come back from the atmosphere as precipitation over the land surface, however its distribution is a complex phenomena influenced by the general circulation of atmospheric conditions and climatic changes taking place around the globe. Over a smaller catchments basis hydrological balance equation is useful to assess the runoff water. Hydrological aspects are regularly monitored over the lake or dam basin.

Example 8:

Determine the evaporation loss over the Bhakra Dam lake basin using the following data during the month of August:

Height of the lake in the beginning of August	= 750.65 m
Height of the lake in the end of August	= 740.81 m
Area of the lake	= 108 m^2
Precipitation over the lake during the month	= 1.50 m
Runoff from land to lake	= 2.5×10^{10} m^3
Inflow from stream above	= 1.0×10^{10} m^3
Outflow from lake below	= 3.61×10^{10} m^3

Solution:

For hydrological balance of the lake in August month Inflow = Outflow

$$750.65 + 1.5 + \frac{2.5 \times 10^{10}}{10^8} + \frac{1.5 \times 10^{10}}{10^8} = 740.81 + \frac{3.61 \times 10^{10}}{10^8} + \text{Evaporation}$$

$$750.65 + 1.5 + 250 + 100 = 740.81 + 361 + \text{Evaporation}$$

Evaporation in August month = 0.34 m

Evaporation rate = 10.97 mm/day

(iii) **Runoff**: Over moderate to large watersheds it is normally assumed that 50 per cent of water surplus is available as runoff,and remaining 50 per cent is held over and added to water surplus in the next month. This delay factor depends on the basin size, soil cover, slope of basin and vegetation conditions. The climatic water balance procedure provides water surplus values in different months. The portion that constitute the surface flow is immediately available as runoff in the same month, but the portion that infiltrates the soil may take enough time to move slowly through the soil back to the surface as base flow. Thornthwate and Mather (1955) assumed 50 per cent of water surplus is available as runoff and remaining 50 per cent is added to water surplus of next month. This figure is adjusted with hit and trial method over a watershed basis till one gets values having good agreement with the measured runoff from the watershed. The procedure is illustrated with the water surplus values of example 1 data.

Example 9:

Compute the runoff values for the example 1 data using 50% criteria of water surplus as runoff in a month.

Solution:

(All values in mm)

Parameter	Jan.	Feb.	Mar.	Apr.	May	Jun.	Jul.	Aug.	Sep.	Oct.	Nov.	Dec.	Year
Water surplus (WS)	9	15	22	63	7	0	0	0	0	164	220	87	587
Surplus carried over from the previous month	114	61	38	30	46	26	13	6	3	(0) 1	62	141	
Total available water surplus	123	76	60	93	53	26	13	6	3	(164) 165	282	228	
Computed Runoff	62	38	30	47	27	13	7	3	2	(62) 63	141	114	587

Measured runoff should match with the computed runoff otherwise criteria may be modified till both are comparable each other. The cycle is repeated till figures are normalized. Repeated values are enclosed in brackets as they are modified.

Example 10:

Compute the runoff values for water surplus data of example 9 to test which of the following criteria is batter (a) 20 percent of available water surplus carried as runoff (b) 30 per cent of available water surplus as runoff in summer months (April to October) and 40 per cent of available water surplus as runoff in winter months (Nov. to March) during seasonal pattern of runoff. (c) 50 per cent of available water surplus as runoff. The measured values of runoff from January to December are 65, 39, 28, 48, 26, 14, 8, 4, 3, 65, 145, and 142 mm respectively.

Solution:

Criteria (a) 20 per cent of available water surplus as runoff.

Parameter	Jan.	Feb.	Mar.	Apr.	May	Jun.	Jul.	Aug.	Sep.	Oct.	Nov.	Dec.	Year
WS	9	15	22	63	7	0	0	0	0	164	220	87	587
WS carried over from the previous month	(294)	(242)	(206)	(182)	(196)	(162)	140	112	90	(0)	(131)	(281)	
	331	272	230	202	212	175				72	189	327	
Total available WS	(303)	(257)	(228)	(245)	(203)	(169)	140	112	90	(164)	(351)	(368)	
	340	287	252	265	219	175				236	409	414	
Runoff assessed	(61)	(51)	(46)	(9)	(41)	(32)	28	22	(18)	(33)	(70)	(74)	
	68	57	50	53	44	35				47	82	83	587
Runoff measured	65	39	28	48	26	14	8	4	3	65	145	142	587

Criteria (b) 30 per cent from April to October and 40 per cent from November to March of available water surplus values.

Parameter	Jan.	Feb.	Mar.	Apr.	May	Jun.	Jul.	Aug.	Sep.	Oct.	Nov.	Dec.	Year
WS	9	15	22	63	7	0	0	0	0	164	220	87	587
WS carried over from previous month	(173)	(109)	(75)	(58)	85	64	45	32	22	(0)	(115)	(201)	
	176	111	76	59						15	125	207	
Total available WS	(182)	(125)	(97)	(121)	92	64	45	32	22	(164)	(335)	(288)	
	185	126	98	122						179	345	294	
Runoff computed	(73)	50	39	(36)	28	19	13	10	7	(49)	(134)	(115)	
	74			37						54	138	118	587
Runoff measured	65	39	28	48	26	14	8	4	3	65	145	142	587

20 per cent criteria is overestimated runoff values from February to September, but underestimated during October to December months. Second criteria is better than first criteria, but χ^2 value is 25.60 which is higher than the tabulated value for 11 degrees of freedom. Third criteria of 50 per cent level is better and closer to measured run off values except in December month with χ^2 value of 8.48 much lower than the tabulated value at 1 per cent level

(iv) **Time trends in runoff**: Human activities on land surface have influenced the surface runoff of water. With increasing urbanization, landuse changes, cutting of forest vegetations the runoff of soil moisture have changed though it may not be apparently visible. Let us illustrate these features with an example of estimating runoff at a gap of 20 to 30 years when adequate landuse changes have taken place in the basin area from which the runoff water is collected. With the help of remote sensing technique the changes in landuse pattern can be assessed over the basin area.

Example 11:

In the Bhakra Dam region basin the following landuse pattern were observed in 1950 and again in 2000 after a gap of 50 years. Assess the runoff changes that have taken place due to landuse changes and deforestation.

Land Cover	Percent in 1955	Percent in 2005	Available Moisture Storage Capacity (mm)
Forests	20.0	16.0	400
Pasture	45.5	35.0	200
Cultivation of crops	25.0	35.0	100
Impervious	9.5	14.0	25

Solution:

Compute the water surplus values using respective 1955 and 2005 precipitation and potential evapotranspiration values with different available water holding capacities of root zone i.e., 400, 300, 100 and 25 mm respectively.

Let us suppose using the data for 1955 four water valance budgets yielded annual runoff values of 700, 650, 600 and 550 mm for 25, 100, 200 and 400 storage capacities respectively. Similarly for the year 2005 again four water balance computations are repeated for same storage capacities to yield values of 325, 265, 210 and 175 mm as annual runoff values for 25, 100, 200 and 400 mm of storage capacities. Multiply these values runoff by their corresponding land use percentages for each type of land cover.

	1955			2005		
	Runoff (mm)	Landuse (%)	Runoff × Landuse	Runoff (mm)	Landuse (%)	Runoff × Landuse
Impervious	700	0.095	66.5	325	0.14	45.5
Cultivation of crops	650	0.25	162.5	265	0.35	92.75
Pasture land	600	0.455	273.0	210	0.35	73.50
Forests	550	0.20	110.0	175	0.16	28.00
			612.0			**239.75**

Compared with actual measured runoff in respective years

	Year 1955	Year 2005
Computed runoff	612.0	239.75
Measured runoff	580.5	252.25
Difference	31.5	-12.5

Therefore over a period of 50 years a value of 44 mm of runoff increase takes place due to land use changes or 0.88 mm per year increase in runoff occurs due to land use changes alone. With rapid increase of urbanization during last 10 years the component will further accelerate and therefore hydrological consequences must be addressed in coming years.

(v) **Soil moisture detention**: Soil moisture content is the important factor in order to assess the soils workability conditions. Water balance computations provide us actual water contents in the soil at a particular time. To assess the moisture detained by the soil in the profile which is the total of the moisture stored within the soil. This value of moisture detained in the soil can be higher than field capacity since it includes both capillary water and gravitational water. Let us illustrate this component with use of data of example 1. It will be the soil moisture status in the soil over which heavy vehicles or transport activities will be in motion with safe precautions.

Example 12:

Compute the soil moisture detention in different months using precipitation and potential evapotranspiration data of example 1.

Solution:

(All values in mm)

Parameter	Jan.	Feb.	Mar.	Apr.	May	Jun.	Jul.	Aug.	Sep.	Oct.	Nov.	Dec.	Year
Storage	200	200	200	200	200	150	128	171	173	200	200	200	
WS	9	15	22	63	7	0	0	0	0	164	220	87	587
Snow	0	0	0	0	0	0	0	0	0	0	0	0	
Carried over WS	119	64	39	30	46	26	13	6	3	(0)1	82	151	
Total available WS for runoff	128	79	61	93	53	26	13	6	3	(164)165	302	238	
Runoff	64	40	31	47	27	13	7	3	2	(82)83	151	119	
Total WS temporarily held	64	39	30	46	26	13	6	3	1	82	151	119	
Moisture Detention (Storage + Total WS Temporarily held + Snow)	264	239	230	240	226	163	134	174	174	282	351	319	

(vi) **Water availability periods**: The smooth curves of precipitation (P) and different levels of potential evapotranspiration (PET) over monthly or weekly periods were compared to quantify the initiation and cessation of different periods when prepitation exceeds PET, PET/2, PET/4 and PET/8 to provide humid, moist, moderately dry and extremely dry periods of water availability. This methodology was first employed by Cocheme and Franquin (1967) for semiaried area of Sahara in West Africa, then by Brown and Cocheme (1969) for high land of Eastern Africa. This method with modification was used in India by Raman and Srinivasamurthy (1971) over monthly periods, Krishnan (1972) and Bishnoi and Virmani (1973) over weekly periods for rainfed crop planning. Different periods were determined using AET values with different levels of PET levels. Weekly analysis of water availability provided by clear picture than over monthly periods for crop planning (Table 3.4).

Table 3.4

Periods of water availability at Hisar (Lat 29°10'N and Long. 75°46'E) India.

Analysis Period	Period without Moisture Stress	Period with Moderate Water Stress	Rainy Season Period (day)	Winter Season Period (day)	Reference
Monthly	July 8 to Sep., 16	June 22 to July 7; Sep. 17 to Oct. 6	107	-	Krishan (1972)
Monthly	July 1 to Sep., 18	June 22 to June 30; Sep. 19 to Oct. 7; Dec. 29 to Jan. 29	108	32	Raman and Srinivasa Murthy (1971)
Weekly	July 8 to Sep. 7	June 18 to July 7; Sep. 8 to Sep. 14; Dec. 10 to Jan. 4; Feb. 9 to Mar. 2.	89	48	Bishnoi and Virmani (1973)

With a rainy season period of 89 days pearl millet (Pennisetum Typhoides (Burn F. Stapf & CE Hubb.), cluster bean (Cyamopsis telragonoloba (L.) Taub.), red gram (Cajanus Cajan (L.) Millsp var. Bicoles D.C.), green gram (Phaseolus avreus Roxb.) varieties are available for successful cultivation within this period. If rainfall water is conserved in a fallow land and used in winter season when PET rate is lowered, can be used for winter crops cultivation. Normally a maximum efficiency of about 30 per cent of rainy season rainfall is possible for soil moisture conservation in a fallow land when rainfall water is conserved after every effective rain and top surface is compacted every time to avoid soil evaporation losses. After tilting the top soil surface layer, lower wet layer is used for cultivation of winter crops. The choice of crop will depend on the total conserved soil water available in the root zone profile which can be quantified and choice of crop can be made as follow:

Available Soil Water in the Profile at Sowing Time	Choice of Crop
<100mm	Taramira (Eruca sativa Mill.)
100 – 125 mm	Mustard (Brassica juneca (L.) Czern & Coss.)
125 – 150 mm	Gram (Cicer arietinum L.)
>200mm	Barley (Hordeum vulgare I.)
>200mm	Wheat (Triticum aestivum L.)

A term called moisture availability index (MAI) was defined by Hargreaves (1971), quantifying the adequacy of monthly precipitation in meeting the crop water requirement and defined the following scale for crop zonation. Adequate precipitation was also termed as dependable precipitation (PD) i.e., expected rainfall at 75 per cent probability level.

MAI (PD/PET)	Classification	Period	Zone	Suitable Crops
0 to 0.33	Very deficient	All months	Very Arid	Nil
>0.34	Moderately deficient to	1 to 2 consecutive months	Arid	Pasture and rangeland grasses
	adequate moisture	3 to 4 consecutive months	Semiarid	Crops with 3 to 4 months growing season.
0.68 tp 1.00	Somewhat deficient	Adequate moisture for four months continuously	Moist	Crops with 4 months growing duration
1.00 to 1.33	Adequate Moisture	Consecutive moisture for more than 4 months	Humid	Crops with moderate to long duration varieties.
>1.34	Excessive moisture	Consecutive moisture for more then 4 months	Wet	Crops with high water requirement needs like rice

Sarkar and Biswas (1980) preferred weekly period with MAI >0.3 and >0.7 for different crop growth period for assessing crop suitability in different parts of India. An index called 'optimum moisture availability index (OMAI)' was defined as the ratio of assurred rainfall (weekly) at 50 per cent probability level to PET level satisfying the criteria for MAI values of 0.3 for early and 0.7 for subsequent growth stages of crops. Therefore using OMAI for 23rd to 42nd week (4 June to 21 October) classified the area into different zones with end of growing season when MAI is 0.3.

Number of Weeks of OMAI at			
>0.3	**>0.7**	**Zone**	**Crop Potential under Rainfed Conditions**
<10	<1	D	Crop prospects at 30 per cent probability level. Low crop potential area, break of 1 to 5 weeks duration, short duration crop is feasible, pasture.
>10	>1	E	medium to long duration crops, long duration crop in once for three years.
>11	>4	F	Medium duration crop once in 2 years, at 40 per cent probability level two short duration crops or a mixed crop is feasible.
>14	>7	G	A crop of 13 to 18 weeks duration is feasible once in 2 yers, crop prospects are high.

Number of weeks with OMAI at <0.3 level are considered as Arid zone with classes of A to C with durations <10 weeks for cultivation of short duration less water requirement crops.

Exercise 2

Ex. 1 Compute monthly climatic water balance of a hilly location using following data of precipitation and potential evapotranspiration using soil moisture retention table of 150 mm (all values in mm). Compute runoff of the location.

Parameter	Jan.	Feb.	Mar.	Apr.	May	Jun.	Jul.	Aug.	Sep.	Oct.	Nov.	Dec.	Total
PET(mm)	20	25	63	123	178	180	166	146	125	83	42	24	1175
P(mm)	59	63	32	17	37	217	668	731	270	32	9	26	2160

Ex. 2 Compute the monthly water balance of the location having following rainfall and potential evapotranspiration with available water holding capacity of 100, 200, 300 and 400 mm in the root zone for two consecutive years.

Parameter	Jan.	Feb.	Mar.	Apr.	May	Jun.	Jul.	Aug.	Sep.	Oct.	Nov.	Dec.
						(First Year)						
PET(mm)	30.4	43.1	91.7	161.1	197.8	198.9	176.7	166.9	151.0	123.6	59.2	24.6
P(mm)	0.0	0.9	14.9	4.9	2.2	78.0	282.0	343.7	253.8	20.5	1.3	1.2
						(Second Year)						
PET(mm)	23.9	43.1	95.8	168.5	194.4	196.7	183.7	169.1	156.1	131.7	68.3	28.2
P(mm)	27.1	12.3	0.0	0.7	47.2	127.4	135.7	218.1	66.4	18.3	28.1	2.6

Ex. 3 Repeat the exercise 2 with two layer model of Palmer's concept with surface layer of 25 mm and remaining available water holding capacity as lower layer of root zone. Compare the soil moisture storage estimation in both the exercises to pinpoint which one is better method to assess the soil moisture variation with time.

Ex. 4 Determine monthly and annual moisture, aridity and humidity indices of the location of exercise 1 data. Represent the location with climatic classification notation.

Ex. 5 Using climatic water balance parameters of exercise 1, determine monthly (a) moisture adequacy indices (b) water availability periods, (c) crop moisture index variations. (d) suggest the feasibility of crops under rainfed conditions.

Ex. 6 Repeat the exercises 1, 3, 4 and 5 for your location using monthly and weekly data of normal precipitation and potential evapotranspiration by combination approach of Penman's equation using agroclimatic observatory data of your location. Use water holding capacity of 1.25 m. soil depth at your location.

Ex. 7 Using data of example 2 determine Palmer's monthly drought and wet spell's (a) intensity (b) beginning and end of spells (c) duration (d) time variation of spells during 2 years period, and (e) impact of these drought and wet spells on rainy crops (June to October) and winter season crops (November to April).

Ex. 8 Using daily precipitation and potential evapotranspiration data of your location during rainy season (June to September), when monsoon rainfall is below normal value, to test the irrigation scheduling criteria of

(a) to irrigate the field when soil moisture storage has reduced to 75 per cent of available water holding capacity with an irrigation amount of 25 mm every time,

(b) to irrigate the field when storage has reduced to 50 per cent of available water holding capacity with an irrigation of 50 mm every time.

(c) to irrigate the field when storage has reduced to 75 per cent of available water holding capacity with an irrigation amount of 75 mm every time.

Which criteria is better for your location soil type?

Ex. 9 Using climatic water balance of past 30 years (year wise) determine temporary climatic shifts taking place at your location?

Ex. 10 Using weekly rainy season data of past 30 years to determine MAI and OMAI values at your location. List the feasibility of crops with availability of crop growing season.

References

Baier, W. and Robertson, G.W. 1966. A new soil moisture budget. Can. J. Plant Sci., 46, p. 299-315.

Benton, G.S., R.T. Blackburn and V.O. Snead 1950. The role of atmosphere in the hydrological cycle. Trans. Am. Geophys. Union, 31, 1, p. 61-73.

Bishnoi, O.P. 1980. A study of climatic balance parameters in relation to crop planning in Punjab and Haryana, Ph.D. Thesis Andhra University, Waltair, A.P., India, p. 1-321.

Bishnoi, O.P. 1984. Crop moisture index in characterising the agricultural drought at Hisar. Mausam, 35, 3, p. 266.

Bishnoi, O.P. and S.M. Virmani 1973. Estiamtion of safe periods for crop planning under dryland agriculture. Indian Jour. of Agri. Science. 43, 5, p. 429-33.

Brown, L.H. and Cocheme, J. 1969. A study of the agroclimatology of the Semiarid Area South of Sahara in West Africa, p. 117-29. FAO/UNESCO/WMO Inter-agency Project, WMO, Geneva.

Budyko, M.I. 1956. The heat balance of the earth's surface. Gidrometeoizdat, Leningrad, Translated by N.A. Stepanova, Office of the Climatology, U.S. Weather Bureau, 1958, 255.

Carter, D.B. and Mather, J.R. 1966. Climatic Classification for Environmental Biology. Publication in Climatology, Vol. XIX, No. 4, Thornthwaite Associates Laboratory of Climatology, Elmer, New Jersey, U.S.A., p. 1-305 to 388.

Cocheme, J. and Fraquin, P. 1967. A study of the agroclimatology of the semiarid area south of Sahara in West Africa, p. 117-29, FAO/UNESCO/WMO Inter-agency Project, WMO, Geneva.

Hargreaves, G.H. 1971. Precipitation dependability and potential for agricultural production in northeast Brazil. Publication No. 74, d-159, EMBRAPA and Utah. State University, USA, p, 123.

Krishan, A. 1972. Water balance parameters and crop growth periods with different degree of water stress. Paper presented at Third All India Dryland Agriculture Workshop held at Hydrabad from Feb. 4 to 8, 1972.

Palmer, W.C. 1965. Meteorological Drought, Weather Bureau Research paper No. 45, p. 58. Washington D.C.

Palmer, W.C. 1968. Keeping track of crop moisture conditions, Nationwide: the new crop moisture index, Weatherwise, 21, 4, p. 156-161.

Paulhus, J.L.H. 1965. Indian Ocean and Taiwan rainfalls set new records. Monthly weather review, 93, 5, p. 331-35.

Penman, H.L. 1948. Natural evaporation from open water, bare soil and grass. Proc. R. Soc. A. 193, p. 120-145.

Raman, C.R.V. and Srinivasamurthy, B. 1971. Water availability periods for crop planning. Prepublished Sci. Rep. No. 173, India Met. Dept. Poona.

Sarkar, R.P. and B.C. Biswas 1980. Agroclimatic classificationfor assessment of crop potential and its application to dry farming tracts of India. In: Proceedings of ICRISAT consultants' meeting on climatic classification held on April 14-16, 1980 at ICRISAT, Patancheru, A.P.

Subrahmanyam, V.P., Subba Rao, B. and Subramanian, A.R. 1963. Proc. of Berkeley (California) Symp. of International Association of Sci., Hydrol. (I.V.G.G.), p. 462.67.

Thonthwaite, C.W. and Mather, J.R. 1955. "The Water Balance", Publication in climatology, laborating of climatology, 8, 1, Elmer, New Jersey, U.S.A.

Thornthwaite, C.W. 1948. An approach towards a rational classification of climate. Geogr. Review., 38, 1, p. 55-94.

4

Agroclimatic Classification

4.1 Introduction

No serious attempts were made before 1960 for agroclimatic classification though the concept of agroclimate appeared as early as 1895 in Thiele's work. Though available energy and moisture have been as active factors of almost all climatic classification in different forms, but systematic appraisal appeared in Thornthwaite and Mather (1955) and Carter and Mather (1966) climatic classification. However, in order to quantify crop's potential for heat, light, moisture and local climatic variation, a systematic appraisal for agroclimatic conditions are essential for agricultural production. For any precise agroclimatic zone it is essential to evaluate the response of crops and their varieties in order to decide best crop, its variety to ensure optimum and stable yields. Since 1960 different form of indices of heat, light radiation, moisture, and soil types have been superimposed in order to delineate homogenous zones/ sub-zones according to the purpose of agroclimatic zoning. These choices are so varied that it is difficult to devise a rational universal method for agroclimatic classification for different applications. Therefore agroclimatic zonings are divisions according to favourability for agriculture.

In equatorial and tropical regions where the temperatures are high and uniform throughout the year, moisture availability is the sole determining factor; whereas in higher latitudes where moisture use is low due to low temperature, the actual available thermal energy limits the growing season for vegetation and crops. The agroclimatic classifications have been attempted with different objectives from general agroclimatic zoning to assessment of agroclimatic resources for specialized zoning for specific enterprise. The efforts have been to utilize agroclimatic conditions as a resource in agriculture with efficient use of inputs in order to increase and stabilize productivity of the zones. The role of agroclimatic variables in crop production have been realized to make best of it. Sometimes it may be necessary to modify the microclimatic conditions, though possible to a limited extent, in order to provide optimum agroclimatic conditions for maximum yield available in zone. It is felt that agroclimatic zonation, with well defined agroclimatic variables and wisely integrated with an objective in view, will ensure as a valuable tool for practical applications in exploiting both the agricultural and agroclimatic potentials in order to maintain ecological balance and scientific farming of value added agricultural products in each zone to raise the economic levels of farming community. Here it will be of interest to have a brief account of the agroclimatic classifications which are useful for different applications in a rational procedure based on agroclimatic indices.

4.2 Agroclimatic Indices

Agroclimatic indices are adopted to express quantitatively the requirements of agricultural plants during the growing season's environment presenting the relationship between a biological entity of agricultural significance i.e., crop growth, development, productivity, and the physical environmental factors. Therefore it will be useful to understand the applications of different agroclimatic indices in characterizing the agricultural climate suitable for different agricultural enterprises. Mischenko (1984) broadly divided these into 5 categorises (Table 4.1).

4.3 Thermal Belts and Sub-belts

The early attempts of (Humboldt, 1817; Koppen, 1884; de Candolle, 1974) climatic zones of

Table 4.1
Indices of agricultural significance with different applications.

Main Category	Indices	Applications	Reference
Light and Radiation Resource Indices	PAR absorption	• Photosynthetic efficiency in crop production • Choice of crops and varieties • Characterising crop geometry • Duration of photosynthetic activity	Tooming (1977), Davitaya (1958), FAO (1978, 1980), Ross (1975)
	Duration of daylight	• Grouping of plants short day (12-14 hrs) Longday (15-18 hrs) and neutral plant	Budyko (1971), Tooming (1977) Monsi and Sekai (1953)
	Heliothermal index	• Rate of development of crops • Phenology • Crop quality assessment	Kanemasu and Arkin (1974)
	Photothermic index	• Growth and development rates • Phenology	Robertson (1968a)
Wintering Conditions of Plant	Mean of absolute annual minimum air and soil temperature	• Quantifying wintering conditions • Boundaries of crop cultivation	Selyaninov (1958, 1966) Gol'tsberg (1967), Zheng and Newman (1986)
	Accumulated negative temperatures and duration of period with temperature less than zero	• Degree of wintering conditions • Depth of snow cover • Low temperature hazards • Frost free period	Zheng and Newman (1986)
	Air temperature of the Coldest month	• Boundary limits of crops cultivation • Low temperature hazards • Cultivation risks in fruits and vegetables	Mischenko (1974) Gol'tsberg (1967)
	Critical Freezing temperature of winter crop species	• Cultivation limits of crops, • Wintering conditions	Moisechik (1975)
Thermal Resource Indices	The date when air temperature passes through 0, 5, 10 and 15°C	• Vegetation period of crops • Frost free and warm period for crop cultivation	USSR (1972) Chirkov (1969) Robertson (1968)
	Accumulated active temperature	• Thermal requirement of plants	Liangzhi et al., (1987)

Table 4.1 Contd.

Main Category	Indices	Applications	Reference
	above threshold level of 0, 5 and 10°C	at different phenophases and maturity • Assessment of heat resources of different agroclimatic belts • Estimation of climate's degree of productivity • Climatic boundaries for maturity of different crop species.	Selyaninov (1958), Shaskho (1967)
	Durnal and extreme temperature level in air and soil biological layer	• Heat resource at a place • Seasonal development of crops	Mischenko (1974) Gol'tsberg (1967)
	Radiation characteristics expressed in heat units	• Assessment of heat resources • Crop growth and development • Evaporative power of air	Budyko (1971)
	Agroclimatic indices of mean daytime and night time temperatures, temperature range, accumulation of these temperatures.	• Assessment of heat resources at micro-, meso- and macro levels • Climate's continentality determination • Agroclimatic regions for specific crops	Robertson (1968a) Mischenko (1977) Gol'tsberg (1967)
	Daytime temperature at 2m above the height of underlying vegetative surface, and their accumulated values above threshold levels.	• Bioclimatic assessment of heat resource of a complex relief, • Heat resource variation due to topography and vegetation types.	Geiger (1960) Budyko (1971) Utaaker (1970) Monteith and Szeiez (1962)
Moisture Resource Indices	Available moisture in soil profile	• Quantification of moisture contents of an area • Agroclimatic assessment for crops/ vegetation at different spatial levels	USSR (1971) Fedoseev (1979)
	Amount of precipitation over	• Dependability of crops on moisture	Azzi (1939)

Table 4.1 Contd.

Main Category	Indices	Applications	Reference
	growing crop/vegetation period	during vegetation period at different risk levels • Choice of crops and cropping patterns • Moisture availability of plants • Drainage and runoff attention needs	Konstantinov (1978), NCA (1976) IRRI (1974)
	Evaporation/evapotranspiration over vegetation period	• Moisture requirements of plants, • Water needs at a place	Penman (1948), Thornthwaite (1948) Budyko (1971)
	Conventional indices of moisture contents called coefficients in terms of ratio of precipitation and potential evapotranspiration or water use and water need ratio.	• Assessment of moisture role in growth and development and yield of crops/vegetation • Water availability periods • Choice of crops and vegetations	Thornthwaite and Mather (1955) Hargreaves (1971) Reddy (1983a and b)
	Use of water balance parameters in terms of evapotranspiration deficit, or moisture adequacy index	• Water availability periods and choice of crops • Irrigation Scheduling, • Drought and Wet spells • Climatic classification • Characterising plant's moisture regime	Yao (1974), Mather (1974) Bishnoi (1980)
Adverse Weather Phenomena	Drought and wet spell indices	• Adverse effects upon adoption of farm technology, • Soil moisture variations for crop growth and development • Management of available soil moisture • Crop growth monitoring and farm advisories • Crop yeild variations.	Palmer (1965), Bishnoi (1980)
	Dates of appearance and	• Frost hazard assessment	Gol'tsberg (1961)

Table 4.1 Contd.

Main Category	Indices	Applications	Reference
	end of frost, and frost intensity	• Delineation of frost affected areas	Sinistina et al. (1973)
	Critical tempeature values	• Degree of frost resistance in plants	Gol'tsberg (1961)
		• Degree of prematuring of crops under high temperature conditions and limited water availability	Selyaninov (1966)
	Hot dry and cold winds	• Damage to plants, • Protection of plants for growth and development under artificial or microclimatic modifications	
	Rainstorms	• Excessive moisture conditions • Damage	
	Hailstroms	• Damage to standing crops	
	Cyclones/thunderstorms/ duststorms	• Wide spread damage.	
	Aridity indices	• Crp conditions, drop in crop yeild • Drought spell.	
	Moisture adequacy index	• Water availability influency crop yield • Drought and wet spells.	

earth's surface were based on temperature distribution to quantify similarities between temperature belts and vegetation zones in order to devise a classification scheme. Koppen (1936) developed a classification in which thermal zones were dominant, but moisture concept was used for separating boundaries between arid/semiarid and semiarid/humid zones. Gaussen (1955) and Meher-Homji (1960) designed bioclimatic types with mean monthly temperatures at more than 20°C as equatorial, 15-20°C for sub-equatorial, 10-15°C for warm temperate and less than 10°C as temperate. Emberger's (1955) pluviothermic quotient based on precipitation and temperature of hottest and coldest month used as bioclimatic types and vegetation zones. Budyko (1958) employed the ratio of net radiation to the heat required to vaporise the local precipitation called energy equivalents as radiational index of dryness. The maps were prepared with index values of greater than 3, 2, 1, 2/3, 1/3 and less than 1/3 to comply with desert, semi desert, steppe, moist, humid and wet forests.

Papadakis (1970, 1975) used average daily maximum and minimum temperatures, winter severity and frost free season, water balance concepts for classifying monthly climates. Using these indices, climates were classified as tropical, terrafria, desert, subtropical, pampean, mediterranean, marine and polar alpine types.

Selyaninov's (1957) work based on thermal conditions, visual composition of crops and vegetations, differences in farming practices, crop productivity and climatic resources presented 5 major thermal belts and several sub-belts, and provided a clear meaningful understanding of thermal climatic classification (Table 4.2). Thornthwaite and Mather (1955) and Carter and Mather (1966) gave five broad thermal regions based on PET values (cm) of >114, 57-114, 28.5 to 57, 14.2 to 28.2 and <14.2 as megathermal, mesothermal, microthermal, tundra and frost types corresponding to Selyaninov's tropical, subtropical, temperate, polar and arctic thermal belts. In terms of comparative scale these are summarized in Table 4.3.

Therefore it appears appropriate to demarcate tropical/subtropical boundary with mean monthly temperature above 18°C or mean annual temperature above 24°/25°C, whereas subtropical/temperate boundary around 16°/20°C, and temperate/cold region around 10°C for the purpose of cultivation of vegetation in the different thermal belts.

In the subtropical belt precipitation varies considerably leading to different types of regional climatic conditions. For example in mediterranean climate, there is dry summer, but winter is moist, and crops are irrigated during summer. In monsoonal climate, dry winter needs irrigation but very moist summer needs drainage. However a few pockets exists with round the year moisture dependability. Perennial plants have well defined vegetation rest period and wood growth is evergreen.

In tropical belt, plants are guaranteed with heat and moisture throughout the year practically without interruption. A succession of dry and rainy periods of varying durations are observed in the belt. Precipitation is highly variable throughout the belt with 100/150 mm in tropical deserts to 10,000/12000 mm in excess moisture zone. Based on moisture conditions, tropical belt is divided into three zones with excess moisture throughout the year,

Table 4.2
Thermal belts for agroclimatic classification

Thermal Belt	Summer Temp. Conditions	Thermal characteristics Effective Temp. $\Sigma T_c > 10°$ (day°c)	Cold Spell with $T_x < 0°C$	Daylength Durations in Summer (hrs)	Moisture Condition	Vegetation Period	Crop Conditions
Arctic	Very low	<300	Long	20-24	No shortage	Short (summer) <40 days	Plant cannot grow in open, but feasible in artificial control conditions
Polar	Low	300-1000	Frost can occur throughout vegetatation period	Uninterrupted long days compensate for low temperature levels, 20-24	No shortage	Short (summer)	Fast growing and low heat requirement plants are feasible such as leafy vegetables, root crops, early potatoes. barley under artificial controlled conditions.
Temperate (1)	Low	1000-2500	Temperature of coldest month ≤20°C	18-20	Excessively moist	Short (summer) very harsh winter (<40 days)	No fruit and winter crops
Temperate (2) Northern belt	Low	1000-2500	Harsh winter with frost damage on crops	16-20	Fairly most region	Short (summer) (40-100 days)	Root crops, potatoes, spring and winter grain crops, sugarbeet, fruit crops, maize for silage, winter and fruit crops destroyed by frost
Temperate(3)	Moderate	2000-4000	Cold winter with frost, temperature of coldest month -5 to 15°C	16-18	Fairly dry region	Medim (summer) (80-150 days)	Sunflower, maize long duration, beet, early maturing rice, soya and cotton, normal wintering of winter crops and perennials.
Temperatue(4)	Moderate with some warm spells	4000-6000	Winter mild with small amount of snow cover, temperature of of coldest month 0 to -5°C	14-18	Dry region	Medium (summer) (150-200 days)	Thermophile crops like grapes, cotton, cilrus fruits, winter crops and wood be frost resistant
Temperate(5)	Warm region	5000-6000	Temperature of coldest month 0°C	14-16	Dry region	Long (summer) (200 days)	Thermophile crops, field work round the year, winter grain crops and wood be frost resistant.

Thermal Belt	Summer Temp. Conditions	Thermal characteristics: Effective Temp. $\sum T_c > 10°$ (day°c)	Cold Spell with $T_x < 0°C$	Daylength Durations in Summer (hrs)	Moisture Condition	Vegetation Period	Crop Conditions
Sub-tropical	Warm region Hot season crops when >10°C	4000-10,000 Subdivided subthermal belts of 4000-6000 6000-8000 8000-10000	No snow, No frost for summer crops, but in winter frost resistent varieties requiring T<10°C during initial stages of development	12-15	Very dry region precitation varies considerably	Very long, Two vegetation periods-summer and winter with harvesing of crops in spring and automn respectively	Themophile crop like cotton & maize (late var.) rice, citrus, grapes, tea, walnuts, detepalm are grown.
Tropical	Mean annual T>5°C	6000 to greater than 10,000	No frost	12-14	Adequate, succession of dry and wet rainy periods of varying durations	Very long	Varied selection of crops, like tea, coffee, tree cocoa, cotton, banana, pineapple, sugarcane, annual crops produces several harvests a year.

T_x - Mean temperature of the coldest month.

(*After:* Salyaninov, 1957)

Table 4.3

Comparison of criteria of thermal belts for agroclimatic classification

Classification	Criteria for Thermal belts of based on mean annual Temperature (T) in °C					
	Tropical/ equatorial	Tropical/sub-equatorial	Subtropical	Temperate	Cold Snow	Arctic Very Cold Polar
Koppen (1936)	Temperature of coldest month >18	Mean annual Temperature >18	mean annual and coldest month Temperature <18	Temperature of coldest month <18, But >-3	Temperature of coldest Month <-3°C, but month 10 Temperature of warmest month <10	Temperature of warmest month <10
Champion and Seth (1968)	>24	>24	17-24	10-17	0-10	-
Gaussan (1955)	T >20	T>20	15-20	10-15	0-10	-
FAO (1980)	30-35	30-35	20-30	15-20	-	-
Krishnan and Mukhtiar Singh (1972)	>28	25-28	20-25	10-20	0-10	-
Selyninov (1957) (Daily, Effective Temperature)*	>27.4	16.4 to 27.4	11.0-27.4	10.5 to 21.4	1-10.5	<1

*The base temperature for tropical and subtropical belt being 10°C, for temperate belt 5°C, and for cold and arctic polar belt 0°C respectively.

excess moisture for most of the year with a dry period and optimal moisture for a short period with a prolonged (more than 6 months) dry period. With greater duration and intensity of dry period, plant growth is retarded and savannas trees drop their leaves. The ecological characteristics of plants in this belt are their inability to endure temperature drops below 5°C, uniform requirements for heat and light during all phases of development, non-stop development cycles in perennial plants (i.e., simultaneous existence of ripe fruit, buds, flowers and ripening fruit), and prolongation of vegetation with the lengthening of the day in annual plants.

4.4 Agroclimatic analogues

With the evolution of climatic classification, it has been a main task to determine areas with similar climatic/agroclimatic conditions for successful borrowing and acclimatization of crops with their cultural technological advancement methods from time to time. Introduction of new crops is based on similarity of general climatic characteristics. The theory of climatic analogues is based on the commonality of the predominant elements of climate which determine the optimal vegetation of plants or farming methods. Therefore it is necessary to detect the predominant climatic elements which limit the establishment and acclimatization of crops to a new areas and to determine, using phenological data, the period for which agroclimatic data is comparable of the area particularly for annual crops. For seasonal crops it is necessary to have the analysis of seasonal conditions. Nuttonson (1965a, 1965b) determined the agroclimatic analogues for grain crops using phenological data on the crop growing period analysis. The climatic conditions were established using available mean monthly temperature and rainfall data. In order to decide about the climatic boundaries for crop cultivation, it is essential to consider the climatic limiting factors such as freezing period, frost period, intensity of frost, drought and wet spell's intensity and duration, dry hot and cold winds, rain storms, excessive moisture, insufficient heat and moisture conditions. Gol'tsberg (1972) developed areas of dry and moist subtropics of first, second and third degree according to the similarity of climatic conditions and their favourability for the introduction of crops in subtropical zone of USSR. The zones were differentiated according to the temperature of coldest month, accumulated temperature, mean summer seasonal temperature for wet sub-tropics, whereas in dry subtropical zone according to duration of drought and dry periods. Agroclimatic conditions for the sprouting of autumn sowing of wheat, three zones were delineated: (i) with a stable winter rest for winter wheat (Table 4.4), with non-stable winter rest and with no winter rest for autumn sowing winter and spring wheat. Within 3 zones, 14 different climatic types were defined varying in seasonal characteristics and agricultural assessment. Gol'tsberg (1967) assessment of microlevel planning at individual farm planning based on actino-climatological work on solar radiation resources, day and night time heat resources, soil climate and topography measurements. For medium and large scale zoning, experimental polygons are prepared to obtain the

Table 4.4

Agroclimatic growing conditions for autumn sowing of wheat for agroclimatic analogue for a zone with stable rest winter wheat

Seasonal Climatic Characteristics	Duration (month) for			Temp. of the Coldest Month (°C)	HTC during heading to waxy ripness	Temp. during harvest (°C)	Daylight in spring (hrs)
	Growing season	of rest when T<5°C	Snow Cover				
Cold winter, moderately warm summer, moist	10 – 11	>6	>4	<-10	1.0 to 1.5	<18	16 – 18
Cold winter, moderately warm summer, fairly dry	10- 11	> 6	>4	<-8	0.8 to 1.0	<18	15 – 16
Moderately cold winter, warm summer, moist	9 – 10	5 – 6	2 – 4	-5 to -10	1.0 to 1.5	18 – 20	14 – 17
Moderately cold winter, warm summer, fairly dry	9 – 10	5 – 6	2 – 4	-1 to -8	0.8 to 1.0	20 – 22	13 – 15
Cold dry winter and spring, warm summer, moisture deficit	8 – 9	4 – 5	Unstable	0 to -8	>2.0	20 – 22	12 – 13
Mild winter, moderately warm summer, moist	8 – 9	4 – 5	2 Unstable	0 to -3	1.0 to 1.5	17 – 20	14 – 16
Mild winter, hot dry, summer	8 – 9	4 – 5	2 Unstable	-3 to -5	0.8 to 1.0	20 – 23	13 – 14

(*After:* Gol'tsberg, 1972)

information on the microclimate, phenology, growth rate, yield and quality. Depending on the importance of local parameters influence on crop productivity parameters of polygons are selected among soil and climatic characteristics, stress conditions, morphology, landuse, adverse weather parameter etc. Table 4.4 shows 6 types of climate varying in spring wheat growing conditions based on climatic indices which can function as climatic analogue values for distribution of wheat varieties.

In order to find the analogue area with similar climatic conditions for different crops, it is desired to have knowledge of optimum climate requirements of different crops or meteorological equivalents in regions which it has yielded best results. A brief summary of important grain crops is given in Table 4.5.

Azzi (1939) criteria of excesses and deficiencies of temperature and rainfall/ evapotranspiration values from optimum values (Table 4.6) provide stable and higher crop yield zones, and therefore can be used as climatic analogues for introduction of new crops/ genotypes according to the adaptability of optimum requirements of crops. Similarly Table 4.7 provides climatic variations data within major climatic belts on the mean temperatures during the growing season, photosynthesis rate at optimum temperatures, maximum photosynthesis at light saturation which can function as climatic analogues for selection and introduction of crop and varieties in new areas for potential productivity. Agroclimatic analogues (Table 4.6) are also useful for the quality of grapes and wine base for quality production and introduction of grapes in different agroclimatic zones (Davitaya, 1948, 58).

4.5 Agroclimatic Classification for General Agricultural Suitability

(i) **When radiation/thermal resources provide limitation**: Simple indices of thermal, moisture, plant wintering conditions and plants requirements are used for delineation of belts/sub-belts under thermal categories and zones/sub-zones under moisture regimes after relating the plant's growing season requirements and analysis of agroclimatic resources. This method is used for regional agroclimatic planning for environmental and typological soil zoning (Sapozhnikova and Shaskho, 1972) and analysis of agroclimatic resources in areas where radiation and thermal resources are limited.

Thermal belts and sub-belts were defined with assessment of heat resources with quantification of $\Sigma T > 10°C$ with boundaries between belts as given in Table 4.2. In each subbelt annual moisture index, as a ratio of precipitation to potential evaporativity or evaporation deficit, was used for demarcation of regions/zones. Further subregions/sub-zones were differentiated with precitation in the warm months and cold months of the year. Now in each sub-zone/sub-region the soil and climatic requirements of each plant species are assessed called agroclimatic zone for the particular crop. Using this approach, Sapozhnikova and Shaskho (1972) divided USSR into 67 provinces or agroclimatic zones of warm and cold belts. In each agroclimatic zone, the bioclimatic potentials (BCP) representing index of climatic productivity as a product of accumulated active air temperature ($\Sigma T > 10°C$),

Table 4.5
Optimum climatic requirements for different crops.

Crop	Optimum temperature needs (°C) at					Optimum evapotranspiration (mm day)				
	Germination	Vegetative	Reproductive	Maturity	Seasonal Break	Vegetative	Reproductive	Maturity	Rainfall Zone (mm)	References
Wheat (spring)	20-25	15-20	25-30	25-28	20-25	2-3	4-6.5	3-5	500-700	Baier (1977)
Wheat winter (Intermediate climates)	15-20	15-20	20-25	22-25	20-25	2-3	4-6.5	3-5	500-750	Mavi (1986)
Pearl millet (warm areas)	33-34	32-34	25-30	30-32	25-30	2-5	5-7	5	500	ICRISAT, (1984)
(Temperature areas)		21-22	22-25	30	22-25	2-5	5-6	5	500	Ong (1983) , Fussell et al. (1980)
Sorghum	30-32	25-35	22-30	25-30	24-30	2-5	4.5-8	5	500-700	ICRISAT, 1984
Maize (warm)	30	25-30	21-30	25-30	25-30	3-5	5-7	4-5	750	WMO, 1977
(temperate)		18-22	21-23	23-25	21-24					Brown (1977), Chirkov (1969)
Cotton (subtropics)	32	27-32	27-33	30-35	27-32	4.5-6	6-7	4.5-5.5	500-700	Bishnoi (1999)
Rice (summer)	30	25-30	22-30	25-30	25-30	4.5-6	6-8	4.5-6	1200-1500	Robertson (1968b)
Rice (winter temperate)	20-24	20-36 daytime 19.5-22.5 night time 18-22	29.5-32.5 daytime 19.0-23.0 night time 20-21 (coldest month >14°C)	22	24.5-31.0 18-22					Nishiyama (1976) Lianghzi Gao et al.(1987) Thomas (1957)
Sugarcane (Tropical)	30 (sprouting of sets)	28-32	28-32	28-30	28-32	High humidity, sultry conditions			1200 to 1650	Mavi (1986)
Gram	20	32 day time 21 night time	22-26	25-30	18-29	2-3	4-6	3-5	300-400	Mavi (1986)
Rapeseed Mustard	22-24	15-20	25	25-30	24	2-3	4-7	3-5	250-350	Mavi (1986)

Table 4.6

Agroclimatic analogues for raw material of grapes and wines

Wine Base	Accumulated active Temperatures ($\sum T_c > 10°C$) (day °C)	Mean temperature of the hottest month (°C)	Total amount of precipitation (mm) during the		Quality of Wine
			Month before harvest	Year	
Champagne	2500-2800	16-18	0-50	400-1200	highly acid,
	2800-3200	18-22	0-100	400-1200	light, good condition
	3200-3600	22-24	0-150	400-1200	No fresh
	>3600	>24	0-200	400 to >1200	Enough heavy
Table wine	2800-3200	18-24	0-170	400-1200	Low strength, fresh
	3200-3600	20-26	0-170	400-1200	Light
	3200-3800	20-26	0-170	400-1200	Heavy
	3800-4100	20-26	0-170	400-1200	Heavy
	>4100	>22	0-200	400 to >1200	Heavy
Strengthened	3600-3800	20-25	0-100	350-800	Low sugar, too thin
Desert and					largely for strong wine
Sweet wines	3800-4100	22-27	0-100	350-800	Sugary, mostly for strong wine
	>4100	>24	0-100	350-800	Highly sugary liquor type
Table, transportable wine	>3800	>82	0-100	500-1100	-
Dry wine	>4000	>25	<20	500-700	

After: Davitaya (1948, 1958)

atmospheric pressure and coefficient of biological productivity characterizing the crop to be assessed to further subdivide into sectors/sub-sectors of similar area which permit local level agroclimatic zoning with different agricultural potentials in the following scale of BCP levels. The conditions tending towards improvement of thermal regime of surface layer of air and soil leads to improve the plant productivity.

BCP value	**Agroclimatic zones at Micro and Meso level**
0.3 to 1.2	Cold belt, least productive, 4.6 to 16.8 centner/ha Least thermal resources
23.0 to 84.0	Temperate belt, good yield potential 11.1 to 56.9 centner/ha
3.1 to 8.0	Warm belt, productivity higher than temperate belt, 41.7 to 113.8
6.3 to 10.0	Hot belt, yield potential of 86.0 to 140 centner/ha

Shaskho (1967) used $\sum T > 10°C$ to delineate four thermal belts of cold (1200 day °C), temperate, warm and hot (>8000 day °C). Then in each belt quantified moisture index on $P/\sum(E - e)$ to quantify as sufficient, insufficient, insignificant moisture with difference in annual water dependability trend. The wintering condition was determined as temperature of the coldest month as shown in Fig. 4.1.

(ii) **When availability of soil moisture provides limitation**: FAO (1978, 1980) is a sound methodology for determination of proper landuse planning and resource management with optimum use of agroecological resources of the agroecological zones for assessing productivity potential and adaptability of crops.

Major climates are delineated based on temperature regimes during the growing period when both soil moisture and temperature permits growth. These major climate's attributes are matched with the crop adaptability groups (Table 4.7) having climatic requirements for photosynthesis and phenology which are functions of crop yield.

Then net biomass and constraint free yield potential of all qualifying crops in respect of the effects of prevailing temperature and radiation regimes on crop photosynthesis and growth for various growing seasons. Finally in the last stage agroclimatic constraints are taken into account resulting in yield losses due to constraints of water stress, effects of pest, diseases, climatic constraints on yield and yield attributes. The amendments in constraint free potential yields by reduction ratings reflect yield losses occurring due to agroclimatic constraints result in crop agroecological zones with suitability characteristics of not suitable, marginally suitable, suitable and very suitable. This classification has been carried out for pearl millet, sorghum, maize, wheat, potato, cassava, and cotton crops.

4.6 Agroclimatic Zoning for Choice of Crops, their Genotypes and Growing Season

Agroclimatic indices based on amount and distribution of rainfall in relation to crop water needs provided favourable climatic factors in the growing reason. The limiting factors of climate such as frost hazard, temperatures below threshold temperature for growth and

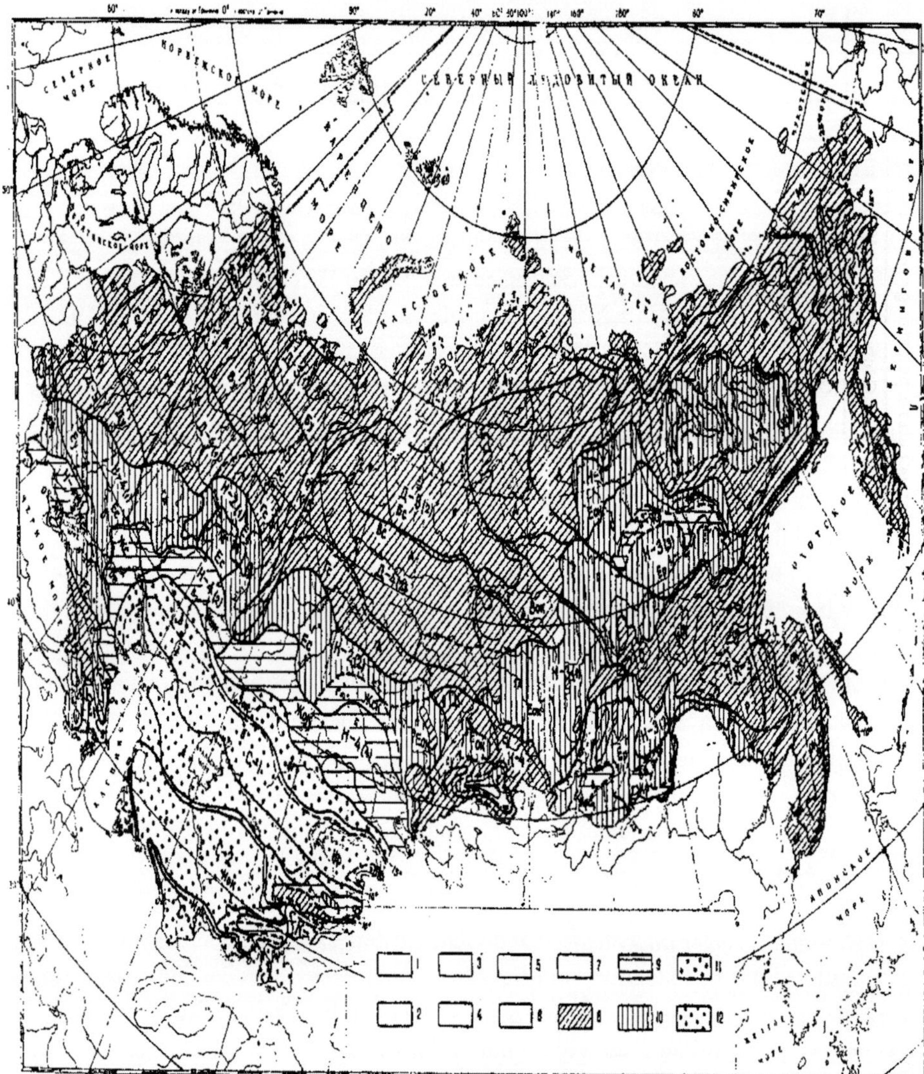

(*After*, Shaskho, 1967; Mischenko, 1984)

Fig. 4.1: Agroclimatic Zones of the USSR (After: Shashko, 1967; Mischenko, 1984)

Cold belt

1. Sub-belt very cold or crops grown in covered/semi-covered ground and fast ripening green vegetables in open ground <400°.
2. Cold sub-belt or early vegetables requiring little heat 400-1200°.

Table 4.7
Characteristics of major climates when water availability, radiation and temperature permits growth during the growing season

Climates	No.	Major Climates during Growing Period	Mean Temperature during Growing Period (°C)	Crop group suitable with			
				Rate of photosynthesis at light saturation (mg CO_2 dm^{-2} hr^{-1})	Optimum Temperature for Maximum Photosynthesis	Radiation Intensity at max. Photosyn-thesis (cal cm^{-2} min^{-1})	Crop Group
Tropics	1	Warm tropics	>20	40-50	20-30	0.3-0.8	II
				>70	30-35	1.0	III
(All months with mean	2	Moderately cool tropics	15-20	20-30	15-20	0.2 - 0.6	I
				>70	20-30	1.0	IV
temperature	3	Cool tropics	5/10-15	20-30	15-20	0.2-0.6	I
>18°C)	4	cold tropics	<5	Not suitable			
Subtropics (one or more months with mean	5	Warm moderately cool subtropics, summer rainfall	>20	40-50	20-30	0.3-0.8	II
				>70	30-35	1.0	III
temperature <18°C, but the remaining months >5°C)	6	Warm moderately cool subtropics, summer rainfall	15-20	20-30	15-20	0.2-0.6	I
				>70	30-35	1.0	III
	7	Warm subtropics summer rainfall	>20	40-50	20-30	0.3-0.8	II
				>70	30-35	1.0	III
	8	Moderately cool subtropics, summer rainfall	15-20	20-30	15-20	0.2-0.6	I
				>70	20-30	1.0	IV
	9	Cool subtropics, summer rainfall	5/10-15	20-30	15-20	0.2-0.6	I
	10	Cool subtropics summer rainfall	<5	Not suitable			
	11	Cool subtropics winter rainfall	5/10-15	20-30	15-20	0.2-0.6	I
	12	Cold subtropics	<5	Not suitable			
Temperate	13	Cold temperate	5/10-15	20-30	15-20	0.2-0.6	I
	14	Cold temperate	<5	Not suitable			

Crop groups
I – Spring wheat, winter wheat, highland P. Bean, white potato (tropical to temperate cultivars)
II – Paddy, rice, lowland P. bean, soybean, cotton, sweet potato, cassava
III – Pearl millet, lowland maize (tropical cultivars), sugarcane
IV – Highland sorghum, highland maize (temperate and tropical high altitude cultivars).

After, FAO (1980)

development i.e., freezing hazard, drought spells, dry winds, rain storms and excess moisture, inadequate thermal heat energy during growing season etc. will limit the crop cultivation.

Having satisfied with these climatic limitations at a location, the water availability periods for different moisture regimes are determined with comparison of precipitation or actual evapotranspiration values against the different levels of potential evapotranspirations i.e., the moisture requirements are met. These are termed as humid, moist, moderately dry, dry and very dry periods when $P \geq PET$, $PET<P \geq PET/2$, $PET/2<P \leq PET/4$, $PET/4<P \geq PET/10$ and $P<PET/10$ respectively. Humid and moist periods are favourable for crop growth, whereas moderately dry periods are adopted as periods for field preparation, sowing, initial growth and maturity (Cocheme and Franquin, 1967). Length of these water availability periods are matched with number of days required for maturity of different crops and their genotypes to fit in the particular location/zone according to the cropping system as

Length of the growing season (days)	**Cropping system**
<75	Low water requirement crops, grasses, shrubs, silviculture,
75 to 125	Monocropping of a medium to long duration cultivar
125 to 150	Intercropping of a short and long duration cultivars combination
150 to 200	Double cropping of short to medium duration cultivars
>200	Multiple cropping

In polar and temperate regions where moisture is not a constraint, the length of growing season is determined by the temperature curve at a place when it crosses the time scale with crop threshold values i.e., for summer crops >10°C and for winter crops above of 5°C i.e., the period without winter conditions.

4.7 Homogeneous Agroclimatic Zoning for Crop Planning and Cropping System

Delineation of agroecological zones with similar agroclimatic conditions have been of special interest to agricultural scientists for adoption of package of practices and transfer of technology for different crops in homogeneous soil-climatic resource zones. Soil and climate being two natural physical resources can be superimposed to demarcate a homogeneous zone for choice of crops and cropping patterns. This methodology can be applied at state level to block level spatial units depending on the quality of input data available. Methodology adopted by Krishnan and Singh (1972), Sly (1970) and Krishnan, 1988, Sehgal et al. (1995) are useful for finer analysis.

Sehgal et al. (1992) defined an agroecological unit as a land unit carved out of agroclimatic zone when superimposed on land form, and soil condition modifies the length of the growing period due to availability of soil moisture to meet plant water needs. FAO (1976) agroecological zone methodology with slight modification in respect of soil physiographic condition was adopted here. The soil-scape map was superimposed using GIS technique on biclimatic map (based on moisture index scale in terms of arid, semiarid, subhumid, humid and perhumid types), and on the length of growing season period map of >210, 180-210, 150-180, 90-150 and <90 days to provide an agroecological regional map. These agroecological zones are used for different applications. Particularly when agroecological zones are further superimposed on detailed soil resource map (1 : 250,000 scale) to provide homogenous agroecological units with finer details of soil, site, land form for landuse planning. These agroecological units are evaluated for growing of pearl millet, sorghum, groundnut, cotton, sugarcane, rice, wheat, maize and other crops using Sys (1985) method to evolve optimum cropping pattern.

Using GIS approach Eswarn and Van Den Berg (1992) pedicted enhancement of length of the growing season which is one index of agricultural production. Therefore with the climate change in different agroecological units we can evaluate the variation in length of the growing period to evaluate the impact of climate change.

4.8 Agroclimatic Zoning for Flexible Crop Planning Covering Risks

Assuming radiation and thermal regimes are not constraints, under rainfed conditions the crop planning is highly dependent on precipitation amount and its distribution during the growing season as against the maximum water need of the crop. Precipitation being highly variable, whereas potential evapotranspiration is a conservative quantity creating imbalance between march of precipitation against potential evapotranspirations. Hargreaves (1971) used 75 per cent probability level of precipitation as an assured precipitation for matching against potential evapotranspiration to quantify moisture availability index. Sarkar and Biswas (1980) further extended this assured precipitation probabilities at 30, 50 and 75 percent levels to cover risk levels for adoption of a flexible crop planning. This methodology is useful in areas where moisture regime is in constraint for meeting the crop water requirements.

Hargreaves (1971) divided the climate into tropical, sub-tropical and temperate boundaries with mean temperature of all months >17°C, of 10 or more months with >10°C in a year. Then in each climatic belt moisture availability index values of <0.33, .34 to 0.67, 0.68 to 1.0, 1.01 to 1.33 and >1.33 to meet crop water requirement conditions of very deficient, moderately deficient, somewhat deficient, adequate moisture, and excessive moisture respectively. Sarkar and Biswas (1980) adopted weekly values of MAI of assured rainfall at 30, 50 and 70 per cent probability levels to cover risks and to meet crop water requirement at different phenophases of crop growing season. In the land preparation, sowing, beginning and end of growing reason were taken with MAI values of 0.3. At 50 per cent probability level the MAI values were called optimum moisture availability index (OMAI). Agroclimatic zones for dry farming tract of India was classified into four agroclimatic zones differing in crop production potentials as per following criteria (Sarkar and Biswas, 1980) (Fig. 4.2).

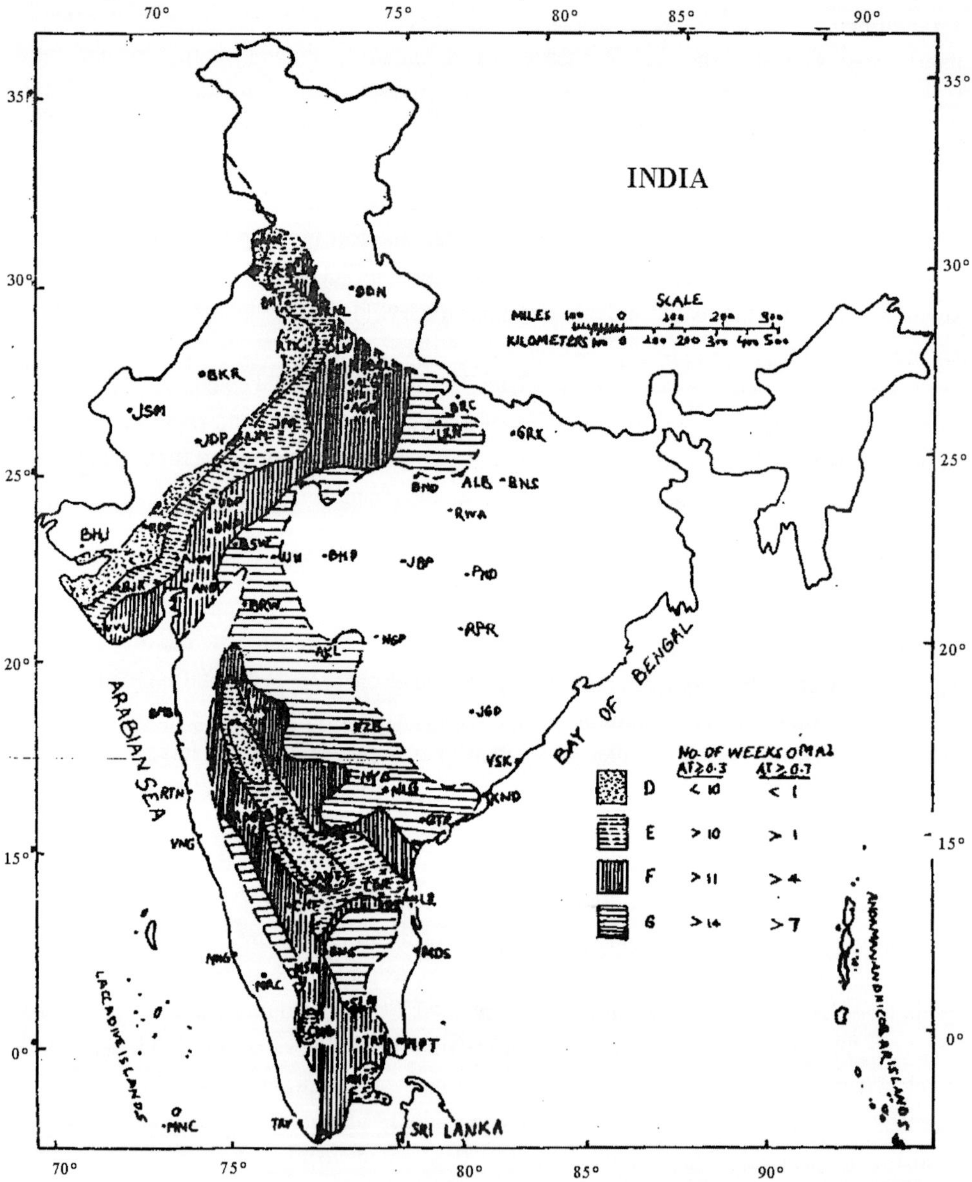

(*After*: Sarkar and Biswas 1980)

Fig. 4.2 Agrolimatic zones of dry farming tracts of India

Agroclimatic Zone	Number of Weeks with OMAI >0.3	>0.7	Av. Daily Temperature Range (°C)	Duration Potential (months)	Crop
D	<10	<1	5 to 10 (Polar)	2 to 4 (Warm month)	Lowest, no crop suitability
			5 to 10 (subpolar)	5 to 7 (warm months)	Short duration crops Once in three year
E	>10	>1	5 to 10 (temperate)	6 to 8 (cooler months)	Potential to raise crops for 40 per cent of years
			10 to 20 (temperate)	4 to 7 (warm months)	
F	>11	>4	10 to 20 (tropical)	4 to 6 (warm months)	Once in two years
			20 to 30	6 to 8 (warm months)	
G	>14	>7	25 to 30 (equatorial)	Throughout year	Crops in 60 per cent of years
A, B, C	<8	0	Arid zone areas	Rainy season	Crop potential of once in 3/4 years

Any stress period within the crop growing season i.e., when OMAI <0.3 is designated with numerical suffixes in the above classification in the ascending order of duration.

4.9 Agroclimatic Zoning for Improving Crop Productivity

Bishnoi (1980) adopted mean crop growing seasonal temperature values for delineation of macroclimatic belts as per following scale

Mean seasonal temperature (°C) (Summer crops)	(Winter crops)	Macroclimatic belts
>35	>25	Very hot
30 – 35	20 – 25	Hot
25 – 30	15 – 20	Moderate/optimum
20 – 25	10 – 15	Cold
<20	<10	Very cold.

In each macroclimatic belt, the crop moisture index values (CMI) are determined during the growing season period for each crop using 3.16 expression.

α is the optimum evapotranspirate of the crop in ith week which varies with different crops and also varies at different growth stages of a crop. The term AET/α.PET regulates the adequacy or anomaly in moisture supply to meet the evapotranspirational needs when it is greater or less than 1.0. The optimum evapotranspiration rate will maintain the optimum growth and development of the crop at different phenophases to achieve maximum crop yields. This optimum evapotranspiration is regulated with the supply of soil moisture from the available soil water stored in the root zone and therefore the moisture regulating factor

M_i is introduced. The optimum evapotranspiration rate values for summer rainfed crops of rice, maize pigeon pea, sorghum/groundnut, pearl millet, green gram/cowpea used were >1.0, 0.9, 0.8, 0.7 and 0.6 respectively, whereas for winter rainfed crops wheat, barley, chickpea rapeseed mustard and grasses the optimum evapotranspiration values were 0.9, 0.8, 0.7, 0.6 and 0.5 respectively. The weekly/monthly variation of CMI_i values will reflect the crop moisture status during the growing season representing the soil moisture status of the crop and crop conditions with different levels of CMI values. In this way 8 different crop conditions are reflected using CMI values, however favourable microclimatic conditions are presented when CMI values are in the range 1 to 4 during growing season for growth, development and reproduction of the crop without deteriorating the crops yield potential.

Using thermal and crop moisture regimes during the growing season, following crop microclimatic zones were formulated from agricultural productivity point of view (Table 4.8) for field crops.

Therefore, for each crop M_3, M_4 and M_5 microclimatic conditions are adopted from short to long duration genotypes meeting the optimum requirements of thermal and moisture regime at different phenophases without deteriorating the yield potential of the crop. However M_2 under irrigation resource facilities and M_6 under excess water management conditions can be adopted. Fig. 4.3 shows the agroclimatic zones for pearl millet crop during rainy season with preferential productivity zones.

In the last step soil resource map is superimposed over the microclimatic zones to take into account of length of growing season with the root zone water availability. Therefore crop agroclimatic zones are delineated for each crop superimposing seasonal temperatures, favourable CMI regime patterns and soil resource map to provide zones for improving and maximizing crop productivity. Varietal differences and introduction of new crops are feasible in different zones according to the thermal and moisture requirements of the crop genotypes. High water requirement and long duration genotypes of crops are feasible in zones with crop moisture regime patterns of M_5 and M_6 and such zones will be preferred for rice and sugarcane crops

4.10 Agroclimatic Zoning for Specific Purposes

The concept of climatic resource for agriculture is closely linked to maximize crop production levels. Davitaya et al. (1973) assessed agroclimate qualitatively as a natural resource in order to evaluate the climatic suitability for agricultural landuse for specific crops or agricultural potential in general. Establishment of agroclimatic indices to evaluate local and regional suitability of climate for specific crops was the first step in this direction. The establishment of climatic types for specific crops (Burgos, 1958a and 1958b; 1982) makes use of algorithms which are nothing but agroclimatic indices.

Specialized agroclimatic zoning have been developed to solve specific crop production

Table 4.8

Microclimatic zones based on crop moisture regime pattern during the growing season.

Criteria adopted for crop moisture regime pattern during		Crop Microclimatic	
Summer Rainy Season	**Winter Rainy Season**	**Zone**	**Features**
Having CMI values in E category for at lest 2 month and one month in any other category	Having 3 months in E category and 2 months in any other category	M_1	Feasibility of crop is not reflected
At least 2 or more months in D category and one month in E category	At least 3 or more months in D category and two months in E category	M_2	Short duration genotypes with external irrigation sources required, rainfed yield potential is low
At least 2 months is C category and one month in D category	At least 3 months in C category and 2 months in D category	M_3	Crop genotypes can be successfully adopted optimum yield potential is achieved with local adopted, genotypes.
At least 3 or more months in C category or one month in B category and 2 months or more in C category	At least 5 or more months in C category and 2 months in B category and 3 or more months in C category	M_4	Optimum potential yield are achieved, composite and hybrid genotypes can be adopted
At least 3 or more months in B category or one month in A category and 2 or more months in B category	At least 3 or more months in B category or two months in A category and remaining in C category	M_5	Crop moisture conditions suitable for long duration hybrid genotypes, maximum yield potential is achieved, pest and diseases control measures needed
At least two months in A category and remaining months in B category	At least three months in A category and remaining months in B category	M_6	Cultivation of crop is restricted for higher water requirement genotypes, soil moisture conservation and recycling of excess rainfed water needed, pest and disease control measures needed.

problems or to grow crops in specific areas with economic advantage. Crop zoning have become popular from seed production quality point of view. A few agroclimatic zones for specific important crops have been presented here to exercise similar problems in different similar agroclimatic analogous areas. Davitaya (1958) study on quality of grapevine zones (Table 4.6) in specific areas with microclimatic differences of rainfall, effective cumulative temperatures, temperature of warmest month for safe harvesting. Moisture conditions during

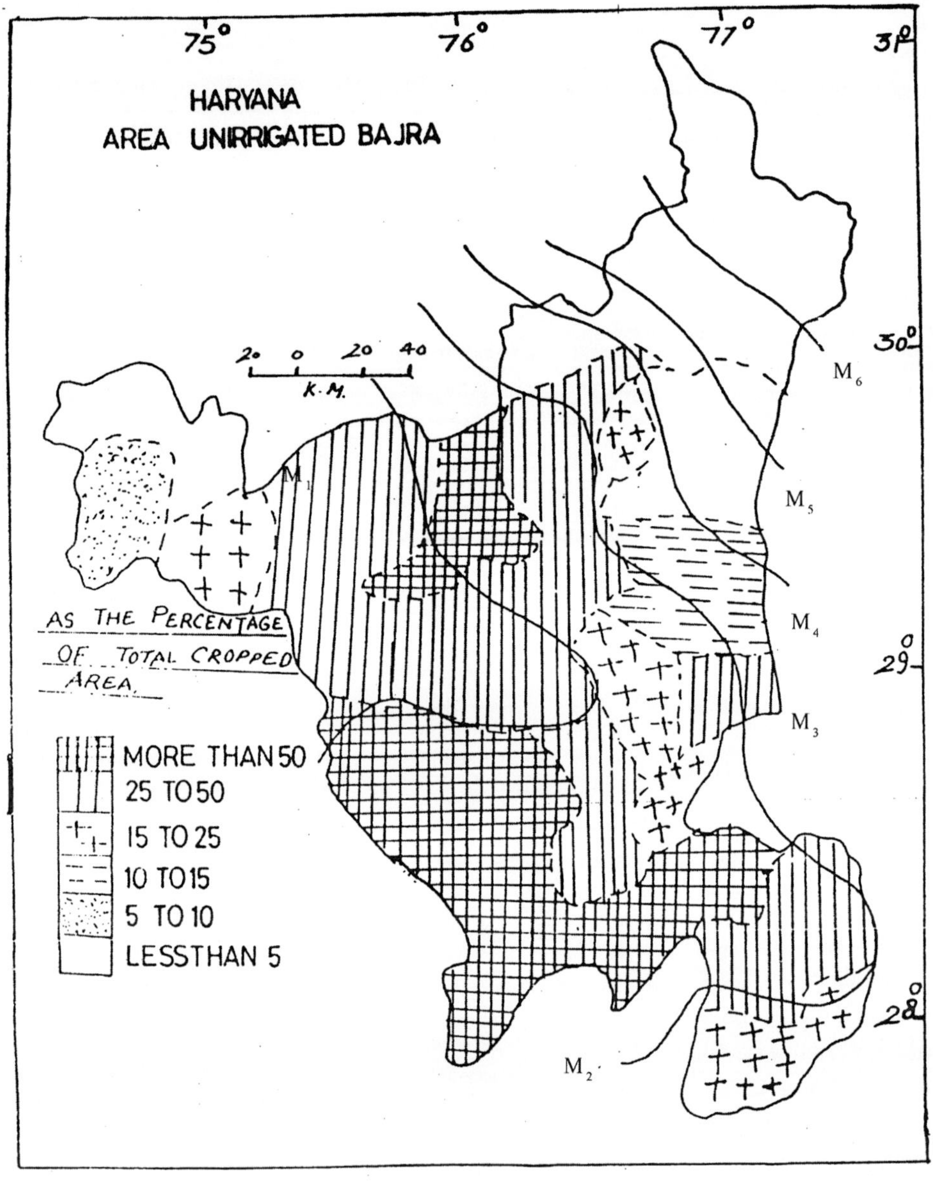

(*After*: Bishnoi, 1989)

Fig. 4.3: Agroclimatic zones for pearl millet in Haryana (India)

growth and reproductive period can also be characterized by Selyaninov's hydrothermal index (HTC) variation between 0.5 to 2.0. HTC values less than 0.5 indicated lack of moisture for grapes and is used as boundary between irrigated and rainfed vine zones, and when HTC is greater than 2.0 represents the excess of moisture. FAO (1978, 1980) concept of potential productivity and adaptability of crops provided sound methodology for agroclimatic classification (Table 4.9) for adoption at field scale to macroclimatic belt scale. As an illustration Fig. 4.4 represent the agroecological zones for cotton in Southeast Asia. A few specialized agroclimatic zonings have been described here to highlight these features.

4.10.1 Assessment of climate's degree of productivity

Sapozhnikova and Shashko (1972) and USSR (1979) assessed climate's degree of productivity with an index (B_K) expressed as product of cumulative effective temperature and moisture index (K_ξ)

$$B_K = K_\xi \,.\, \Sigma T_C > 10°C \qquad \ldots (4.1)$$

K_ξ moisture index characterizes the relationship of crop yield to moisture regime expressed as

$$K_\xi = \frac{0.5P_x + P_r}{0.18\ \Sigma T_C > 10°C}, \qquad \ldots (4.2)$$

P_x and P_r are precipitation during cold and warm periods. $0.18\ \Sigma T_c > 10°C$ represents the evaporative power of the year.

B_k quantitatively assess the climate's degree of productivity to achieve optimum potential productivity with a given combination of heat and moisture regime B_k values varying from 100 per cent in south of 60°N to 145 per cent in Bulgaria confirming the practical importance.

4.10.2 Potential productivity of wheat

Zheng and Newman (1986) delineated potential productivity areas of wheat using agroclimatic resource parameters of seasonal temperature and water balance parameters during crop growing season. A monthly mean temperature of 9-12°C in January with extreme winter minimum of ≥ –20°C was used as a criteria for separating winter and spring wheat. When accumulated negative temperature is more than 700 day °C and mean winter minimum of < –22°C, there is no possibility of winter wheat. Winter wheat survival depends on cloud cover, snow cover, drought period in the growing season. N – S extent in China is determined by temperature, where E – W extent by annual and seasonal precipitation for wheat productivity. With use of temperature conditions, China was divided into 5 geographical zones for wheat production—one for spring and 4 for winter wheat. Using water balance, agroclimatic potential of wheat yields were delineated into sub-zones of arid, semiarid, dry subhumid, moist subhumid, moist, humid and perhumid with moisture index values of < - 60, -30 to -60, -15 to -30, 0 to -15, 0 to 40, 40 to 80 and > 80 respectively. Agroclimatic potential of wheat yield (Ym) was determined in each subzone using the expression.

$$Y_m = Y \,.\, k\ HI, CT, GS, CEC, \text{ where} \qquad \ldots (4.3)$$

Y is dry matter in kg/ha/day, k is crop constant, HI is harvest index, CT is mean temperature crop response constant, GS is growing season (days) from planting to harvest, CEC is correction factor for effect of climate.

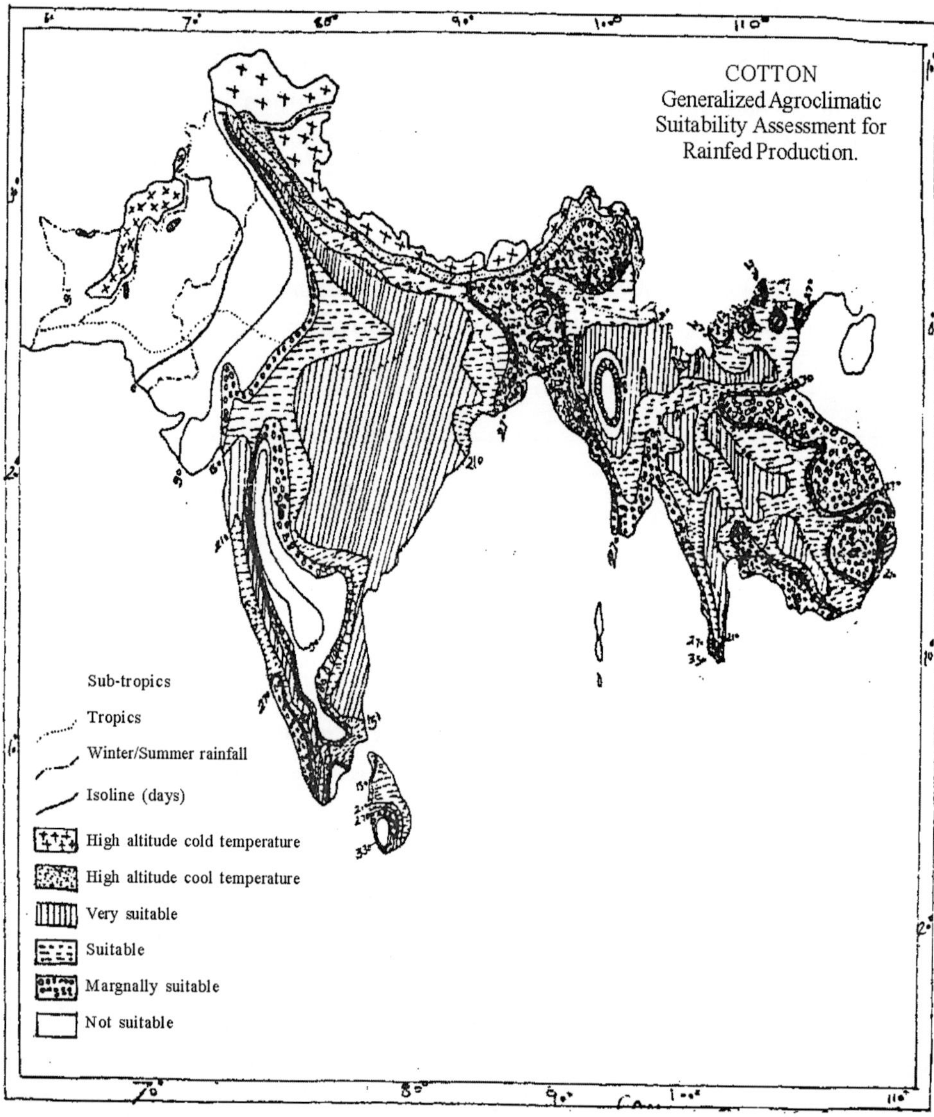

(*After*: FAO, 1978, 1980)

Fig. 4.4: Generalized agroclimatic suitability for cotton production in SE Asia

4.10.3 Potential productivity of rice:

Liangzhi Gao et al. (1987) utilized land and water resource indices for delineation of rice production potential zones in China with selection of suitable crops, cropping systems and rice genotypes (Fig. 4.5). Temperature indices were used for demarcating major potential rice production areas for growth and development of rice Japonica variety. For early maturing

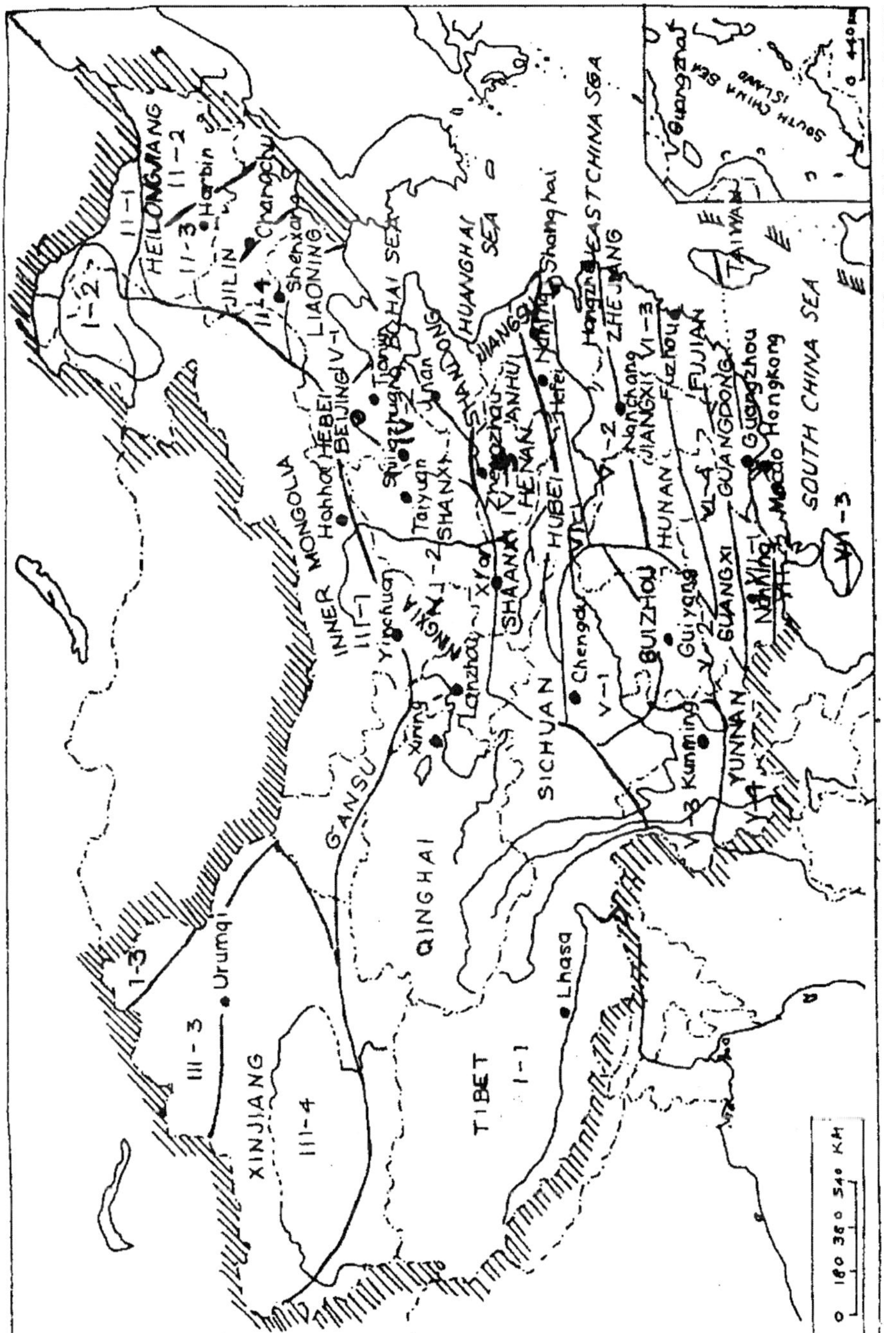

(*After*: Liangzhi Gao et al., 1987)

Fig. 4.5: Classification of rice growing areas based on climatic factors and cropping systems in China

Table 4.9
Major rice agroclimatic characteristics for different rice climatic belts and subbelts in China (Liangzhi Gao et al., 1987)

Rice Climatic Belt	Rice climatic Subbelt	Agroclimatic Classification Based on HI	RAI	RCI	RGS (days)
1. Belt without rice	Tibet, Xinman mount, Alati mount	Unsatisfied	-	-	-
2. NE China subhumid and single crop early rice belt.	i) Heilung R.	Satisfied	0.9-2.0	<1 for early Japonica with early maturity	120
	ii) Mudan R.	-	0.9 – 2.0	>1 for Japonica	120
	iii) Songhua R.	-	0.9 to 2.0	>1 for medium maturity	130 – 140
	iv) Yanbian	-	0.9 to 2.0	>1 for late maturity Japonica	140 – 160
3. NW China dry and single crop early rice belt	i) Yellow R.	Satisfied	2 to 5	≥ 1 for early maturity Japonica	140 – 160
	ii) Less plateau		2 to 5	≥ 1 for medium Japonica	140 – 160
	iii) N. Xinjiang		2 to 18	≥ 1 for late maturity Japonica	160 – 190
	iv) S. Xinjiang		18 to 24	≥ 1 for early Japonica with late maturity	160 – 180
4. N. China sub-humid and for single rice belt	i) Lianning	Satisfied	1 to 2	≥ 1 for medium Japonica with early maturity	160 – 180
	ii) Beijing Ticanjin		1.3 to 2	≥ 1 for medium maturity	160 – 180
	iii) Yellow R.		0.5 to 1	≥ 1 for hybrid indica	190 – 200
	iv) Huei R.				
5. SW. China humid and double rice belt	i) Sichuan basin	Satisfied	0.5-1.0	≥ 1 for wheat and hybrid indica	200 – 240
	ii) Guizhan plateau		0.4-0.5	≥1.0 for hybrid indica	200 – 220
	iii) N. Yumnan & S. Sichuan		0.6-0.8	≥1.0 for wheat medium Japonica	200 – 240
	iv) Yunnan		0.5 – 0.7	≥1.0 for early triple cropping	200 – 240
6. Central China humid single and double	i) Jiang Hua irrigated	Satisfied	0.9-1.0	≥1.0 for wheat plus hybrid indica	200 – 210
	ii) Yangzi R.		0.5 – 0.6	≥ 1.0 for early triple cropping	210 – 220
	iii) Jiangnan		0.5-0.6	≥1.0 for medium triple cropping	220 – 240
	iv) Middle south China		0.5-0.8	≥1.0 for late triple cropping	240 – 260
7. S. China humid and double cropping	i) South of Nauling	Satisfied	0.4 – 0.5	≥1.0 for late triple cropping	260 – 280
	ii) Pearl R.		0.5-0.9	≥1.0 for late triple cropping	280 – 290
	iii) Hainan Island		0.7-1.0	≥1.0 for late triple cropping	Throughout the year

(*After*: Liangzhi Gao et al., 1987).

Japonica rice varieties, mean daily temperatures greater than threshold value of 10°C for 110 consecutive days or more are required, whereas in hilly area mean temperature of 20°C is preferred. Topography, isothermal lines of 18°C and 14°C mean temperatures are also drawn over the region as January temperature of 18°C is optimum temperatures of rice booting, heading and flowering of winter rice, whereas temperature of coldest month January should be more than 14°C to provide suitable conditions for rice, rice-sweet potato cropping pattern.

After delineating major rice belt based on temperature values, moisture index expressed in terms of Rice Aridity Index (RAI) divide it into sub-zones of arid, sub-humid and humid with RAI values >2, 1 to 2 and <1 respectively.

$$RAI = \frac{ET}{P} = k_r \frac{PET}{P} \qquad \ldots (4.4)$$

ET and PET are total actual and potential evapotranspiration for rice crop during the growing season, P is precipitation during the growing season. k_r is rice crop coefficient having values of 1.05 in S. China and 1.25 in N. China.

To analyse rice cropping systems a compound index was also expressed as:

$$RCI = \frac{RGS}{RGD + T} \quad \text{(for single rice crop)} \qquad \ldots (4.5)$$

$$RCI = \frac{ARGS}{RGD + \Sigma T_i} \quad \text{(for multiple cropping systems)} \qquad \ldots (4.6)$$

where ARGS is actual rice growing season (days) for a specific region under double cropped area, ΣT_i is time interval (days) for second crop, RGS is length of rice growing season based on temperature and RGD is length of rice growth duration from sowing to maturity.

Two major thermal belts of rice were divided into 6 rice climatic zones using moisture index. In arid zone production is limited by water stress and needs irrigation resource. In subhumid zone, production depends on irrigation to a large extent. Humid zone receives adequate rainfall to meet rice water requirement. These six rice climatic zones were further splitted into 12 subbelts based on supplementary indices and suitability of different cropping system (Table 4.9).

When RCI is >1.0, 0.9 to 1.0 and < 0.9 zone is suitable for rice cultivation, partial planting of rice varieties and unsuitable for rice cultivation.

4.10.4 Potential productivity of maize:

Chirkov (1969) developed agroclimatic zoning for maize species and hybrids using agroclimatic indices of heat requirements of various species and hybrids of maize (Fig. 4.6). A comparative assessment of the heat requirements of various species and hybrids of maize by varying the ripening rates characterised with the number of leaves on the main stem and the accumulated effective temperature for the initiation of main development

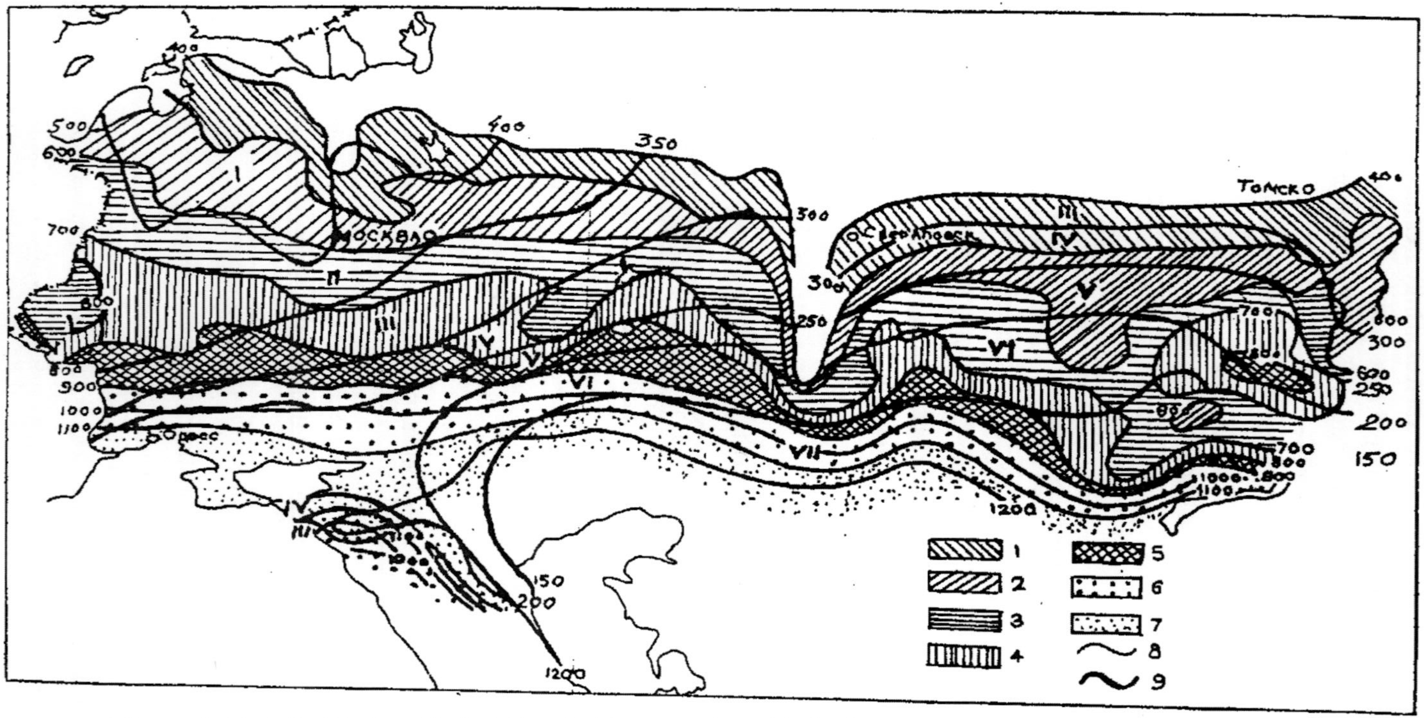

1-7 as per Table 4.10, 8-Isoline of $\sum Tc>10°C$ with 80 per cent dependability for vegetation period, 9-Isoline for precipitation amount during vegetation period.

(*After*: Chirkov, 1969)

Fig. 4.6: Agroclimatic zoning of species and hybrids of maize in the USSR.

phenophases. With 80 per cent risk coverage of heat requirement for the main development phenophases of maize, thermal zones were delineated within the accumulated temperature limit of 400-1200 day °C (Table 4.10).

In each thermal zone, the dependability of moisture regime in maize was made according to available soil moisture or assessed through modified hydrothermal coefficient (K) based on rainfall, soil moisture and accumulated temperature during the vegetation period.

K	moisture dependability of maize
1.2 to 1.3	5 degrees
0.8 to 0.9	3 degrees, irrigation essential
0.5 to 0.7	cultivation uneconomical without irrigation

Chirkov observed that fairly good available moisture in the root zone of 140-180 cm and optimum temperature regime 18-22 °C provided agroclimatic conditions for stable high yields of maize. Similar relationship between yield and heat/water dependability have been developed for wheat distribution with varying growing probabilities and stable yield potential (William, 1969) in Canada.

4.10.5 Winter wheat:

Agroclimatic zoning for winter wheat were delineated with the assessment of autumn vegetation conditions, the wintering period and the agroclimatic conditions of crop formation in spring—summer period. Sapozhnikova and Shashko (1972) developed agroclimatic maps for autumn vegetation period for winter wheat with consideration of the period between the passing of mean temperatures from 12°C to 5°C, mean amount of precipitation for the period 20 days before to 20 days after sowing, and HTC for the given accumulated temperatures and different amounts of precipitation. Coefficient of tillering, during autumn vegetation period for winter wheat, is expressed by the relationship between the tillering rate and accumulated active temperatures for the period from tillering to end of vegetation. Moisechik (1975) developed a comprehensive index (CI) for wintering conditions of winter wheat expressed as

$$CI = \overline{H} \cdot \frac{\overline{T_m}}{\overline{N}} \cdot T_X \text{, where} \qquad \ldots (4.7)$$

$\overline{H}$ is mean maximum for the winter of the depth to which soil freezes, $\overline{N}$ is mean duration of snow cover, $\overline{T}_m$ is mean of absolute minimum temperature for the plants and T_X is critical temperature for the plant freezing. Index reflects the winter's severity for wheat liability to damage.

4.10.6 Agroclimatic zoning for cash crops

Le Quang Huynh (1987) presented agroclimatic zoning for cereals, fruits, vegetables, and industrial crops after evaluating the agroclimatic resources for agrocultural development.

Table 4.10

Heat and moisture dependability for main phenophases development of maize species and hybrids of different ripening rates.

Thermal Zone	Cumulative $\Sigma Tc>10°C$ with 80 per cent Risk Coverage	Phenophase	Ripening Rates of Species and Hybrids and Characteristic Number of Leaves, Moderate 13-14	Moderate 15-16	Moderate 17-18	Late ripening 15-20	Moisture Reserves (mm)
1.	400-500	Heading	80-90	55-80	45-70	35-55	400
		Milky ripeness	14-15	8-20	3-10	5	
		Waxy ripeness	5-10	5	-	-	
2.	500-600	Heading	90-100	90-98	70-90	55-79	350-400
		Milky ripeness	40-60	20-40	8-25	8-10	
		Waxy ripeness	15-35	5-10	5	-	
3.	600-700	Heading	100	98-100	90-100	75-95	300-350
		Milky ripeness	60-80	40-65	25-50	10-30	
		Waxy ripeness	35-60	10-35	5-15	5	
4.	700-800	Heading	100	100	100	95-100	250-300
		Milky ripeness	80-95	60-80	50-70	30-55	
		Waxy ripeness	80-90	35-60	15-40	5-15	
5.	800-900	Heading	100	100	100	100	200-250
		Milky ripeness	95-100	80-98	70-90	55-75	
		Waxy ripeness	80-95	60-82	40-55	15-40	
6.	900-1100	Heading	100	100	100	100	150-200
		Milky ripeness	100	100	90-100	75-100	
		Waxy ripeness	100	82-100	55-90	40-80	
7.	>1100	Heading	100	100	100	100	150
		Milky ripeness	100	100	100	100	
		Waxy ripeness	100	100	90	80	

(*After*: Chirkov, 1969)

Agroclimatic indices in terms of heat, light and moisture were evaluated for humid and tropical complicated relief area of Vietnam. Four thermal belts were delineated with use of accumulated temperature greater than 10°C to meet the thermal requirement of crop with values of <7000, 7000-8000, 8000-9000 and >9000 day °C termed as less hot, medium hot, hot and very hot belts. Average absolute minimum temperature is used to assess the wintering conditions of perennial crops. In each thermal belt, Selyaninov hydrothermal coefficient (HTC) was evaluated for assurance of moisture to meet water requirements of crops and also related it to maximum crop yields. Corn yields was maximum when HTC is between 1-2 and tea yield was maximum when HTC >1.0. HTC became less than 1.0 for dry rainy season. Decreasing values of HTC indicated drier conditions.

Accumulated heat units can be determined for maturity of different crops and cropping patterns. Isotherms of 4, 9 and 15°C were drawn to assess areas of low temperature causing damage to banana, pepper and rubber respectively to provide limits of crop cultivation in the area.

Exercise 3

Q. 1. Develop agroclimatic zones of your state based on moisture and thermal indices, soil types and specify these zones w.r.t. the major cropping system being followed.

Q. 2. Develop agroclimatic zones for individual specific major crops of your state and demarcate the potential productivity zones for seed production.

Q. 3. Suggest alternate cropping/farming system in different zones to improve productivity minimizing risk due to climatic variability and assess the climate's potentiality for different crops/cropping systems.

References

Azzi, G. 1939. Trattato di ecologia agraria. Torino. Soc. Ed. Internationale, p. 594.

Baeir, W. 1977. Crop weather models and their use in yield assessments. WMO Tech. Note 151, WMO, No. 458, Geneva, p. 1-48.

Bishnoi, O.P. 1980. A study of climatic water balance parameters in relation to crop planning over Punjab and Haryana. Thesis submitted to Andhra University, Waltair, p. 250.

Bishnoi, O.P. 1989. Agroclimatic zoning. CAgM Report No 30, Agril. Meteorology, WMO/TD No. 238, Geneva, p. 1-134.

Bishnoi, O.P. 1999. Performance of COTTAM simulation model in Haryana, In: Proc. of national workshop on dynamic crop simulation modelling for agrometeorological advisory services, NCMWRF, DST, Ch. 10, p. 133-39.

Brown, D.M. 1977. Response of Maize to Environmental Temperature: A Review, p. 12-27. In: Proc. of Symp. on Agrometeorology of the Maize Crop. WMO, No. 481, WMO, Geneva.

Budyko, M.I. 1958. The heat balance of earth surface. U.S. Deptt. of Commerce, Washington, D.C.

Budyko, M.I. 1971. Climate and Life. Gidrometeoizdat, L. p. 19-71.

Burgos, J.J. 1958a. Tipos agroclimaticos de la palma datilifersa en la Republica Argentina. Constitucion No. 101. al Congraso, Fruticola, Argentina.

Burgos J.J. 1958b. Agroclimatic classifications and representations. Report, on the application value of climatic and agroclimatic classifications for agriculture purposes. WMO commission for Agric. Meteo., Second Session, Warsaw, p. 4.

Burgos, J.J. 1982. Bases climatolo'gicas para la planification de recursos naturales Y artificiales renovables en la Amazonia Brasilera. OMM – PNUD. Sudan, p. 93.

Carter, D.B. and Mather, J.R. 1966. Climatic classification for environmental biology Pub. in climatology, XIX, 4, p. 1-136.

Champion, H.G. and S.K. Seth 1968. A revised survey of the forest types of India. Delhi: Manager of Publications, GOI.

Chirkov, U.I. 1969. Agroclimatic Conditions and Maize Productivity, Gidrometeoizolat, L. p. 238.

Cocheme, J. and Franquin, 1967. An agroclimatological survey of a semiarid area in Africa south of Sahara. WMO Tech. Note No. 86.

Davitaya, F.F. 1948. Climatic zones for grape cultivation in USSR. M. Pishipromizdat, p. 1-192.

Davitaya, F.F. 1958. Temperature forecasting and some problems of seasons development in the nature. Gidrometeoizdat, L.

Davitaya, F.F., Saposzhnikova, S.A. and Shashko, D.I. 1973. Agroclimatic resources of the USSR. Karta l: Glavnoie upravienic goodezii karlografii pri soc. Min. SSSR, M.

de Candolle 1974. Geographie botanique. Archives de la Bibl. Univ. de Geneve.

Emberger, L. 1955. Une classification biogeographique des climates. Recucil des travaux des laboratories de botanique ge'ologic de la faculte' des sciences de l'Universite' de Montpellier.

Eswarn, H. and E. Van Den Berg (1992). Impact of doubling atmospheric CO_2 on length of the growing season in the Indian sub-continent. Pedologic 42, p. 289-96.

FAO, 1976. Report on the agroecological zones project. Vol. I Methodology and results for Africa, FAO, Rome, p. 15.

FAO, 1978. Report on the agroecological zones Vol. 1, Meteorology and results for Africa. Report, No. 48, p. 1-157.

FAO, 1980. Report of the agroecological zones. Vol. 4, Results of SE Asia. Report No. 48/4, p. 1-39.

Fedoseev, A.P. 1979. Agrotechnique and weather. L., Gidrometeoizdat.

Fussell, L.K. C.J. Pearson, and M.J.T. Norman 1980. Effect of temperature during various growth stages on grain development and yield of Pennisetum americanum. Jour. or Expt. Botany, 31, p. 621-633.

Gaussen, H. 1955. Les climates analogues a l'e'chelle due monde. Compterendus hebdomadaires des, sources de l'Academie d'Agriculture de France, 41.

Geigher, R. 1960. Climate near the earth atmosphere. M. Publication of foreign literature.

Gol'tsberg, I.A. 1961. Agroclimatic characteristics of hard frosts (Ultra low temperature) in USSR, and dealing methods with them, Gidrometeoizdat.

Gol'tsberg, I.A. 1967. Microclimates of the USSR. Gidrometeoizdat, L.

Gol'tsberg, I.A. 1972. Microclimatic Atlas of World, GUGK. Gidrometeoizdat.

Hargreaves, G.H. 1971. Precipitation dependability and potentials for agricultural production in NE Brazil. Rep. 74-D159, EMBAPA Utah.

Humboldt, A. 1817. Memoires Soc. D'Aroveil., 3, p. 469-602.

ICRISAT, 1984. Agrometeorology of sorghum and pearl millet in the semiarid tropics. Proc. Of Intern. Symp. held at Hyderabad, 15-20, Nov. 1982, p. 1-317.

IRRI, 1974. An agrocliamtic classification for evaluating cropping system potentials of SE Asian rice growing regions. IRRI report, p. 1-5, Los, Banos, Phillipiness.

Kanemasu, E.T. and Arkin, G.F. 1974. Radiant energy and light environment of crops. Agril. Meteorolo., 14, NNI-2.

Konstantinov, A.R. 1978. Climate, soil and yield of winter wheat, Gidrometeoizdat, L., p. 24.

Koppen, W. 1884. Die Warmezonen der Erde Met., 2, 1, p. 215-26.

Koppen, W. 1936 Das geograpische system der climate handbuch der klimatologic, Band, Teil C. Koppen – Geiger, Berlin.

Krishnan, A. 1988. Delineation of soil-climatic zones of India and its practical applications in agriculture. Fertilizer News, 33, 4, p. 11-19.

Krishnan, A. and Singh, M. 1972. Soil-climatic zones in relation to cropping patterns. Proceedings of symp. on cropping patterns. ICAR, New Delhi, pp. 172-85.

Le Quang Huynh 1987. Agroclimatic Zonation of Vietnam, Hydrometeorological Service of Vietnam, p. 1-11.

Liangzhi, Gao, Li Lin, and Jin Zhiging 1987. A climatic classification for rice production in China. Agril. & Forest Meteorol., 39, p. 55-67.

Mather, J.R. 1974. Publ. in climatology, Drex. Inst. of Tech., Centerton, N.J., 7, 1.

Mavi, H.S. 1986. Introduction to Agrometeorology. Oxford and IBH Publishing Co., New Delhi.

Mehar – Homji 1960. Carte des types bioclimatiques due sub-est asiatique. Toulouse, France, Faculte' des Sciences, Service International des Cartes.

Mischenko, Z.A. 1974. About the methods of day and night temperature recording and their accuracy, Tradi. GGO, Edi., 339, L., p. 139-51.

Mischenko, Z.A. 1977. Meso and microclimatic division of winter temperatures. J. of Meteorology and Hydrology, No. 10.

Mischenko, Z.A. 1984. Agroclimatic mapping of the continents. CAgM VII, Rapporteur on agroclimatic zoning. CAgM Report No 23, WMO, Geneva, p. 1-83.

Moisechik, V.A. 1975. Agrometeorological conditions and hibernation of winter crops. Gidrometeoizdat, L., p. 294.

Monsi, M. and Sakai, T. 1953. Uber den lichtfactor in den planzenge-selzchaften und seine bedeutung fur die staffproduktion. Japanese Jour. of Botany, Vol. 14, 1.

Monteith, J.L. and Szeicz, C. 1962. Radiative temperature in the heat balance of natural surface. Q.J. Roy. Met. Soc., 88, 378.

NCA, 1976. National commission on agriculture report, Vol. IX, Krishi Bhavan, New Delhi, India.

Nishiyama, I. 1976. Effect of Temperature on the vegetative growth of rice plants, p. 159-185. In: Climate and Rice, IRRI Publication, LOS Banos, Philippines.

Nuttonson, M.Y. 1965a. Agroclimatic analogues for the intermountain region of the United States and an outline of its physiography, climate and farm crops. Washington, p. 36.

Nuttonson, M.Y. 1965b. Agroclimatic analogues for the northern plains regions of the United States and an outline of its physiography, climate and farm crops. Washington.

Ong, C.K. 1983. Response to remperature in a stand of Pearl Millet, I. Vegetation Development. Jour. of Expt. Botany, 34, 140, p. 322-36.

Palmer, W.C. 1965. Meteorological drought. Research Note No. 45, US Weather Bureau, Washington, D.C.

Papadakis, J. 1970. Agricultural potentialities of world climates Ac., Cordoba 4564, Buenos Aires, Argentina.

Papadakis, J. 1975. Climates of the world and their potentialities. Buenos, Aires.

Penman, J.L. 1948. Natural evaporation from open water, bare soil and grass Proc. Roy. Soc., Ser. A. 193, London.

Reddy, S.J. 1983a. Agroclimatic classification on the semiarid tropics I. A method for computation of classification variables. Agri. Meteo., 30, p. 185-200.

Reddy, S.J. 1983b. Agroclimatic classification on the semiarid tropics II. Identification of classificatory variables. Agril. Meteol., 30, p. 201-19.

Robertson, G.W. 1968a. A biometeorological time scale for a cereal crop involving day and night temperature and photoperiod. Int. Jour. Biometeorol., 12, 3, p. 191-223.

Robertson, G.W. 1968b. Rice and Weather. WMO Tech. note No. 144, WMO. No. 423, Geneva, p. 1-39.

Ross, J.K. 1975. Radiation regime and plant vegetation. Gidrometeozdat, L., p. 341.

Sapozhnikova, S.A. and Shaskho, D.I. 1972. Agroclimatic division of the USSR. Ed. Davitaya, D.D., Karta-11, 40.000, GUGK.

Sarkar, R.P. and Biswas B.C. 1980. Agroclimatic classification for assessment of crop potential and its application to dry farming tracts of India. In 'Climatic Classification' UCRISAT, a consultant meeting 14-16, April, 1980. ICRISAT, Patancheru, A.P., op. 89-100.

Sehgal, J. and Mandal, D.K. Mandal, C., and S. Vadivelu 1992. Agroecological regions of India NBSS & LUP Tech. Bull No. 24, Oxford IBH, NBSS & LUP Publication, p. 73.

Sehgal, J., Mandal, D.K., Virmani, S.M. and Mandal, C. 1995. Agroecological subregions of India NBSS and LUP Publication, No. 35, p. 365.

Selyaninov, G.T. 1957. Prinzipy agroklimatichestogo rationito bania SSSR. (Basis for the agroclimatic divisons of USSR). Izvestia Akademiv Nauk SSSR, 4, 97-101.

Selyaninov, G.T. 1958. Prospects of subtropical management in USSR in connection with the natural conditions. L. Gidrometeoizdat, p. 193.

Selyaninov, G.T. 1966. Agroclimatic map of the world. L. Gidrometeoizdat.

Shaskho, D.I. 1967. Agroclimatic divisions of the USSR. M. Publisher, Koles, p. 333.

Sinistina, N.I., Gol'tsberg, I.A., Strunnikov, Z.A. 1973. Agroclimatology, L. Gidrometeoizdat, p. 343.

Sly, W.K. 1970. A climatic moisture index for land and soil classification in Canada.. Canadian Jour. of Soil. Sci., 50, p. 291-300.

Sys, C. 1985. Land evaluation, part I, ITC for postgraduate soil scientists. State Univ. of Ghent, Belgium, p. 141.

Thiele 1895. Deustchlands landwirtschaffliche klimatographic, Bonn.

Thomas, J.E. 1957. Rice in Spain, World Crops, 9, p. 247-50.

Thornthwaite, C.W. 1948. Topoclimatology Proc. towards a rational classification of climate. Geogr. Review, 38, p. 55-94.

Thornthwaite, C.W. and J.R. Mather 1955. The water balance, Pub. in climatology, 8, 1, p. 1-104.

Tooming, K.G. 1977. Solar radiation and yield formation. Gidrometeoizdat, L., p. 1-199.

USSR 1971. Agroclimatic resources of European socialist countries. Ind. Bulgaria, Sci, Acad. Sofia., p. 33.

USSR 1979. European socialist countries. Izd. Bulgaria, Scie, Acad., Sofia, p. 120.

USSR, 1972. Agroclimatic Aflas of world, GUGK, Gidrometeoizdat.

Utaaker, K. 1970. Approaches in agricultural meteorology on the mesoscale in Norway. Publ. Dep. A.N., p. 17-35.

W.M.O., 1977. Agrometeorology of the Maize (Corn) Crop. WMO, No. 481, Proc. of Symp., WMO, Geneva, p. 1-451.

William, G.D.V. 1969. Applying estimated temperature normals to the zonation of the Canadian great plain for wheat. Can. J. Soil Sci., 49, p. 263-76.

Yao, A.Y.M. 1974. Agril Meteorology, 13, p. 405-17.

Zheng, J.F. und Newman, J.E. 1986. The climatic resources for wheat production in China. Agril. Meteorol., 38, p. 205-15.

5

Thermal Time Bioclimatology

5.1 Introduction

Plant's interaction with thermal environment lies with the phenomena of growth and development, respiration and transpiration relations. Temperature, an index of sensible heat component of heat balance equation at the surface, is simultaneously influenced by the hydrological balance components surface at the vegetative particularly evapotranspiration. Therefore available heat and availability of moisture at the transpiring surface influence the thermal environment surrounding the vegetative surface to simulate an effective temperature influencing the growth and development of the plant. Similar to moisture requirement of a plant to meet daily evapotranspirational needs, plants also require a some quantity of thermal heat from the environment for accelerating its development and to sustain growth balance of different organs of the plant. At optimum thermal environment morphological and development changes in the plant will take place at most rapid rate. Heat exchanges, at the plant's surface are in the form of adsorption of heat, and heat loss by conduction, transpiration and reradiation influencing the heat balance at the surface to provide net available heat assessed in terms of temperature index. Therefore accumulated effective temperature have been used as an useful agroclimatic index of heat requirement of plants in addition to the quantification of heat resources at place. Temperature, as an index of thermal environment, strongly influence the development rate in terms of cardinal temperature levels of threshold, optimum and inversion or upper threshold points for the all living organism including plants. Therefore plant's response to environment not only involves in the direct modification of the organism (increase of size or mass), but also sets limits to the adaptability of the organism, and provide a thermal environment force in plants to trigger a set of sequential steps in organism's life cycle.

During the active vegetation period it is essential to quantify the thermal heat available to provide optimum condition for growth, development and periodic occurrences of different phenophases in different plant species from germination to reproduction. This would also enhance the better understanding of bioclimatic activity to characterise link between plant development rate, productivity and thermal factors. Availability of thermal heat in humid to perhumid and temperate to frigid regions is a major constraint for the biological activity.

The concept of growing degree days (GDD) or heat units (HU) or thermal time (TT) originated with observations by Reaumur (1735) that plant development is more closely related to the temperature accumulated to a given stage than with time alone. However, this concept became popular in scheduling plantings for the harvest of rapidly maturing vegetables. The calendar of growing degree days have proved useful for evaluating climate relative to crop and variety adaptability and for crop scheduling. Breeders and seed producers are using GDD concept to predict time of flowering for controlling pollination and successful cropping of hybrid genotypes of crops.

Went's phenomena of thermal periodism with diurnal variation of temperature is well known. Modification of thermal environment to achieve optimum conditions for higher growth and rapid development of crop growth stages in temperate and colder environments have proved very beneficial. Therefore optimum use of thermal environment data for different applications of phenological development, predicting flowering and harvesting/maturity, and adaptability of different cropping schedules and rotations, crop phenological zonings and modification of thermal environment are useful applications for the benefit of developing crop weather advisories to raise the economic status of farming community engaged in raising value aided agricultural products.

5.2 Growth and Temperature

Growth refers to the accumulation of biomass of an organism with increase of its mass or size/shape, whereas development is the sequential periodic occurrences of different phenophases of the organism to reproduce it. Normally growth is photochemical light sensitive reaction, but the dark reaction i.e., respiration is sensitive to temperature influencing the net growth and its distribution in different plant organs sharing at that time. Low temperatures limits respiration, but high temperature will enhance respiration reducing net biomass generation. At low light intensity, the net photosynthesis rate is linearly related, but at higher light intensities, the net photosynthesis rate become light limited provided CO_2 and heat are available in the environment. When CO_2 becomes inadequate in the environment, the rate of dark reaction is limited and net photosynthesis is insensitive to both light and heat availability. With adequate CO_2 levels in a warmer environment and higher light intensities, the net photosynthesis saturates at a higher level than in a cooler environment. The net biomass generated due to light interception is distributed in plant organs to increase their mass and size in terms of plant height, total leaf area, total length/mass of the root system. Since agrometeorological interaction of plant with their surrounding environment takes place at the leaf surface, there is close correspondence between development index and both leaf area development and interception of photosynthetically active radiation indicating a significant climatic control of plant's productivity with the development index. The stage of the plant's development determines the amount and relative activity of leaf surface exposed to incoming PAR, CO_2, and transpirational water loss. Baker and Musgrave (1964) demonstrated that maximum green leaf area and root development at heading stage of corn and sorghum when the transpiration rate becomes maximum for peak vegetative growth. Therefore it is useful to develop climatic indices (development heat units) to predict leaf area development to depict the plant's capacity to exploit environment during the crop growing season attaining maximum leaf area and root growth around heading in field grain crops.

A simple method of evaluating the rate of biochemical process due to temperature is temperature coefficient commonly represented by Q_{10} i.e., the ratio of reaction rate at any temperature to the rate at 10°C higher. For growth enzymatic reactions it is around 2. The relative larger value of Q_{10} for enzymatic reactions shows that with increase of temperature, the enzymatic activities also increases, but at lower temperature the enzymatic activity may cease. Therefore a temperature response curve for the growth of the plants will follow the shape of the enzyme response curve, rising rapidly at linear rate at lower temperature ranges and less rapidly to intermediate optimum temperature, and then falling at higher temperatures. Growth of different plant components leaves, stems, roots etc. follow a similar pattern. The increase of growth rate at higher temperature within the growth range would result in shortening of growth duration period of enzyme activities.

Went (1944, 48) work on growth rate of tomato plants exhibited the phenomena of thermoperiodism when the day temperature (26°C) was altered diurnally with lower night temperature of 17°C. Went (1957) also observed that the lower temperature at night had beneficial effect on plants such as stem growth, quality of growth, earliness, flowering and fruiting rate process, increase in sugar contents and carbohydrate translocation. Enhanced root growth over stem growth at lower night temperature was useful for the growth of root crops for obtaining higher yield of potato at lower night temperature (Went, 1959) and increase of sugar contents of sugarbeet (Ulrich, 1952). Therefore annual cyclic pattern of day and night temperatures at a location or in a region are useful to select the adaptability of different crop species in particular months where optimum day and night temperature are experienced for beneficial effects of lower night temperatures on the growth rate and fruit setting in horticultural crops. The thermal tolerance of crops also changed with phenological stages from planting to harvest indicating a variation of optimum temperature needs at different phenophases (Table 5.1).

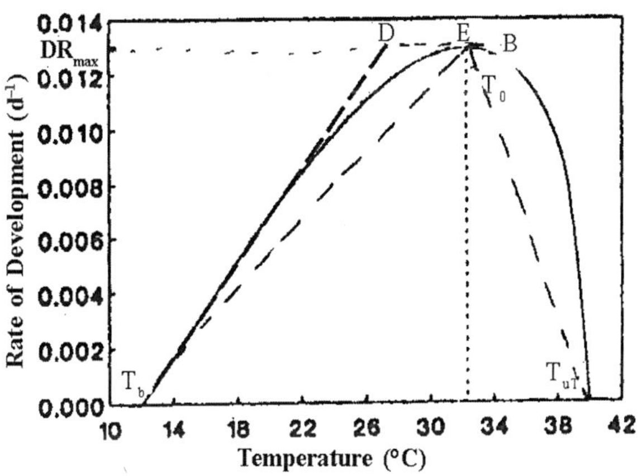

Fig. 5.1: Relation between temperature and rate of vegetative development

Therefore thermoperiodic adaptation of plants within specific range of day and night temperatures are useful for higher productivity and seed production programmes. During daytime growth rate is faster following a polynomial curve, whereas during the night time growth rate response is linear.

5.3 Crop Development

Development of crops is primarily influenced by temperature in a linear manner between threshold and optimum levels. This response of temperature on crop development rate (DR) is usually expressed with thermal time or growing degree days concept within cardinal temperature limits. Crop development rate is linearly related to increase of temperature between base to optimum temperature levels and with a decreasing linear response between optimum to upper threshold limit of temperature (Fig. 5.1). If temperature beyond optimum is not achieved in the environment or it is attained only for a very short period, then linear thermal time concept (ADEB) is adequate to describble

Table 5.1
Threshold, optimum and upper limit temperature values of some field crops

Crop	Cardinal Temperature (°C)			References
	T_{base}	$T_{op.}$	T_{UT}	
Winter wheat	0	15	25.0	McMaster and Smika, 1988
Corn	10	25	30.0	Cross and Zuber, 1972
Tomato	7	15-18° night,		Witter and Aung, 1969
	7	18-27 day	27	Owens and Moore, 1974
	4.3			Warnok and Isaacs, 1969
	6 to 10			Calado and Portas, 1987
Cassava (sowing to	14.8	30	36.6	Keating and Evenson, 1979
emergence)	12.5	29.5	39.8	
Maize (leaf development)	12.8	37.5	40	Watts, 1971
Maize (sowing to tasseling) photoperiod 12 hors	10-11.8	21-29	37.0-37.2	Ellis et al., 1992
Rice (sowing to flowering) photoperiod 8 hours	8-10	24-25		IRRI, 1977
Wheat (winter)	3 to 4.5	25	30-32	Nuttonson, 1955
Wheat (spring)	5	25-30	30-32	Nuttonson, 1957; Bierhuizen, 1973
Barley	3 to 4.5	15-20	38-40	Mavi, 1986
Maize (warm)	8 to 10	32-35	40-44	Anonymous, 1988
Maize (temperate)	8 to 10	21-25	30-35	Gilmore and Rogers, 1958
Rice (summer)	10 to 12	30-32	36-38	Anonymous, 1988
Rice (temeprate)	8 to 10	21-24	30-32	Gao et al., 1992
Tobacco	13 to 14	28	35	Mavi, 1986
Sugarbeet	4 to 5	25	28-30	Mavi, 1986
Cotton	10	27.5	40	Anonymous, 1988
Sorghum/Pearl Millet	8 to 11	32-35	40	Anonymous, 1988
Potato	3	18	-	Mavi, 1986
Peas	4 to 5	30	35	Wang and Bryson, 1956
Gram	5	20 to 28	30-35	Anonymous, 1988
Rapeseed and Mustard	5	24	30-35	Anonymous, 1988
Sugarcane	8-10	28 to 32	35-40	Anonymous, 1988
Pearl Millet	10.9	33.4	43.8	Anonymous, 1987
Sorghum	10.1	35.6	47.4	Anonymous, 1987
Groundnut	9.7	33.6	43.4	Anonymous, 1987

$$DR \propto (T - T_b)$$

$$DR = \frac{(T - T_b)}{k} \qquad \ldots (5.1)$$

When a particular phenological event is completed in n number of days, DR is 1.0 and constant k attains the value of total thermal time or growing degree days (GDD) (Hodges, 1991; Tollenaar et al., 1979; Kiniry and Keener, 1982) expressed it as

$$GDD = \sum_{i=1}^{n} (T_i - T_b) \qquad \ldots (5.2)$$

However, work of Lehenbauer (1914) on elongation of maize seedlings showed a decline rate when temperature was raised above optimum level and Gilmore and Rogers (1958) presented a bilinear model which accounted for a reversal linear function between optimum to upper threshold ceiling temperature level (AEC path) (Roberts and Summerfield, 1987). Performance of this model was better than linear model, but still it does not describe the pattern accurately (Solid line response). It was observed that in different crop genotypes, the DR_{max} and slope of increasing and decreasing linear responses differed, and therefore in order to describe these limitations Gao et al. (1952) presented a Rice Clock Model (RCM) to describe the response of DR to temperature expressed as

$$DR = \exp(k)\left\{\frac{T - T_b}{T_0 - T_b}\right\}^{\alpha}\left\{\frac{T_{uT} - T}{T_{uT} - T_0}\right\}^{\beta}, \qquad \ldots (5.3)$$

where k, α and β are model parameters, and DR_{max} occurs at $T = T_0$. However Gao et al (1992) observed that maxima DR in indica rice cultivar occurred at 2.3°C higher based on equation above than the prespecified value of T_0(30°C). Therefore above equation was modified to achieve DR_{max} at T_0 expressed as

$$DR_{max} = \exp(k) \text{ when } \left\{\frac{T - T_b}{T_0 - T_b}\right\}^{\alpha}\left\{\frac{T_{uT} - T}{T_{uT} - T_0}\right\}^{\beta} > 1 \qquad \ldots (5.4)$$

Yin et al (1995) developed the Beta function to describe the effect of temperature on the rate of crop development with use of three cardinal temperatures base (T_b), optimum (T_0) and upper threshold or ceiling (T_{uT}) levels. The three model coefficient μ, α and β provides a flexible family of non-symmetric, unimodel probability density functions with fixed end points of T_b and T_{uT} to describe the response of DR to temperature expressed as

$$DR = \exp(\mu)\{T - T_b\}^{\alpha}\{T_{uT} - T\}^{\beta}, \qquad \ldots (5.5)$$

The DR_{max} corresponding to T_0 is determined from the equation by expressing its first order derivative to zero to get T_0 expressed as

$$T_0 = \frac{\alpha T_{uT} + \beta T_b}{\alpha + \beta} \qquad \ldots (5.6)$$

Substituting T_0 in equation of DR, we get DR_{max} as

$$DR_{max} = \exp(\mu)\alpha^{\alpha}\beta^{\beta}\left\{\frac{T_{uT} + \beta T}{\alpha + \beta}\right\}^{\alpha+\beta} \quad \ldots (5.7)$$

Yin et al. (1995) used Beta model to describe the crop development with temperature accurately with superiority over thermal time approaches in predicting rice flowing time. α defines the curvature of the relationship, with temperature between T_b and T_0, whereas β characterises the curvature between T_0 and T_{uT}. Table 5.1 provides data of temperature on cardinal points for some field crops.

Example 1:

Develop the rice clock model (RCM) for describing the development rate of indica rice variety with temperature, using following limiting parameters from sowing to flowering.

$T_b = 8°C$, $T_{uT} = 42°C$

DR(day^{-1})	Temperature (°C)
0.004	17
0.012	26
0.013	30

Solution:

$$DR = \exp(k)\left\{\frac{T - T_b}{T_0 - T_b}\right\}^{\alpha}\left\{\frac{T_{uT} - T}{T_{uT} - T_0}\right\}^{\beta},$$

$$\text{Ln } DR = k + \alpha \text{Ln}\left(\frac{T - T_b}{T_0 - T_b}\right) + \beta \text{Ln}\left(\frac{T_{uT} - T}{T_{uT} - T_0}\right),$$

$DR_{max} = \exp(k)$ at $T = T_0$ Around $T_0 = 30°C$, $k = \text{Ln}(DR_{max})$, therefore $k = -4.3428$

$$\text{Ln } 0.004 = k + \alpha \text{Ln}\left(\frac{9}{22}\right) + \beta \text{Ln}\left(\frac{25}{12}\right)$$

$$\text{Ln } 0.012 = k + \alpha \text{Ln}\left(\frac{18}{22}\right) + \beta \text{Ln}\left(\frac{16}{12}\right)$$

Solving these equations for α and β after substituting k value, we get

$$\alpha = \frac{+1.17866 + .73397\beta}{+.8938}$$

$\beta = 1.5022$, $\alpha = 2.5512$

$$DR = \exp(-4.3428)\left(\frac{T - T_b}{T_0 - T_b}\right)^{2.5522}\left(\frac{T_{uT} - T}{T_{uT} - T_0}\right)^{1.5022}$$

$$T_0 = \frac{\alpha T_c + \beta T_b}{\alpha + \beta}$$

$$T_0 = \frac{2.5512 \times 42 + 1.5022 \times 8}{4.0522}$$

$$= \frac{107.15 + 12.0176}{4.0522} = 29.4\ °C$$

$T_0 = 29.4$ °C

Example 2:

Using the following data on development rate of rice determine the development rate function with temperature using Beta function model from sowing to flowering using.

$T_b = 8°C$,	$T_{uT} = 42°C$
R(day^{-1})	**Temperature (°C)**
0.004	17
0.012	26
0.013	30

Solution:

β function response of DR on temperature is

$$DR = \exp(\mu)\,(T - T_b)^{\alpha}\,(T_{uT} - T)^{\beta}$$

$$DR_{max} = \exp(\mu)\alpha^{\alpha}\beta^{\beta}\left(\frac{T_c - T_b}{\alpha + \beta}\right)^{\alpha+\beta}$$

$$T_0 = \frac{\alpha T_c + \beta T_b}{\alpha + \beta}$$

In $DR = \mu + \alpha Ln(T - T_b) + \beta Ln(T_{uT} - T)$

Therefore

$$Ln(0.004) = \mu + \alpha Ln(9) + \beta Ln(25)$$
$$Ln(0.012) = \mu + \alpha Ln(18) + \beta Ln(16)$$
$$Ln(0.013) = \mu + \alpha Ln(22) + \beta Ln(12)$$
$$-5.215 = \mu + 2.1972\alpha + 3.2189\beta$$
$$-4.4228 = \mu + 2.8904\alpha + 2.7726\beta$$
$$-4.3428 = \mu + 3.0910\alpha + 2.4849\beta$$
$$-1.0987 = -0.6932\alpha + 0.4463\beta$$
$$-0.0800 = -0.2006\alpha + 0.2877\beta$$

$$\alpha = \frac{+1.0987 + 0.4463\beta}{+0.6932}$$

$$\therefore \quad -0.08 = -0.5006\left(\frac{1.0987 + 0.4463\beta}{+0.6932}\right) + 0.2877$$

$$-0.08 = -0.3179 - 0.1292\beta + 0.2877\beta$$

$$0.2379 = 0.1585\beta$$

$$\beta = 1.5009, \alpha = 2.5513$$

$$T_0 = \frac{\alpha T_c + \beta T_b}{\alpha + \beta}$$

$$T_0 = \frac{2.5513 \times 42 + 1.5009 \times 8}{4.0522}$$

$$T_0 = \frac{107.1546 + 12.0072}{4.0522} = \frac{119.1618}{4.0522} = 29.4°\,C$$

$$-4.4228 = \mu + 7.37428 + 4.16139$$

$$\mu = -15.9585.$$

$$DR_{max} = \exp(-15.9585)\ 2.5513^{2.5513}\ 1.5009^{1.5009}\left(\frac{34}{4.0522}\right)^{4.0522}$$

$$= 1.73 \times 10^{-7} \times 10.9086 \times 1.8394(5{,}538.2673)$$

$$DR_{max} = 0.013035\ day^{-1}.$$

5.4 Growing Degree Days/Thermal Time

The concept of heat units (Reaumer, 1735) was successfully used in assessing crop phenology and plant development. The simple form of heat units, measured in growing degree day (GDD), is

$$GDD = \left[\frac{T_{Max} + T_{Min}}{2}\right] - T_{base} \qquad \ldots (5.8)$$

where T_{max} and T_{min} are the daily maximum and minimum temperatures, and T_{base} is threshold temperature at which the development is initiated. T_{base} varies with crops, their varieties, growth stages and seasonal conditions (Wang, 1960). This expression is intended to describe the heat energy received by the crop over a given time period, and accumulated heat energy is related to the growth or development processes. There are several interpretations of this equation when put to practical use, as how these parameters are incorporated. For example, two common interpretations (McMaster and Wilhelm, 1997) are (i) most common method for grain cereals when $[(T_{Max} + T_{Min})/2] < T_{base}$, then $[(T_{Max} + T_{Min})/2] = T_{base}$. (Davidson and Campbell, 1983; Kirby, 1995; Masle et al., 1989; Gallagher, 1979); (ii) When $T_{max} < T_{base}$ and $T_{min} < T_{base}$, then $T_{max} = T_{base}$, $T_{min} = T_{base}$—a most common method used for crops such as corn (Baker et al., 1986; Baver et al., 1988; Swanson and Wilhelm, 1996; Wilhelm et al., 1989). The basic difference is T_{base} is compared with average temperature in first approach, whereas in the second approach T_{base} is compared with T_{max} and T_{min} before calculating the average temperature.

Example 3:

Compute GDD (°C days) with basic equation of GDD with two interpretations in use of daily temperature values using the following data for a wheat crop with base temperature of 0°C.

Day	T_{max} °C	T_{min} °C
1	15.5	5.0
2	12.7	4.1
3	9.4	-2.0
4	10.0	-1.0
5	12.4	0.0
6	10.0	5.0
7	15.8	-5.0
8	5.0	-7.0
9	-2.0	-10.0
10	-5.0	-15.0
11	0	-6.0
12	10.0	1.0
13	15.0	5.0
14	128	3.0
15	10.0	0.0
16	9.0	-1.6
17	8.0	-2.0
18	12.0	-4.0
19	15.0	6.0
20	20.0	4.0

Solution:

Day	T_{max}(°C)	T_{min}(°C)	$T_{av.}$(°C)	First Interpretation		Second Interpretation		
				Adjusted $T_{av.}$(°C)	GDD (°C day)	Adjusted T_{max}(°C)	Adjusted T_{min}(°C)	GDD (°C day)
1	15.5	5.0	10.0	10.0	10.0	15.5	5.0	10.0
2	12.7	4.1	8.4	8.4	8.4	12.7	4.1	8.4
3	9.4	-2.0	3.7	3.7	3.7	9.4	0.0	4.7
4	10.0	-1.0	4.5	4.5	4.5	10.0	0.0	5.0
5	12.4	0.0	6.2	6.2	6.2	12.4	0.0	6.2

6	10.0	5.0	7.5	7.5	7.5	10.0	5.0	7.5
7	15.8	-5.0	5.4	5.4	5.4	15.8	0.0	7.9
8	5.0	-7.0	-1.0	0.0	0.0	5.0	0.0	2.5
9	-2.0	-10.0	-6.0	0.0	0.0	0.0	0.0	0.0
10	-5.0	-15.0	-10.0	0.0	0.0	0.0	0.0	0.0
11	0	-6.0	-3.0	0.0	0.0	0.0	0.0	0.0
12	10.0	1.0	5.5	5.5	5.5	10.0	1.0	5.5
13	15.0	5.0	10.0	10.0	10.0	15.0	5.0	10.0
14	128	3.0	7.9	7.9	7.9	12.8	3.0	7.9
15	10.0	0.0	5.0	5.0	5.0	10.0	0.0	5.0
16	9.0	-1.6	3.7	3.7	3.7	9.0	0.0	4.5
17	8.0	-2.0	3.0	3.0	3.0	8.0	0.0	4.0
18	12.0	-4.0	4.0	4.0	4.0	12.0	0.0	6.0
19	15.0	6.0	10.5	10.5	10.5	15.0	6.0	10.5
20	20.0	4.0	12.0	12.0	12.0	20.0	4.0	12.0
Total					**107.3**			**117.6**

Second interpretation estimates more GDD than first interpretation because the adjusted $T_{av.}$ is greater due to resetting of T_{Max} and T_{Min} to T_{base}.

5.4.1 Modifications of GDD/TT expression

When T_{Min} falls below T_b, GDD values are not accurate. Growth of plants is retarded when temperature increases above optimum temperature which frequently happen in temperate to tropical environment. Several modifications have been put forward to understand the biological meaning of upper temperature threshold (Gilmore and Rogers, 1958; McMaster and Smika, 1988); converting to photothermal units by adding a photoperiod variable (Nuttonson, 1948; Masle et al., 1989; using only maximum or minimum temperature or portion of the day (Cross and Zuber, 1972; Masle et al., 1989), incorporating sunshine for heliothermal time units for thermosensitive genotypes (Smith, 1920.), energy development units when daily radiation (total or net) is incorporated with GDD.

Soybean development units (SDU) as average of daytime and night time growth development units (Brown, 1960; Brown and Chapman, 1961), development units in terms of 100 growth units with one cm of transpiration of water (Thornthwaite and Mather, 1955) and physiological index based on temperature efficiency for plant growth (Lehenbauer, 1914) and for elongation of corn plants (Livingston, 1916). The upper limit of maximum temperature is set at 26.6/30/32.2°C and a minimum temperature below T_{base} is set at T_{base} depending on crop species. Newman et al. (1967) adjusted GDD based on the maximum temperature and GDD daily values for maize crop were assessed as

(i) When $T_{Max} > 32.2°C$ and GDD ≥ 30, then

GDD (Adjusted) = GDD – (T_{Max} – 32.2°C), and ...(5.9)

(ii) When $T_{Max} > 26.6$ °C and GDD < 5, then

GDD (Adjusted) = GDD + (T_{Max} – 26.6°C) ... (5.10)

Brown (1960) method treated day and night time temperatures separately to determine their contribution towards phenological development (Y) expressed as

$$Y_{Max} = 1.85(T_{Max} - 10°C) - 0.026\,(T_{Max} - 10°C)^2 \quad ...\,(5.11)$$

$$Y_{Min} = T_{Min} - 4.5°C \quad ...\,(5.12)$$

$$GDD = (Y_{Max} + Y_{Min})/2 \quad ...\,(5.13)$$

The summation of interdiurnal temperature was introduced by Wang and Bryson (1956) for pea yield and flowering of native plants where effective diurnal spread was converted into degree hours (DH) as

$$DH = \frac{12(T_{Max} - T_b)}{T_{Max} - T_{Min}} \quad ...\,(5.14)$$

Therefore upper threshold temperature is adjusted in several ways for computing GDD from the basic equation 5.8, however commonly used approaches are, (i) before entering the temperature data in equation 5.8, $T_{av.}$ is set equal to T_{base} if less than T_{base} and is set equal to T_{UT} when greater than T_{UT}; (ii) before entering the temperature data into equation 5.8, T_{max} and T_{min} are set equal to T_{base} if less than T_{base} and are set equal to T_{UT} when greater than T_{UT}.

Example 4:

Use the following temperature data for computing GDD by two different modified methods for winter maize with use of a upper threshold value of 30°C and base temperature of 10°C.

Day	T_{max}(°C)	T_{min}(°C)
1	30	14
2	32	15
3	35	18
4	40	22
5	38	20
6	35	18
7	30	15
8	27	12
9	25	10
10	20	8
11	18	5
12	10	3
13	16	9

14	20	8
15	22	10
16	25	8
17	32	10
18	35	11
19	37	15
20	30	10

Solution:

Method 1 if $T_{av.} < T_{base}$, then $T_{av.} = T_{base}$; If $T_{av.} > T_{uT}$, then $T_{av.} = T_{uT}$.
Method 2 if T_{max} or $T_{min} < T_{base}$, then T_{max} or $T_{min} = T_{base}$; If T_{max} or $T_{min} > T_{uT}$, then T_{max} or $T_{min} = T_{UT}$

Day	T_{Max}(°C)	T_{Min}(°C)	$T_{av.}$(°C)	First Modified Method			Second Modified Method			
				Adjusted $T_{av.}$(°C)	GDD (°C day)	ΣGDD (°C day)	Adjusted T_{Max}(°C)	Adjusted T_{Min}(°C)	GDD (°C day)	ΣGDD (°C day)
1	30	14	22.0	22.0	12.0	12.0	30	14	12.0	12.0
2	32	15	23.5	23.5	13.5	25.5	30	15	12.5	24.5
3	35	18	26.5	26.5	16.5	42.0	30	18	14.0	38.5
4	40	22	31.0	30.0	20.0	62.0	30	22	16.0	54.5
5	38	20	29.0	29.0	19.0	81.0	30	20	15.0	69.5
6	35	18	26.5	26.5	16.5	97.5	30	18	14.0	83.5
7	30	15	22.5	22.5	12.5	110.0	30	15	12.5	96.0
8	27	12	24.5	24.5	14.5	124.5	27	12	14.5	110.5
9	25	10	17.5	17.5	7.5	132.0	25	10	7.5	118.0
10	20	8	14.0	14.0	4.0	136.0	20	10	5.0	123.0
11	18	5	11.5	11.5	1.5	137.5	18	10	4.0	127.0
12	9	4	6.5	10.0	0.0	137.5	10	10	0.0	127.0
13	16	9	12.5	12.5	2.5	140.0	16	10	3.0	130.0
14	20	8	14.0	14.0	4.0	144.0	20	10	5.0	135.0
15	22	10	16.0	16.0	6.0	150.0	22	10	6.0	141.0
16	25	8	16.5	16.5	6.5	156.5	25	10	7.5	148.5
17	32	10	21.0	21.0	11.0	167.5	30	10	10.0	158.5
18	35	11	23.0	23.0	13.0	180.5	30	11	10.5	169.0
19	37	15	26.0	26.0	16.0	196.5	30	15	12.5	181.5
20	30	10	20.0	20.0	10.0	206.5	30	10	10.0	191.5

First modification estimates more GDD than second modification due to lowering of T_{Max} to T_{UT}.

Perry et al. (1986) and Perry and Wehner (1990) observed that use of upper threshold temperature method was better among 14 different heat unit methods used for assessing harvesting of cucumber. Perry et al. (1997) compared 11 methods of computing heat units for tomato harvest at different locations and seasons, and reported the most reliable method with least coefficient of variation (Arnold, 1959) as the reduced ceiling method (Perry et al. 1986) multiplied by day length (DL) resulted in least variations expressed as

$$GDD = \Sigma[\{T_{UT} - (T_{Max} - T_{UT})\} - T_{base}]DL \text{ when } T_{Max} > T_{UT}, \text{ and} \quad \ldots (5.15)$$

$$GDD = \Sigma[T_{Max} - T_{base}]DL \quad \text{when } T_{Max} < T_{UT} \quad \ldots (5.16)$$

Owens and Moore (1974) adopted a different procedure for computing GDD as

$$GDD = [\{(T_{Max} + T_{Min})/2 - (T_{max} - T_{UT})\} - T_{base}] \text{ where } T_{Max} > T_{UT}, \text{ and} \quad \ldots (5.17)$$

$$GDD = [(T_{Max} + T_{Min})/2 - T_{base}] \text{ when } T_{Max} \leq T_{UT}, \quad \ldots (5.18)$$

Tyldesley (1978) expressed GDD as

$$GDD = [(T_{Max} + T_{Min})/2 - T_{base}] \text{ when } T_{Min} > T_{base}, \quad \ldots (5.19)$$

$$GDD = [\tfrac{1}{2}(T_{Max} - T_{base}) - \tfrac{1}{4}(T_{base} - T_{Min})] \text{ when } (T_{Max} - T_{base}) > (T_{base} - T_{Min}) > 0, \quad \ldots (5.20)$$

$$GDD = [\tfrac{1}{4}(T_{Max} - T_{base})] \text{ when } 0 < (T_{Max} - T_{base}) < (T_{base} - T_{Min}), \text{ and} \quad \ldots (5.21)$$

$$GDD = 0 \text{ when } T_{Max} < T_{base} \quad \ldots (5.22)$$

Inclusion of daylength improved the accuracy of the model for daylength sensitive genotypes. Similarly to threshold temperature (T_b) there exists a threshold photoperiod (P_b) during the preanthesis growth period of a crop. Photoperiod plays critical role in case of photosensitive genotypes of crops particularly beans and the accumulated heat units are expressed as below when temperature and daylength values are between threshold and optimum levels.

$$PTT = \sum_{i=1}^{n}(T_i - T_b)(P_i - P_b) \quad \ldots (5.23)$$

$$HTT = \sum_{i=1}^{n}(T_i - T_b)(S_i) \text{ where} \quad \ldots (5.24)$$

PTT is photothermal time for day length sensitive genotype, HTT is heliothermal time for thermsensitive genotypes, S_i is sunshine hours or radiation amount values.

Thermal time variations take place due to variations in threshold temperature with crop development (increases at reproductive stages), variations in photoperiod, vernalization effect for winter crops, and apical development behaviour of different genotypes.

Example 5:

Compute GDD for winter wheat using base temperature of 5°C and upper threshold temperature of 25°C during tillering phase using following data for the two modified approaches.

Day	T_{Max}(°C)	T_{Min}(°C)
1	25	5
2	22	0
3	26	2
4	30	6

5	36	10
6	38	14
7	36	16
8	28	6
9	26	10
10	25	9
11	36	10
12	35	9
13	26	8
14	19	-5
15	21	-5
16	20	-6
17	18	-10
18	20	0
19	22	6
20	25	11

Solution:

				Method 1_b	Method $1_{b\&UT}$		Method $_{2b}$			Method $_{2b\&UT}$		
Day	T_{max}(°C)	T_{min}(°C)	$T_{Av.}$(°C)	GDD (°C day)	Adj. $T_{av.}$(°C)	GDD (°C day)	Adj. T_{max}	Temp. T_{min}	GDD (°C day)	Adj. T_{max}	Temp. T_{min}	GDD (°C day)
1	25	5	15	10	15	10	25	5	10.0	25	5	10.0
2	22	0	11	6	11	6	22	5	8.5	22	5	8.5
3	26	2	14	09	14	9	26	5	10.5	25	5	10.0
4	30	6	18	13	18	13	30	6	13.0	25	6	10.5
5	36	10	23	18	23	18	36	10	18.0	25	10	12.5
6	38	14	26	21	25	20	38	14	21.0	25	14	14.5
7	36	16	26	21	25	20	36	16	21.0	25	16	15.5
8	28	6	17	12	17	12	28	6	12.0	25	6	10.5
9	26	10	18	13	18	13	26	10	13.0	25	10	12.5
10	25	9	17	12	17	12	25	9	12.0	25	9	12.0
11	36	10	23	18	23	18	36	10	18.0	25	10	12.5
12	35	9	22	17	22	17	35	9	17.0	25	9	12.0
13	26	8	17	12	17	12	26	8	12.0	25	8	11.5
14	19	-5	7	2	7	2	19	5	7.0	19	5	7.0
15	21	-5	8	3	8	3	21	5	8.0	21	5	8.0
16	20	-6	7	2	7	2	20	5	7.5	20	5	7.5
17	18	-10	4	0	5	0	18	5	6.5	18	5	6.5
18	20	0	10	5	10	5	20	5	7.5	20	5	7.5
19	22	6	14	9	14	9	22	6	9.0	22	6	9.0
20	25	11	18	13	18	13	26	11	13.5	25	11	13.0
Total				216		214			245.0			211.0

Method 1: When $T_{av.} < T_{base}$, then $T_{av.} = T_{base}$; when $T_{av.} > T_{UT}$, then $T_{av.} = T_{UT}$

Method 2: When T_{max} or $T_{min} < T_{base}$, then T_{max} or $T_{min} = T_{base}$; T_{max} or $T_{min} > T_{UT}$, then T_{max} or $T_{min} = T_{UT}$

b = base temperature correction only

b & UT = base and upper threshold temperature corrections.

Method 2 when base and upper threshold corrections are applied estimated lowest cumulative GDD values over other procedures.

Therefore several modification have been made for upper and lower temperature limits in order to quantify the stable heat unit system with least coefficient of variation. The ideal heat unit system with limiting conditions is of the form.

$$TT / GDD = \frac{[T_{Max} - (T_{Max} - T_0)] + [T_b + (T_{Min} - T_b)]}{2} - T_b \qquad \ldots (5.25)$$

Brown (1960) developed soybean development units (SDU) as a photothermal unit based on the rate of development for the number of hours of darkness as a function of mean daily temperature. The rate of development unit (Y) at each temperature is defined as reciprocal of night hours from planting to flowering as about same number of night hours were required to reach flowering for all photoperiods at any temperature level.

$$Y = \frac{10{,}000}{\text{Night hours to flowering}} \qquad \ldots (5.26)$$

The rate of development and temperature proved to be a curve linear and parabolic in nature as the base temperature of 50°F and SDU was expressed as

$$SDU = 4.39T - 0.0256\,T^2 - 155.18 \qquad \ldots (527)$$

where, T is average temperature from planting to flowering. Daily mean temperature values are used to assess daily SDU contribution. Brown (1960) compared various heat indices with increasing temperatures and reported that SDU and physiological index systems had similar optimum temperatures between 86 to 89°F respectively.

5.5 Transpirational Needs for Plant Development

Since dry matter production is strongly related to transpiration, therefore accumulated transpiration values will quantify the amount of dry matter which is used for plant's development through its distribution in different plant organs. Thus plant's development is directly influenced by the transpirated soil water which is easy to quantify to translate plant development.

5.5.1 EPIC model for transpiration

Similar to Ritchie (1972) EPIC model used following expressions for plant (Pt_p) and soil (PE_s) evaporation expressed as (Sharpley and Williams, 1990).

$$Pt_p = PE_0 \times \frac{LAI}{3} \text{ for } 0 \leq LAI \leq 3 \quad \ldots (5.28)$$

$$Pt_p = PE_0 \text{ for } LAI > 3.0 \quad \ldots (5.29)$$

where PE_0 is Penman's potential evaporation under no shortage of water. However when soil water was limited, Pt_p will decrease according to the availability of water in the root zone depth. The actual transpiration from the plant surface is determined as product of potential transpiration and ratio of actual water available as fraction of potential available water in the root zone depth.

5.5.2 Ritchie method for potential transpiration

Ritchie and Burnett (1971) assessed the daily transpirational needs for a row crop of sorghum with no shortage of water as

$$Pt_p = PE_0 \, (-0.21 + 0.7\sqrt{LAI}\,) \text{ for } 0.1 \leq LAI \leq 2.7 \quad \ldots (5.30)$$

$$Pt_p = PE_0 \text{ for } LAI > 2.7 \quad \ldots (5.31)$$

, where PE_0 and Pt_p are Penman's potential evapotranspiration and potential transpiration rate on a day having LAI. Ritchie (1972) model assessed potential transpiration also as

$$PT_p = PET - PE_s \quad \ldots (5.32)$$

, where PET is potential evapotranspiration of a developing canopy for short uniform sorghum crop with no shortage of water and PE_s is potential soil evaporation from soil surface saturated with soil water.

It can be expressed as

$$PT_p = PET - \frac{\Delta}{\Delta + \gamma} R_{ns}, \quad \ldots (5.33)$$

, where R_{ns} is net radiation over the soil surface saturated with water. Ritchie (1972) expressed R_{ns} as a function R_{no} for several plants with different LAI values as

$$R_{ns} = R_{no} \exp(-0.398 \, LAI) \quad \ldots (5.34)$$

, where R_{no} is net radiation values over short crop with saturated water surface. Therefore, the evapotranspiration from plant surface at saturated conditions or wet surfaces are expressed assuming that acrodynamic factors do not influence the evaporation process.

$$PT_p = PET - \frac{\Delta}{\Delta + \gamma} R_{no} \exp(-0.398 \, LAI), \quad \ldots (5.35)$$

However, when surface soil water contents decreased below a threshold value so that E_s depends on the flux of water available at the evaporating site. E_s from soil surface below the canopy is identical with evaporation from a bare soil surface depending on hydraulic properties of soil expressed as

$$E_s = \alpha\sqrt{t} - \alpha.\sqrt{(t-1)}, \quad \ldots (5.33)$$

, where α dependent on hydroaulic property of soil determined as the slope of the linear relation between cumulative soil evaporation for a drying cycle and square root of time for different soil types varying from 3 to 5 mm day$^{-½}$ in sandy to clay loam soils respectively.

$$T_p = ET - E_s \quad \ldots (5.37)$$

5.5.3 Driessen's method for maximum transpiration:

Normally a crop under optimum water supply transpires at maximum rate (Mt_p) and will be a function of all transpiring leaves. Potential transpiration under prevailing environment for the well watered, closed, short green standard crop (grass) was defined by Penman as $PE_0 \simeq PT_0$ (Ritchie, 1972). Driessen (1986) expressed maximum transpiration as

$$Mt_p = [1 - \exp(-0.8\ LAI)]\ PT_p \quad \ldots (5.38)$$

Potential transpiration rate can be expressed for a crop having fully shaded crop leaf area

$$PT_p = PET - PE_s \quad \ldots (5.39)$$

$$PT_p = PET - PE_0 \exp(-0.4\ LAI) \quad \ldots (5.40)$$

Therefore

$$Mt_p = [1 - \exp(-0.8\ LAI)]\ [PET - PE_0 \exp(-0.4\ LAI)] \quad \ldots (5.41)$$

when there is no shortage of water in the crop root zone. However, when soil water decreases in the root zone, the upper soil layer a dry mulch layer is formed and transpiration from plant decreases below PT_m depending on availability of water in the root zone depth. The available water in the root zone depth depending on critical level of soil moisture (below wilting point) expressed by Doorenbos et al., 1978 and Doorenbos and Pruitt, 1977 as

$$SM_c = (1 - p)\ (SM_{fc} - SM_{pwp}) + SM_{pwp} \quad \ldots (5.42)$$

, where p soil water depletion fraction which depends on the maximum transpiration rate from the plant and type of crop as given below (Table 5.2).

Table 5.2

p values at different maximum transpiration rates of field crops.

Crop Type	Mt_p (mm day^{-1})								
	1	2	3	4	5	6	7	8	9
Leafy vegetables	0.50	0.44	0.36	0.30	0.26	0.24	0.22	0.20	0.17
Peas, potato orchards	0.75	0.65	0.55	0.45	0.40	0.38	0.33	0.30	0.25
Beans, grass, tomato, sunflowers	0.80	0.70	0.60	0.50	0.45	0.43	0.38	0.34	0.30
Field cereal crops, cotton, wheat, rice	0.90	0.80	0.70	0.60	0.55	0.51	0.46	0.43	0.40
Sorghum, sugarcane, soybeans, fast growing crops	0.92	0.85	0.75	0.65	0.60	0.55	0.50	0.48	0.45

(*After*: Doorenbos et al., 1978)

Therefore $T_p \sim MT_p$ when moisture content in root zone is optimum between SM_{fe} and SM_{Cr}. When $(SM_0 - 0.05) \geq SM_x > SM_{fe}$ a linear relation is expressed between uptake of water for transpiration and soil air content.

$$T_p = 0 \text{ when } SM_x > SM_0 - 0.05 \quad \ldots (5.43)$$

$$T_p/Mt_p = \frac{\{(SM_0 - 0.05) - SM_x\}}{\{(SM_0 - 0.05) - SM_{fe}\}} \text{ when } (SM_0 - 0.05) \geq SM_x \geq SM_{fc} \quad \ldots (5.44)$$

$$T_p = MT_p \text{ when } SM_{fe} > SM_x > SM \quad \ldots (5.45)$$

$$T_p = MT_p \frac{(SM_x - SM_{PWP})}{SM_{Cr} - SM_{PWP}}, \text{ when } SM_{Cr} \geq SM_x \geq SM_{pwp} \quad \ldots (5.46)$$

$$T_p = 0 \text{ when when } SM_{pwp} > SM_x \quad \ldots (5.47)$$

5.5.4 Monteith's concept of transpirational demand

In a given environment, the amount of dry matter produced by a crop is proportional to the amount of water transpired (de Wit, 1958). Therefore with thermal time accumulation for a crop development the ratio of amount of dry matter produced to water transpired (T.E.) is constant (Monteith, 1986).

Monteith (1986) showed that ratio of dry matter to transpired water is conservative and behaves in a linear manner (AC) with time when there is no shortage of water and crop function at a maximum growth rate under optimum conditions of radiation and temperature (Fig. 5.2). However under limiting conditions of availability of soil water or temperature conditions, the crop will grow and develops at a rate governed by water uptake by the expansion of root system (ABDF). However with availability of rains or soil water at any stage (D), the growth rate will improve, but will come down with shortage of water with time. Therefore transpiration from a well watered, short, uniform crop stand covering ground following Penman's assumptions will approach potential evapotranspiration rate, but in a drier environment there will be less production of dry matter due to decrease of transpiration rates. The transpiration rate influenced the temperature of the crop's stand and its rate of development with the growing season. The transpirational demand of water by the crops in essential for production of growth and development in any environment. Monteith (1986) assessed transpiration demands by expressing it in terms of the dry matter production determined by incoming solar radiation (R_s), fraction of this intercepted by crop (f) and conversion coefficient (e) and dry matter – transpiration water ratio i.e., transpiration efficiency (T.E.) expressed as (Monteith, 1986).

$$\text{Transpirational Demand} = \frac{R_s.f.e}{T.E.} \quad \ldots (5.48)$$

In moist conditions the efficiency of radiation conversion into dry matter (e) and transpiration efficiency (T.E.). are conservative and constant during the crop growing season. However under dry conditions (when saturation deficit > 2 kPa) more transpirational water will be required to produce maximum amount of dry matter as determined by solar radiation and temperature, and the dry air will reduce the size of the canopy, conversion coefficient (e) and fraction of (f) radiation intercepted limiting potential growth. This negative effect of dry air on e and f will limit the rate at which water is required for maximum growth. f value can be simulated with the progress of leaf area development of the crop using Beer's law or Monteith's expression of radiation interception. It can also be developed with the progress of crop development with thermal time during the crop growing season as Marshall and Terry (1981) expressed this relationship for groundnut as

$$f = \frac{0.92}{1 + \exp\{-0.0092(TT - 634)\}} \quad \ldots (5.49)$$

where, TT is thermal time accumulation from sowing.

For Pearl Millet (MHB-110) leaf area index can be simulated with thermal time after sowing ($T_b \sim 10°C$) as

$$LAI = -2.8675 + 0.011879\,TT - 0.00000615\,(TT)^2, \quad \ldots (5.50)$$

$$(5.71) \qquad (-4.74) \qquad (R^2 = 0.94, F = 29.99)$$

The conversion coefficient, e, remained constant during the expansion of crop canopy where response of photosynthesis to solar radiation is almost constant with time. However e was sensitive to the environment conditions of the location. e varied with temperature and saturation deficit condition of the location, but remained constant in moist temperature conditions for a crop. If optimum temperature exist during the crop growing season then the variation of e with saturation deficit can be regulated by an decreasing exponential function.

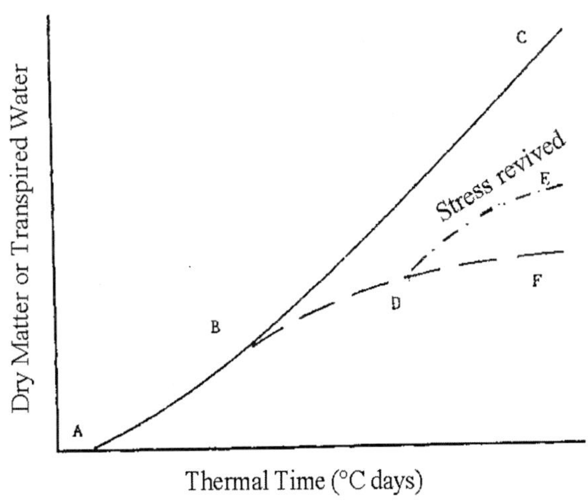

Fig. 5.2: Dry matter or transpirated water with thermal time (After: Monteith, 1986).

$$e = e^* \, 2.065 \exp(-0.546 \text{ S.D.}), r = -0.92 \qquad \ldots (5.51)$$

where e* is the value of conversion efficiency of the crop under wet temperature and optimum temperature conditions at 1.5 kPa saturation deficit level.

Transpiration efficiency (TE) is inversely related to saturation deficit (S.D.) such that TE × S.D. is constant (Table 5.3).

Example 6:

Estimate the transpirational demand for a sorghum variety using following data and determine the development units for sorghum maturity.

Daily solar radiation = 16MJm^{-2}

Mean temperature during crop growing season = 25°C

Saturation deficit = 2 kPa

Base temperature = 10°C

Initial slow phase of crop establishment = 200°C days

Sowing to the end of canopy formation = 650°C days

Sowing to physiological maturity = 1000°C days

Sowing to harvest = 1100°C days

Conversion coefficient before senescence = 2.5 gMJ^{-1}

Transpiration efficiency = 5 g kg^{-1}

Conversion coefficient after senescence = 1.25 g MJ^{-1}

$$f = \frac{0.95}{1 + \exp\{-0.009(TT - 650)\}}.$$

Solution:

Compute f values at different thermal time values.

	Thermal time (°C days)										
	100	**200**	**300**	**400**	**500**	**600**	**700**	**800**	**900**	**1000**	**1100**
f	0.007	0.016	0.07	0.09	0.16	0.37	0.58	0.76	0.86	0.91	0.93
e(gMJ^{-1})	2.5	2.5	2.5	2.5	2.5	2.5	2.5	2.5	2.5	2.5	1.25
Demand (kg m^{-2})	-	-	0.056	0.72	1.28	2.96	4.64	6.08	6.88	7.28	3.72
T_p(mm day^{-1})	-	-	0.06	0.72	1.28	2.96	4.64	6.08	6.88	7.28	3.72
Days after sowing (days)	10	20	30	40	50	60	70	80	90	100	110
Total T_p(mm)	-	-	0.6	7.2	12.8	29.6	46.4	60.8	68.8	72.8	37.2
Total T_p(cm)	-	-	0.06	0.72	1.28	2.96	4.64	6.08	6.88	7.28	3.72

Total transpiration during growing season = 33.60 cms

Development units for sorghum maturity = 3360 growth units.

Table 5.3

Conversion coefficient and transpiration efficiency and transpired water by vegetation crops in different environments.

Crop	Conversion Coefficient Before Flowering (e) gMJ^{-1}	After Flowering	T.E. ($g\ kg^{-1}$)	Total Transpiration Vt(mm)	(q×S.D.) (g kPa kg^{-1})	Conditions	Reference
Pearl Millet	2.5	0.8	3.3	200	10	Wet, temperate,	Anonymous, 1987
	1.3			150		Dry, semiarid	-do-
						tropics	-do-
Groundnut	1.9	0.5	1.7	340	5	Wet, temperate	-do-
	0.5-0.8			220		Dry, semiarid	-do-
						tropics	-do-
Sorghum	2.9	0.8	2.4	336	9	Wet, temperate	Hammer and Muchow, 1994
	0.5			250		Dry, semiarid tropics	-do-
Oats			1.63				Arkley, 1963
Barley			1.86				Briggs and Shantz, 1914
Sorghum			6 to 7.5				Hammer et al., 1997
Well watered			8.09				Mortlock and Hammer, 2004
Water limited			7.33				Hammer et al., 1997
C_3 crops	1-1.5	0.8				Temperate species	Monteith and Elston, 1983
C_3 crops	1.5-1.7	0.8				Tropical species	Kiniry et al., 1989; Monteith, 1990
C_4 crops	2.5	0.8				Tropical species	Squire, 1990

Example 4:

Maize crop with a leaf area index of 4.0 is grown on a clay loam soil. Assuming potential evapotranspiration rate of 3.5 mm day^{-1} and potential evaporation rate of 4 mm day^{-1} over a weekly period. Calculate the actual transpiration rate T_p of the crop when matric suction attained 625 cm.

Solution:

$$PT_p = PET - PE_0 \exp(-0.4\ LAI)$$

$$PT_p = 3.5 - 4 \exp(-0.4 \times 4) = 3.5 - 0.2 = 3.3 \text{ mm day}^{-1}.$$

$$T_{pm} = 3.3\{1 - \exp(-0.8\ LAI)\} = 3.3\{1 - \exp(-0.8 \times 4)\}$$

$$T_{pm} = 3.16 \text{ mm day}^{-1}.$$

$$SM_x = SM_0 \exp\{-Y(Ln\ x)^2\}$$

For a clay loam soil at field capacity level, the maximum available water is about 0.445 cm^3, Y is 0.0058

Therefore $SM_x = 0.445 \exp\{-0.0058(Ln\ 625)^2\}$

$SM_x = 0.35$ cm^{3}cm^{-3}

$SM_{cr} = (1 - p)(SM_{f.c.} - SM_{PWP}) + SM_{PWP}$

$= (1 - 0.8)(0.393 - 0.258) + 0.258$

$SM_{cr} = 0.285$ cm^{3}cm^{-3}

Threfore $SM_x < SM_{f.c.} < SM_{10}$,

and $> SM_{cr}$ and $> SM_{16000}$

Thus $T_p = T_{pm} = 3.16$ mm day^{-1}.

$SM_{f.c.} = 0.445 \exp\{-0.0058(Ln100)^2\}$

$= 0.393$

$SM_{PWP} = 0.445 \exp\{-0.0058(Ln(16000)^2)\}$

$= .258$ cm^{3}cm^{-3}

$SM_{10} = 0.445 \exp\{-0.0085(Ln\ 10)^2\}$

$= 0.432$cm^{3}cm^{-3}

5.6 Applications of Growing Degree Days

5.6.1 Phenological and morphological development

Phenology is the periodic occurrence of event of the life cycle of any living organism and is related to climatic variables particularly accumulation of temperature, daylength, sunshine and radiation variables depending on sensitivity of phenophases to the climatic variables. Development of the plant refers to the qualitative changes at the different stages development cycle from germination of seed to maturity. Development in other words refers to the timing of critical events of organism's progress towards maturity. Development is the series of record of observation of morphological and phasic development changes that takes place in the life cycle of the plant. Therefore phenological development of each organism needs to be studied in terms of detailed morphological characters so that plant life cycle can be modelled with the use of climatic variables. Phasic development is the occurrence of peridic changes in growth stages linking with biomass partitioning patterns, whereas morphological development refers to beginning and end of organ's development within the plant's life cycle. Temperature and photoperiod are main controlling factors for phasic durations, whereas morphological features are sensitive to crop input variables. Plants have been considered as an integrator of impact of weather variables so that phenological observations act as ecological indicators for different applications over seasonal/annual periods. Three group of plant are normally studied i.e., wild growing plants (trees, bushes, herbage), field crop plants and horticultural and vegetable fruit plants depending on the location and environment. Fig. 5.3 Shows the appearance of different morphological and phenological characters in pearl millet genotype sown HHB-50 on 7th June, 1987 at Hisar (latitude 29° 10'N and Longitude 75°76'E)—a semiarid tropical monsoonal location (Bishnoi, 1998). With growing season in terms of accumulated thermal time/heat units scale the appearance of leaves on main stem and first four effective tillers; tillers, nodes on main stem and first four effective tillers; ear emergence on main stem and tillers; appearance of seed setting on main ear and first four effective tiller ears, appearance of physiological maturity of main ear and effective tiller ears have been shown with successive upper head arrows. Dry matter accumulation per plant and leaf area index have been also shown with dotted lines in the graph to present the overall plant growth with the thermal time. Thus appearance of different morphological and phenological characters in the life cycle of pearl millet shows the phenological development. The threshold heat units required at different phenophases increased with growth stages reflecting its tolerance limits to thermal environment. The temperature response of individual plant organs will change when integrated into the whole plant, and even the plant will have shifting optimum characteristics at different developmental stages (Table 5.5).

Phenological observations have been of special interest all over the world with the recommendation of World Meteorological Organization (WMO) with regard to the uniformity of observations at large number of adaptive locations of different species of food grain, horticultural crop and forest vegetations. Neog and Chakravarty (2005) developed model for Brassica Juncea using thermal time. Brown (1963) developed a numerical scale of 0 to 9 for different stages of bud development in apricot flower for developing temperature relationships to predict harvesting. Similarly Higgin (1952) divided the phenological phases of garden peas between nodes into 10 subphases and successive nodes with increasing numerals of 1, 2, 3, 4, . . ., etc., for first, second, third etc. for the nodel number with development of figures of minute morphological and growth changes for vegetative, phenological development of peas. Hanway's (1971) 10 point scale for successive development stages of corn and growth stages of grain sorghum by Vanderlip (1972) became popular in field cereal crops for prediction of successive stages and maturity dates for harvesting. Therefore for each crop these 9/10 successive critical stages can be identified for developing their relationship with thermal time scale, Table 5.4 below presents the Hanway (1971) scale for corn and sorghum grain crops.

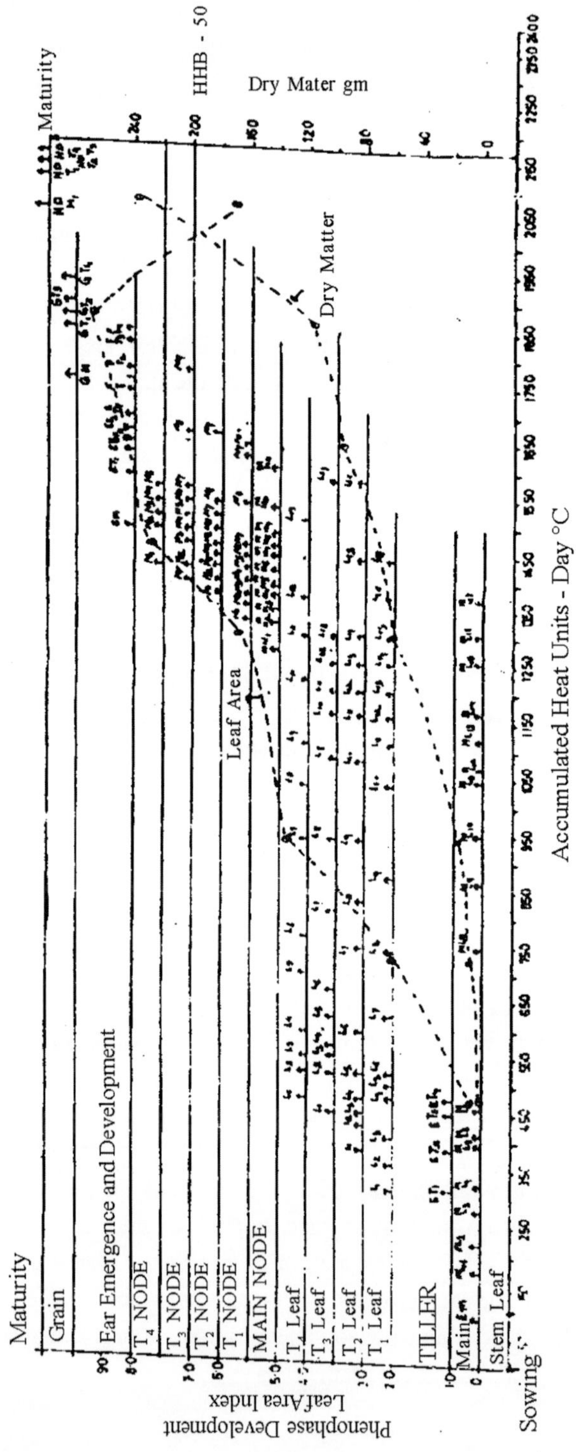

Fig. 5.3: Phenological development, leaf area and dry matter accmulation in Peal Millet with thermal time

Table 5.4(a)

Hanway scale and growing degree days for different crops

Stage of Development	Stage Number (scale)	Corn GDD (day °C) for Corn NEB611 V_1	A619 × A632 V_2	N28× Mo17 V_3	Sorghum NB505 V_1	Sorghum RS626 V_2	Sorghum RS671 V_3
Emergence – Coleoptile visible	0.0	0.0	0.0	0.0	0.0	0.0	0.0
2 leaves fully emerged	0.5	213	193	213			
4 leaves fully emerged	1.0	345	312	349	254	264	294
6 leaves fully emerged	1.5	476	431	484	-	-	-
8th leaves fully emerged	2.0	608	550	619	498	527	580
10th leaves fully emerged	2.5	739	669	754	-	-	-
12th leaves fully emerged	3.0	871	788	889	741	790	866
14th leaves fully emerged	3.5	1003	907	1024	-	-	-
16th leaves fully emerged	4.0	1134	1026	1159	985	1053	1151
Silk emergence/anthesis	5.0	1397	1264	1430	1229	1317	1437
Kernals in blister stage/ seed setting	6.0	1661	1502	1700	1473	1580	1723
Kernals in soft dough stage	7.0	1924	1740	1970	1717	1843	2008
Kernals beginning to dent	8.0	2187	1979	2240	839	1970	2151
Kernals fully dented	9.0	2450	2217	2511	1961	2106	2294
Physiological maturity	10.0	2713	2455	2781	2205	2369	2580

Corn	A	B	R^2	Sorghum	A	B	R^2
V_1	-0.246	0.00376	0.99	V_1	-0.266	0.00459	0.99
V_2	-0.246	0.00416	0.99	V_2	-0.240	0.0042	0.99
V_3	-0.230	0.00367	0.99	V_3	-0.258	0.00283	0.99

Stage number = A + B $\sum$GDD

Table 5.4(b)

For cotton variety H-777 for different dates of sowing in 1992 at Hisar location

Stage of Development	Stage Number	Growing degree days (day °C) for 29/5	9/6	19/6
Sowing	0.0	0	0	0
Emergence	1.0	116	311	200
Square/Bud appearance	2.0	1021	913	788
Flowering/initiation	3.0	1404	1316	1172
Green Bods initiation	4.0	1490	1420	1270
Square/Bull maxima	5.0	1576	1529	1363
Flowering maxima	6.0	1719	1739	1527

Green Bods maxima	7.0	1913	1810	1659
Boll opening initiation	8.0	2018	1952	1818
Boll's 1st picking	9.0	2391	2293	2142
Boll's 2nd picking	10.0	2452	2345	2175

Stage Number = -0.69 + 0.0039 ΣHU, R^2 = 0.90 for 29th May sowing ...(5.52)

Stage Number = -1.02 + 0.0042 ΣHU, R^2 = 0.93 for 9th June sowing ...(5.53)

Stage Number = -0.77 + 0.0045 ΣHU, R^2 = 0.94 for 19th June sowing

Table 5.4 (c)

Growing degree days for wheat (WH-147) for different dates of sowing

Stage of Development	Stage Number	Accumulated Growing Degree Days (day °C) for Wheat Sown on				
		29/10	13/11	29/11	13/12	29/12
Sowing	0.0	0.0	0.0	0.0	0.0	0.0
Germination	0.5	115.6	107.0	96.6	94.2	78.1
Leaf initiation	1.0	154.6	188.0	184.2	194.4	141.3
Tiller initiation	1.5	376.6	315.5	379.4	344.8	310.1
Nodes initiation	2.0	520.7	476.3	400.5	376.8	371.4
Last tiller appearance	2.5	692.1	542.5	422.2	392.3	449.8
Flag leaf	3.0	708.7	622.7	513.5	523.8	583.5
Last node appearance	4.0	947.0	763.5	749.3	724.7	753.9
Ear emergence	5.0	1015.7	991.1	784.2	849.5	773.9
Anthesis	6.0	1027.1	1122.2	939.5	996.3	875.4
Seed setting	7.0	1119.7	1199.9	985.8	1075.9	950.5
Milking	8.0	1418.1	1360.5	1151.6	1167.0	1118.6
Hard dough	9.0	1612.7	1531.2	1286.3	1328.4	1280.7
Maturity	10.0	1722.4	1636.6	1524.8	1490.3	1369.1

Stage Number = -0.56 + 0.0059 ΣHU, R^2 = 0.96 for 29th October ...(5.54)

Stage Number = -0.46 + 0.0061 ΣHU, R^2 = 0.99 for 13th November

Stage Number = -0.49 + 0.0071 ΣHU, R^2 = 0.98 for 29th November ...(5.55)

Stage Number = -0.44 + 0.0069 ΣHU, R^2 = 0.99 for 13th December

Stage Number = -0.51 + 0.0074 ΣHU, R^2 = 0.98 for 29th December ...(5.56)

The regression developed for the Hanway's scale of data shows the strong linear relationship between Hanway scale and accumulated growing degree days for different crops. It is clearly indicated that corn (var. A619 × A632), sorghum NB 505, cotton (H-777) sown on 19th June, and wheat (WH-147) sown on 29th December are faster growing accumulating the biomass with absorption of available heat in the environment. Using similar responses between Hanway's scale and accumulated heat unit values, we can evaluate the responses of a crop grown at different locations or responses of different crop hybrids at a location sown on same date (Table 5.4a) or response of crop sown on different dates at a location (Table 5.4b and Table 5.4c). Knowing the rate of development, the crop stage at a future date can be predicted with the prevailing environment temperature data. In a particular abnormal year (wet or drought situation) it is easy to assess the number of days the growing is ahead or behind the normal schedule. Ultimately crop itself is the ultimate integrator of the impact of all environmental variables, therefore crop's day to day observations be compared with predictions due to GDD where temperature and daylength values are mainly used. A crop grown on warm site (sandy soil with south facing slope) can attain quicker accumulation of heat units as compared to other sites having variations in soil texture and slopes.

Example 8:

Assuming wheat WH-147 sown on 13th November has accumulated heat units of 400 °C days by 10th January when will flag leaf appear if

(i) Normal temperature of 20°C prevail

(ii) Temperature prevails 5°C above or below normal.

Solution:

Hanway scale number and accumulated growing degree days from 13th November are related as

Stage number = -0.46 + 0.0061 ΣGDD, $R^2 = 0.99$

With GDD values of 400°C days,

Stage Number = -0.46 + 0.0061 × 550

Stage Number = 1.98

For flag leaf to appear (stage number 3) the accumulated GDD required are 3 = -0.46 + 0.0061 ΣGDD

Therefore ΣGDD (flag leaf) = 567.2 °C days

Therefore for flag leaf to appear there is requirement of GDD accumulation of 167.2°C days or growth of stage number from 1.98 to 3.0

(i) with prevailing normal temperature of 20°C, it would need [(167.2)/(20 – 5)] days or 11 days approximately.

(ii) when temperature is 5°C above or below normal it would require 167.2/(20 or 10) days or approximately 8 or 17 days for flag leaf to appear.

Example 9:

Use 2006 monsoon season temperature data to assess the number of days that growing season of kharif crops is ahead or behind the normal schedule.

Solution:

Progress of 1976 growing season at a monsoonal location compared with normal.

Date from (July 1st)	Accumulated Normal	GDD(°C days) 2006	GDD Difference (°C days)	Normal GDD per day	Days Difference
August 1	677.0	666.0	-11.0	21.8	-0.5
September 1	1309.0	1242.0	-67.0	20.4	-3.3
October 1	1884.0	1782.0	-102.0	19.1	-5.3
November 1	2250.0	2012.0	-238.0	11.8	-20.2

Therefore the growing season in 2006 is behind the normal schedule for different dates of monsoon rainy season crops and in daily crop weather advisories a quantitative measure of adverse crop weather situation needs to be reflected with the prevailing environment.

Therefore with the degree day concept it is feasible to predict anthesis, maturity of crops, and to synchronize the flowing time for breeding and seed production of cross pollinated crops, preparation of phenological maps for different crop species at different times is useful for time sequential and adaptability analysis to optimize productivity with heat utilization concept.

5.6.2 Phenological zonation

Based on phenological and meteorological data over space and time scales, phenological zonation have been developed with agroclimatic differences. Geiger (1951) considered plants as a biochemical integrators of their surrounding meteorological environment and therefore plant's characters can be used as a climatic difference indicators over spatial and time scales for bioclimatic zonation (Huard and Chevrette, 1973). Castonguay and Dube (1985) used data of 300 phenological and meteorological stations in Quebec area (Canada) to delineate 5 phenological zones having significant climatic differences with significant thermal gradients. Dube and Chevrette (1978) used pheenological data to describe specific periods of growing season of three vegetation indicator plants—Lilac, Zabeli and Arnold Red to develop phonological indices in the form of

Earliness Index (EI) = A/6, where

$$A = [(P_1 + P_2)\text{ Lilac} + (P_1 + P_2)\text{ Zabeli} + (P_1 + P_2)\text{ Arnold Red}], \quad \ldots (5.57)$$

Summer Index (SI) = B/9, where ... (5.58)

$$B = [(P_3 + P_4 + P_5)\text{ Lilac} + (P_3 + P_4 + P_5)\text{ Zabeli} + (P_3 + P_4 + P_5)\text{ Arnold Red}], \text{ and} \quad \ldots (5.59)$$

$$\text{Growth Season Index (GSI)} = (A + B)/15, \quad \ldots (5.60)$$

, where P_1, P_2, P_3, P_4 and P_5 are phenophases for appearance of first leaf, full leaves, first bloom, full bloom and end of bloom from March 1st.

Phenological zones were distinguished based on phenological indices particularly GSI values, thermal indices (heat degree days accumulation in tropical to subtropical areas, and cold degree days accumulation in temperate to colder areas). The comparsion of accumulated degree days in each zone with respect to a reference zone in terms of percentage of monthly values or monthly difference will indicate the thermal gradients availability in each zone (Fig. 5.4). Phenological zonation is indirectly related to precipitation through rain and snow distribution due to the influence of thermal potential the zone. These distinguishing features of climatic differences in different phenological zones can be used as a tool to classify meteorological records into specific phenological groups. On yearly basis such analysis will assess the impact of climatic variability on phenoclimatic zonation. Such phenological variability studies can be used to refine the existing agroclimatic zonations and the package of practices adopted for different crops in the zone. Phenological classification having thermal gradient differences, will also assess the influence of climatic variability on plant's adaptation and bioclimatic potential of the zone. Inversely phenological data can be used to assess the climatic conditions at locations where meteorological data are not available since meteorological and phenological observations are complementary to each other.

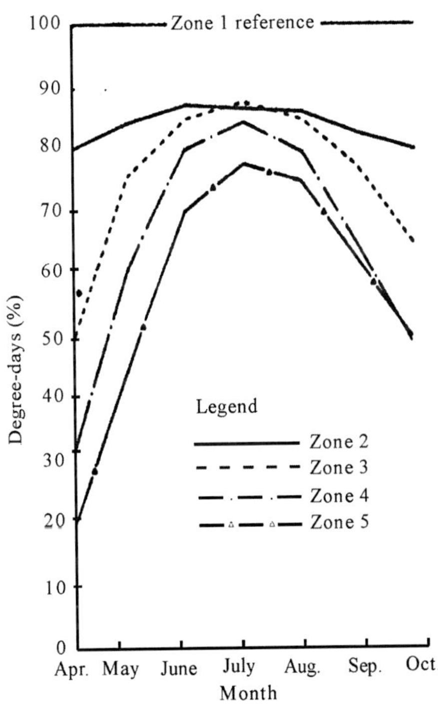

Fig. 5.4 Heat degree days accumulation (5°C) in each phenological zone, expressed as a percentage of those observed in zone 1. (Values in parentheses are the mean monthly accumulations in zone 1). (After Dube and Chevrette, 1978).

Therefore phenological indices will permit us to depict the agroclimatic zonation of plant species as it integrates the effect of all productive environmental variables (temperature, daylength, vernalization and light intensity) at different phenophases.

Phenologial maps for the initiation of a development phase will reflect a differential behaviour of temperature of the plants. Schnelle (1965, 1953a and 1953b) produced the phenological maps for apple blossoming in Europe and wheat harvest. Production of such phenological maps for different growth stages particularly sowing, flowering and physiological maturity alongwith meteorological data will provide a spatial coherence of different plant growth stages to present the relative environmental control of plant growth and development with time. Harrison et al. (2000) evaluated ARCWHEAT-2 model for assessing phenological development of wheat in Europe to reproduce the spatial variability of sowing dates, calibration of several cultivars and compared with observed phenological dates.

5.6.3 Climatic calender

Thornthwaite (1952) using daily development data of pea plant (Higgin, 1952) developed the climatic calendar with use of successive nodes in pea plant. The total number of nodes were numbered successively termed plant development as consisting of growth units. The pea plant was thought of a climatic yardstick in terms of node's scale. Thornthwaite and Matter (1954) related the plant's transpiration with the development units as both are related to the availability of environmental thermal energy, and defined a development unit (growth unit) as the interval between successive nodes on a green peas plant or the amount of water needed for transpiration during the growth period. Inversely a development or growth unit can be defined as the amount of development that will occur with the transpiration of unit amount of water. In metric system 100 development units are normally taken for 1 cm of transpiration of soil moisture. Since it is easy to quantify the plant's water need, therefore development index for a location can be computed and matched with the crop phenological observations for refinement. If will be very useful to compute daily development units for a location to quantify the accumulated values during the growing season for simulated phenological stages. Normally a Julian calendar is translated into a climatic calender where each day is expressed in terms of the number of accumulated development units. Therefore transpirational needs and growth or development index of a plant are strongly related, and with assessing of transpirational water needs of a plant we can quantify development rate by this method.

Mather (1974) developed a climatic calendar to form crop growing zones based on the availability of development or growth units over the yearly period, and demarcated 10 crop meter growing zones of USA for growing of vegetable crops such as peas, potatoes, tomatoes, blossoming of apples, cherries, peaches, beans and sweet corn. The methodology is useful to develop climatic calendar for growing of different crops during the year.

Example 10:

Green peas require 1680 growth units for maturity. Determine the harvest date when peas are grown with dates of sowing on March 1, March 15, March 31, April 15, April 30, May 15, May 31 and June 15 when development units accumulates at the rate given below in different months.

Month	Development rate (day^{-1})
Jan.	0
Feb.	0
Mar.	5
Apr.	15
May	30
June	40
Jul.	53
Aug.	40
Sep.	25
Oct.	18
Nov.	10
Dec.	5

Solution:

Dates of Sowings	Accumulated Development Units	Date of Maturity	Maturity Days
March 1	$5\times31+15\times30+30\times31+40\times3=1665$	June 4	96
March 15	$5\times16+15\times30+30\times31+40\times5=1660$	June 6	83
March 31	$1\times5+15\times30+30\times31+40\times7=1665$	June 8	68
April 15	$15\times15+30\times31+40\times13=1675$	June 14	60
April 30	$1\times15+30\times31+40\times18=1665$	June 19	51
May 15	$16\times30+40\times30=1680$	June 30	46
May 31	$1\times30+30\times40+53\times8=1654$	July 9	40
June 15	$15\times40+53\times20=1660$	July 21	36

Therefore peas mature from 36 to 96 days depending on development rate availability in different months.

Example 11:

Use the climatic calendar data to determine the development units for maturity of following crops at different times of the year. Also indicate which crop develops faster?

Crops	Planting Date	Harvesting Date	Development Rate (day^{-1})	
Lima beans	April 20	July 27	April	16
Sweet corn	May 15	Aug. 15	May	28
Radish	April 5	May 25	June	40
Spinch	August 25	October 15	July	35
			Aug.	35
			Sept.	40
			Oct.	20

Solution:

Crop	Duration (days)	Accumulated Development Units	Rate of Development (day^{-1})
Lima beans	98	3173	32.38
Sweet corn	92	3258	35.41
Radish	50	1100	22.00
Spinch	51	1710	33.53

Sweet corn develops faster than other vegetables, whereas radish rate of development is least.

5.6.4 Forecasting quality of agricultural products

Davitaya (1948, 1958) adopted accumulated heat units for the distribution of grapes for champagne, table wines and strong and dessert wines of different qualities. The microclimatic differences in radiational heating influenced the different species to mature at different timings. The accumulated heat units and rainfall amount both influenced the availability of heat energy and transpriational needs to stimulate the grapes quality for different types of wines (Table 4.6).

Temperature regime not only helps to delineate the course of plant's development, but also functions as an index to exploit the insolation and moisture regimes in controlling the photosynthesis and respiration rates particularly in reproductive phenophase influencing the quality of the produce. For example in cotton, lint quality is based on temperature during reproductive phenophase and is determined in terms of fibre strength (g/tex.1/8 inch), fibre length (inches) and ginning percentage (Hesketh and Low, 1968) as

$$\text{Fibre strength} = 56.603 - 2.91\,\overline{T} + 0.059\,\overline{T}^2 \quad \ldots (5.61)$$

$$\text{Fibre length} = 1.219 - 0.0065\,\overline{T} \quad \ldots (5.62)$$

$$\text{Ginning percentage} = 50.54 - 0.6755\,\overline{T} \quad \ldots (5.63)$$

where $\overline{T}$ is average temperature through the boll period provided water and nutrients are fully met.

Exercise 4

Q. 1. Compute GDD for the months with lowest temperature at your location for wheat crop with base temperature of 0, 1, 2, 3, 4 and 5°C with basic equation or GDD having two interpretations of daily average temperature.

Q. 2. Develop a Beta response development rate function to temperature in maize genotypes with $T_b = 10°C$, $T_{uT} = 37°C$ using following data of development rate between sowing and tasselling.

Development Rate (day^{-1})	Temp. (°C)
0.00	10
0.04	20
0.06	30
0.03	36

(Ans. $\mu = -5.7$, $\alpha = 0.75$, $\beta = 0.31$, $T_0 = 29.1°C$, $RD_{max} = 0.058\ day^{-1}$)

Q. 3. Compute GDD for your location in the coldest month for wheat crop to test the following forms of GDD expressions:

(i) $GDD = T_{Min} - T_b$

(ii) $GDD = \frac{T_{min} + (T_{max} - T_{max} \text{ degree above } T_0)}{2} - T_b$

(iii) $GDD = \frac{(T_b + T_{min} \text{ above } T_b) + T_{max}}{2} - T_b$

(iv) $GDD = \frac{(T_b + T_{min} \text{ above } T_b) + (T_{max} - T_{max} \text{ Degree above } T_0)}{2} - T_b$

Given that T_b is 5°C, T_0 is 25°C, T_{uT} is 35°C

Show that C.V. is least for the last (iv) expression.

Q. 4. Use observatory temperature data at your location to quantify the

(i) behaviour of different genotype planted on same date

(ii) behaviour of a hybrid genotype planted on different dates

(iii) assess the number of days that the growing season is ahead or behind the normal schedule (use normal values of climatic data) in a rainy or dry season.

Q. 5. Use All India Coordinated Project data to examine the expected response of a crop hybrid at different locations?

Q. 6. Use the climatic data at your location to examine which form of GDD have least coefficient of variation for maturity of the most important crop grown at the location.

(i) GDD or $TT = \Sigma\left(\frac{T_{max} + T_{min}}{2}\right) - T_b$

(ii) PTT = Σ(GDD × daylength)
(iii) HTT = Σ(GDD × sunshine hours)
(iv) EDU = Σ(GDD × R_s)
Where TT is thermal time (°C days), PTT is photothermal time (°C days hours), HTT is heliothermal units (°C days hrs.) and EDU is energy development units (°C days Wm^{-2}), Rs is total daily solar radiation at the earth's surface.

Q. 7. A marriage ceremony requires following fresh vegetables on 15th September in large quantity. Determine on which date the farmer should plant the following selected vegetables so as to be ready on September 15 with use of following data.

Vegetables	Total Development Units Required	Development Rate in the Months (day^{-1}) month	DR
Sweet corn	3500	September	25
Lima beans	3250	August	40
Green beans	1850	July	50
Peas	1500	June	40
Spinch	1340	May	25

(Ans. June 22, July 30, July 27, Aug. 3 and Aug. 7)

Q. 8. A farmer has a contract of supplying tomatoes on each successive days from June 1. Determine the date of planting for each day's harvest of tomatoes when tomatoes require about 1525 development units for maturation. Development rates in the months of February, March, April, May and June are 1, 5, 16, 28 and 40 day^{-1} respectively.
Ans.

Maturity date	Planting date
June 1	Feb. 6
June 2	Mar. 3
June 3	Mar. 11
June 4	Mar. 19
June 5	Mar. 28
June 6	April 2

Q. 9. Quantify the transpiration rates of different field/vegetable crops grown at your location throughout the year and develop the development scale for predicting maturity of different crops.

Q. 10. Develop the climatic calendar for developing growing season's availability over the yearly period and demarcate cropmeter growing zones based on thermal development data or transpiration based growth units for your state.

Q. 11. Determine the relationship between heat units and phenological development of different important field crops including horticultural and vegetable crops at your location?

Q. 12. Develop a map for phenological development of flowering and maturity of different important field crops including horticultural and vegetable crops in different agroecological zones of your state?

References

Anonymous, 1987. Microcimatology in tropical agriculture. Vol. 1 & 2. Overseas development administration, Research Schemes Report R3208 and R3819, Eland House, Stag Place, London, SWIE, 5DH.

Anonymous, 1988. Agroclimatic survey of Haryana. Agrimet. Publ. on Deptt. of Agricultural Meteorology, Haryana Agril. Univ., Hisar, India.

Arkley, R.J. 1963. Relationship between plant growth and transpiration. Hilgardia, 34, p. 559-84.

Arnold, C.Y. 1959. The determination and significance of the base temperature in a linear heat unit system. Proc. Am. Soc. Hort. Sci., 74, p. 430-45.

Baker, D.N. and R.B., Musgrave, 1964. Photosynthesis under field conditions V. Further plant chamber studies of the effects of light on corn. Crop. Sci., 4, p. 127-31.

Barker, J.T., Pinter, P.J., Reginato, R.J., Kenemasu, E.T. 1986. Effects of temperature on leaf appearance in spring and winter wheat cultivations. Agron. J., 78, p. 605-13.

Baver, A., Garcia, R., Kanemasu, E.J., Blad, B.L., Hatfield, J.L., Major, D.J., Reginato, R.J., Hubbard, K.G. 1988. Effects of latitude on phenology of winter wheat. Agric. For. Meteorol., 44, p. 131-40.

Bierhuizen, J.E. 1973. The effect of temperature on plant growth, development and yield. Proceeding of uppasala symposium plant response to climatic factors, UNESCO, p. 88-98.

Bishnoi, O.P. 1998. Predicting crop yields through phonological and plant biophysiological parameters. Jour. of Applied Hydrol., IX, 2 & 3, p. 1-10.

Briggs L.J., H.L., Shantz 1914. Relative water requirement of plants. J. Agric. Res., 3, p. 1-63.

Brown, D.M. 1960. Soybean ecology I. Development temperature relationships from controlled environmental studies. Agron. J., 52, p. 493-96.

Brown, D.M. 1963. A heat unit system for corn hybrid recommendations. Paper presented at 5th National Conf. on Agril. Meteorology, Lakeland, Florida, Ont. Res. Foundation, tomato, Canada.

Brown, D.M. and L.J. Chapman, 1961. Soybean ecology III soybean development units for zones and varieties in the great lakes region. Agron. J. 53, p. 306-308.

Calado, A.M. and C.M. Portas 1987. Base temperature and date of planting in processing tomatoes. In: W.L. Sims (ed.) Second International Symp. on Processing Tomatoes. Acta. Horticulture, 200.

Castonguay, Y. and P.A. Dube, 1985. Climatic analysis of a phenological zonation. A multivariate approach. Agril. And for Meteorol., 35, p. 31-45.

Cross, H.Z. and M.S. Zuber 1972. Prediction of flowering dates in maize base on different methods of estimating thermal units. Agron. J., 64, p. 351-55.

Davidson, H.R. and C.A. Campbell, 1983. The effect of temperature, moisture and nitrogen on the rate of development of spring wheat as measured by degree days. Can. J. Plant Sci., 63, p. 833-46.

Davitaya, F.F. 1948. Climatic zones for grape cultivation in USSR, M. Pishipromizdat, p. 192.

Davitaya, F.F. 1958. Temperature forecasting and some problems of seasons development in the nature. Gidrometeoizdat, L.

de Wit, C.T. 1958. Transpiration and Crop Yield. Institut. Voor Biologisch en Scheikundig Onderzock van Landouwassen. Wageningen. Modedeling, No. 59.

Doorenbos, J., A.H., Kassam, C. Bentvelder, and G. Vittenboogaard, 1978. Yield response to water. U.N. Economic Commission, West Asia, Rome.

Doorenbos, J. and W.O. Pruitt 1977. Guidelines for predicting crop water requirements. Irrigation and Drainage paper 24, FAO, Rome, p. 144.

Driessen, P.M. 1986. The waer balance of the soil. In: Modelling of agricultural production: Weather, soils and crops (ed.), H. van Keulen and J. Wolf. Pudoc Wageningen, Netherland.

Dube, P.A. and J.E. Chevrette, 1978. Phenology applied to bioclimatic zonation in Quebec. In: Phenology, an aid to agricultural technology, Vermont Agric. Exp. Stat. Bull., 684, p. 33-42.

Ellis, R.H., Summerfield, R.J., Edmeades, G.O. and E.H. Roberts 1992. Photoperiod, temperature and the interval from sowing to tassel initiation in diverse cultivars of maize. Crop. Sci., 32, p. 1225-32.

Gallagher, J.N. 1979. Field studies in cereal leaf growth: Initiation and expansion in relation to temperature and ontogeny. J. Exp. Bot., 30, p. 605-36.

Gao, L.Z., Jin, Z.Q., Huang, Y. and L.Z., Zhang, 1992. Rice clock model—a computer simulation model of rice development. Agric. For. Meteorol., 60, p. 1-16.

Geiger, R. 1951. Micrometeorology. In: T.F. Malone (ed.), Compendium for Meteorology. Am. Meteorol. Soc., Boston, p. 993-1003.

Gilmore, E.C. and G.S. Rogers, 1958. Heat units as a method of measuring maturity in corn. Agron. J. 50, 611-50.

Hammer, G.L. and R.C. Muchow, 1994. Assessing climatic risk to sorghum production in water limiting subtropical environments. I. developing and testing a simulation model. Field Crops Res., 36, p. 221-34.

Hammer, G.L., G.D. Farquhar and I. Broad, 1997. On the extent of genetic variation for transpiration efficiency in sorghum. Anstrolian Jour. of Agril. Res., 48, p. 649-55.

Hanway, J.J. 1971. How a corn plant develops. Iowa State Univ. of Sci. and Technology, Spec. Rep. No. 48.

Harrison, P.A., J.A. Porter, and T.E. Downing 2000. Scaling up the AFRCWHEAT-2 model to assess phenological development for wheat in Europe. Agril. for Meteorology, 101, p. 167-86.

Hesketh, J.D. and A. Low, 1968. Effect of temperature on components of fibre quality on cotton varieties of diverse origin. Cott. Gro. Rev. 45, p. 243-57.

Higgins, J.J. 1952. Instructions for making phonological observations of garden peas. Pub. Climatology, Laboratory of Climatology, 5, 2, p. 11.

Hodges, T. 1991. Temperature and water stress effects on phenology. In: T. Hodges (ed.), Predicting crop phoenology. CRC Press, Boca Raton, F.L., p. 7-13.

Huard, R. and J.E., Chevrette, 1973. La phenologie au Quebec. Agriculture, 30, 2, p. 49-51.

IRRI, 1977. Annual report for 1976, Los. Banos, Philippines, p. 113-20.

Keating, B.A. and J.P. Evenson, 1979. Effect of soil temperature on sprouting and sprout elongation of stem cuttings of cassava. Field Crops Res., 2, p. 241-51.

Kiniry, J.R., C.A. Jones, J.C. O'Toole, R. Blanchet, M. Cabelgenne, and D.A. Spannel, 1989. Radiation use efficiency in biomass accumulation prior to grain filling for five grain crops species. Field Crops Res., 20, p. 51-64.

Kiniry, J.R. and M.E. Keener 1982. An enzyme kinetic equation to estimate maize development rates. Agron. J., 74, p. 115-19.

Kirby, E.J.M. 1995. Factors effecting rate of leaf emergence in barley and wheat. Crop Sci., 35, p. 11-19.

Lehenbauer, P.A. 1914. Growth maize seedling in relation to temperature. Physiol. Res. I: 247-88.

Livingston, B.E. 1916. Physiological temp indices for the study of plant growth in relation to climatic factors. Physiol. Res. I, 399-420.

Masle, J., Doussinalut, G., Farquhar, G.D. and Sun, B. 1989. Foliar stage in wheat correlates better to photothermal time than to thermal time. Plant, Cell Environ., 12, p. 235-47.

Marshall, B. and A.C. Terry, 1987. A comprehensive set of data on leaf area, intercepted radiation and dry weight of groundnut grown at five temperatures (19, 22, 25, 28 and 31°C) with and without irrigation, in the controlled environments glasshouses. In: Microclimatology in Tropical Agriculture. Vol. I & Vol. II. ODA, Eland House, Stag Place, London, SWE, SDH.

Mather, J.R. 1974. Climatology: Fundamentals and Applications. McGraw Hill Book Co., N.Y., p. 412.

Mavi, H.S. 1986. Introduction to agrometeorology. Oxford and IBH Publications. New Delhi.

McMaster, G.S. and D.E. Smika, 1988. Estimation and evaluation of winter wheat phenology in the central great plains. Agric. For. Meteorol., 43, p. 1-18.

McMaster, G.S. and W.W. Wilhelm, 1997. Growing degree days: one equation, two interpretations. Agril. For. Meteorol., 87, p. 291-300.

Monteith J.L. 1965. Evaporation and environment. Symposia of the society for experimental biology, 19, 205-34.

Monteith, J.L. and J. Elston, 1983. Performance and productivity of foliage in the field. In: Dale, J.E. and Milthorpe, L. (eds.). The growth and function of leaves, Cambridge University Press, Cambridge, U.K., p. 199-518.

Monteith, J.L., 1986. How do crops manipulate water supply and demand? Philosophical transactions of the Royal Society, London, A 316, p. 245-59.

Monteith, J.L., 1990. Conservation behaviour in response of crops to water and light. In: R. Rabbinage, J. Goudriaan, Van keulen, F.W. T. Penning de Vries, H. Van Laar (eds.). Theortical Production Ecology: Reflection and Prospects. Pudoc. Wageningen, p. 3-16.

Mortlock, M.Y. and G.L. Hammer 2004. Genotype and water limitation effects on transpiration efficiency in sorghum. In: Water Use in Crop Production (ed.). M.B. Kirkham, p. 287-99. IBDC, Lucknow, India.

Neog, Prasanta and N.V.K. Charavarty, 2005. Thermal time and phonological model for Brassica Juncea. Jour. of Agrometeorology, 7, 2, p. 174-81.

Newman, J.E., Cooper, W.C., Reuther, W., Cahoon, G.A. and A Peynado, 1967. Orange fruit maturity and net heat accumulation. Ground Level Climatology, p. 127-47.

Nuttonson, M.Y. 1948. Some preliminary observations of phonological data as tool in the study of photoperiodic and thermal requirements of various plant material. In: Murneed, A.E. and Whyte, R.D. (eds.) Vernalization and Photoperiodism – a symposium. Chronica Botanica. Pub., Waltham, M.A.

Nuttonson, M.Y. 1955. Wheat climate relationships and the use of phenology in ascertaining the thermal and photothermal requirements of wheat. Amer. Inst. Crop. Ecol., Washington, D.C.

Nuttonson, M.Y. 1957. Barley climate relationships and the use of phenology in ascertaining the thermal and photophermal requirement of barley. Amer. Inst. Crop Ecol., Washington, D.C.

Owens, T.O. and E.L. Moore 1974. A comparison of various methods of calculating heat units requirements of tomato. Tech. Bull. No. 70, Mississippi Agric. and Forestry Experiment Station, Mississippi State University, M.S.

Perry, K.B. and T.C., Wehner, 1990. Prediction of cucumber harvest date by using a heat unit model. Hort. Sci., 25, p. 405-06.

Perry, K.B., Wehner, T.C. and G.L. Johnson, 1986. Comparison of 14 methods to determine heat unit requirements for cucumber harvest. Hort. Sci., 21, p. 419-23.

Perry, K.B., Y.Wu., Sanders, D.C., Garrett, J.J., Decoteau, D.R., Nagata, R.T., Dufautt, R.J., Batal, K.D., Granberry, D.M. and W.J. Mclaurin, 1997. Heat units to predict tomato harvest in the southeast USA. Agric. For. Meteorol., 84, p. 249-54.

Pinter,, J.T., P.J., Reginato, R.J., Kanemasu, E.T. 1986. Effects of temperature on leaf appearance in spring and winter wheat cultivars. Agron. J., 78, p. 605-13.

Reaumur, R.A.E de 1735. Observation du thermometer, faites a paris pendent 1'anne'e 1735, Compare'es evec celles qui ont ete faites sous la ligne, a 1'Isle de France, a Alger et en quelquesunes de nos isles de 1'Amerique. Mem. Acad. Sci., Paris, 1735: 545.

Ritchie, J.T. 1972. Model for predicting evaporation from a row crop with incomplete cover. Water Resour. Res., 8, p. 1204-13.

Ritchie, J.T. and E. Burnett, 1971. Dryland evaporative flux in a subhumid climate 2. Plant influences. Agron. J., 63, p. 55-62.

Roberts, E.H. and R.J. Summerfield, 1987. Measurement and prediction of flowering in annual crops. In: J.G. Atherton (ed.). Manipulation of flowering. Prutterworths, London, p. 17-50.

Schnelle, F. 1953a. Contribution to the phenology of apple blossoming in Germany. Ber. D.W.D. 1.

Schnelle, F. 1965. Contribution to phenology of apple blossoming in Europe. I Ber. D. W.D. 101.

Schnelle, F. 1953b. Phenological world map: beginning of wheat harvest. Geogr. Pocket book.

Sharpley, A.N. and J.R. Williams, 1990. EPIC-Erosion-productivity impact calculator 1. Model documentation U.S. Deptt. of Agri., Tech Bull No. 1768, p. 1-235.

Smith, J.W. 1920. Agricultural Meteorology. The effect of weather on crops. The Macmillan Co., New Your, p. 304.

Squire, 1990. The physiology of tropical crop production. CAB International, Wallingford, U.K., p. 236.

Swanson, S.P. and W.W., Wilhelm, 1996. Plant date and residue rate effects on growth, partitioning, and yield of Cron. Agron. J., 88.

Thornthwaite, C.W. and J.R. Mather 1954. Climate in relation to crops. Meteorol. Monograph, 2, 8, p. 1-10.

Thornthwaite, C.W. and J.R. Mather, 1955. The water balance. Publ. Climatology, Laboratory of Climatology, 8, 1, p. 104.

Thornthwaite, C.W., 1952. Climate in relation to planting and irrigation of vegetable crops. Lab. of Climatology, Publ. in Climatology, V, 5, p. 1-6.

Tollenaar, M., Daynard, T.B. and R.B. Hunter, 1979. Effect of temperature on rate of leaf appearance and flowering date in maize. Crop Sci., 19, p. 363-66.

Tyldesley, J.B. 1978. A method of evaluating the effect of temperature on an organism when the response is non-linear. Agric. Meterol., 19, p. 137-53.

Ulrich, A. 1952. The influence of temperature and light factors on the growth and development of sugarbeets. Agron. J., 44, p. 66-73.

Vanderlip, R.L. 1972. How a sorghum plant develops. Kansas State Univ. Bull, C-447.

Wang, J.Y. 1960. A critique of heat unit approach to plant response studies. Ecology., 41, p. 785-90

Wang, J.Y. and R.A. Bryson, 1956. A study of phytometeorological effects on the growth and development of peas. Dept. Meteorol., University Wisonsin, Madison, p. 165.

Warnok, S.J. and R.L. Isaacs, 1969. A linear heat unit system for tomatoes in California. J. Am. Soc. Hort. Sci., 94, p. 677-78.

Watts, W.R. 1971. Role of temperature in the regulation of leaf extension in zea mays. Nature, 229, p. 46-47.

Went, F.W. 1944. Plant growth under controlled conditions XI. Thermoperiodicity in growth and fruiting of tomato. Amer. J. Bot. 31, p. 135-50.

Went, F.W. 1948. Thermoperiodicity, p. 145-57. In: Verification and Photoperiodism by A.E. Murneek and R.O. Whyte (eds.). Chrom'ca Bofanica. Co., Waltham, Mass.

Went, F.W. 1957. The experiemental control of plant growth. Chronica. Bofanica. Co., Waltham, Mass, p. 343.

Went, F.W. 1959. Effects of environment of parent and grandparent generations on tuber production by potatoes. Amer. J. Botany, 46, p. 277-82.

Wilhelm, W.W., Bouzerzour, H., Power, J.F. 1989. Soil disturbance – residue management effect on winter wheat growth and yield. Agron. J., 81, p. 5, 81-88.

Witter, S.H. and L.H. Aung 1969. Lycopersicon esculentum mill. In: L.T. Evans (ed.). The induction of flowering. Macmillan of Australia.

Yin, Xinyou, Marin, J. Kropff, Graham, McLaren and Romeo, M. Visperas, 1995. A non-linear model for crop development as a function of temperature. Agril. and For. Meteorol., 77, p. 1-16.

6

Human and Animal Bioclimatology

6.1 Introduction

Atmosphere is a resource in terms of weather or climate which supports a range of biological and human activities. Agriculture is an economic symbiosis of crop and livestock production where cattle have been the backbone of the farmer to provide nutritious and balanced diet in developing countries. Genetic diversity of different animals exist all over the world with different unique characters. The low productivity of Asian cow of 987 kg/lactation as against the world average of 2038 kg/lactation (Kadervel, 2002) shows the importance of microclimatic conditions for their adaptability. The importance of stress of physical environment on animal production in tropics have been well recognized so as to influence the feed intake, production and efficiency of utilization of energy for milk/meat production (McDowell, 1976; Smith and Taneja, 1986). Therefore, Bioclimatology is a science to study these biophysiological effects of climatic conditions on the living organisms and quantification of their interaction effects. Even the crossbreeds of tropical cattle with elite exotic breeds are having a potential for high milk production, but the harsh tropical environmental conditions acts as a constraint in the phenotypic expression of their genetic character. However, production can be improved if proper environmental conditions are provided with use of hygienic ventilated animal houses to reduce the stress factors.

Weather variables play an important role in human adaptability and reactions for their discomfort. Abnormal and sudden change in weather conditions lead to various forms of human physical ailments, therefore comfort indices have been devised for extreme warm and winter conditions separately so as to maintain human efficiency in his daily endures to generate maximum qualitative output to serve the humanity all over. Energy budget expression is the basic aspect which describes the interacting role of meteorological variables with the human/animal biological system (Lowry, 1969).

6.2 Adaptability

Carlson (1958) defined acclimatization as the ability of an organism to adjust for survival within its lifetime to changes in climate, whereas Ladell (1957) expressed it as the physiological changes that result in improved performance with successive exposure in a hot or cold environment.

The basic physical principles and laws of exchange of radiation and energy also holds true over human and animal surfaces, but due to mobility they are in a position to adjust for favourable weather conditions. Climatic factors interact to produce a resultant microclimatic optimum environment where the organism thrives comfortably well to produce maximum output.

Those animals which provide warm feeling to human touch are called warm blooded, whereas cold touch feeling are called cold blooded. When cold blooded animals were perceived not to be perpetually cold, but to warm up and cool down with their environment, they are known as poikilotherm expressing changeability. But warm blooded animals are known to maintain a constant stable and relatively higher body temperature. Cold blooded animals have little capacity to generate internal metabolic heat and maintains warming of body from external radiant sources called ecototherms. For ex., lizards, snakes, bettles. Warm blooded have higher metabolic rates and better insulation and also generates much of heat internally called endotherms (Hardy, 1979). For ex. birds and mammals.

The interaction mechanism are different for warm blooded animals (Homeotherms) in comparation with cold blooded animal (Poikilotherms). The homeotherms can maintain a constant body temperature by physiological thermoregulation and metabolic mechanisms, whereas piokilotherms cannot modify their body temperature with behavioural responses, and therefore varies widely. All courtyard cattle and animals belong to warm blooded group and maintains a constant body temperature range of 39.0 ± 0.5°C by heat balance with a thermoregulatosy mechanism located in the body. A dysfunction of this mechanism seriously effects the health of the animal causing disease and reduction in production. The constant body temperature also varies from animal to animal such as 36°C for elephants to 43°C for small birds.

Bianca (1976) and Primault (1979) defined 5 thermal zones based on the responses of homeotherms (Fig. 6.1) as

A. A zone of hypothermia where animal cannot supply adequate heat to maintain body temperature.
B. A broad range within which the optimum body temperature is maintained by produced heat.
C. A zone of thermal indifference where optimum body temperature is maintained with normal metabolism activity providing sufficient energy.
D. A zone of neutral range where conditions are not ideal, but with cooling effect body temperature is maintained.
E. A zone of hyperthermea where temperature control mechanisms cannot provide sufficient cooling to maintain body temperature.

Bianca (1976) divided the animal's response to heat stress in four categories termed as

(i) Thermal indifference: a zone within which compensatory responses by an organism are absent.
(ii) Mild heat: where thermoregulatory mechanisms of body completely compensate for the extra heat load to maintain body temperature.

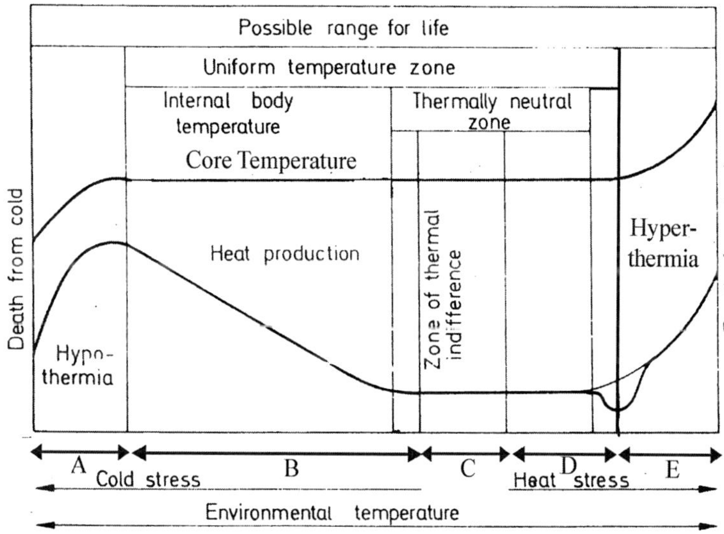

Fig. 6.1: Temperature and critical ranges. (From Bianca, 1976)

(iii) Moderate heat: where thermoregulatory mechanisms work at greater intensity to stabilize the body temperature within the limited range.

(iv) Severe heat: where thermoregularity mechanisms are overloaded and body temperature rises leading to exhaustion and ultimately ends with death.

6.2.1 Optimum environmental conditions

Depending on the original climatic conditions, the animal species have been adapted to heat or cold temperature ranges. Therefore the livestock and poultary production are widely influenced by the climatological variables (Table 6.1). For example, the European cattle have a comfort zone between -1.1 to 15.6°C, whereas Indian cattle are comfortable between 10 to 26.7°C (Appleman and Owens, 1970). However new born animals have more narrowly defined range of comfort at a higher temperature than those of their parents. A calf of 50 kg is comfortable at 18°C, whereas a calf of 200 kg is comfortable at 12°C. A breeding cattle of 50 kg will develop faster at 20°C, whereas a 500 kg will develop faster at 5°C which are intended for reproduction. The situation in sucking pigs is more peculiar as at birth they requires higher temperature of 33°C (artificial control to provide warmth), whereas mother is kept between 8 to 15°C (weighting between 350 to 150 kg). The optimum temperature for the young ones reduces as they gain weight with time. At 20 kg stage sucking pigs are comfortable at 22°C, whereas at 60 kg stage are comfortable at 15°C. Fattening pigs are very sensitive to humidity levels (50 to 70 per cent). Similarly chicks need a higher temperature of 34°C at hatching, but a decrease of 2°C over weekly growth is ideal so as to attain 300 gm weight at 21°C. However a low humidity of 55 to 60 per cent is desired. Therefore

Table 6.1
Optimum environmental conditions for animals.

Type	Temperature Range (°C)			Humidity	Conditions	Reference
	Threshold	Optimum	Maximum	Range (%)		
Calves for breeding	5	10-20	15-21	50-80	Temperature	Primault, 1979
Calves while flattening		18→12		50-60	-	-do-
Young breeding cattle	-6.5	5-15	21	50-80	-	-do-
Young cattle while flattening	-6.5	10-15	21	50-80	-	-do-
Milk cows	-6.5	0-15	21	50-80	-	-do-
Pigs	0	0-15	30	50-80	-	-
Sucking pigs	15	33→15	33	50-80	-	-
Young pigs	5	22→15	29	50-80	-	-
Sows	5	10-15	24	50-80	-	-
Indian cattle	10	10-27	30/35	50-80	Tropical	Appleman and Owen, 1970
Sheep for slaughter or wool		5-15/20		50-80	-	-
Horses		8-15		50-80	-	-
Chicks		34→21		50-70	-	-
Egg laying hens	5	15-22/28	28	50-70	-	-
German cows		0-16		70-85	Temperate, better fat percentage	Renner, 1976
Karan fries and Karan swiss cows		7-27				Shinde and Taneja, 1986

→ Indicates the decrease of temperature with growth of animal

(*After*: Primalt, 1971, 1979)

proper controlled micrclimatic conditions are needed for their brought up. The optimum temperature for egg production lies in the range of 10 to 16°C, however egg production can be maintained up to 25°C provided nutritional foods are not limiting. The optimum temperature for egg laying hens lies between 15 to 22°C with relative humidity around 50 to 70 per cent only.

The optimum temperature for milk production in temperate breeds is around 10°C, whereas the critical temperature after which the milk production steeply declines is 21 to 27°C in Jersey and Holstein Freisian; 29 to 32°C for Brown Swiss and in case of tropical cattles. Yousef (1985) cited the upper critical temperature for milk production in Holstein as 21°C and for Brown Swiss and Jersey as 24°C. Alhassen and Buvanendran (1985) enumerated that milk yield declined sharply at temperatures above 27°C for early and mid lactations in Friesian – Bunaji Cross in Nigeria. Optimum temperature of the barns air for European dairy cattle was between 10 to 18°C with relative humidity range of 60 to 70 per cent (Ozhan et al., 2001). Mandal et al. (2002) observed optimum milk production in Frieswal cows when maximum temperature was less than 25°C and minimum temperature varied between 5 to 15°C with relative humidity less than 60 per cent, however milk yield declined when mean temperature was above 27°C with humidity being 48 per cent but milk yield showed an upward trend when temperature was below 27°C and humidity 67 per cent. University of Wisconsin Cooperative Extension (2002) published information that when Temperature – Humidity Index (THI) values were above 75, there is heat stress on high milk producing cows; when THI was above 80, severe heat stress occurred for cows on pasture, and when THI was above 83-85, there was danger of fatal heat stress. Gerret Rietveld (2002) also reported that at 37.7°C and RH of 20 per cent the environment induces heat stress in cattle, whereas at 37.7°C and 50 per cent RH induces a danger zone, and at 37.7°C and 80 per cent RH becomes a lethal temperature for cattles.

In tropical conditions high environmental temperature and desire to produce high milk production (high metabolic rate) are incompatible. It is possible that cattle of Indian origin may have contributed to the heredity of Swiss cattle for Brown Swiss and bred Zebu and 3/4 bred cattle of appropriate varieties are alike and have been shown to have a compatible degrees of inherent resistance to piroplasmosis (Mahadevan and Hutchison, 1964).

Indian cattle having well developed sweat glands with high density which helps in forming a fine water barrier under skin. The short coat and white colour also help in minimising the solar effect similar to water cooled engines. Comparatively European cattle have less developed sweat glands, dark colours and thick coat, and therefore helps them to conserve body heat in cold climates. These animals produce higher milk yield with liberation of more heat dissipated in cool environment. Thus they need a large quantity of food to generate metabolic heat to maintain balance of heat loss and can be compared to air cooled engines.

In tropical environment, animals cannot get rid of excess heat due to high environmental temperature, so a dairy cow even at rest produces 446 W of heat energy, whereas a lactating cow produces about double this amount. Burton and Edholm (1955) reported that at 18°C a lactating Holstein cow required a metabolic heat of 1047W, whereas at 40.5°C required metabolic heat of 756 W only to maintain body's core temperature. Therefore in animals thermal effects are most important for maintaining the balance of heat production and heat loss regulated by thermoregulatory mechanism located in the hypothalmus in the brain for their survival. These mechanisms differ for different animals, but its dysfunction will deteriorate health of animal leading to reduction in production and stimulation of health diseases. Both heat and cold conditions affect their physiological mechanisms in different ways. The excessive heat in the body is reduced by evaporative thermoregulation and respiratory thermoregulation with sweat glands. Another phenomena known as panting also exists in animals, whereby the increase in volume of air exhaled per unit time functioning as respiratory ventilation by keeping animal's mouth open (dog, lion) or closed (ox and sheep). In such excessive heating effects, the quantity and quality of productivity decreases. Heat stress in combination with high humidity reduces food consumption and produce less output. Therefore animals (cattle) are better adapted to lower temperature (0 to 15°C) comfort zone where heating requirement are supplemented with metabolic heat.

Pigs, being nonsweating fatty animal, are strongly influenced by excessive heat balance conditions. The major difference in heat regulation between a tropical and temperate cattle lies in the respiratory frequency which increases steeply in temperate cattle with rise of temperature as compared to gradual increase in tropical cattle due to lower ratio of surface area to body weight. The rectal temperature rises above 30°C with steep increase of heart rate, and also above 25°C in temperate cattle. Therefore stress also influence the blood composition of cattle with increase of creatinine contents and decrease of CO_2 combining capacity, ascorbic acid and cholesterol levels. Heat stress also reduces the reproduction rate of animals. A white, yellow or red coat with a smooth texture will considerably minimise the direct effect of solar radiation.

6.3 Energy Balance Equation and its Components

Energy budget or heat balance is the basic principle which is regulated by thermoregulatory mechanism to maintain thermal equilibrium between metabolic heat produced and the total heat loss due to evaporation, conduction and sensible heat transfers (Fig. 6.2). The shaded area in Fig. 6.2 presents the increased maintaince requirements by metabolism in relation to prevailing environmental temperatures. However, when it is not maintained the production is affected both in quantity and quality. The temperature values corresponding to C and E levels varies with species, breeds and individuals. The temperature range (CE) is the ability to regulate evaporative heat loss. C level is influenced by insulation and heat production. Normally species, adapted to higher environmental heat, will have low insulation, well developed evaporative loss mechanism and low basel metabolism. But cold adapted species results to have a higher metabolic rate and greater thickness of insulation. Sweating is the

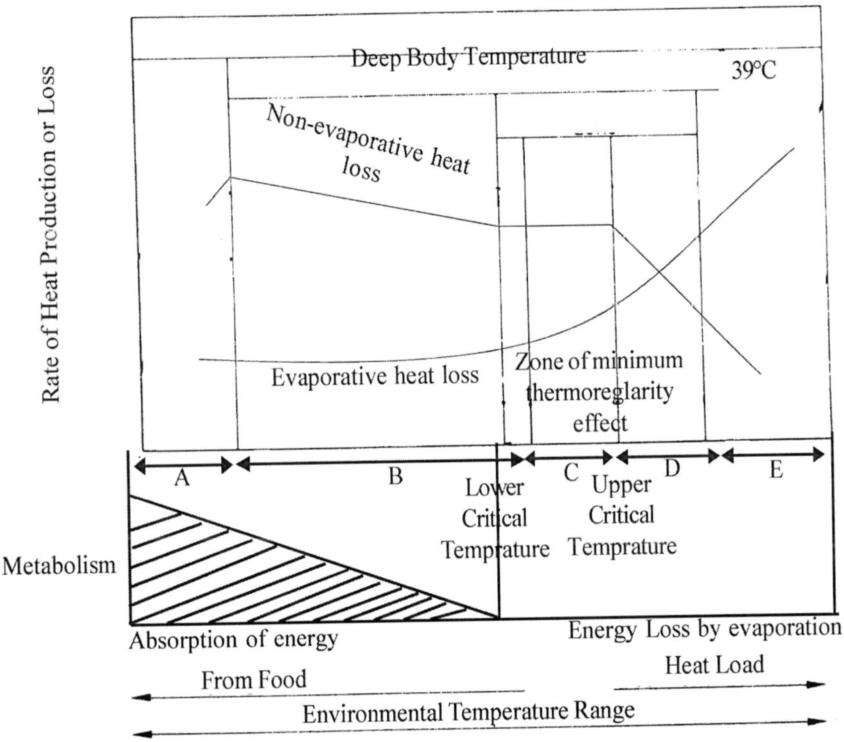

Fig. 6.2 Heat balance over environment temperature range.

main thermoregulatory mechanism in hot environments (Hobb, 1980), but in cold climates animals constrict the blood vessels so that less blood flows to the surface and also due to the development of insulating properties of skin, hairs and body fat.

6.3.1 Radiant energy fluxes

The short and longwave fluxes are computed for various environments and converted into average values by multiplying with the ratio of total body surface area of the animal to the surface area exposed to the radiant energy source. Since the longwave radiations are emitted by the total surface area of animal, therefore longwave emitted is $\in\sigma T_{bs}^4$ having T_{bs} as average body surface temperature. Incoming longwave radiations and reflected longwave from earth are absorbed by animal's surface and its average value will depend on the animal's shape-spherical or horizontal cylinder type. If the animal is inside a shed, then surrounding wall environment temperature may be uniform to give R_{LWi} as $\in\sigma T_w^4$ with $\in_w$ and T_w as emissivity and temperature of the surrounding wall surface. However, if the animal is in open environment, then upper hemisphere is radiating at sky temperature and emissivity, and lower hemisphere may be radiating at earth's temperature of $\in_e\sigma T_e^4$ with earth temperature (T_e) and emissivity ($\in_e$). An average value may is determined to provide R_{LWi} (Fig. 6.3).

The direct shortwave radiation (R_{si}) for animals is determined by the projection of the animal on a plane perpendicular to the radiation beam i.e. shadow of animal. The average direct shortwave radiation is the ratio of projected area to the total area multiplied by the radiation perpendicular to he solar beam (R_{sp}). Here shape of animal plays critical role. For a spherical shape it is 0.25. Therefore $a_s R_{si}$ is a_s 0.25. R_{sp} or 0.25 $a_s R_{si} \sin \phi$ or 0.25 $R_{si} \cos \theta$, where ϕ and θ are sun elevation angle from horizon and zenith angle between solar beam and the normal to the surface. The absorption coefficients for zula cattle, red sussex cattle, Aberdeen angus cattle, sheep, newly shorn sheep are 0.49, 0.83, 0.89, 0.74 and 0.58 respectively (Monteith, 1973). Similarly diffuse radiation is absorbed as ½ $a_s R_{sd}$ from the upper half hemisphere and reflected radiations from ground or lawn surface i.e., lower hemisphere is absorbed as ½ $a_5(1 - \alpha_g) R_{st}$ where R_{st} is the total incoming solar radiation direct and diffuse components both. Albedo for white and black Holstein is 61 and 8 per cent; light and dark brown Jersey being 32 and 27 per cent; light tan, medium red and dark red Herefold being 35, 24 and 18 per cent; red sussex 17 per cent, Aberdean Angus being 11 per cent, sheep at 26 per cent with newly born at 42 to 50 per cent respectively.

The energy budget over a unit area of the animal's surface can be expressed as (Fig. 6.3).

$$a_s R_{si} + a_s R_{sd} + a_s(1 - \alpha_g) R_{st} + a_L R_{LWi} - \epsilon_b R_{LWbs} + M - LE - A - G \pm S = 0 \quad \ldots (6.1)$$

where a_s and a_L are absorption coefficients for short and longwave radiations falling on the surface; R_{si}, R_{sd} and R_{LWi} are the incident direct shortwave, diffuse shortwave and incoming longwave radiation including reflactance from grove at the surface, ϵ_b is emissivity of the animal's surface, R_{LWb} is longwave emittance, M is metabolic heat per unit surface area supplied to the skin surface; LE, A and G are rate of loss due to evaporation of water from respiration and skin, convection and conduction phenomenon, ±S is rate of heat storage in animal.

In homeotherm animals a steady state is attained by acquiring a constant body temperature of 39 ± 2°C, then ±S approaches zero. If animal is standing over a surface then G can be neglected otherwise in sitting or touching surrounding surface conduction takes place significantly.

Each of the components of equation (6.1) can be quantified as an average value since each different parts of the animal are exposed to different flux densities. Appendix VII and VIII can be used for evaluating different components depending on temperature and radiation data.

6.3.2 Metabolic heat production (M)

Metabolic production of heat depends on the animal's body weight, its thermal insulation, activity level, nutritional diet with the season, and the surrounding physical environment.

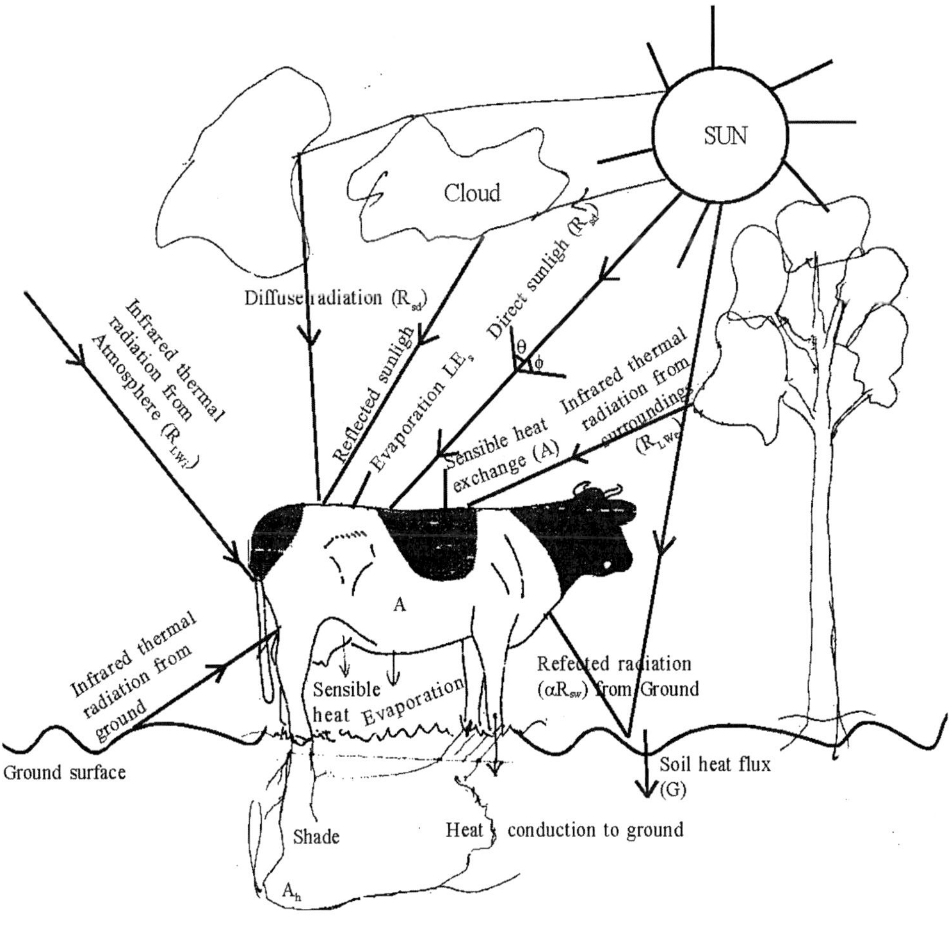

Fig. 6.3: Energy balance components influencing cow standing in an open environment.

Animals obtain energy through oxidation of foodstuff; therefore oxygen consumption can represent the rate of energy metabolism. Even at rest animal requires a minimum metabolic rate to maintain the body functions and constant body temperature around 39°C in homeotherms. Kleiber (1932) expressed the relationship between basal metabolic rate (Mb) in terms of body mass (m, kg) as

$$M_b = cm^{0.75} \qquad \ldots (6.2)$$

where, M_b is in Watt and c is constant of 3.4 for homeothems (Kleiber, 1965) and for poikilothems it is 5 per cent of homeotherms values (Hemmingsen, 1960). M_b ranged between 30 to 50 Wm^{-2} (Fig. 6.4)

Energy exchange of animal's body with the surrounding environment is proportional to the surface area of the animal. For a spherical body it is proportional to $(radius)^2$ and in terms of animal's volume it is proportional to $(volume)^{2/3}$, and for a constant density of animal's body it is proportional to $(mass)^{2/3}$. Therefore metabolic rates can be expressed in terms of body's mass or volume or density. Table 6.2 shows the metabolic rates for different homeotherms.

Table 6.2
Metabolic rates for different homeotherms

Animal	Body Mass (kg)	Oxygen Consumption (lt. O_2 hr^{-1})	Approx.** Body Surface Area (m^2)	Oxygen Consumption* (lt. O_2 $kg^{-1}hr^{-1}$)	Metabolic Rate W kg^{-1}	Metabolic Rate Wm^{-2}
Small birds	0.1	0.12	0.022	1.20	6.69	30.4
Rat	0.3	0.26	0.045	0.88	4.91	32.6
Cat	2.5	1.70	0.184	0.68	3.79	51.5
Dog	12.0	3.84	0.525	0.32	1.79	40.9
Sheep/Goat	45.0	10.35	1.267	0.23	1.28	45.5
Man	70.0	15.40	1.701	0.22	1.23	50.6
Horse/Cattle	700.0	70.00	7.901	0.10	0.56	49.6
Elephant	4000.0	240.0	25.268	0.06	0.33	53.0

(*After*: Schmidt Nielsen, 1969).

* One litre of oxygen consumption at S.T.P. equals 5.5798 W of energy release.

** Area = 0.1 $m^{2/3}$.

1 Watt = 0.8598 Kcal hr^{-1}

1 Wm^{-2} = 2.388 Kcal $cm^{-2}s^{-1}$

M will vary 50 to 500 Wm^{-2} depending on activity of the animal or man. Man at rest (60 Wm^{-2}), standing (90 Wm^{-2}), doing light work (95-120 Wm^{-2}), doing moderate work (180 Wm^{-2}), during sustained hard work (300-350 Wm^{-2}), and during heavy activity climbing or sports activity (600 Wm^{-2}) respectively (Landsberg, 1969).

6.3.3 Latent heat exchange

The evaporation of water from a human or animal occurs through insensible perspiration from lungs or from respiratory tracts and skin surfaces, and evaporation of sweat from the body. It is essential process in homeotherms to prevent from excess heat available in warm environmental conditions. The evaporation rate will depend on vapour pressure gradients between the evaporative surface and surrounding aerial environment, and the resistances offered during the diffusion pathway. Wind force will also augment the maintaince of saturation deficit at the evaporative surface. The evaporation of sweat will also depend on the activity of sweat glands. The sweating rate of humans often exceeds the maximum evaporative capacity, but not in animals (Robertshaw, 1976).

Sweating and panting both dissipate the body heat through evaporation process. Evaporation will depend on the vapour pressure gradients developed while panting and respiratory ventilation (Mclean, 1974). Therefore respiratory latent heat loss is a function of metabolic rate due to increased breathing and oxygen consumption. Monteith (1973)

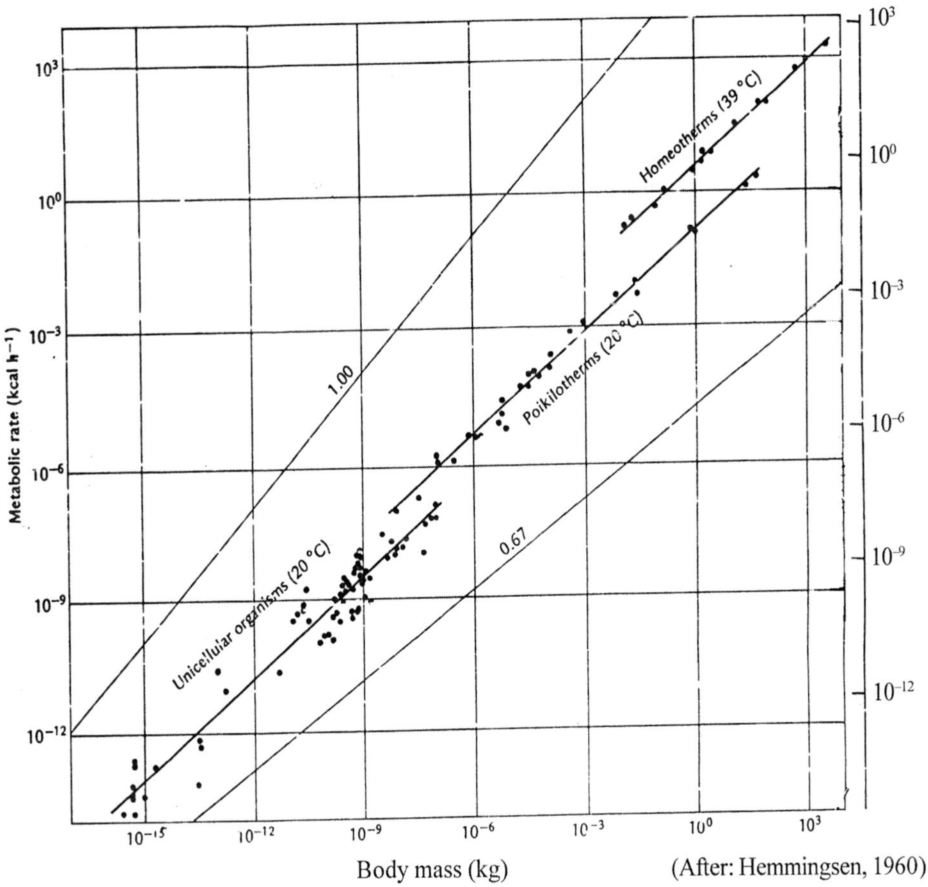

(After: Hemmingsen, 1960)

Fig. 6.4 Metabolic rates for different organisms and animals.

reported this ratio of 0.08 to 0.1 for wet to dry air, and for kangaroo it is 0.02 (Campbell, 1977). Monteith (1973) quantified this insensible perspiration heat loss from dry human skin (LE_s) as of 20 to 30 per cent of metabolic rate (Monteith, 1975) expressed as

$$LE_s = \frac{L(\rho_{vs} - \rho_{va})}{r_{vs} + r_{vc} + r_{va}} \qquad \ldots(6.3)$$

where, ρ_{vs} and ρ_{va} are saturated water vapour density at skin surface and actual water vapour density of surrounding air, r_{vs} and r_{vc} are resistances to diffusion of vapour through the skin and coat, and r_{va} is aerial resistance. When skin is dry (non sweating animals) r_{vc} and r_{va} are much less than r_{vs} (Table 6.3), however when skin is wet (moist skins) r_{va} contributes largely.

The ratio of respiratory evaporation to metabolic heat production is expressed as:

$$\frac{LE_R}{M} = \frac{(\rho_{ae} - \rho_{ai})\overline{L}}{(\rho_{0i} - \rho_{oe})\Gamma} \qquad \ldots(6.4)$$

Table 6.3
Skin resistance to vapour diffusion for non-heat stressed animals and human

Animal	r_{vs}(sm^{-1}) (dry skin)	Coats r_c Thermal resistance (s m^{-1})	Coat/cm depth (s m^{-1})		Referebce
Human	7700		Air	470	Kerslake (1972)
Camel	13000		Red fox	330	Schmidt-Nielsen (1969)
Desert tortoise	120000		Huskydog	290	-do-
			Merino sheep	280	-do-
Zebra	10200		Down	190	Bernstein (1971)
Ostrich	56000		Block fox	150	Schmidt-Nielson et al. (1969)
Calf (Vasocontricted)	1100		Ayshire cattle		Blaxter (1967)
(Dilated)	500		Flat coat	120	-
Pig	1000		Voiterect	80	-
(Dilated)	600		Gallowary	90	-
Human	1200			90	-
(Dilated)	300				-
Downsheep (Dialated)	900	1900			
Dallsheep		1450			
Galloway Cattle		1000			
Marino Sheep		2800			
Sparrow	7600				Robinson et al., (1976)

where ρ_a is water vapour density in air, ρ_0 is oxygen density in air, *i* is inspired concentration, *e* is expired concentration. Γ is heat produced per kg of O_2 consumed about 15 MJ kg^{-1}.

However, some animals with small nasal passages exhale air at temperatures below body temperature.

Evaporative heat losses in sweating animals are much greater than the losses through insensible perspiration. The sweat rate of man is much higher attaining maximum evaporative capacity, but in animals sweat rate is much lower (Robertshaw, 1976). Panting is an compensatory mechanism for increasing evaporative losses in animals with low capacity for sweating (Gates, 1980) such as birds, dogs and cats exhibiting high panting rates with low sweating rates as compared to man, horses and cattles which sweats freely. Evaporation from wet skin is controlled by environmental factors with the conditions of r_{vs}.

The metabolic heat production rate will rise with the activity level attaining maxima. The maximum metabolic rate that an animal can sustain is about 10 times their basal rates expressed as (Campbell, 1977)

$$M = M_b\left(1 + 9\frac{a}{a_m}\right), \qquad \ldots (6.5)$$

where a/a_m is ratio of actual activity to the maximum sustainable activity.

For poikilotherm animals LE approaches M in the energy budget equation and (M – LE) is taken as zero.

Campbell (1977), Johnson (1965) showed the variation in the ratio of latent heat loss to heat production for different animals (Fig. 6.5). The resistances to heat and water vapour losses in animals depends on coat depth, wind speed, vascometer action and respiratory rate. With rise of temperature the latent heat loss also increases sharply.

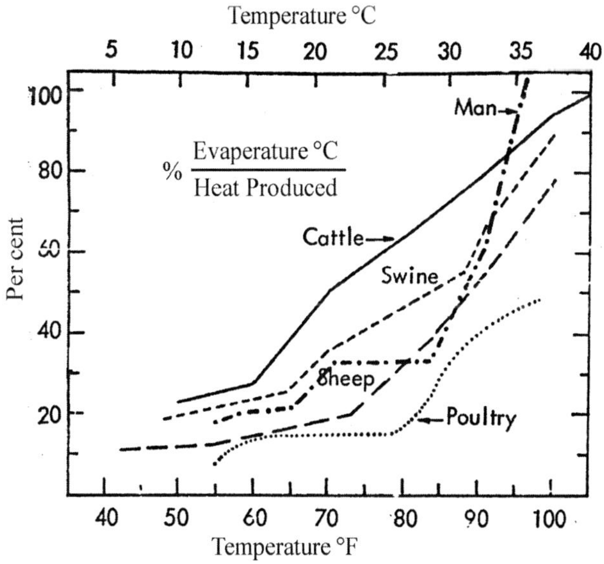

(*After*: Johnson, 1965)

Fig. 6.5 Ratio of total evaporative heat loss to heat production for cattle, swine, sheep, poultry and man.

6.3.4 Conductive heat exchange

Conditions of heat exchange in animals takes place through conduction of heat to or from air through skin and coat or clothing, and to or from the ground surface through body contact while sitting or quenching body heat in water pool. The conduction heat flux on the basis of outer body surface of animal is of form (Cena and Monteith, 1975) as

$$G = \frac{k(T_b - T_{bs})}{r_2 Ln(r_2 / r_1)} \qquad \ldots (6.6)$$

where k is total thermal conductance of the fat (0.205 $Wm^{-1}\,°C^{-1}$), skin (0.502 $Wm^{-1}\,°C^{-1}$), fur etc., T_b is internal body temperature, T_{bs} is temperature of outer surface of the body Monteith (1973), preferred to use the term insulation which when multiplied with $\rho_a C_p$ to provide resistance. Campbell (1977) suggested that the coat resistance decreases with square root of wind speed in thin hairs, but will be independent in thick coats (for ex sheep). Insulation for sheep fur (6/7 cm thickness) is 1.1 to 1.25 Wm^{-2} °C, for dog fur (3/5 cm thickness is 0.6 to 0.95 Wm^{-2}°C), black surface is 0.1 Wm^{-2}°C, and cotton of thickness 6 cm as 1.5 Wm^{-2} °C. Heat transfer to the skin surface of an animal depends on blood flow and regulatory mechanism of vasoconstriction or vasodilation to control body temperature. The tissue resistance is variable and average values can be used to quantify the thermal resistance offered with use of following data.

Animal/man	Vasocontriction Resistance ($s\ m^{-1}$)	Vasodilation Resistance ($s\ m^{-1}$)	Reference
Calf	110	50	Monteith (1973)
Pig (60 days)	100	50	Kerslake (1972)
Sheep	90	30	-do-
Man	90	15	Monteith (1973)
	120	30	-do-

6.3.5 Sensible heat exchange

Sensible heat exchange takes place at the outer surface of the body. At low wind speeds free convection dominates, but at higher wind speeds (>0.2 to 0.5 m sec^{-1}) forced convection predominates. The sensible heat exchange (A) is expressed as

$$A = h_c (T_a - T_{bs}) \qquad \ldots (6.7)$$

where, h_c is convective heat transfer coefficient from animal's body surface temperature (T_{bs}) with the environment (T_a). h_c is expressed is terms of constants in different forms as

$$h_c = \frac{Nu\,k}{D}, \qquad \ldots (6.8)$$

$$h_c = 2.631 \times u^{1/3} D^{-1/2}$$

where, Nu is Nusselt number, k is thermal conductivity of air ($2.57 \times 10^{-2}\ Wm^{-1}\ °C^{-1}$), u is wind speed in m sec^{-1} and D is diameter in m., h_c in $Wm^{-2}\ °C^{-1}$. h_c depends on the shape of the body (vertical or horizontal cylindrical shape) and free or forced convective conditions in terms of $D^3(T_a - T_{bs})$ as (Mitchell, 1974; Monteith, 1975; Gates, 1980) given below:

Shape	$D^3(T_a - T_{bs})$	Free Convection h_c ($Wm^{-2}°C^{-1}$)	Forced Convection
Vertical cylinder	1.5×10^{-4} to 15	$1.4[(T_a - T_{bs})/D]^{0.25}$	$0.78\ R_e^{0.5}.k/D$
(D-height in m.)	15 to 1.5×10^{4}	$1.5(T_a - T_{bs})^{0.33}$	
Horizontal cylinder	1.5×10^{-4} to 15	$1.2[(T_a - T_{bs})/D]^{0.25}$	$0.65\ R_e^{0.5}.k/D$
(D-diameter in m.)	15 to 1.5×10^{4}	$1.2(T_a - T_{bs})^{0.33}$	

During respiratory process, air enters and exits through body orifices there is always respiratory convective heat losses (Ar) expressed as

$$A_r = C_1 V(T_i - T_e) \qquad \ldots (6.9)$$

where A_r is respiratory heat loss (Wm^{-2}), V is ventilation rate l min^{-1}, T_i and T_e are temperatures of inhaled and exhaled air, C_1 is constant factor. A_r is very small at low temperatures about 25 Wm^{-2} (Mitchell, 1974). Convection takes place both by the upward passage of heated air as well as by wind action. Radiation is reflected by the fleece of animal. However, the energy absorbed induces temperature. For example in a Merino sheep energy absorbed in outer 1 cm. of fleece induced temperature of about 85°C in calm air and this energy is lost as a longwave radiation to sky. Fleece functions as an insulator (9:1 air and wool fibre). Heat regulation by a sheep is different in shade and exposed area to the sun when incident and diffuse radiation of 800 and 120 $Kcal/m^2/hr$ is falling on the body surface. Temperature varied from 40 to 85°C in different parts of the body. Tight wool raises temperature higher than loose one. Temperature gradient is 40 to 50°C between fleece and skin surface in desert sun. Tip is about 92°C, whereas 1 cm below about 70°C with exponential fall across the insulator. Skin temperature 42°C. Thus sheep withstand the heat of summer desert by physical rejection of most of energy falling on it.

Heat storage: It accounts for the change in heat contents of the animal. For homeotherms the body temperature being constant, heat storage is least. However in case of desert animals such as camel whose body temperature rises to 40°C in afternoon and drops to 34°C in the morning amounting to 40 Wm^{-2} and this heat may help in conserving body water.

6.3.6 Energy budget over human or animal system

Equation (6.1) can be expressed for human (a vertical cylindrical system) or animal surfaces (a horizontal cylindrical system) as

$$[a_S A_{Si} \cdot R_{Si} + a_S A_{Se} R_{Se} + a_L A_{Se} R_{SLi} - \epsilon\sigma A_{Se} T_S^4]$$

$$- h_c(A - A_c)\,\Delta T - k A_c \frac{dT}{dz} - LE + M \pm S = 0 \qquad \ldots (6.10)$$

where, R_{Si}, R_{Se} and R_{SLi} are flux of direct shortwave incoming, scattered including reflected shortwave energy from surrounding environment and flux of longwave energy from surrounding environment; T_s is surface temperature of human or animal's body surface, A's are areas of transfer surface appropriate for different components of radiation, A_{Si} and A_{Se} are for direct beam of shortwave given by shadow area for different orientations and extended source areas seen by spherical surroundings of the system for shortwave; A is total area of the system man or animal, A_c is the area of system in contact with the floor.

System	A_{Si}	A_{Se}	Characteristics of the Surface		Q	Reference
A man standing on ground or a vertical cylinder	Dh cosθ	πDh	diameter (D)	height (h)	Altitude angle of the sun (θ)	Lee and Vaughan (1964)
Horizontal cylinder or an animal standing or sitting on surface	Dl sinθ	πD*l*	diameter (D)	length (*l*)	Altitude of sun	Lee and Vaughan (1964)
Standing man	A/π	(Ac/A) (A – Ac) <0.10			Perpendicular solar beam	Gates (1980) Lowry (1969) Campbell (1977)
Conifer fir trees	0.32A	0.94A			Branches perpendicular to solar beam	Gates (1980) Lowry (1969), Campbell (1977)
Spruse trees	0.34A	0.88A			-do-	

a_S and a_L are absorption coefficient for shortwave and longwave spectrums, a_L will equal $\in$ of the body system, h_c is convective heat transfer coefficient, ΔT is the difference between the system surface with surroundings as proportional to 1/3 power of wind speed, k is thermal conductivity of the system, LE is latent heat dissipated and S and M are stored and utilized energy.

Example 1:

Examine the energy budget equation of an animal surrounded with a open spherical environment with the following data.

$R_{Si} = 1116.8\ Wm^{-2}$	$A = 0.0075\ m^2$
$R_{Se} = 69.8\ Wm^{-2}$	$a_S = 0.9$
$R_{SLi} = 174.5\ Wm^{-2}$	$a_L = \in = 0.97$
$T_a = 27°C$ or $300°K$	$A_{Si} = 0.0018\ m^2$
$dT/dz = 0.2°C\ m^{-1}$	$A_{Se} = 0.0060\ m^2$
$u = 0.27\ ms^{-1}$	$Ac = 0.0010\ m^2$
$\sigma = 5.67 \times 10^{-8}\ Wm^{-2}°K^{-4}$	$T_s = 31°C$ or $304°K$
$k = 0.035\ Wm^{-2}\ °K^{-4}$	$hc = kv^{1/3} = 10.523\ Wm^{-2}°K^{-1}$
	$LE = 0.035\ W$
	$M = 0.063\ W$
	$S = 0$

Solution:

Energy budget equation over animal's surface from surroundings as

$$[a_S A_{Si} . R_{Si} + a_S A_{Se} R_{Se} + a_L A_{Se} R_{SLi} - \in\sigma A_{Se} T_S^4]$$
$$- h_c(A - A_c)\ \Delta T - k A_c \frac{dT}{dz} - LE + M \pm S = 0$$

$$0.9 \times 0.0018 \times 1116.8 + 0.9 \times 0.006 \times 69.8 + 0.97 \times 0.006 \times 174.5 - 0.97 \times 5.67 \times 10^{-8} \times 0.006 \times (304)^4$$
$$- 10.523 \times 0.0065 \times 4 - 0.035 \times 0.001 \times 0.2 - 0.035 + 0.063 = 0$$
$$1.8092 + 0.3769 + 1.0156 - 2.8184 - 0.2736 - 0.07 - 0.035 + 0.063 = 0$$
$$0.07 \simeq 0$$

6.4 Applications of Energy Budget

6.4.1 Body surface temperature

In order to use energy balance equation it is essential to quantify the average body surface temperature. Heat transfer from animal's core to the outer exchange environment can be expressed assuming that all latent heat loss takes place from inside of the non-sweating animals.

$$M - LE = \frac{\rho c_p (T_b - R_s)}{r_{Hc} + r_{Ht}} \qquad \ldots (6.11)$$

where, T_b is the animal's deep body temperature, T_s is body surface temperature, r_{Hc} and r_{Ht} are average coat and tissue thermal resistances.

$$T_{bs} = T_b - \frac{(M - LE)(r_{Hc} + r_{Ht})}{\rho c_p} = T_b - \frac{(M - LE) r_{Hb}}{\rho c_p} \qquad \ldots (6.12)$$

Tissue resistance is variable within limits in order to balance the energy budget. Computation of T_{bs} from energy budget equation will be more complex as several parameters are involved. However we will solve the energy budget equation for T_{bs} with certain assumption in order the arrive its values with negligible errors. Tissue fats provide adequate energy of the order of 40 kJ g^{-1}. A man doing heavy sports exercises would consume about (at the rate of M of 600 Wm^{-2}) 54 gm of fat. Table 6.4 provides the body temperature of different homeotherms. The core temperature of body also depends on metabolic heat production expressed as (Kerslake, 1972).

$$T_b = 36.5 + 4.3 \times 10^{-3} M \qquad \ldots (6.13)$$

where, M is in Wm^{-2} and T_b is core temperature of human body in °C.

Example 2:

Given that thermal resistance for vasoconstriction and vasodialation varied between 100 to 20 sm^{-1} for a human body of diameter 0.17m, wind speed 3m sec^{-1}, LE_R is 0.1M, LE_s is 9 Wm^{-2}, T_b is 37°C, M is 60 Wm^{-2}, r_{Hc} is 100 sm^{-1}. Compute the body's surface temperature.

Solution:

$$T_{bs} = Tb - \frac{(M - LE)(r_{Hc} + r_{Ht})}{\rho c_p} \qquad LE = LE_R + LE_S$$

Table 6.4
Body temperature of different animals

Animal	Body Temperature °C	Lethal Core Temperature °C	Reference
Marsupials	36 ± 2	40 – 41	Robinson and Morrison (1957) Dawson and Hulbert (1970)
Eutherian mammals	38 ± 2	42 – 44	Morrison and Ryser, (1952)
Lower montremes	31 ± 2	37	Schmidt – Nielsen et al. (1966)
Higher eutherians	37 ± 1	42 – 44	Adolph (1947)
Birds	40 ± 2	42 – 46	Dawson and Hulbert (1970) Robinson and Lee (1946)
Human	37	43	
Insects	35 ± 1	41	Skholnik and Schmidt-Nielson (1976)
Mammals arctic	38.6		Irving and Krog (1954)
Sheep/Goat	39 ± 2	41.5	
Buffalow	38.0 ± 1.5	42.0	Yousef (1972)
Camel	37.5 ± 1.5	45.0	Yousef (1972)
Cow	39.0 ± 1.0	42.0	Yousef (1972)
Bull	38.5 ± 0.5	42.0	Yousef (1972)
Donkey/Ass	38.0 ± 2.0	42.0	Yousef (1972)

$$T_{bs} = 37 - \frac{(60-15)(73.4)}{1200}$$

$$= 37 - \frac{45 \times 73.4}{1200}$$

$$= 37 - 2.8$$

$$T_{bs} = 34.2°C$$

$$T_{bs} = T_b \text{ when } r_{He} + r_{Ht} = 0$$

$$= 0.1M + 9$$

$$LE = 6 + 9 = 15 \text{ Wm}^{-2}$$

$$r_{He} = r'_{He} - 50\sqrt{3}$$

$$= 100 - 50\sqrt{3}$$

$$r_{He} = 13.4 \text{ sec m}^{-1}$$

$$r_{Ht} = \frac{100+20}{2} = 60 \text{ sm}^{-1}.$$

6.4.2 Climate – space diagrams for animals

A climate space diagram is a graphical representation of the impact of varying conditions of temperature and radiation on the animal. The acceptable environmental limits on the diagram represents the climate space. Assuming no heat exchange in conduction, heat storage and work done by animal, the energy balance equation can be expressed as

$$Rn - A - LE + M = 0 \qquad \ldots (6.14)$$

The average body surface temperature (T_{bs}) will quantify A and R_{nabs}, therefore

$$\left(R_{ab} - \epsilon_b \,\sigma T_{bs}^{\;4}\right) + \frac{\rho_a c_p}{r_{Ha}}(T_a - T_{bs}) - LE + M = 0 \qquad \ldots (6.15)$$

where R_{ab} in amount of shortwave and longwave absorbed by the animal's body surface. Assuming all latent heat loss is from inside the animal (non-sweating animals), then

$$M - LE = \frac{\rho_a c_p}{r_{Hc} + r_{Ht}}(T_b - T_{bs}) \qquad \ldots (6.16)$$

where, r_{Hc} is resistance to heat transfer through the fat and coat skin

$$T_{bs} = T_b - \frac{(r_{Hc} + r_{Ht})}{\rho_a c_p}(M - LE) = T_b - \frac{r_{Hb}}{\rho_a c_p}(M - LE)$$

$$T_{bs} = T_b - I(M - LE) \qquad \ldots (6.17)$$

Substituting T_{bs} in equation above, we get on simplification

$$R_{ab} - \epsilon_b \,\sigma\left\{T_b^{\;4} + \frac{4T_b^{\;3} r_{Hb}(M - LE)}{\rho c_p}\right\} + \frac{\rho c_p}{r_{Ha}}(T_a - T_b) + \frac{r_{Hb}}{r_{Ha}}(M - LE) + M - LE = 0$$

$$(R_{ab} - \epsilon_b \,\sigma T_b^{\;4} + M - LE) - \frac{\rho c_p}{r_{Ha}}(T_b - T_a) + r_{Hb}(M - LE)\left\{\frac{1}{r_{Ha}} + \frac{4\,\epsilon_b \,\rho T_b^{\;4}}{\rho c_p}\right\} = 0$$

$$(R_{ab} - \epsilon_b \,\sigma T_b^{\;4} + M - LE) - \frac{\rho c_p}{r_{Ha}}(T_b - T_a) + r_{Hb}(M - LE)\left\{\frac{1}{r_{Ha}} + \frac{1}{r_{LWr}}\right\} = 0$$

$$(R_{ab} - \epsilon_b \,\sigma T_b^{\;4} + M - LE) - \frac{\rho c_p}{r_{Ha}}(T_a - T_b) + \frac{r_{Hb}}{r_{eq}}(M - LE) = 0 \qquad \ldots (6.19)$$

In terms of insulation (I)

$$(R_{ab} - \epsilon_b \,\sigma T_b^{\;4} + M - LE) - \frac{\rho c_p}{r_{Ha}}(T_b - T_a) + \frac{\rho c_p I}{r_{eq}}(M - LE) = 0 \qquad \ldots (6.20)$$

In simple terms first T_{bs} is determined and then it is substituted in the equation to find out T_a for different values of R_{ab}.

Here $\frac{1}{r_{eq}} = \frac{1}{r_{Ha}} + \frac{1}{r_{LWR}}$ and the resistance to longwave radiative transfer as

$$\frac{1}{r_{LWR}} = \frac{\rho c_p}{4\,\epsilon_b \,\sigma T_b^{\;3}}$$

Equations (6.18) or (6.19) can be expressed for T_a by rearranging the terms as

$$T_b - T_a = \frac{r_{Ha}}{\rho c_p}\left(R_{ab} - \epsilon_b \sigma T_b^{\,4} + M - LE\right) + \frac{r_{Hb}\, r_{Ha}}{\rho c_p\, r_{eq}}(M - LE)$$

$$T_a = T_b - \frac{r_{Ha}}{\rho c_p}\left(R_{ab} - \epsilon_b \sigma T_b^{\,4} + M - LE\right) + \frac{r_{Hb}\, r_{Ha}}{\rho c_p\, r_{eq}}(M - LE) \qquad \ldots (6.21)$$

$$T_a = T_b - \frac{r_{Ha}}{\rho c_p}\left(R_{ab} - \epsilon_b \sigma T_b^{\,4} + M - LE\right) + \frac{r_{Ha} I}{r_{eq}}(M - LE) \qquad \ldots (6.22)$$

These expressions represent animal's energy budget and for a given values of M, LE, T_b, r_{Hb} or I graphical relations are presented between R_{abs} and T_a in astraight line relationship providing average slope value of $\frac{\rho c_p}{r_{Ha}}$ called climate space diagram for animals. Three more lines are depicted in climate space diagram to describe the physically attainable environments for the animal such as blackbody cavity expressed as $R_{backbody} = \sigma T_a^{\,4}$, where wall temperature equals air temperature. Secondly, the clear night sky line presents the minimum radiant energy environment available to the animal expressed as

$$R_{ab}\ \text{night} = a_L\, \overline{\epsilon}_e\, \sigma T_a^{\,4} \qquad \ldots (6.23)$$

where e_e is average emissivity of clear sky and ground.

Thirdly – the bright sun line represents the maximum energy available to the animal under a clear sky expressed as

$$R_{ab}\ \text{sun} = R_{ab}\ \text{night} + a_s \frac{A_p}{A} R_{sb} + \frac{1}{2} a_s R_{sd} + \frac{1}{2}(1 - \alpha_g) \in_b R_{st}$$

$$R_{ab}\ \text{sun} = R_{ab}\ \text{night} + a_s \frac{A_p}{A} R_{sp} + a_s R_{sd} \qquad \ldots (6.24)$$

Climate space diagram for poikilotherm is also of similar type where body temperature increases instead of M and r_{Hb}. (M – LE) also approaches zero for such species. Therefore, using simplified energy budget equations for different body temperatures lowest to highest lines are drawn. Climate space diagrams are useful in comparing a high temperature environment with little radiant energy input with a low temperature environment having high radiant energy.

Example 3:

Develop a climate space diagram for sheep using following data. Given that at maximum temperature for survival (i.e., Thermal maxima)

M = 102 Wm^{-2}

LE = 100 Wm^{-2}

$T_b = 44.0\ °C$

$I = 1.25\ m^2\ °C\ w^{-1}$ or $r_{Hb} = 1450\ s\ m^{-1}$

At thermal minimum:

$M_b = 65\ Wm^{-2}$

$LE = 20\ Wm^{-2}$

$T_b = 39°C$

$I - 1\ m^2\ °C\ W^{-1}$ or $r_{Hb} = 1200\ s\ m^{-1}$

r_{Hb} (sheep) = 1200 to 1450 s m^{-1}

$$\frac{A_p}{A} = \frac{1 + \frac{4 \times 1 \times 7}{22 \times .4}}{2 + 4 \times 2.5} = \frac{1 + \frac{280}{88}}{12} = 0.26$$

Diameter of sheep = 0.25m, length = 1m, θ = 90°, then $A_p/A = 0.26$, wind speed = 0.1 m sec^{-1} at free convection

$R_{sp} = 100\ Wm^{-2}$, $R_{Sd} = 250\ Wm^{-2}$, $a_S = 0.6$, $\alpha_b = 0.4$, $\alpha_g = 0.23$, $\in_b = .98$, $\in_g = .97$

Wind speed = 2m sec^{-1} at forced convective conditions.

Solution:

Assuming horizontal cylindrical shape of sheep, r_a is calculated using Gates (1980) expression as

$$ra = \frac{\rho_a c_p}{k\left(\frac{u}{D}\right)^{0.5}}$$

Therefore during free convection

$\rho_a = 1.208\ kg\ m^{-3}$

$c_p = 1008\ J\ kg^{-1}\ °K^{-1}$

$k = 3.89\ Wm^{-2}\ sec^{½}\ °C^{-1}$

$$r_a = \frac{313.0}{\sqrt{0.4}} = 495\ s\ m^{-1}$$

During forced convection

$$r_a = \frac{1.208 \times 1008}{3.89\sqrt{\frac{2}{0.25}}} = \frac{1217.66}{3.89 \times 2.828}$$

$r_a = 110.7\ s\ m^{-1}$.

Determine T_{bs} at thermal minima and maxima using

$T_{bs} = T_b - I\ (M - LE)$

T_{bs} at thermal maxima = 44.0 – 1.25 (102 – 100)

= 44.0 – 1.25 × 2 = 44.0 – 2.5

$T_{bs}max$ = 41.5°C or 314.5 °K

$T_{bs}min$ = 39.0 – 45 × 1

$T_{bs}min$ = –6°C or 266°K

Curves for clear night, back body and full sun on the diagram will help to define the limits of tolerable radiation environments for the sheep as shown in Fig. 6.6. These are described with use of following expressions

$R_{abs} = \sigma T_a^4$ (Blackbody curve)

$R_{abs} = \epsilon_b \sigma T_a^4$ (Clear night curve)

$R_{abs} = 317.2\ Wm^{-2} + \overline{\epsilon}\,\epsilon_b\, \sigma T_a^4$ (Full sun curve)

Now use the expression of energy balance at thermal maxima and minima separately to determine T_a values at different values of R_{abs}.

$\alpha_b = 0.4$, $a_s = 0.6$, $a_{grass} = 0.23$, $\epsilon_b = 0.98$, $\overline{\epsilon}_g = 0.97$, $\overline{\epsilon}\,\epsilon_s = .94$, $\overline{\epsilon} = .97$

$$\text{Full sun } R_{ab} = \text{Night } R_{ab} + \frac{A_h}{A}(1-\alpha_b)\,R_{sb} + \frac{1}{2}(1-\alpha_b)\,R_{sd} + \frac{1}{2}(1-\alpha_b)\,\alpha_g\,R_{st}$$

$$= \text{Night } R_{ab} + 0.6 \times 0.26 \times 1000 + \frac{1}{2}(.6)\,250 + \frac{1}{2}(.6)(0.23)\,1250$$

$$= \text{Night } R_{ab} + 156 + 75 + 86.2$$

$$\text{Full sun } R_{ab} = \text{Night } R_{ab} + 317.2$$

Case I: At thermal maxima or upper lethal temperature

T_{bs} = 41.5°C or 314.5°K

$$T_a = \frac{r_{Ha}}{\rho c_p}(\epsilon_b\, \sigma T_{bs}^4 - R_{abs} - M + LE) + T_{bs}$$

$$= \frac{495}{1.208 \times 1008}\left\{0.98 \times 5.67 \times 10^{-8} \times (314.5)^4 - R_{abs} - 102 + 100\right\} + 41.5$$

$$T_a = 0.4065\,\{541.6 - R_{abs}\} + 41.5$$

$$= 0.4065\,R_{ab} + 220.2 + 41.5$$

$$T_a = -0.4065\,R_{ab} + 261.7 \text{ for free convection}$$

Under forced convection

$$T_a = \frac{110.7}{1.208 \times 1008}\{541.6 - R_{ab}\} + 41.5$$

$$T_a = 0.0909\,\{541.6 - R_{ab}\} + 41.5$$

$$T_a = -0.0909\,R_{abs} + 49.2 + 41.5$$

$$T_a = -0.0909\,R_{ab} + 90.7 \text{ for forced convection}$$

Case II: At thermal minima or lower lethal temperatures for free convection

$$T_a = \frac{495}{1.208 \times 1008}\left\{0.98 \times 5.67 \times 10^{-8} \times (266)^4 - R_{abs} - 65 + 20\right\} + T_{bs}$$

$$= 0.4065\ \{278.19 - R_{ab} - 45\} - 6.0$$

$$= -0.04065\ R_{ab} + 0.4065 \times 233.19 - 6$$

$$T_a = -0.4065\ R_{ab} + 88.79 \text{ for free convection}$$

Under forced convections

$$T_a = 0.0909\ \{233.19 - R_{ab}\} - 6 = -0.0909\ R_{ab} + 15.2$$

Using these lines expression between T_a and R_{ab}, Ta value are plotted for different values of R_{ab} in Fig. 6.6.

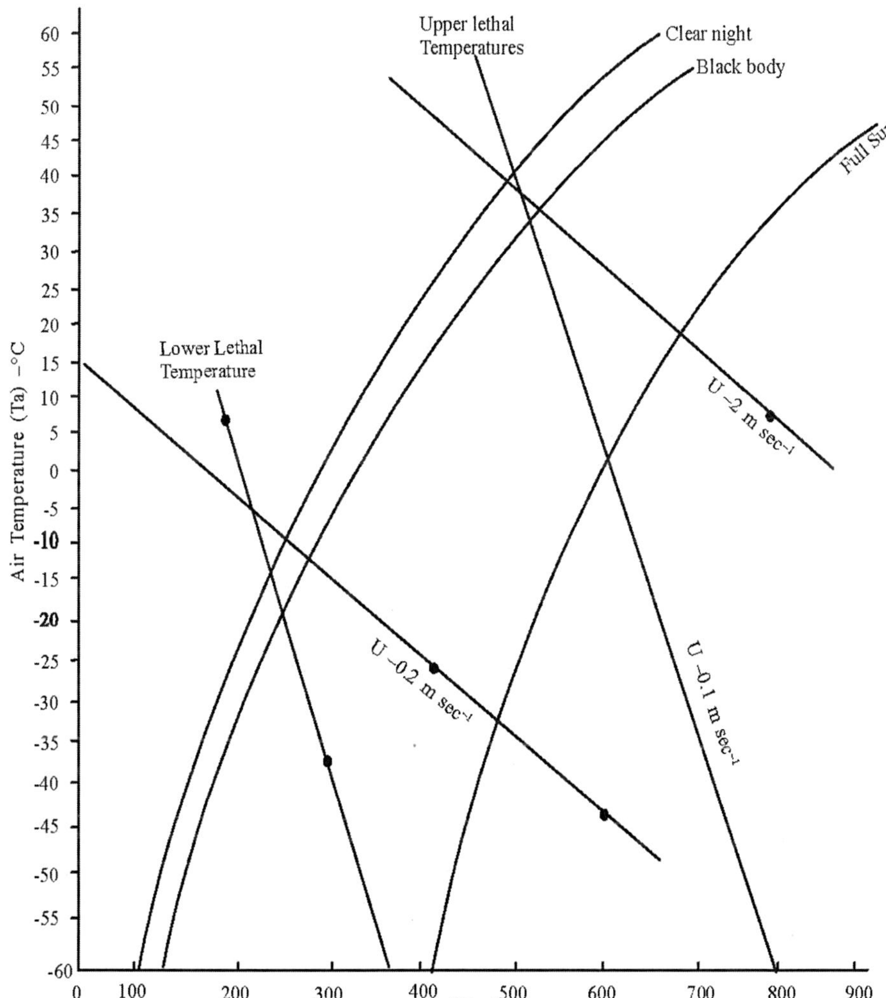

Fig. 6.6: Climate space diagram for sheep.

6.4.3 Equivalent blackbody temperature

The equivalent blackbody temperature T_{eq} is that temperature which will provide same radiant and convective heat exchange as is present in the animal's natural environment. Climate space diagram provides T_{eq} value as the intersection of animal's energy budget line with the blackbody line. T_{eq} combines two environmental variables of T_a and R_{ab}.

The energy budget equation (6.19) for an animal in a blackbody cavity can be expressed as

$$\in\sigma T_{eq}^{4} - \in\sigma T_b^{4} + M - LE - \frac{\rho c_p}{r_{Ha}}(T_b - T_{eq}) + \frac{r_{Hb}}{r_e}(M - LE) = 0 \qquad \text{. . . (6.25)}$$

The energy budget equation of the animal in natural environment is of the form (Equation 6.19).

Equating 6.19 and 6.22 equations, we get

$$R_{ab} = \frac{\rho c_p}{r_{Ha}}(T_{eq} - T_a) + \in\rho T_{eq}^{4} \qquad \text{. . . (6.26)}$$

Solving this equation for T_{eq} assuming that $T_{eq} = T_a + \Delta T$

$T_{eq}^{4} = T_a^{4} + 4T_a^{3}(T_{eq} - T_a)$ neglecting other smaller terms

$$R_{ab} = \frac{\rho c_p}{r_{Ha}}(T_{eq} - T_a) + \in\sigma\{T_a^{4} + 4T_a^{3}(T_{eq} - T_a)\}$$

$$R_{ab} = \in\sigma T_a^{4} + (T_{eq} - T_a)\left(\frac{\rho c_p}{r_{Ha}} + 4\in\sigma T_a^{3}\right)$$

$$R_{ab} = \in\sigma T_a^{4} + (T_{eq} - T_a)\,\rho c_p\left(\frac{1}{r_{Ha}} + \frac{1}{r_{LWR}}\right)$$

$$R_{ab} = \in\sigma T_a^{4} + \frac{\rho c_p}{r_{eq}}(T_{eq} - T_a)$$

$$T_{eq} = T_a - \frac{r_{eq}}{\rho c_p}\left(R_{ab} - \in\sigma T_a^{4}\right) \qquad \text{. . . (6.27)}$$

where, r_{eq} is equivalent parallel resistance of r_{Ha} and r_{LWR}

Case I: When $R_{ab} = \in\sigma T_a^{4}$, $T_{eq} = T_a$ i.e., when animals are in burrow or under dense crop/vegetation stands or cloud cover or in sheltered home.

Case II: When animals are exposed to a clear sky daytime solar radiation, $\frac{r_{eq}}{\rho c_p}\left(R_{ab} - \in\sigma T_b^{4}\right)$ term will represent as a change of radiant energy w.r.to the surroundings. T_{eq} will vary with the surrounding environment.

6.4.4 Food requirement of animals

Energy budget equations for animals can be used to quantify food requirements expressing sensible and latent heat losses of animal in a blackbody cavity with temperature (T_{eq}) as

$$M - LE = \frac{\rho c_p (T_b - T_{eq})}{r_{Hb} + r_{eq}} \quad \ldots (6.28)$$

where, (r_{Hb} + r_{eq}) is total heat exchange resistance the sum of r_{Hb} and equivalent parallel resistance of r_{Ha} and r_r. M value is determined knowing other variables which will quantify the energy requirements. Heat produced and O_2 consumed in metabolism in common foodstuff are given below. Ratio of CO_2 produced to O_2 consumed is known as respiratory quotient (RQ).

Food	Kcal/gm^{-1}	$lt\ O_2 gm^{-1}$	Kcal $lt\ O_2^{-1}$	RQ
Carbohydrate	4.2	0.84	5.0	1.0
Fat	9.4	2.0	4.7	0.71
Protein (urea)	4.3	0.96	4.5	0.81
Protein (uric acid)	4.25	0.97	4.4	0.74

(After: Schmidt – Nielson, 1972).

Example 4:

Compute the food requirement of a sheep of 50 kg in an environment with T_{eq} of 0°C with wind speed of 2ms^{-1}, diameter 40 cm, r_{Hb} of 1500 sm^{-1}, assuming LE = 0.1M and body temperature of 39°C.

Solution:

$$r_{Ha} = 0.7 \times 307\sqrt{\frac{d}{u}}$$

$$r_{Ha} = 0.7 \times 307\left(\frac{0.4}{2}\right)^{1\backslash 2} = 96 \text{ s m}^{-1}$$

r_r at °C = 260 s m^{-1},

$$r_{eq} = \frac{96 \times 260}{96 + 260} = 70.1 \text{ s m}^{-1},$$

r_{Hb} = 1500 s m^{-1}.

Using the expression

$$M - LE = \frac{\rho c_p (T_b - T_{eq})}{r_{Hb} + r_{eq}}$$

$$M - 0.1M = \frac{1200(39-0)}{1500+70.1}$$

$$.9M = \frac{1200 \times 39}{1570.1}$$

$$M = 33.1\ Wm^{-2}$$

Body area of sheep of 50 kg $= 0.1m^{2/3}$

$= 0.1\ (50)^{2/3}$

$= 1.36\ m^2$

Therefore energy requirement of sheep for thermoregulation $= 1.36 \times 33.1 = 45W$

Assuming caloric content of glucose as 15.7 MJ/kg, a kilogram

of glucose would be sufficient for $= \dfrac{15.7 \times 10^6}{45}$ or 4 days

6.4.5 Stored energy utilization

Animals also experience short periods of intense activity such as running, chasing, climbing hills etc. during which stored energy is utilized. Stored energy can be expressed as

$$S = \frac{V}{A}\rho_b c_{pb}\frac{dT_b}{dt} \qquad \text{. . . (6.27)}$$

where, V and A are volume and surface area of animal's body, ρ_b and c_{pb} are body density and specific heat.

In such a transient state the energy balance becomes important where sensible and radiant heat loss from animal takes place

$$M - LE = \frac{\rho c_p (T_b - T_{eq})}{r_{Hb} + r_{eq}} = \frac{V}{A}\rho_b c_{pb}\frac{dT_b}{dt} \qquad \text{. . . (6.28)}$$

Homeotherm animals control the body temperature to satisfy this expression, whereas for poikilotherms M – LE is zero, T_b is constant, and expression reduces to

$$\frac{dT_b}{dt} = -\frac{A\rho c_p (T_b - T_{eq})}{(r_{Hb} + r_{eq})V\rho_b c_{pb}}$$

$$\frac{dT_b}{(T_b - T_{eq})} = -\frac{A\rho c_p}{(r_{Hb} + r_{eq})V\rho_b c_{pb}}dt$$

$$\int_{T_{b_1}}^{T_{b2}} \frac{dT_b}{(T_b - T_{eq})} = \int_0^t \frac{A\rho c_p}{V(r_{Hb} + r_{eq})\rho_b c_{pb}}\, dt = -\int_0^t \frac{dt}{\Gamma}$$

where, Γ is called time constant of the animal.

$$\left|\text{Ln}\left(T_b - T_{eq}\right)\right|_{T_{b_1}}^{T_{b_2}} = -\frac{t}{\Gamma}$$

$$\text{Ln}\left(T_{b_2} - T_{eq}\right) - \text{Ln}\left(T_{b_1} - T_{eq}\right) = -\frac{t}{\Gamma}$$

$$\text{Ln}\frac{\left(T_{b_2} - T_{eq}\right)}{\left(T_{b_1} - T_{eq}\right)} = -\frac{t}{\Gamma}$$

$$\frac{T_{b_2} - T_{eq}}{T_{b_1} - T_{eq}} = e^{-\frac{t}{\Gamma}} \qquad \text{. . . (6.31)}$$

This equation will quantify the time required for the poikilotherm animal to change from T_{b_1} to T_{b_2} when T_{eq} is changed during transient state activity. Γ is index of response time and animals with higher values of Γ can survive better in outside exposure. Also animals with large V, small A, higher thermal resistance will have higher Γ. On the other hand the animals with less weight will respond quickly to the environmental changes in order to attain equilibrium.

For homeotherms under transient conditions, M, LE and r_{Hb} becomes constant at their extreme values (maxima or minima) so as to provide the final temperature of T_{eq} as

$T_{eq} + \dfrac{(M - LE)(r_{Hb} + r_{eq})}{\rho c_p}$ due to metabolic heat and evaporation.

Example 5:

How much food does a 600 kg animal needs (kg/day) to survive in arctic winter with an average T_a of –20°C and wind speed of 3m s^{-1}. T_{eq} varies day and night and its average value is T_a. Diameter of animal 1m., r_{Hb} is 1200 sm^{-1} and LE is 0.1M.

Solution:

Determining the thermal resistances

$$r_{Ha} = 0.7 \times 307\sqrt{\frac{d}{u}}$$

$$r_{Ha} = 0.7 \times 307\sqrt{\frac{1}{3}} = 123.5 \text{ s m}^{-1}$$

$$r_r = \frac{\rho c_p}{4 \in \sigma T_a^{\,3}} = \frac{1200}{4 \times 0.98 \times 5.67 \times 10^{-8} \times (253)^4} = 333.4 \text{ s m}^{-1}.$$

$$r_{eq} = \frac{r_{Ha}\, r_r}{r_{Ha} + r_r} = \frac{123.5 \times 333.4}{123.5 + 333.4} = 90.1 \text{ s m}^{-1}.$$

Using energy budget equation

$$M - LE = \frac{\rho c_p (T_b - T_{eq})}{r_{Hb} + r_{eq}}$$

$$M - 0.1M = \frac{1200(39 + 20)}{1200 + 90.1}$$

$$M = \frac{1200 \times 59}{1290.1 \times .9}$$

$$M = 61\ Wm^{-2}$$

Surface of the animal (A) is given by

$= 0.1\ m^{2/3}$

$A = 0.1\ (600)^{2/3} = 7.13\ m^2$.

Therefore energy requirement of animal = M × A

$= 61 \times 7.13$

$= 434.8$ W

Assuming the caloric content of glucose as 15.7 MJ/kg,

Therefore for one day the food requirement will be $\frac{434.8 \times 60 \times 60 \times 24}{15.7 \times 10^6} = 2.39$ Kg of glucose

Example 6:

Determine the equivalent temperature of a sunbather on a beach at noon on a clear day with T_a of 28°C, wind speed of 2.5 s m^{-1}, absorption coefficient of 0.75. Height of man is 1.65m, radius of man is 0.117m, R_s is 1000 Wm^{-2}, R_{sd} is 250 Wm^{-2}.

Solution:

Equivalent temperature (T_{eq}) expression is

$$T_{eq} = T_a + \frac{r_{eq}(R_{abs} - \in \sigma T_a^{\ 4})}{\rho c_p}$$

$$R_{ab} = a_s . \frac{A_p}{A} R_{sp} + a_s R_{sd} + \in_s \sigma T_a^{\ 4} \quad \text{(No clothing)}$$

$$R_{ab} = 0.75 \times 0.297 \times 1000 + 250 \times 0.75 + 0.98 \times 5.67 \times 10^{-8} \times (301)^4$$

$$R_{ab} = 222.75 + 187.50 + 456.1$$

$$R_{ab} = 866.36\ Wm^{-2}$$

$$R_{eq} = \frac{r_{Ha}\ r_r}{r_{Ha} + r_r} \qquad \left[\frac{A_h}{A} = \frac{x}{\pi(x+1)} \text{where } x = \frac{h}{r} = 14.1\right] = \frac{14.1 \times 7}{22 \times 15.1} = 0.297$$

r_r at 28°C = 194 s m^{-1}

$$r_{Ha} = 0.7 \times 307 \times \sqrt{\frac{d}{u}}$$

$$= 0.7 \times 307 \times \left(\frac{2.34}{2.5}\right)^{½}$$

$$r_{Ha} = 65.75 \text{ s m}^{-1}$$

$$r_{eq} - \frac{65.75 \times 194}{259.75} = 49.1 \text{ s m}^{-1}.$$

Therefore $T_{eq} = 28 + \frac{49.1}{1200} [866.36 - 0.98 \times 5.67 \times 10^{-8} \times (301)^4]$

$$= 28 + \frac{49.1 \times 410.26}{1200}$$

$$= 28 + 16.79 = 44.8°C.$$

Example 7:

Calculate respiratory heat loss in a cow with use of following data:

ρ_{ae} = 44gm^{-3}
ρ_{ai} = 7 gm^{-3}
ρ_{oi} = 260 gm^{-3}
ρ_{oe} = 200g m^{-3}
M = 100 Wm^{-2}
L = 2.4 × 10^6 J kg^{-1}
Γ = 1.5 × 10^6 J kg^{-1}.

Solution:

$$\frac{LE_R}{M} = \frac{(\rho_{ae} - \rho_{ai})L}{(\rho_{oi} - \rho_{oe})\Gamma}$$

$$= \frac{(44-7)2.4 \times 10^6}{(260-200)15 \times 10^6} - \frac{37 \times 2.4}{60 \times 15} = 0.098$$

$$LE_R = 0.098 \times M$$

$$= 0.098 \times 100 = 9.8 \text{ Wm}^{-2}.$$

Example 8:

Compute the respiratory evaporation for a house sparrow at 20°C with use fo following data:

Temperature of expired moist air = 18°C
ρ_{ae} = 15 gm^{-3}
ρ_{ai} = 7 gm^{-3}
ρ_{oi} = 260 gm^{-3}
ρ_{oe} = 200 m^{-3}
M = 100 Wm^{-2}
L = 2.4 × 10^{6} J kg^{-1}
Γ = 1.5 × 10^{6} J kg^{-1}.

Solution:

$$\frac{LE_R}{M} = \frac{(\rho_{ae} - \rho_{ai})L}{(\rho_{oi} - \rho_{oe})\Gamma}$$

$$= \frac{(15-7)2.4\times10^6}{(260-200)15\times10^6} = \frac{8\times2.4}{60\times15} = 0.021$$

$$LE_R = 0.021 \times 1 = 0.021 \text{ Wm}^{-2}.$$

Example 9:

Assess the rate of skin surface water loss in a camel with use of following data:
ρ_a = 7 gm^{-3}
ρ_{ss} = 44 gm^{-3}
r_{as} = 13000 sm^{-1}.

Solution:

L = 2440 J g^{-1}

$$LE_S = \frac{2440\times(44-7)}{13000}$$ (neglecting r_c and r_a as compared to r_s)

$$LE_S = \frac{2440\times37}{13000} = 6.94 \text{ Wm}^{-2}.$$

Example 10:

Compute the thermal resistance offered by 1 cm thick layer of cotton clotting and calm air over human body surface (Given k_a = 0.0258 Wm^{-1} °K^{-1}, k_c = 0.59 Wm^{-1} °K^{-1}, p_{aCp} = 1200 J m^{-3} °K^{-1})

Solution:

Thermal conductivity of cotton cloth (K_{cl}) = 0.059 Wm^{-1} °K^{-1}

$$r_{hc} = \frac{l}{D_h} = \frac{l\,\rho c_p}{k_{cl}}$$

$$= \frac{1 \times 1200}{0.059 \times 100} = 0.203 \times 10^3 \text{ sm}^{-1}$$

$r_{hc} = 203 \text{ sm}^{-1}$.

Thermal resistance of calm air of 1cm thickness

$$r_{ha} = \frac{l\,\rho c_p}{k}$$

$$= \frac{1 \times 1200}{100 \times 0.0258} = 0.0465 \times 10^4$$

$r_{ha} = 465 \text{ s m}^{-1}$.

Calm air offers more thermal resistance than 1 cm cotton layer in reducing heat loss.

Example 11:

Determine the coat thermal resistance for an animal having 4 cm thick coat with negligible free convection. Given that radiative conductivity of 0.02 $Wm^{-1}\,°K^{-1}$.

Solution:

Coat conductivity comprises of conductivities of still air, radiation and free convection. Therefore coat thermal conductivity will be

$k_c = 0.0258 + 0.02 = 0.0548 \text{ Wm}^{-1}\,°K^{-1}$

Thermal resistance of coat (r_{hc}) $= \dfrac{4 \times 1200}{0.0458 \times 100} = 1048 \text{ s m}^{-1}$.

Example 12:

Determine the equivalent temperature for a person standing in an open environment at full sun with wind speed of 2m s^{-1} and temperature of 35°C. Shadow area ratio is 0.3, a_s is 0.75 and e is 0.98, R_{sp} is 1100 Wm^{-2} and R_{sd} is 300 Wm^{-2}, diameter of man is 0.18 m.

Solution:

$$R_{ab} = a_s \times 0.3 \times R_{sp} + a_s R_{sd} + \epsilon_s \sigma T_a^4$$
$$= 0.75 \times 0.3 \times 1100 + 0.75 \times 300 + 0.98 \times 5.67 \times 10^{-8} \times (308)4$$
$$= 247.5 + 225.0 + 500.0$$

$R_{ab} = 972.5 \text{ Wm}^{-2}$

$r_{Ha} = 0.7 \times 307\sqrt{\dfrac{d}{u}}$ (For outdoor resistance is about 70 per cent of values under laminar flow)

$$r_{Ha} = 0.7 \times 307\left(\frac{0.18}{2}\right)^{½} = 64.5 \text{ s m}^{-1}.$$

$$r_r \text{ at } 35°C = \frac{\rho c_p}{4 \in \sigma T_a^{\,3}} = 181 \text{ s m}^{-1}.$$

$$T_e = \frac{r_{Ha}\, r_r}{r_{Ha} + r_r} = \frac{64.5 \times 181}{64.5 + 181} = \frac{64.5 \times 181}{245.5} = 47.5 \text{ s m}^{-1}$$

$$T_{eq} = T_a + \frac{r_e(R_{ab} - 4 \in \sigma T_a^{\,4})}{\rho c_p}$$

$$= 35 + \frac{47.5}{1200}(972.5 - 500)$$

$$= 35 + \frac{472.5 \times 47.5}{1200}$$

$$T_{eq} = 35 + 18.7 = 53.7°C$$

with a high equivalent temperature, the man will look for shade or shelter.

Example 13:

If an adult person is doing 1 hours of strenuous sports exercise, sleeps for 10 hours, 6 hours working at desk, and remaining resting awake. Assess the daily food intake required to maintain the body weight of 77 kg.

Solution:

Daily Activity	Hours	Metabolic Heat Production (M) (W/m^2)	Food Intake Required per hr (KJ)	Daily Food Intake (KJ)
Sports	1	600	4284.0	4284.0
Sleep	10	50	357.3	3570.0
Desk work	6	95	678.3	4069.8
Resting awake	7	60	428.4	2998.8
	24 hrs			14922.6

Daily food intake required = 14922.6 KJ

Knowing the caloric content of different food items, the desired amount can be determined to maintain the body weight.

Example 14:

Determine the lowest temperature at which a human can survive with use of following data at different activity levels and use of clothings.

Diameter of human = 0.22 m, wind speed = $3 ms^{-1}$.
$LE_R = 0.1$ M,
$LE_S = 9\ Wm^{-2}$
$T_b = 39°C$
$T_a = 0°C$.

Solution:

The lowest equivalent temperature is given by

$$M - LE = \frac{\rho c_p}{r_{Hb} + r_{eq}}(T_b - T_{eq})$$

r_r at 0°C = 260 s m^{-1}

$$r_{Ha} = 0.7 \times 307\left(\frac{0.22}{3}\right)^{½} = 58.2 \text{ s m}^{-1}$$

$$r_{eq} = \frac{58.2 \times 260}{318.2} = 47.5 \text{ s m}^{-1}$$

Therefore

$$M - 0.1M = \frac{1200}{(r_{Hb} + 47.5)} = 39 - T_{eq}$$

$$T_{eq} = 39 - \frac{0.9M}{1200}(r_{Hb} + 47.5)$$

(i) For no clothing, r_{Hb} = 90 s m^{-1}

$$T_{eq} = 39 - \frac{0.9M}{1200}(90 + 47.5) = 39 - \frac{0.9M \times 137.5}{1200}$$

$$T_{eq} = 39 - 0.103 \text{ M}$$

(ii) For a wool business suit, r_{Hb} = 300 s m^{-1}.

$$T_{eq} = 39 - \frac{0.9M}{1200}(347.5)$$

$$T_{eq} = 39 - 0.26 \text{ M}$$

(iii) For a thick winter clothing, r_{Hb} = 1000 s m^{-1}

$$T_{eq} = 39 - \frac{0.9M}{1200}(1047.5)$$

$$T_{eq} = 39 - 0.7856 \text{ M}$$

T_{eq} (°C) at Different Activity Levels

Clothing Resistance (r_{Hb})	Sleeping (M = 50)	Resting (60)	Standing (90)	Desk Working (95)	Light Work (120)	Moderate Work (180)	Heavy Work (250)	Sustained Hard Work (350)
No clothing (90 sm^{-1})	33.9	32.8	29.7	29.2	26.6	20.5	13.3	3.0
Wool business Suit (300 sm^{-1})	26.0	23.4	15.6	14.3	7.8	-7.8	-26.0	-52.0
Thick winter Clothing (1000 sm^{-1})	-0.28	-8.1	-31.7	-35.6	-55.3	-102.4	-157.4	-236.0

Example 15:

A person wants to undertake a climbing activity with metabolic heat production of 350 Wm^{-2} during noontime with clear sunshine at T_{eq} temperature of -20°C. Determine the clothing insulation required in this environment.

Diameter of man = 0.22 m
$LE_R = 0.1m$
$LE_S = 5\ Wm^{-2}$
$T_b = 37°C$
$T_{eq} = 20°C$
$u = 3\ m\ s^{-1}$.

Solution:

$$r_{Ha} = 0.7 \times 307\sqrt{\frac{0.22}{3}} = 58.2\ s\ m^{-1}$$

r_r at 0°C = 260 s m^{-1}

$$r_{eq} = \frac{260 \times 58.2}{318.2} = 47.55\ s\ m^{-1}$$

$$LE = LE_R + LE_s = 0.1\ M + 5$$

$$M - LE = \frac{\rho c_p}{r_{Hb} + r_{eq}}(T_b - T_{eq})$$

$$M - 0.1M - 5 = \frac{1200(37 + 20)}{r_{Hb} + 47.55}$$

$$0.9 \times 350 - 5 = \frac{1200 \times 57}{r_{Hb} + 47.55}$$

$$315.0 - 5 = \frac{1200 \times 57}{r_{Hb} + 47.55}$$

$$310 = \frac{1200 \times 57}{r_{Hb} + 47.55}$$

$$r_{Hb} + 47.55 = 220.64$$

$$r_{Hb} = 220.64 - 47.55$$

$$r_{Hb} = 173.1 \text{ s m}^{-1}$$

Therefore clothing plus tissue resistance required is 173.1 s m^{-1}.

6.4.6 Survival in hot environment

During hot environments heat balance is maintained with increase of metabolic or equivalent temperature with dissipation of heat keeping the body temperature constant. With dissipation of heat skin temperature increases, but is maintained with increase of sweating rate. However, we cannot allow increase of skin temperature to create uncomfortness beyond 36°C in human or 39°C in homeotherms. Therefore the sustainability of maximum environmental heat is related to sweating rate depending on acclimatization and duration of exposure. A commonly used value of maximum sweat rate during heat stress for human is 380 Wm^{-2} for dry skin conditions or LE_S values when skin is wet (or $r_{vs} = 0$) whichever is smaller.

Robinson et al (1976) used energy balance equation under heat stress conditions expressed as

$$\frac{\rho c_p (T_b - T_{eq})}{(r_{Hb} + r_{eq})} = M - LE_R - LE_S \frac{(r_{Hc} + r_{eq})}{(r_{Hb} + r_{eq})} \qquad \ldots (6.32)$$

when LE_S approaches zero or $r_{Hc} >> r_{Ht}$, expression reduces to heat loss expression in a blackbody cavity with temperature T_{eq}. Impact of clothing can be assessed with this equation which can be tolerated by a person. Normally LE_s comprises of clothing, boundary layer resistance and atmospheric vapour density. The boundary layer resistance for vapour transport is given by i.e., for air at 100 kpa at 20°C.

$$r_{va} = 283\sqrt{\frac{d}{u}} \qquad \ldots (6.33)$$

Similarly for CO_2 and O_2 gases

$$r_{ca} = 395\sqrt{\frac{d}{u}} \qquad \ldots (6.34)$$

$$r_{oa} = 319\sqrt{\frac{d}{u}} \qquad \ldots (6.35)$$

In clothings, heat ad water vapour are diffused in same manner, therefore $r_{vc} \sim r_{Hc}$ under free or forced convections.

Example 16:

Determine the effect of clothing on maximum equivalent temperature tolerated by a person with the use of following data:

$\rho_{va} = 5\ g\ m^{-3}$
$T_b = 38°C$
$r_{eq} = 40\ s\ m^{-1}$
$r_{va} = 47\ s\ m^{-1}$
$r_{vs} = 200\ s\ m^{-1}$
$r_{Ht} = 15\ s\ m^{-1}$
$r_{Hc} = r_{vc}$
$LE_R = 0.1\ M$
$T_{eq} = ?$

Solution:

Using the equation (6.32) under stress conditions in hot environment

$$\frac{\rho c_p(T_b - T_{eq})}{(r_{Hb} + r_{eq})} = M - LE_R - LE_S \frac{(r_{Hc} + r_{eq})}{(r_{Hb} + r_{eq})}$$

$$\frac{\rho c_p(T_b - T_{eq})}{(r_{Hb} + r_{eq})} = M - LE_R - L\frac{(\rho_{sv} - \rho_{va})}{(r_{vs} + r_{Hc} + r_{va})}\frac{(r_{Hc} + r_{eq})}{(r_{Hb} + r_{eq})}$$

Substituting the different values, we get

$$\frac{1200(38 - T_{eq})}{55} = M - 0.1M - \frac{2420(44-5)}{(200+47+r_{Hc})}\frac{(r_{Hc}+40)}{(15+40)}$$

$$\frac{1200(38 - T_{eq})}{55} = 0.9M - \frac{2420 \times 39(r_{Hc}+40)}{55(247+r_{Hc})}$$

$$T_{eq} = 38 - \frac{0.9 \times 55M}{1200} + \frac{2420 \times 39 \times 55(r_{Hc}+40)}{55 \times 1200(r_{Hc}+247)}$$

$$T_{eq} = 38 - 4.125 + 78.65\frac{(r_{Hc}+40)}{(r_{Hc}+247)}$$ (For $M = 100\ Wm^{-2}$ for light work)

$$T_{eq} = 33.875 + 78.65\frac{(r_{Hc}+40)}{(r_{Hc}+247)}$$ (For $M = 100\ Wm^{-2}$)

$$T_{eq} = 29.75 + 78.65\frac{(r_{Hc}+40)}{(r_{Hc}+247)}$$ (For $M = 200\ Wm^{-2}$)

T_{eq} (°C)	r_{Hc}(s m^{-1}) When M is 100 Wm^{-2} (Light Work)	T_{eq} (°C)	r_{Hc} (s m^{-1}) When M is 200 Wm^{-2} (Moderate to Heavy Work)
46.6	0	42.5	0
65.6	100	61.5	100
76	200	72.0	200
82.8	300	78.7	300
87.4	400	83.3	400

6.4.7 Equivalent wet blackbody temperature (T_{weq})

In order to derive a stress index parameter with use of environmental variables energy budget equations is expressed in terms of physiologic and environmental variables as equation (6.32).

Substituting $LE_S = \dfrac{L(e_{vs} - e_{va})}{(r_{vs} + r_{vc} + r_{va})}$

and $T_{eq} = T_a + \dfrac{r_{eq}}{\rho c_p}(R_{ab} - \in \sigma T_a^{\,4})$ in equation (6.32) we get . . . (6.35)

$$e_{vs} - e_{va} = (e_{vs} - e_{sa}) + (e_{sa} - e_{va})$$
$$e_{vs} - e_{va} = \Delta(T_b - T_a) + (e_{sa} - e_{va})$$

where Δ is slope of s.v.p. curve at skin temperature which is closer to body temperature (T_b).

$$\frac{\rho c_p}{(r_{Hb} + r_{eq})}\left[T_b - T_a - \frac{r_{eq}}{\rho c_p}(R_{ab} - \in \sigma T_a^{\,4})\right] = (M - LE_R) - \frac{L[\Delta(T_b - T_a) + (e_{sa} - e_{va})]}{(r_{vs} + r_{vc} + r_{va})}\frac{(r_{Hc} + r_{eq})}{(r_{Hb} + r_{eq})} \quad \text{. . . (6.36)}$$

Put $r_v = (r_{vs} + r_{vc} + r_{va})$

$$\gamma = \frac{\rho c_p}{L}$$

$$\gamma^* = \frac{\gamma_{rv}}{(r_{Hc} + r_{eq})}$$

$$(T_b - T_a) - \frac{r_{eq}}{\rho c_p}(R_{ab} - \in \sigma T_a^{\,4}) = \frac{(M - LE_R)(r_{Hb} + r_{eq})}{\rho c_p} - \frac{[\Delta(T_b - T_a) + (e_{sa} - e_{va})](r_{Hc} + r_{eq})L}{r_v\, \rho c_p}$$

$$(T_b - T_a) + \frac{\Delta}{\gamma^*}(T_b - T_a) = \frac{r_{eq}}{\rho c_p}(R_{ab} - \in \sigma T_a^{\,4}) + \frac{(M - LE_R)(r_{Hb} + r_{eq})}{\rho c_p} - \frac{(e_{sa} - e_{va})}{\gamma^*}$$

$$(1 + \frac{\Delta}{\gamma^*})(T_b - T_a) - \frac{r_{eq}}{\rho c_p}(R_{ab} - \in \sigma T_a^{\,4}) = \frac{(M - LE_R)(r_{Hb} + r_{eq})}{\rho c_p} - \frac{(e_{sa} - e_{va})}{\gamma^*} \quad \text{. . . (6.37)}$$

when a person is in an environment with saturated blackbody cavity, then

$T_a = T_{weq}$ and $\rho_{va} = \rho_{sa}$, therefore energy budget equation (6.37) reduces to

$$(1+\frac{\Delta}{\gamma *})(T_b - T_{weq}) = \frac{(M-LE_R)(r_{Hb}+r_{eq})}{\rho c_p} \quad \ldots (6.38)$$

Subtracting equation (6.38) from (6.37), we get

$$(1+\frac{\Delta}{\gamma *})(T_b - T_a) - (1+\frac{\Delta}{\gamma *})(T_b - T_{weq}) - \frac{r_{eq}(R_{ab} - \in\sigma T_a^4)}{\rho c_p} + \frac{(e_{vs} - e_{va})}{\gamma *} = 0$$

$$(1+\frac{\Delta}{\gamma *})(T_{weq} - T_a) = \frac{r_{eq}(R_{ab} - \in\sigma T_a^4)}{\rho c_p} - \frac{(e_{vs} - e_{va})}{\gamma *}$$

$$T_{weq} = T_a + \frac{\gamma *}{\Delta + \gamma *}\left\{\frac{r_{eq}(R_{ab} - \in\sigma T_a^4)}{\rho c_p} - \frac{(e_{vs} - e_{va})}{\gamma *}\right\} \quad \ldots (6.39)$$

Equation 6.39 combines are the variables into a single T_{weq} parameter representing heat stress index. As with dry equivalent temperature, T_{weq} represents the temperature of a poikilotherm (M – ME = 0) exposed to the same environment as a person with identical radiative convective and latent heat exchange properties.It is measured experimentally with use of copper sphere covered with black moistened cloth, filled with water and is used to relate to human comfort in hot environments called wet globe temperature which is related to T_{weq} to quantify human comforts.

Case I: When r_{vs} becomes large under dry conditions, γ^* increases and expression reduces to dry equivalent temperature.

Case II: When there is air movement and skin surface is wet, then clothing vapour and heat transfer resistances are almost equal, and $\gamma^* = \gamma$. Therefore $\frac{\gamma}{\Delta + \gamma}$ becomes 0.17 and second term is quite small, therefore $T_{weq} \simeq T_a$.

Case III: When vapour and heat transfer resistances are equal and skin surface is wet, we can determine the maximum equivalent temperature that can be tolerated for various body heat transfer resistances and the metabolic rates.

Example 17:

Determine the maximum equivalent temperature tolerated by a man at 0, 100, 200, 300, 400 and 500 Wm^{-2} metablic rates and body heat transfer resistances of (r_{Hb}) 15, 100 and 200 s m^{-1} assuming that skin is wet, and vapour and heat transfer resistances are equal. Given that T_b = 37°C, r_{eq} is 40 sm^{-1}, LE_R is 0.06m.

Solution:

When skin is wet, using energy budget equation in a saturated blackbody cavity i.e., T_{weq} = T_a, $e_{va} = e_{sa}$ and $r_{vc} = r_{Hc}$ then $\gamma^* = \gamma$, $\frac{\Delta}{\Delta + \gamma} = 0.17$.

Therefore using equation (6.38)

$$\frac{\gamma^*}{\Delta + \gamma^*}(M - LE_R) = \frac{\rho c_p (T_b - T_{weq})}{(r_{Hb} + r_{eq})}$$

$$0.17(M - 0.06M) = \frac{1200(37 - T_{weq})}{r_{Hb} + 40}$$

$$\frac{0.17 \times 0.94M(r_{Hb} + 40)}{1200} = 37 - T_{weq}$$

$T_{weq} = 37 - 0\ 0.001332\ M\ (r_{Hb} + 40)$
$T_{weq} = 37 - 0.005328\ M$ (for $r_{Hb} = 0$)
$T_{weq} = 37 - 0.018648\ M$ (for $r_{Hb} = 100$ s m^{-1})
$T_{weq} = 37 - 0.032\ M$ (for $r_{Hb} = 200$ s m^{-1}).

Equivalent Wet Temperature (°C)	**M(Wm-²)**					
	0	**100**	**200**	**300**	**400**	**500**
T_{weq}	37.0	36.5	35.9	35.4	34.9	34.3
T_{weq}	37.0	35.1	33.3	31.4	29.5	27.7
T_{weq}	37.0	33.8	30.6	27.4	24.2	21.0

The wet equivalent temperature is always lower than the body temperature to maintain energy balance under such conditions.

Example 18:

Determine the comfortness of a human being using energy budget equation when inside wall temperature is 15°C lower than air temperature of 22°C. Given that r_{eq} is 100 s m^{-1}.

Solution:

Using energy budget equation

$$T_{eq} = T_a + \frac{r_e}{\rho c_p}\{R_{ab} - \epsilon\sigma T_a^4\}$$

$$= 22 + \frac{100}{1200}\{\epsilon\sigma T_w^4 - \epsilon\sigma T_a^4\}$$

$$= 22 + \frac{1}{12}\{391 - 430\} \qquad \text{assuming } \in = 1$$

$$= 22 - \frac{39}{12} = 22 - 3.25$$

T_{eq} = 18.8°C.

This is an uncomfortable temperature for a person resting under the wall cover. Therefore wall temperature should be increased artificially so as to provide a comfortable temperature of 37°C for which a radiant heat of R_{ab} = 610 Wm^{-2} will be desired.

6.5 Human Comfort

Every person spends half of its life in the indoor microclimate of the building which is different from the external environment conditions. Therefore interrelationship between internal and external weather conditions is desired. It is quite common that with the arrival of northwesterly cold winds with snowfall, rains and atmospheric turbulence, the inner surface of walls and windows suddenly have a drop in temperature with the change of weather conditions outside the buildings. Reverse situation is experienced when outside environment is experienced with the flow of hot summer winds. Relative humidity plays an important role. In winter season the cold air masses are undesirable indoors, but any such wind during summer season are most welcomed due to low absolute humidity indoors. But lower relative humidity in artificially controlled heated buildings may cause an unpleasant dryness of throat to contribute in multiplication of viruses and bacteria, whereas very high humidity leads to dampness of clothes and walls in stimulating fungus attacks. Therefore a humidity level around 50 per cent with a slight higher absolute humidity inside is required if ventilation rate are adequate. Therefore comfort indices have been devised by several workers for winter and summer conditions based on meteorological variables and physiological responses to thermal stresses and metabolic heat production. Since it is difficult to quantify the role of individual weather variables among radiation, temperature, wind, global radiation temperature, and heat balance over body surfaces; empirical relationships have been developed to seek the resultant effect of all these variables in a given environment to seek the level of human comforts in terms of indices.

In man, the excessive heat in the body is reduced by stimuli from the hypothalamus – a principal heat regulatory mechanism through pores in skin and evaporation of sensible (sweat) and insensible perspiration. The energy exchange between human and the surrounding environment can be expressed as

$$\underbrace{\left[a_s R_{si} A_{sd} + a_s R_{se} A_{se}\right]}_{\text{Shortwave}} + \underbrace{\left[\in\sigma(T_r^4 - T_s^4)A_{se}\right]}_{\text{Longwave}} + \underbrace{\left[\frac{LK_v(e_m - e_a)}{r_m + r_a}\right]}_{\text{Evaporative}} + \underbrace{\left[h_c A_{se}(T_a - T_s)\right]}_{\text{Convective}} + M_b + M_{HP} = 0 \qquad \ldots (6.40)$$

where, T_r is mean radiant temperature of environment, $\in$ is $\in_b \in_e$ product of emissivity of body and environment.

Body area of an adult is approximated with the expression of the form as A = 0.026m, where m is mass in kgm and A in m^2.

Olgyay (1960) developed a bioclimatic chart to translate the climatic data for comfort zone for architectural applications based on relation of wind with high temperature, air movement on vapour pressure, effect of evaporative cooling, and radiation effect on low temperatures as shown in Fig. 6.7. The comfort zone lies in the middle with climatic boundaries needed for comfort feeling at different temperatures for a inhabitant for a moderate climatic zone. Brooks (1950) showed that the British comfort zones lie around 14.5 to 20.5°C, in U.S.A. from 20.5 to 26.6°C and in tropics from 23.3 to 29.4°C between 30 to 70 per cent of relative humidity. Fig. 6.8 presents temperature-relative humidity relationships with summer comfort zone in middle, and a little lower winter comfort zone is indicated.

At higher temperatures, the wind effects can retrace the comfort feelings with nearly parallel lines presenting the upper limit of comfort perimeter with desired wind velocities. Wind force counteracts the high temperatures. Also the vapour pressure is counteracted by the winds above 60 per cent relative humidity level. At 15 mm of Hg vapour pressure a depressed feeling is noticed. Siple (1960) stated that for over 15 mm vapour pressure, each additional mm of pressure can be counteracted with 1 mph wind effect (88 ft/min) presenting boundaries from 15 to 23 mm vapour pressure counteracted by wind velocities from calm (20 to 30 ft/min) to 700 ft/min air movements to provide a reduction of 8°F. At low relative humidities, evaporative cooling reduces high temperatures as shown with dotted line in Fig. 6.8 with required amount of grains of moisture per pound air which brings down the temperature to upper comfortable perimeter. Higher dotted line shows the limit of the combined effect of wind and evaporative cooling. At higher temperatures, there are curves for indicating the limit of work of moderate intensity and probability of heat stroke as the danger zone.

At the lower perimeter of comfort zone is the line from which shading is needed towards higher temperatures, and conversely at lower temperature radiation is needed to counteract the cold temperature feelings. Radiations are expressed in BTU to represent outside conditions of mean radiant temperature (MRT) values which can bring comfort zone feeling in a comfort zone range. Olgyay (1960) reported that a 50 BTU of sun's radiation can counteract 3.85 °F drop in dry bulb temperature as the amount of radiation needed as a relative index depending on exposure and the possible absorption of the radiant energy.

The results of the above processes can be interpreted on a yearly time table. Time areas for various needs of radiation, shading and winds can be described indicating underheated period, over heated periods, and wind required for design conditions with extreme temperature conditions.

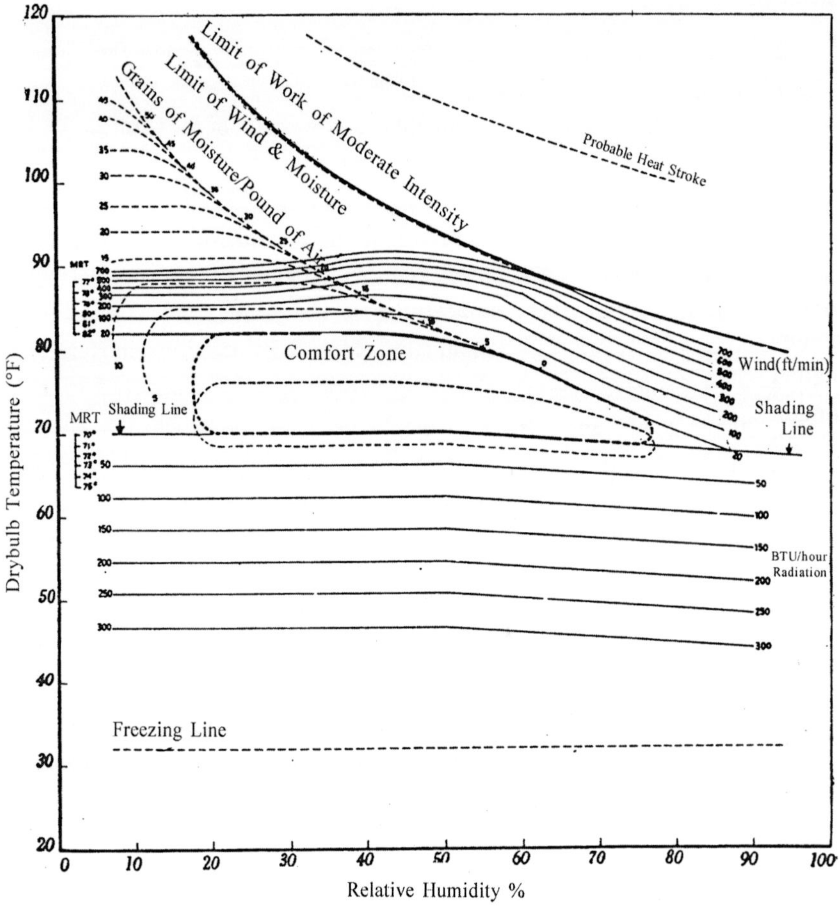

Fig. 6.7: Bioclimatic chart for a moderate zone inhabitants (After: Olgyay, 1960)

It is useful to define 'h_r' a radiant transfer coefficient based on T_r values. Net longwave radiative transfer can be expressed similar to net convective transfer processes of $h_c A_{se} (T_a - T_s)$ as

$$\in\sigma(T_r^4 - T_s^4) \propto (T_r - T_s)$$

$$\in\sigma (T_r^2 + T_s^2) (T_r^2 - T_s^2) = h_r(T_r - T_s)$$

$$\in\sigma(T_r^2 + T_s^2) (T_r + T_s) (T_r - T_s) = hr(T_r - T_s)$$

$$\in\sigma(T_r^3 + T_s^3 T_r + T_s^2 + T_s . T_r^2) (T_r - T_s) = h_r(T_r - T_s)$$

$$2\in\sigma(T_r^3 + T_s^3) (T_r - T_s) = hr(T_r - T_s) \quad \text{(approximation)}$$

$$\therefore \quad h_r = 2\in\sigma(T_r^3 + T_s^3) \qquad \text{. . . (6.41)}$$

In cloudiness or night time net shortwave, evaporation component are negligible with thermal equilibrium ($M_{Hp} = 0$) equation (6.39) reduces to

$$\in\sigma(T_r^4 - T_s^4)\,A_{se} + h_c A_{se}(T_a - T_s) + M_b = 0$$
$$h_r A_{se}(T_r - T_s) + h_c A_{se}(T_a - T_s) + M_b = 0$$
$$h_r A_{se} T_r - hr\,A_{se} T_s + h_c A_{se} T_a - h_c A_{se} T_s + M_b = 0$$
$$A_{se} T_s(h_r + h_c) = M_b + A_{se}(h_r T_r + h_r T_a)$$

$$T_s = \frac{M_b}{A_{se}(h_r + h_c)} + \frac{h_r T_r + h_c T_a}{h_r + h_c} \quad \ldots (6.42)$$

First term is constant, whereas second term is a weighted average of the air temperature and effective radiant temperature of the environment. Neglecting first term T_s is called operative temperature a resultant effect of wind, radiation, air temperature and humidity.

It is difficult to solve the equation (6.42) with use of all the components, therefore it can be used for specific situations at any temperature so that algebric sum of all terms is zero.

Example 19:

Determine the surface temperature and operative temperature of a man of diameter 20 cm standing in a cloudy midday environment using following data

M_b = 50 Kcal m^{-2} hr^{-1} or 58.1 Wm^{-2}

$h_c = h_r$ = 10^{-2} cal cm^{-2} min^{-1} °K^{-1} or 6.97 Wm^{-2} °K^{-1}

T_a = 32°C or 305 °K

T_r = 32.5°C or 305.5°K

Solution:

$$T_s = \frac{M_b}{A_{se}(h_r + h_c)} + \frac{h_r T_r + h_c T_a}{h_r + h_c}$$

$$= \frac{58.1}{6.97 + 6.97} + \frac{6.97 \times 305.5 + 6.97 \times 305}{6.97 + 6.97}$$

$$= 4.17 + 305.25$$

T_s = 309.4°K

T_0 = 305.3°K.

6.5.1 Role of clothing in energy budget of human

Let us express the energy budget equation at the outer surface of clothing where $T_a = T_r$ as

$$a_s A_x R_{st} + A_x(h_c + h_r)(T_a - T_x) + A_x LE + A_x \times \frac{k_{cl}}{L_{cl}}(T_s - T_x) = 0 \quad \ldots (6.43)$$

where, A_x is the area of clothing surface area, R_{st} is total incoming s.w. radiation per unit area per unit time, T_x is temperature of outer surface of clothing, LE is the latent heat transfer per unit area per unit time, k_{cl} is thermal conductivity of clothing and L_{cl} is thinkness of clothing.

Conduction between inner surface of clothing and skin interface, and between body core and skin must attain equilibrium so that

$$\frac{k_{cl}A_s(T_s - T_x)}{L_{cl}} = \frac{k_b A_s(T_c - T_s)}{L_b} \qquad \ldots (6.44)$$

where k_b is effective conductivity of body and L_b is normally taken as 0.1 m, T_c is the core temperature (usually body or rectal temperature). Metabolic and stored heat will determine core temperature expressed as

$$(M_b + M_{HP}) = \frac{k_b}{l_b} A_{se}(T_c - T_s)$$

$$T_s = T_c - \frac{l_b(M_b + M_{HP})}{k_b A_{se}} \qquad \ldots (6.45)$$

Solving equation for $(T_s - T_x)$ using (6.44)

$$(T_s - T_x) = \frac{l_{cl} k_b A_{se}(T_c - T_s)}{k_{cl} l_b A_{se}}$$

$$= \frac{l_{cl} k_b A_{se}}{k_{cl} l_b A_{se}} \frac{l_b(M_b + M_{HP})}{k_b A_{se}}$$

$$T_s - T_x = \frac{l_{cl}(M_b + M_{Hc})}{k_{cl} A_{se}} \qquad \ldots (6.46)$$

Solving equation (6.43) for $(T_c - T_a)$ using equations (6.45) and (6.46)

$$(a_s R_{st} + E) + (h_c + h_r)(T_a - T_x) + \frac{k_{cl}}{l_{cl}}(T_s - T_x) = 0$$

$$\frac{(a_s R_s + E)}{(h_c + h_r)} + (T_a - T_x) + \frac{k_c(T_s - T_x)}{l_c(h_c + h_r)} = 0$$

$$\frac{(a_s R_s + E)}{(h_c + h_r)} + \frac{(M_b + M_{Hc})}{(h_c + h_r)A_{se}} = (T_x - T_a) \quad \text{[using eq. 6.46]}$$

$$T_s - \frac{l_{cl}(M_b + M_{Hc})}{k_{cl} A_{se}} - T_a = \frac{(a_s R_s + E)}{(h_c + h_r)} + \frac{(M_b + M_{Hc})}{(h_c + h_r)A_{se}}$$

(Substituting T_x from equation 6.46)

$$T_c - \frac{l_b(M_b + M_{Hc})}{k_b.A_{se}} - \frac{l_{cl}(M_b + M_{Hc})}{k_{cl}.A_{se}} - T_a = \frac{(a_s R_s + E)}{(h_c + h_r)} + \frac{(M_b + M_{Hc})}{(h_c + h_r)A_{se}}$$

(Substituting T_s from equation 6.45)

$$T_c - T_a = \frac{(M_b + M_{Hc})}{A_{se}}\left\{\frac{l_b}{k_b} + \frac{l_{cl}}{k_{cl}}\right\} + \frac{(a_s R_s + E)}{(h_c + h_r)} + \frac{(M_b + M_{Hc})}{(h_c + h_r)A_{se}}$$

$$T_c - T_a = \frac{(a_s R_s + E)}{(h_c + h_r)} + \frac{(M_b + M_{Hc})}{(h_c + h_r)A_{se}} + \frac{(M_b + M_{Hc})R}{A_{se}} \quad \text{where } R = \left\{\frac{l_b}{k_b} + \frac{l_{cl}}{k_{cl}}\right\}$$

$$T_c - T_a = \frac{(a_s R_s + E)}{(h_c + h_r)} + \frac{(M_b + M_{Hc})}{(h_c + h_r)A_{se}} + \frac{(M_b + M_{Hc})R}{A_{se}} \quad (\text{when } A_s = A_{se})$$

Case I: Hot environment: In hot environment $T_a > T_c$, therefore M_{HC} becomes of larger negative magnitude value than M_b. With a cooling rate E, a large T_a will result in a large negative withdrawal rate and body will have increase in temperature. M terms needs to be expressed in same units as R_s and E i.e., per unit area.

Case II: Cold Environment: In cold environment T_c is much greater than T_a as all terms in the expression are positive. Here clothing plays an important role. Clothing increases the resistance (R) with adequate clothing to maintain T_c in colder conditions. Coat thickness (l_{cl}) and coat colour absorption (a_s) influence strongly the contribution for ($T_c - T_a$) values. Normally a coat of 1 cm thickness leads to a contribution of 19°C than the maxima possible by darkening the coat ($a_s = 1$).

Values of k_{cl} are 0.02, 0.03, 0.10, 10.0 kcal m^{-1} hr^{-1} $°C^{-1}$ for still air, dry wool, wet wool and air in gentle motion (0.167 cal cm^{-1} min^{-1} $°C^{-1}$).

Example 20:

Compute the temperature difference ($T_c - T_a$) for a clothed and unclothed man standing in an open environment using following data.

R_s = 698 Wm^{-2}

h_r = 6 Wm^{-2}

u = 10^4 cm min^{-1} or 1.667 m s^{-1}

k_c = 0.03487 Wcm^{-1} $°C^{-1}$

l_c = 0.01m

M_{HP} = 0

Solar altitude = 30°

Diameter of man = 0.2

l_b = 0.1m

M_b/A_s = 58 Wm^{-2}

$h_c = 6.98\ Wm^{-2}\,°C^{-1}$
$a_s = 0.7$, albedo = 0.3
$k_b = 2.44\ Wm^{-1}\ °C^{-1}$
$E = -23.034\ Wm^{-2}$

Solution:

(a) For a clothed person:

Using the equation

$$T_c - T_a = \frac{(M_b + M_{HP})R}{A_{se}} + \frac{(a_s R_s + E)}{(h_c + h_r)} + \frac{(M_b + M_{HP})}{A_{se}(h_c + h_r)}$$

with a convective transfer coefficient of

$$h_c = 2.2 \times 10^{-3}\ w^{1/3}\ D^{-1/2}$$
$$h_c = 2.2 \times 10^{-3}\ (10^4)^{1/3}(20)^{-1/2}$$
$$= 2.2 \times 10^{-3} \times 21.557 \times 0.2236$$
$$h_c = 0.0106\ \text{cal cm}^{-2}\ \text{min}^{-1}\ °C^{-1}.$$

$$R = \frac{l_b}{k_b} + \frac{l_c}{k_c} = \frac{0.1}{2.44} + \frac{0.01}{0.03487}$$
$$= 0.041 + 0.287 = 0.328\ \text{s m}^{-1}.$$

$$(T_c - T_a) = (58 + 0)\ 0.328 + \frac{\left(\dfrac{0.7 \times 698 \times 0.87 \times 7}{22} - 23.034\right)}{12.98} + \frac{(58 + 0)}{12.98}$$
$$= 19.024 + (135.2 - 23.034)/12.98 + 58/12.98$$
$$= 19.024 + 8.645 + 4.468$$
$$(T_c - T_a) = 32.1\ °C$$

If core temperature is 36°C for a man, then Ta is 3.9°C. Therefore a wool suit will bring in equilibrium with the environmental temperature of 3.9°C.

(b) For a unclothed person:

Thermal resistance (R) = 0.041 s m^{-1}

$$(T_c - T_a) = 58 \times 0.041 + 8.645 + 4.468$$
$$= 2.378 + 8.645 + 4.468$$
$$(T_c - T_a) = 15.5°C$$

Therefore for a core temperature of 36°C, for equilibrium T_a is 20.5°C.

Example 21:

A person is comfortable at 37°C and r_{Ht} of 40 s m^{-1}. Thermal resistance for normal indoor conditions r_{Ha} of 200 s m^{-1}, r_r of 200 s m^{-1} and r_{eq} of 100 s m^{-1} is adequate. If thermal resistance for normal indoor clothing is 100 s m^{-1}, show that

(a) When skin and expired air temperature are 33.8°C, LE = $(2.286 \times 10^{-3}M + 0.31)(37 - \rho_{va})$

(b) Determine the comfortable temperature for a normal metabolic activity (M = 120 Wm^{-2}) when ρ_{va} is 15 gm^{-3}.

Solution:

$$LE = LE_R + LE_S$$

$$LE = \frac{(\rho_{ve} - \rho_{vi})LM}{(\rho_{oi} - \rho_{oe})\Gamma} + \frac{L(\rho_{vs} - \rho_{va})}{(r_{vs} + r_{vc} + r_{va})}$$

$$LE = \frac{(37 - \rho_{va})2.4M}{70 \times 15} + \frac{2430(37 - \rho_{va})}{(7700 + 100 + 100)}$$

$$LE = \left(\frac{2.4M}{70 \times 15} + \frac{2430}{7900}\right)(37 - \rho_{va})$$

$$LE = (2.286 \times 10^{-3} M + 0.31)(37 - \rho_{va})$$

For a normal condition $T_w = T_a$, then $T_{eq} = T_a$,

$$\frac{\rho c_p (T_b - T_{eq})}{r_{Hb} + r_{eq}} = M - LE$$

$$= M - (2.286 \times 10^{-3}M + 0.31)(37 - 15)$$

$$= M - .05M - 6.82$$

Using ρ_{ve} at 33.8°C = 37 gm^{-3}

$\rho_{vi} = \rho_{va}$

$\rho_{oi} = 240$ gm^{-3}

$\rho_{oe} = 210$ gm^{-3}

$r_{vs} = 7700$ s m^{-1}

$r_{vc} = 100$ s m^{-1}

$$\frac{1200}{100 + 40}(37 - T_{eq}) = 0.95M - 6.82$$

$$T_{eq} = 37 - \frac{.95}{1200} \times 140M + \frac{6.82 \times 140}{1200}$$

$$= -0.1108M$$

$$T_{eq} = 37.7957 - 0.1108M$$
$$= 37.7958 - 1108 \times 120$$
$$T_{eq} = 24.5°C.$$

Therefore comfort is maintained at lower temperature when air humidity is increased. A small variation in activity (M value) results in change in comfortable temperature.

Example 22:

Compute the temperature difference ($T_c - T_a$) for an unclothed and clothed person standing on a cloudy day noon time in the open environment using following data.

$R_s = 0.0$	Diameter of man = 0.2 m
$h_r = 6\ Wm^{-2}\ °C^{-1}$	$L_b = 0.1$ m
$u = 1.667\ ms^{-1}$	$M_b/A_s = 58\ Wm^{-2}$
$k_{cl} = 0.03487\ Wm^{-1}\ °C^{-1}$	$h_c = 6.98\ Wm^{-2}\ °C^{-1}$
$L_{cl} = 0.01m$	$a_s = 0.7$
$M_{HP} = 0$	$k_b = 2.44\ Wm^{-1}\ °C^{-1}$
Solar altitude = 90°	E = -23.034 Wm^{-2}.

Solution:

(a) For a clothed person:

$$\text{Thermal resistance} = \frac{l_b}{k_b} + \frac{l_c}{k_c} = \frac{0.1}{2.44} + \frac{0.01}{0.03487}$$
$$= 0.041 + 0.287 = 0.328\ s\ m^{-1}.$$

$$(T_c - T_a) = 58 + 0.328 + \frac{(0 - 23.034)}{12.98} + \frac{58 + 0}{12.98}$$
$$= 19.024 - 1.775 + 4468$$
$$(T_c - T_a) = 21.7°C$$

For a core temperature of 36°C, air temperature needed is 14.3°C.

(b) For an unclothed person:

$$(T_c - T_a) = 58 \times 0.041 - 1.775 + 4.468$$
$$= 2.378 - 1.775 + 4.468$$
$$(T_c - T_a) = 5.1°C$$

For a core temperature of 36°C, air temperature needed for comfort is 30.9°C a value closer to thermal neutrality.

6.6 Comfort Indices

In order to quantify living beings (men or animals) reaction to weather conditions, comfort indices have been used for physical adaptation (Herrington, 1954). These are of two categories-one based on meteorological data only and the other based on physiological informations in addition to meteorological data. The meteorological indices are based on dry (T_d) and wet (T_w) bulb temperatures, effective (Yaglou, 1926), black globe temperatures (T_g) scales (Bedford, 1946). Effective temperature was derived based on dry and wet bulb temperatures as temperature of still air saturated with water vapour which induced the same outstation of warnith as that expressed in a given environment. Minard et al. (1957) used T_d, T_w and T_g values to derive human comfort indices.

McArdle (1947) developed a nomogram for the production of sweat loss by fitting a acclimatized man exposed to heat for 4 hours based on T_d, T_w, wind velocity (u) and rate of heat production to give an excellent physiological index for a human comfort. Belding and Hatch (1955) developed an 'Index of heat stress' using heat balance equation based on T_d, T_g, T_w, u and metabolic heat production for a normal human. Several indices are available in literature based on these parameters. There are two groups of indices – one which delineated safe limits of heat stress for acclimatized men (A1 to A3 in Fig. 6.8) and the other which delineated safe limits for unacclimatized men (UA1 to UA4 in Fig. 6.9) below which effective physical and mental work can be carried out safely for a long period of exposure (Olgyay, 1960). Smith (1955) derived A1 curve for conditions under which men can comfortably and safely do light work (100 – 150 Kcal/hr) with a sweat loss of 2.5 *l.* in 4 hours. Brunt (1943) provided A2 curve for separating very warm conditions from intolerably hot conditions (heat stroke), and Largent and Ashe (1958) and Lee (1957) evolved A3 curve as a safe limit for workers in mines and textile mills representing thermal stress where deterioration of mental function begins. Similarly curves for unacclimatised men—UA1 (Brunt, 1943) separating warm conditions from very hot conditions, UA2 (Schickele, 1947) as a heat death line below which no cases of heat stroke were noticed, UA-3 (Ladell, 1955) as a line for an effective temperature of 80°F with an air motion of 500 ft/min separating warm from too hot environmental conditions, and UA4 (Lee, 1957) representing the upper limit of thermal stress for daily work by an unacclimatized men. Ladell et al. (1944), Schickele (1947), Borden et al (1945), Hefferman et al. (1953) and Austin and Berry (1956) observed morbidity and mortality from heat disease developed between these group lines among the personal in relation to limit of thermal stress.

There are two types of human comfort indices – warm season and cold season indices where environmental variables reacts in different manner. In a warm environment there is dilation of blood vessels which increases the flow of blood near the skin, and thermoregulation response through sweating, panting and evaporative cooling. However in cold season wind and temperature are more crucial affecting human comfort feeling than humidity and solar radiation. Therefore the indices have been developed accordingly.

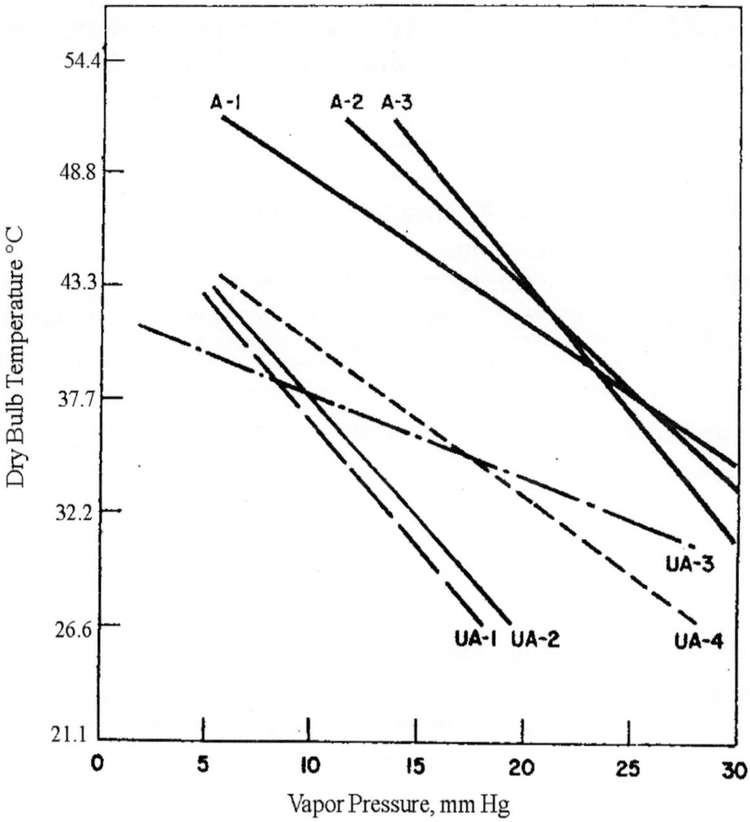

Fig. 6.8: Limits of thermal stress for acclimatized (A–1 to A–3) and unacclimatized (UA–1 to UA–4) men. (*After*; Olgyay, 1960)

6.6.1 Warm season indices

(i) Effective Temperature Indices (ETI)
(ii) Temperature and Humidity Indices (THI)
(iii) Humidex Index (HUI)
(iv) Heat Stress Index (HSI) based on evaporation and sweat rate
(v) Relative Strain Index (RSI) based on resistance of clothing to heat and vapour transfer
(vi) Thermal Stress Index (TSI) based on heat exchange of body with environment
(vii) Equivalent Temperature Index ($T_{eq}I$) using surface temperatures of human body
(viii) Equivalent Blackbody Temperature Index ($T_{eqb}I$)
(ix) Equivalent Wet Blackbody Temperature Index ($T_{eqw}I$)

These indices have been described in Table 6.5.

Table 6.5
Human comfort indices during a warmer season

Index	Definition	Meteorological and Physiological Parameters	Performance for Physological Sensory Responses	References
(A) Based on meteorological variables				
Effective Temperature Index (ETI)	Temperature of air saturated with water vapours to produce same degree of warmth or comfort as in open environment	Air temperature (Ta), humidity (RH), wind speed (u), clothing, globe thermometer temperature (Tg)	Least reliable, uncomfortable and heat stress conditions	Jaglou and Miler (1925)
Radiation - Convection Temperature Index (RCI)		Tg	Internal microclimates	Vernon (1930)
Resultant Temperature Index (RTI)	Mcdified ETI	Ta, Tg, RH, U	Satisfactory for people at rest or engaged in sedentary activity	Missenard (1948)
Ecessive Heat Stress Index or (THI)	Temperature in combination with humidity at higher values attaining permissible maximum discomfort	Td, Tw, Tdp (dew point) related as THI = 0.4(Td + Tw) + 15 THI = 0.55 Td + 0.2 T_{dp} + 17.5 THI = Td – (0.55 – 0.55 RH) (Td – 58) when temperature are in °F. THI = 0.4(Td – Tw) + 4.78 when temperature in °C	when THI < 70 – No discomfort THI > 75, 50 per cent people feel discomfort THI > 80, almost 100 per cent people feel discomfort THI > 85 – No work without airconditioning	Thom (1957) Hobb (1980)
Equivalent Temperature Index (T_{eq}I)	Surface temperatrue of an instrument (eupatheosfat) representing human body temperature for comfortable conditions.	T_{eq} = 0.522 T_a + 0.478 T_s – 0.01474 $\sqrt{v}$(100–T_a), where T_a in °F, T_s is median temperature of the surroundings in °F, U in ft/min.		Dufton (1932)
Humidity Index (HI)		HI = T_a + 5/9(e_a – 10), where e_a is actual vapour pressure in kρa×10	HI beween 20 to 29 – comfortable HI between 30 to 39 – varying degree of discomfort HI between 40 to 45 – everyone uncomfortable HI > 46 – labour work restricted.	Anderson (1965)

Index	Definition	Meteorological and Physiological Parameters	Performance for Physological Sensory Responses	References
Climatic Tourist Index (CTI)		Monthly Index = (S + T - 5D)/5, where S is number of hours solar insolation during one month, T is average monthly temperature in °C, D is number of hours of rainfall during one month. Weekly Index = (30S + 7T − 150D)/35 where S, T, D are values during one week.		Rivolier (1975)
Aggressive Biometeorological Index (ABI)		Daily ABI = k_1 + ΔT + U + k_2 + k_3 − 2I, where k_1 (humidity coefficient) = (Relative humidity above 50 per cent level + mm of rainfall + duration of rainfall in hours)/5, ΔT is number of °C below 15°C and above 25°C U is wind speed in m sec^{-1}, k_2 is the sum of each sudden pressures change counted as 5, k_3 is the sum of each abnormal meteorological penomenon counted as 5 such as fog, lightening, Thunderstorm, squall etc.		Rivolier (1975)
(b) Based on meteorological and physiological parameters:				
Predicted 4 hours Sweat Rate Index (P_4SRI)	Sweat rate with four hours exposure to the environment	T_a, U, Clothing, metabolic level	Satisfactory under light to heavy heat stresses to predict sweat rate response, but not suitable around the comfort zone and under cold exposure	McArdle (1947)
Heat Stress Index (HSI)	It is the ratio fo heat produced by metabolism for any activity or required evaporation to the maximim evaporation capacity of the environment	Radiation, convection, vapour pressure, wind velocity and metabolic levels to quantify external heat stress on man.	Suitable for thermal stress conditions, but not for physiological responses to the stress	Belding and Hatch (1955)
Relative Strain Index (RSI)	It is the ratio of total heat load i.e., metabolism and heat transfer between a clothed body in the environment to the evaporative capacity of air	-do-	Suitable for thermal stresses under different clothing conditions, but not for working people to quantify physiological response to heat stress.	Lee and Henschel (1963)

Index	Definition	Meteorological and Physiological Parameters	Performance for Physological Sensory Responses	References
Thermal Stress Index (TSI)	It is the total thermal stress on the body (metabolic and environmental) and sweat rate required to maintain thermal equalibrium	$S = [M - W] \pm (C \pm R)(1/f)$ where S is required sweat rate (kcal hr^{-1}), W is metabolic rate converted into mechanical work (kcal hr^{-1}), M is metabolic rate (kcal hr^{-1}), C is convective heat exchange (kcal hr^{-1}), R is radiative heat exchange (kcal hr^{-1}), and F is cooling efficiency of sweating	Satisfactory for working or resting persons between comfort and severe stress conditions when thermal equalibrium is maintained only	Givoni, (1963, 1969)
Total Temperature Index (TTI)	Fluctuations in temperature and pulse frequencies	Temperature curve based on morning and evening temperatures, pulse frequencies	Post operative pneumonia and thromboembolism complications occur when high peak of temperature is followed by a deep fall.	Boerema (1941)
Pulse Breathing Index (PBI)	It is the ratio of pulse rate to the breathing rate		When PBI = 4 for perfect autonomic balance When PBI > 4 for liability to autonomic conditons, When PBI < 4 for very high stability.	Hildebrandt (1962)
Vegetative or Nervous System (VI)		$VI = (1 - d/p) \times 100$, where d is diastolic pressure in mm, p is pulse rate for minute	During ergotropic phase (increased metabolism and blood flow and increased oxygen consumption) d is small, d/p becomes <1 and VI is positive. During trophotropic phase (domination of parasympathetic stimulation) d is higher, d/p becomes >1 and VI is negative.	Kerdö (1967)

6.6.2 Cold season indices

(i) Wind Chill Index (WCI)
(ii) Wind Chill Equivalent Temperature Index ($T_{weq}I$)
(iii) Clothing Insulation.

These indices have been described in Table 6.6 and are computed based the on interrelated temperature, humidity, dew point temperature, vapour pressure as given in appendix IX to XIII.

Szokolay (1985) differentiated these empirical and physiologically based methods using social survey approach such as comfort vote technique using Bedford Scale (Bedford, 1936), physiological responses (Winslow et al., 1937) and ASHRAE scale (Gagge et al., 1971) as give below in term of their corresponding numerical code.

Numerical Code	ASHRAE Scale	Bedford Scale	Physiological Response	Comfort Level
+3	Hot	Much too warm	Profuse sweating	Unacceptable
+2	Warm	Too warm	Onset of sweating	Uncomfortable
+1	Slightly warm	Comfortably warm	Vasodialation	Acceptable
0	Neutral	Comfortable	Minimal	Maximum acceptable
–1	Slightly cool	Comfortably cool	Vasocontriction	Acceptable
–2	Cool	Tool cool	Onset of body cooling	Uncomfortable
–3	Cold	Much too cold	Shivering	Unacceptable

Auliciems (1983) developed regression between the group comfort values (T_{cf}) and outdoor mean monthly temperature (T_{Mo}) and indooor mean monthly temperature values (T_{Mi}), and reported that T_{cf} appeared to be some 10°C in width, ranging above 17°C expressed as

$T_{cf} = 0.31\,T_{Mo} + 17.6$, $r = 0.8$
$T_{cf} = 0.73\,T_{Mi} + 5.41$, $r = 0.92$
$T_{cf} = 0.48\,T_{Mi} + 0.14\,T_{Mo} + 9.22$ $r = 0.95$

However with increasing thresholds for humans in Asia and Africa the physiological responses are activated in a curvelinear manner than linear relationship as above (Cena, 1974).

Most livestock animals do not easily adjust to heat stress. Temperature in combination with high humidity (about 26.5°C in temperate locations) leads to animal's death. U.S. National Weather Service (1976) is using Humidity-Temperature Index (HTI) for issuing alert warnings as

$$HTI = 0.55(1.8\,T_a + 32) + 0.2(1.8\,T_d + 32) + 17.5$$

where, when HTI ranges between 75-78, livestock producers are alerted to avoid losses; when HTI ranged between 79-83, dangerous conditions exist for livestock safety; and when HTI > 84 emergency conditions are declared for livestock safety and care must be provided with shade, proper food and water.

Table 6.6
Human comfort indices during a cooler season

Index	Definition	Meteorological and Physiological Parameters	Performance for Physological Sensory Responses	References
Windchill Index (WCI)	It is quantity of heat that atmosphere could absorb in an hour from an exposed one sq. m. surface as a measure of cooling effect created due to temperature and humidity.	$WCI = \Delta T(9+10.9\sqrt{U}-U)$ where WCI is in kcal m^{-2} hr^{-1}, ΔT is difference of body temperature with the surrounding environment in °C, U is wind speed in m sec^{-1}. However, in polar area $WCI = (\sqrt{100U} + 10.45 - U)(33 - T_a)$, where T_a is in °C	WCI < 200 - Hot sensation, 200 to 399 - pleasant sensation, 400 to 599 - cold sensation, 600 to 799 - very cold sensation 800 to 999 - cold sensation 1000 to 1199 - very cold sensation >1200 - Bitterly cold sensation WCI = 50 - Excessive heat WCI = 2500 - Enormous cold	Bedford (1948), Siple (1945), Court (1948), Thomas and Boyd (1957) Wilson (1962)
Winddchill Equivalent Index (T_{weq})	It is temperature under a light wind (2.22 m sec^{-1}) which will produce same cooling effect as produced in an open environment.	$T_{weq} = \frac{WCI}{23.08} + 33$ WCI, U and T_a for prevailing environment is determined first.	T_{weq} is a sensational temperature temperature experienced inside	Siple and Passel (1945)
Clothing Insulation (CLO)	It is the thickness of wearing clothing to insulate and counteract windchill index under different temperature and windspeed conditions.	One CLO insulation allows a heat flow of 1 kcal $m^{-2}hr^{-1}$ with a temperature gradient of 0.18°C across the fabric.	Thickness of clothing (CLO) / Comfortable temperature under light wind (°F): Nil – 86; 1 – 70; 2 – 54; 3 – 38; 4 – 22	Steadman (1971)
Equivalent Blackbody Temperature Index ($T_{eq.b}I$)	Energy balance equation is used to express comfortable conditions using physiological parameters	$T_{eqb}I = \frac{\rho c_p}{(r_c + r_{eq})} T_b - \frac{(r_c + r_{eq})}{\rho c_p}(M - LE)$ $LE = (2.3 \times 10^{-3}M + 0.31)(31-\rho_{va})$ where ρ_{va} is water vapour density in air in kg m^{-3}, M is metabolic rate	When room wall temperature equals T_a, $T_{eqb} = T_a$ for normal metabolic rate of 90 Wm^{-2} at low vapour	Campbell (1977)

Index	Definition	Meteorological and Physiological Parameters	Performance for Physological Sensory Responses	References
		for human varying from 50 to 200 Wm^{-2}.	density of 0.005 kg m^{-3}, T_{eq} is 22°C - comfortable level, however if humidity is high 19°C is comfortable.	
Equivalent Wet Blackbody Temperature Index ($T_{eqw}I$)	Environmental variables are combined into a single parameter representing heat stress called equivalent wet blackbody temperature index. The wet equivalent temperature must always be lower than body temperature to maintain energy balance	Eq. (6.38) represents expression for equivalent wet blackbody temperature. However, when r_{vs} becomes large, γ^* becomes higher and expression reduces to dry equivalent temperature equations. And with air movement over wet skin surface, the resistances for water vapour and heat transfer becomes equal so that $\gamma^* = \gamma$	T_{eqw} I is determined for different values of r_{Hb} and M values. $T_{eqw}I$ decreases with increase of M and r_{Hb}	Campbell (1977)

6.7 Characteristics of Animal's Shelters

Environmental need of livestock and their shelters must be compatible each other to provide a optimum microlimatological conditions for their development and higher productivity (Landsberg, 1954). The prevailing outside environment strongly influences the climate inside through the shape, exposure, material used, ventilation and hygienic conditions of the shelter. The design of shelter are mainly influenced by radiation and wind keeping in view the due comforts. Therefore a climatic data of past at least 30 years record at a location is analysed in order to decide whether the winter or summer conditions are more stressful. If winter season is more stressful, then it will be desirable to weaken the impact of cold winds and to increase the winter receipt of solar radiation for heating purposes. Conversely, if summer's heat is most stressful, then summer shading and increased cooling breezes are necessary for evading excess heat load. If both winter extreme cold and summer intense heat are problematic at a location, then a compromise design may have to be adjusted keeping in view the seasonal march of climatic conditions. A bioclimatic comfort chart based on temperature and humidities, can be plotted to highlight the maximum or minimum conditions of climatic stresses.

(i) Radiation and shading: Olgyay (1963) developed a procedure to quantify the radiation load on a shelter/house and on a body placed inside the shelter. It will depend on the exposure of the wall i.e., on its azimuth, solar angle and on the surface ratio between outer walls and roof. Excessive heating takes place due to radiation received by building porches facing south (northern hemisphere) or north (southern hemisphere), maximum heat loss takes place through roof and ventilation. Therefore thermal insulation is very important in cold months. It is possible to use reflecting materials (painting roof and white walls) and to use ventilated double walls to prevent excess heating in daytime. Use of a material with lower thermal conductivity will provide better insulation such as glass wood, polystyrene slabs. The heat loss through a complete animal structure is called its U-value which is 0.1 or less for roof, 0.1 to 0.2 for walls, and 0.1 for floor in a animal's house. In order to provide a roof insulation a 2" mineral layer is covered both sides by corrugated asbestos or tiles. Panting helps to some extent in getting rid of heat through hot air blowing on moist mucous membranes of lungs leading to cooling effect. However panting itself produces some heat through muscular activity. Therefore ultimately the animal milk productivity will go down. Also the major source of energy in animals is through grasses and straw which undergoes fermentation and produces heat. Therefore, in summer during daytime cattle do not eat as it worsens their body temperature.

Normally for high milk production in tropics, cross-breeding between low producing local cattle and European breeds is practised. The level of European blood is to be limited to 62.5 per cent and local blood to 37.5 per cent which is ideal as they can have better adoptability to tropical climate, disease resistance, and fairly higher productivity compared to local breeds. Buffaloes are poorly adopted to tropical climate as they have less sweat glands only 1/10 to

1/6 compared to cattle. Therefore they prefer marshy lands with water for cooling their body. They are also poorly adopted in cold climates being having no body coat and are limited to pockets in the globe.

Net radiation variations over the shelter provides available energy to produce outside environment which varies throughout day and night expressed as

$$R_n = (1 - \alpha_r) R_s + \in_a R_{LWa} - \in_r \sigma T_r^4 + \in_e R_{LWe} \quad \ldots (6.48)$$

where, R_n is net radiation over the roof surface, α_r is shortwave reflectivity of roof material, R_s is total incoming solar radiation, $\in$ is emissivity with subscripts a, r, and e for atmosphere, roof and environment surroundings, R_{LWa} is longwave radiation emitted from atmosphere towards the roof surface and R_{LWe} is longwave radiation coming from surrounding environment falling towards building or roof, T_r is temperature of roof surface contributing for longwave radiations emitted by the roof in the atmosphere. However, if an animal is grazing in an open lawn surface, then radiation balance is

$$R_n = (1 - \alpha_b) (R_s)(1 + \alpha_g) + \in_a R_{LWa} + \in_e R_{LWe} - 2\sigma \in T_b^4 \quad \ldots (6.49)$$

where α_b is reflectivity from animal's body, T_b is temperature of animal's body, α_g is reflectivity from ground lawn surface.

Walls with cavity brickwork or insulated concrete blocks or hollow blocks in dense concrete are used to provide insulation. Floor insulation is done with use of hard core concretion, damping of insulation material to provide hygienic conditions. The degree of insulation will depend over the climatic needs of the livestock. Calves, poultry and pig accommodation needs insulated shelter particularly.

6.7.1 Radiation analysis

Olgyay (1963) used noon time radiation record received on a vertical wall facing various directions during both the underheated and overheated periods (Fig. 6.9). Keeping Fig. 6.9 in view, it could be desirable to orient a house/shelter rectangular in shape so that its walls would receive minimum radiation in the summer overheated period and the maximum during the winter underheated period. The maximum radiation received during underheated period is on a wall facing S75°W, whereas during overheated period on a vertical wall facing E75°S. Therefore we can have a rectangular dwelling with long side facing S75°W to maximize winter heating and short side facing E75°S to create less heating. In some cases, it may not be feasible to have a rectangular shelter, but may take a irregular shape.

Temperature, being sensible heat index of the surrounding air, also influence the shelter depending on the thermal gradient existing between environment and the outer surface.

6.7.2 Wind rose analysis

The overall temperature of the shelter is influenced by the wind. When wind is calm, a zone of thermal balance is formed around the shelter, whereas with flow of air the zone of thermal balance is diffused with direct impact on the roof and walls. Therefore at each

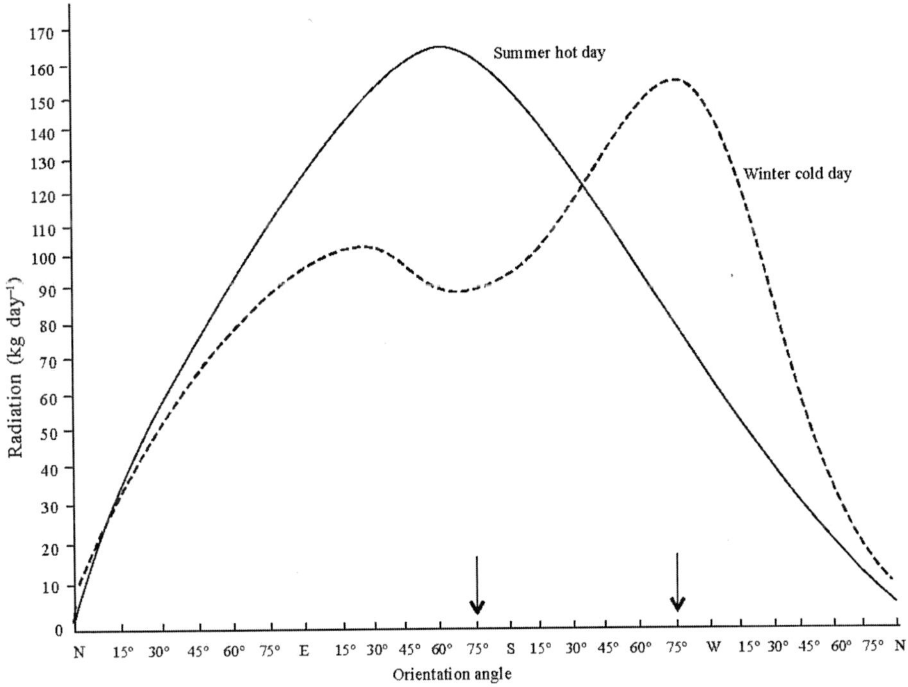

Fig. 6.9: Radiation receipt on a vertical wall during noontime in different directions.

location seasonal wind analysis is done with formation of wind roses. If summer is hot and dry or humid will require shade and ventilation by natural breezes, whereas in cold winter winds will require the breakdown of wind force. Therefore wind analysis is done during all seasons particularly during most stressful periods. Fig. 6.10 shows wind roses diagrams during the months of January (winter) and July (summer) to highlight the distribution of wind speed and direction, and its impact on the surrounding environment of the shelter. Concentric circles are marked to cover the range of frequencies observed in wind data. Average values of long term data, converted in percentage, are plotted in each wind direction frequencies in different shapes for each wind category ranging between 0 – 3.99, 4 – 7.99, 8 to 11.99 and more than 12 mph. Wind rose data shows seasonal shift from northwest in the winter period and from southwest in the summer period. For maximum winter and dry summer protection, and utilization of wind in the summer humid period, it as necessary that a windbreak to the northwest of shelter be erected in winter and southwest in dry simmer period and there should be no interference with the wind in the southwest during humid summer period. Since northwest winds are also quite frequent in summer to bring relief from heat, a sliding window in northwest can be used to have benefit of both the seasons. A wind break of mixed (deciduous and coniferous) belt of trees of proper density be adopted so as to plant them at right angles to the direction of wind to be protected. Normally the barrier is designed to ward off the cold winter winds and use the cooling summer breezes for the animal's

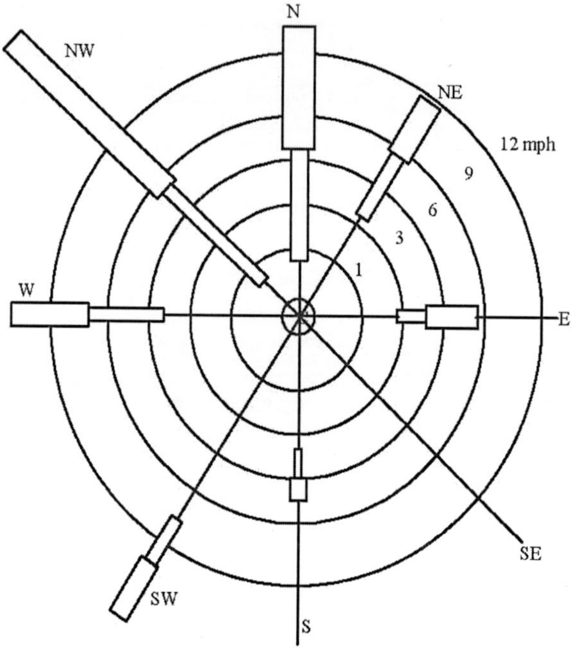

Fig. 6.10(a): Wind rose during January - a winter month C represents calm.

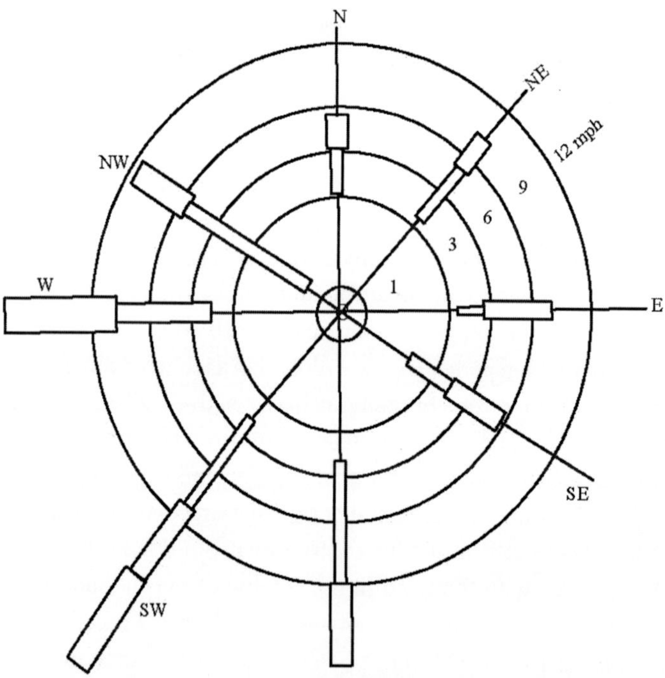

Fig. 6.10(b): Wind rose during July - a summer rainy month.

shelter. The prevailing winds during the rainy months are useful in hot dry areas, but occurrence of rains during winter alongwith snow lowers down the temperature of the shelter which is further aggravated if winds are strong. In temperate areas precipitation also occurs in the form of snow and this snow layer on shelter will form a screen against the effects of outside radiation and temperature with insulating and reflecting properties reducing diurnal variations of temperature.

6.7.3 Climatic environment inside the shelter

Climatic conditions outside the shelter have an immediate effect on the temperature of the walls and microclimatic conditions prevailing inside the shelter (Olgyay, 1954; Primault, 1979). Radiation balance inside the shelter with be influenced by radiated heat from the walls and roof. Therefore heat balance must be maintained for proper inside environment expressed in energy processes as

$$M - LE \pm F \pm G \pm A \pm R_n \pm St = 0 \qquad \ldots (6.50)$$

where, M is metabolic heat production by the animal in the shelter, LE is heat loss from skin and respiratory organs, F is heat loss or gain due to food or water, G is the heat loss or gain by direct contact between skin and surrounding surfaces, A is the heat loss or gain by convection due to contact between the air, skin and respiratory passages, R_n is heat loss or gain by radiation and St is stored energy.

In ideal situation the heat loss from the animal's body must be compensated by the radiation from the surrounding walls towards the animal so that metabolism of animals is used for maintaining normal processes, digestive heat production, muscular heat production and productive processes like milk, growth, work etc. (Oosterlee, 1959).

Animals continuously exchange latent heat from the environment through evaporation of water from the skin and respiratory organ (about 150 g hr^{-1} for a 300 kg milk cow). This saturated air is exhaled continuously raising the humidity inside the shelter. If environment is dry, then air inside will absorb these water vapours ultimately attaining saturation level and condensation of water vapours will take place inside the shelter. Therefore thermal insulation of walls inside is necessary. In a temperate or winter season thermal insulation for temperature and humidity both becomes almost essential. Therefore temperature and humidity in combination are important microclimatic parameters in order to create a suitable thermal hygenic environment. Thermal insulation properties of the shelter are decided according to the thermal gradients between inside and outside environment of the location.

Since animals release both heat and water vapour with metabolism process, therefore microclimatic conditions will continuously change with increase of temperature and water vapour concentrations and it will become necessary to have a protective measures by means of natural (wind rose) or forced ventilation (artificial control) to maintain optimum microclimatic conditions (Bianca, 1961). In winter due to continuous low temperature and cold northwesterly winds there is negative radiation balance and shelter becomes unsuitable area unless it is artificially warmed or cold winds are protected by a wind break or shelterbelt erection. Longwave radiations emitted by walls are protected by thermal insulation. However in tropical arid/semiarid areas a reverse situation is commonly experienced which is further aggravated by the release of metabolic heat. A cooling mechanism is desired to maintain optimum microclimatic conditions. Alongwith release of water vapours there is considerable release of harmful gases such as carbon dioxide, methane, ammonia, etc. due to chemical reactions in metabolism and degradation of stable litter. Therefore proper ventilation becomes essential to provide hygienic conditions. Temperature and wind speeds plays important role in ventilation process (Smith, 1964a). Table 6.7 shows the role of temperature and wind velocity for comfortable management of pigs.

In poultry the microclimatic requirements are very peculiar in order to maintain day and night time comfortable microclimatic conditions. Adult birds can tolerate to the variations in climate change, but the young chicks needs a constant starting temperature of 35°C, however with growth (gaining weight) up to 45 days the optimum temperature of adults of the order of 12 to 15°C. Diurnal variations of day and night temperatures is unsuitable for these young chicks, and therefore a artificial heating or insulation for loss of longwave radiation is necessary to maintain the optimum levels in the nigh time. High relative humidity also does not help in stimulating feathers, therefore a lower humidity between 40 to 70 per cent is

Table 6.7
Temperature and wind velocity for proper ventilation of pigs in a sheltered environment

	Wind Velocity (m sec^{-1})		
Temperature °C	**<0.15**	**0.15 – 0.25**	**0.25 to 0.37**
21	Pigs are comfortable	Pigs are comfortable	Young piglets are uncomfortable (less than 60 days old)
18	Pigs below 7 days are uncomfortable	Pigs below 35 days are uncomfortable	Pigs below 85 days are uncomfortable
15	Pigs below 10 days are uncomfortable	Young piglets are uncomfortable	Piglets below 85 days are uncomfortable
12	Piglets below 60 days are uncomfortable	Pigs below 85 days are uncomfortable	-
10	Pigs below 100 days are uncomfortable	Pigs below 115 days are uncomfortable	-
7	Pigs below 115 days are uncomfortable	Pigs below 125 days are uncomfortable	Pigs below 140 day are uncomfortable
5	Pigs below 140 days are uncomfortable	Pigs below 140 days are uncomfortable	Pigs below 140 days are uncomfortable
2	All flattening pigs are uncomfortable	All flattening pigs are uncomfortable	All flattening pigs are uncomfortable

maintained. However for hatching the fertile eggs at 12-13°C at 80 to 90 per cent relative humidity provided better results (Winn and Godfrey, 1967).

Prevailing winds at low temperature are susceptible, but at higher temperature wind dissipate heat to maintain the heat balance. Therefore a ventilating system design is necessary to maintain correct air movement. With a wind speed of 1m/sec at enterance of hatchery will provide a U value (heat loss) of 0.35 m per sec. to reduce the mortality rate by increasing temperature from 8/10 to 18°C.

Therefore ventilation is an essential aspect which needs to be decided depending of choice of animals, their age and number of livestock and weather conditions of the location.

6.7.4 Reduction of climatic stresses

High environmental temperatures have deleterious effects on milk production as the animal cannot balance the heat production and heat losses. The situation is further aggravated with prevailing of high humidity as the animals fail to get body cooled through evaporation of sweat. Also when a period of high temperature (daytime) is followed by a period of low temperature (night time), the animals can maintain the productivity if diurnal variation of temperature is greater by adjusting eating in night time to maintain milk production. But if such differences does not occur in cloudy sky during pre-monsoon period the milk production will fall steeply. Heat stroke and sun stroke can occur in European breeds if exposed to sun during summer. Therefore it is necessary to reduce the climatic stresses through

(i) Shade – shade will reduce the impact of solar radiation. Tree shade is preferred as it will cool the environment through evapotranspiration.
(ii) Breeze – wind flow will drift the air close to the body saturated with water vapours and getting dry air will help in cooling the body.
(iii) Water spray – it will help in cooling the body temperature of animals.
(iv) Green fodder – it will help in providing high water contents.

Milk biosynthesis depends on the mammary gland receiving a continuous supply of various metabolites and harmones from the blood. In cattle yield is uneffected in 0 to 21°C, but at lower temperature below 5°C as well as above 21 to 27°C yield decreased slowly, however above 27°C declined rapidly. Decline also takes place with increase of humidity as compared to lower values. A decrease of 1 kg per °C rise in rectal temperature is normally observed. However the optimum temperature for lactation is dependent on the species, breed and degree of tolerance to heat or cold. For example, decrease in milk yield in Holstein cattle at 21°C, in Brown Swiss and Jersey at 24 to 27°C, and in Brahma at about 32°C. Minimum critical temperature for Jersey cattle is around 2°C, Holstein is about -13°C.

6.8 Livestock Productivity and Weather Relationships

Bianca (1976) stated that most efficient production of meat, milk and eggs occur when meteorological elements are within a certain range. Hancock (1954) concluded that temperature in the range of 5.0 to 21°C have no effect on milk yield and below which there is significant decrease in milk yield. Das and Balaine (1980) observed that Haryana Cows calving in spring and winter are having higher yields and longer lactation than those calving in autumn and summer. Whittemore (1980) enumerated that as the animal grows, lactations yield increases, then as aging occurs, lactation yield is reduced forming a hump-back response for lactation yield to lactation number. Dalal et al. (1991) reported in Haryana Cattle, (NW India) caving in March to June period had higher production capacity, whereas in crossbreds of Haryana, the autumn calvers were superior to other season calvers (Singh and Yadav, 1999), and higher persistency was observed in summer and rainy season calvers than winter calvers in crossbred cattle (Shiv Prasad et al., 1999). Deokar and Ulmek 2001; Odera et al. (1978) observed the season of calving had significant effect on lactation length in Jersey Cows.

Johnson (1958) observed that gradual increase in temperature and humidity unlike sudden changes does not cause an immediate reduction of milk yield in dairy cattle. Galvano (1968) found significant correlation between milk yield and maximum environmental temperature (0.93), minimum temperature (–0.9), mean temperature (–0.93) and mean humidity (0.57) in Friesian cows. Thomas and Acharya (1981) reported that milking average had a positive relationship with average maximum temperature and negative relationship with vapour pressure. Palia and Arora (1983) observed that season of calving had a significant effect on peak yield in Jersey Cows. February – May calving period had a higher peak yield as compared to other seasons. Panda and Sandhu (1983) observed that winter calvers produced more milk followed by rainy and summer calvers, but this effect was non-significant in crossbreds of Holstein and Jersey with Haryana and Desi Bengal cows. Shinde (1984) reported significant and negative correlation of temperature and humidity with milk yield per day in Brown – Swiss and Holstein crossbreds cows in Northern India. Shinde et al. (1990) explained the largest variations in daily milk yield with the stage of lactation. Kabuga (1991) reported that temperature was significant in early and late lactation in milk yield, but not in middle lactation. Guerreno et al. (1998) found that Friesian cows calving in late summer and autumn period produced less than those calving in winter and spring. Depression in milk yield above 21°C proved to be 1.1 kg (31 per cent), whereas above 86 per cent relative humidity the daily average milk yield decreased by 1 kg (30 per cent) (Szues et al., 2001). Maximum temperature, minimum temperature and relative humidity significantly affected the dynamics of total daily milk production (Mandal et al., 2002). Genetic group and season of calving had a significant effect on exotic as well as crossbreds of exotic with Indian cattle (Das et al., 2003).

Smith (1967) used June rainfall and April production data to predict milk production in Great Britain eleven months ahead. Shinde et al. (1990), Ramachandran and Murli (1992) enumerated the importance of temperature-humidity-sunshine index which had a negative and significant correlation with milk yield. Kathilankal (2003) reported that in crossbreds December – January calvers had higher average lactation yields and longer lactations, whereas among Desi/indigenous cattle October – November and February – March calvers had higher lactation yields, but longer lactation was observed in April – June calvers in NW Indian environmental conditions (Table 6.8). High persistency was observed in crossbred animals calving in July – September, whereas among Desi/indegrenous breeds February – March and April – June calvers were most persistent. Higher persistency was observed in Desi/indigenous breeds than in crossbred indicating their adaptability.

Kahilankal (2003) also analysed the milk yield and weather variables data for past six years (1995 – 2000) in crossbreds and Desi/Indigenous cattles to quantify optimum environmental variables for their maximum milk yields and adaptability levels (Table 6.9). It was observed that optimum temperature for crossbreds varied between 13 to 22°C, whereas for Desi it was between 16 to 22°C. The optimum relative humidity range were 52 to 66 per cent and 63 to 67 per cent for crossbreds and Desi cattles. The crossbred cattle adaptability in NW India required optimum temperature of 11°C, relative humidity 66 per cent, and THI value of 55 and preferred high wind speeds (5.85 kmph). Crossbreds gave higher milk yield, but there is wide variation in milk yield throughout the lactation period and lactation persistency is lower than desi cattles. On the other hand adaptability of desi/indigenous cattle required optimum temperature for peak milk yield of 18.5 °C and relative humidity of 66 per cent and THI value of 64 and preferred lower wind speed values (3.62 kmph). Though milk yield was lower, but the variation was less throughout the lactation period presenting higher persistency (Table 6.9).

Milk yield was predicted with the use of multiple regressions developed between milk yield and significant weather variables expressed as (Kathilankal, 2003).

$$\underset{}{Y = 9.786} - \underset{(-3.15)}{0.6888\ WS} + \underset{(+2.27)}{0.625\ SH} + \underset{(3.05)}{5.01 \times 10^{-2}\ R} - \underset{(3.05)}{6.7 \times 10^{-2}\ HTU}$$

$$+ \underset{(+2.18)}{4.6 \times 10^{-2}\ PET} - \underset{(-3.1)}{6.5 \times 10^{-2}\ AET}, \quad (R^2 = 0.69, F = 14.54) \quad \ldots (6.51)$$

where, Y is milk yield (kg/day) for crossbred calving in October – November months.

In case of Desi cattle calving in October – November month the milk and weather variables regressions explaining 80 per cent milk yield variations.

$$Y = 2.548 + \underset{(2.56)}{.288\ T_{max}} - \underset{(-2.04)}{2.29\ VP_m} - \underset{(1.22)}{0.061\ RHE} - \underset{(-2.35)}{4.36\ WS} - \underset{(-2.34)}{0.377\ SH}$$

$$+ \underset{(-3.85)}{0.69\ E} - \underset{(-3.46)}{0.063\ R} + \underset{(3.6)}{0.00331\ K_0} - \underset{(-3.46)}{0.252\ VPD_e} + \underset{(2.18)}{0.0507\ AET}$$

$$+ \underset{(1.21)}{1.598\ MAI}, \quad (R^2 = 0.80, F = 5.06) \quad \ldots (6.52)$$

where, Y is milk yield (kg/day), T_{max} is maximum temperature in °C.

VP_m is mean monthly vapour pressure in mm of Hg in morning, RHE is relative humidity

Table 6.8

Milk yield, lactation length and rate of decrease in milk yield after attaining peak yield in crossbreds and Desi/indigenous cows.

Month of Calving	Lactation Length (Months) and Milk Yield (kg/day) and Rate of Decrease (kg/day)					
	Crossbreds			Desi/Indigenous		
	Group I	Group II	Decrease Rate	Group I	Group II	Decrease Rate
April to Jun.	8.13±2.36 (5.83)	10.08±2.13 (6.21)	-0.024	7.53±2.80 (2.01)	9.50±1.96 (2.55)	-0.004
July to Sep.	11.00±2.08 (5.85)	10.11±1.45 (6.74)	-0.019	6.50±1.45 (1.77)	8.33±1.76 (3.02)	-0.015
Dec. to Jan.	10.58±2.63 (6.42)	11.21±2.65 (7.85)	-0.027	7.40±1.02 (2.23)	6.83±1.21 (2.79)	-0.013
Oct. to Nov.	11.19±1.88 (6.03)	9.67±1.49 (7.38)	-0.024	6.64±1.84 (2.09)	7.15±2.21 (2.79)	-0.013
Feb. to Mar.	9.88±1.85 (6.00)	11.19±2.70 (7.78)	-0.020	7.17±2.22 (2.14)	8.17±2.41 (2.46)	-0.004
Average (kg/day)	10.16 (6.03)	10.45 (7.19)		7.04 (2.05)	8.00 (2.78)	

Figure in brackets are the average milk yield (kg/day).

Group I – First, second and third lactations.

Group II – Fourth, fifth and sixth lactations.

(*After*: Kathilankal, 2003)

Table 6.9

Optimum weather variables for peak yield in crossbred and desi/indigenous cattle.

Calving Period	Peak Milk Yield (kg/day)	Temperature (°C)	Relative Humidity (%)	THI	Wind Speed (kmph)	Sunshine (hrs.)
Crossbred						
Ap. – Jun.	8.5±0.73	30.9±1.1	67.7±7.3	83.8±2.84	7.8±1.6	6.1±1.4
Jul. – Sep.	7.57±0.86	25.1±2.5	59.5±9.0	71.4±4.4	4.0±1.0	7.4±1.0
Octo. – Nov.	9.26±0.88	13.4±1.8	71.1±7.3	56.4±2.8	3.8±0.9	6.3±1.6
Dec. – Jan.	10.85±1.90	17.1±2.6	65.4±7.2	61.9±4.0	4.5±0.6	7.6±0.9
Feb. – Mar.	10.22±1.61	22.9±3.7	52.9±12.7	70.6±4.6	5.1±0.6	8.6±1.1
Desi/Indigenous						
Apr. – Jun.	2.25±0.74	21.6±3.2	63.4±8.4	69.8±5.3	2.8±0.8	7.3±1.0
Jul. – Sep.	3.32±1.00	17.1±2.8	65.8±7.5	62.5±4.5	4.5±0.6	7.7±1.0
Oct. – Nov.	3.53±1.24	16.2±3.2	64.9±10.7	60.9±5.3	2.7±0.6	6.4±2.0
Dec. – Jan.	3.34±1.07	22.9±3.9	52.9±12.6	70.6±4.6	5.1±0.6	8.6±1.1
Feb. – Mar.	2.68±0.56	22.9±3.9	51.8±12.4	70.4±4.8	5.1±0.7	8.6±1.1

(*After*: Kathilankal, 2003)

in evening in percent, WS is mean wind sped in kmph, SH is mean sunshine hours, E is mean monthly pan evaporation in mm/day, R is total monthly rainfall in mm, k_0 is is wind chill factor, VPD_e is mean monthly vapour pressure deficit in mm of mercury in the evening, AET and PET are actual and potential evapotranspiration in mm, and HTU is heliothermal units with 10°C of base temperature in °C days hours.

Exercise 5

Q. 1. The small animal in the example 1 had a surface with zero albedo, calculate (i) the resulting increase in its heat load (ii) the resulting increase in its surface temperature at equilibrium assuming all other factors are unchanged.

[**Ans.** (i) 0.2743 W, (ii) 6.5°C]

Q. 2 Determine the equivalent temperature for a man standing in an open environment at full sun with temperature of 35°C (example 12 parameters). (a) wearing white clothing with $a_s = 0.3$ (b) when wind speed in open environment increases to 35 m s^{-1}. (c) when (a) and (b) both situations exists. (**Ans.** 42.1°C, 50.1°C, 40.7°C)

Q. 3. Depict the climate space diagram for a small homeotherm animal with metabolic rates of 50 and 150 Wm^{-2} as lowest and highest values, latent heat of 0.2 M, under wind speeds of 0.1 and 1.0 m sec^{-1} in extreme wind environments. The body temperature for these two environments are 41 and 37.5°C as survival limits of the mouse.

r_{Hb} is 300 sm^{-1} at 37.5°C and 230 sm^{-1} at 41°C, a_s is 0.8, R_{sp} is 1000 Wm^{-2}, R_{Sd} is 225 Wm^{-2}, diameter of mouse is 2 cm, length of mouse is 3.5 cm, ϕ is 30°, R_{ab} sun = R_{ab} night + 424 Wm^{-2}.

Q. 4. Depict the climate space diagram for a poikilotherm desert animal using following data.

M – LE = 0

T_b = 3, 30, 40 and 45°C

Wind speed (u) = 0.3 ms^{-1}

Diameter of animal = 1.5 cm

$a_s = 0.8$

Ap/A = 0.3

R_{ab} sun = R_{ab} night + 550 a_s (in Wm^{-2}).

Q. 5. In the example 19 if the environment is without sunlight with a radiant temperature of 0°C and air temperature of 10°C calculate

(i) Operative temperature change as wind changes from 3.7 to 185 mph.

(ii) The operative temperature change with increase of wind sped from 3.7 to 18.5 mph and the cylindrical diameter is 0.5 m.

[**Ans.** (i) 5.40°C to 6.5°C as h_c varied from 0.01 to 0.0183, (ii) No change almost].

Q. 6. Calculate the temperature difference ($T_c - T_a$) for a clothed man standing during the noon environment with the following data.

Diameter of man = 0.30 m
M_b/A_s = 60 Wm^{-2}
R_s = 700 Wm^{-2}
h_r = 6 Wm^{-2} °C^{-1}
u = 2 ms^{-1}
k_c = 0.035 Wm^{-1} °C^{-1}
l_c = 0.01 m
Solar angle = 60°
l_b = 0.1m
h_c = 7 Wm^{-2} °C^{-1}
a_s = 0.85
k_b = 2.6 Wm^{-1} °C^{-1}
E = –25 Wm^{-2}

(**Ans.** 29°C)

Q. 7. Compute the temperature difference ($T_c - T_a$) for a clothed person standing in an open environment using following data.

R_s = 698 Wm^{-2}	l_b = 0.1 m
h_r = 6 Wm^{-2} °C^{-1}	Diameter of man = 0.2 m
k_c = 0.03487 Wm^{-1} °C^{-1}	h_c = 6.98 Wm^{-2} °C^{-1}
M_{HP} = 0	k_b = 2.44 Wm^{-1} °C^{-1}
Solar altitude = 30°	E = -23.03 Wm^{-2}.

(a) When a_s = 0.7 and l_c is 0.0, 0.005, 0.01 and 0.02

[**Ans.** ($T_c - T_a$) = 15.5, 23.8, 32.1, 45.7°C]

(b) When l_c = 0.01m, a_s is 0.0, 0.2, 0.5 and 1.0

[**Ans.** $T_c - T_a$ = 21.7, 24.7, 29.2, 36.6°C].

Q. 8. A camel of 500 Kg body weight, surface area of 6 m^2 is capable of adjusting body temperature from 40°C in the afternoon to 34°C in the morning by adjusting the metabolic rate. Determine how much saving of body water is conserved. sp. heat is 3.4 J gm^{-1}.

[**Ans.** 40 Wm^{-2}].

Q. 9. Estimate the food requirement of a cow of 700 Kg body weight in a environment with equivalent temperature of 5°C with wind speed of 3m sec^{-1}, given that cow's diameter is 3m, and r_{Hb} of 2000 s m^{-1}, LE as 0.1M and body temperature -39°C.

[**Ans.** 14.86 MJ daily].

Q. 10. Draw a climate space diagram for a cattle (white cow) with following data

At thermal maxima	**At thermal minima**
T_b = 40.5°C	T_b = 42.0°C

$LE = 410\ Wm^{-2}$ $M = 100\ Wm^{-2}$

$M = 400\ Wm^{-2}$ $LE = 0.2\ M$

$r_{Hb} = 1420\ s\ m^{-1}$ $r_{Hb} = 1200\ s\ m^{-1}$.

Diameter of cow = 0.75 m

Length of cow = 1.80 m

Wind velocity for free convection = 0.1 m sec^{-1}

Wind velocity for forced convection = 2 m sec^{-1}

Direct solar radiation = 800 Wm^{-2}

Diffuse solar radiation = 100 Wm^{-2}

Total solar radiation = 900 Wm^{-2}

Albedo of cow's body surface = 0.3

Emissivity of cow's body surface = 0.98

Albedo of Lawn surface = 0.23

Emissivity of Lawn surface = 0.96.

Q. 11. Using the meteorological and physiological data available at your location, compare the different methods of human comfort indices to draw sensatory conclusions during the hottest and coldest month's daily data.

Q. 12. Develop milk yield and weather variable relationships at your location using past 10 years seasonal data.

References

Adolph, E.F. 1947. Tolerance to heat and dehydration in several species of mammals. Am. J. Physiol., 151, p. 564-75.

Alhassen, W.S. and V. Bavenendran, 1985.Weather variables and milk production in Friesian – bunagi crosses reared outdoors in the dry tropics. Agricultural and Forest Meteorology, 34, p. 163-71.

Anderson, S.R. 1965. Humidex calculation. Atmospheric environment service of Canada. C.D.S. No. 24-65.

Appleman, R.D. and F.G. Owens 1970. Relationship of the environment to calf health. ASAE Annual Meeting paper No. 70: 355, Minneapolis, July, 7-10.

Auliciems, A. 1983. Psycho – physiological criteria for global zones of building design. In: O. Oerdieck, J. Muller, and H. Leith (eds.). Proceeding of 9th Intern. Soc. of Biom. Conf. Part 2, Bimeteorology, 8, Lisse, Swets and Zetlinger, p. 69-86.

Austin, M.G. and J.W. Berry, 1956. Observations on one hundred cases of heat stroke. J. Amer. Med. Ass., 161, p. 1525-29.

Bedford, T. 1946. Environmental warmth and its measurement. Med. Res. Council, War Memorandum No. 17, Lodnon.

Bedford, T. 1948. Basic principles of ventilation and Heating, Lewis, London.

Belding, H.S. and T.F. Hatch, 1955. Index for evaluating heat stress in terms of resulting physiological strains. Heating, piping and air-conditioning, 27, p. 129-36.

Bernstein, M.H. 1971. Cutaneous water loss in small birds. Condor, 73, p. 468-69.

Bianca, W. 1961. Heat tolerance in cattle. Int. Jour. of Biometeor., 1, p. 5-30.

Bianca, W. 1976. The significance of meteorology in animal production Intern. Jour. of Biometeorology, 20, p. 139-56.

Bedford, T. 1936. The warmth factor in comfort at work. Report of Industrial Health Res. Board, 76, London, HMSO.

Blaxter, K.L. 1967. The energy metabolism of ruminants. Hutchinson.

Boerma, L. 1941. Mededelingen Chirug. Klin. v. Prof. Gerland, Groningen, Netharlands.

Bordon, D.W., J.F. Waddill and G.S. Grice, 1945. III. Statistical study of cases of heat disease. J. Amer. Med. Ass. 128, p. 1200-5.

Brooks, C.E.P. 1950. Climate in everyday life. Ernest Benn., London.

Brunt, D. 1943. The reactions of the human body to its physical environment. Quart. J. Roy. Meteorol., 69, p. 77-114.

Buettner, K.J.K. 1951. Physical aspects of human biometeorology, p. 1112, In: The compendium of Meteorology Amer Meteorol. Soc., Boston.

Burton, A.C. and O.G. Edholm, 1955. Man in cold environments. London, Edward Arnold Ltd., p. 273.

Campbell, G.S. 1977. Radiative heat loss from animals and man. Heat loss from animals and man (Monteith and Mount Eds.) Buttorworths, London, p. 33-58.

Carlson, L.D. 1958. Symposium on metabolic aspects of adaption of warm blooded animals in cold environment. Fed. Proc., 17, p. 1044-45.

Cena, K. and J.L. Monteith, 1975. Transfer processes in animal coats. II conduction and convection. Proc. R. Soc. Lond. B., 188, p. 395-411.

Cena, K. 1974. Radiative heat loss from animals and man. In: Heat loss from animals and man (eds.), J.L. Monteith and L.C. Mount. Butterworths, London, p. 33-58.

Court, A. 1948. Bull Amer. Meteorol. Soc., 29, p. 487.

Dalal, D.S., K.C. Arora and S.S. Rathi, 1991. Evaluation of performance traits in half bred cattle. Indian Jour. of Dairy Science, 44, 7, p. 401-5.

Das, D. and D.S. Balaine, 1980. Non-genetic factors affecting economic traits in Haryana cattle. India Jour. of Dairy Science, 33, 3, p. 406-08.

Das, G.C., D. Das, R.N. Goswami, T.C. Roy and N. Nahardeka, 2003. Factors affecting some production traits in Jersey, Red Sindhi and their crosses in Assam. Indian Veterinary Journal, 80, p. 126-27.

Dawson, T.J. and A.J. Hulbert, 1970. Standard metabolism, body temperature and surface areas of Australian marsupials. Am. J. Physiol., 218, p. 1233-38.

Deoker, D.K. and B.R. Ulmek, 2001. Studies on lactation length in Jersey Cattle. Jour. of Maharastra Agril. University, 26, 1, p. 104-06.

Dufton, A.F. 1932. Bldg. Res. Techn., paper 13.

Gagge, A.P., Stolwijk, J.A.J., and Y. Nishi, 1971. An effective temperature scale based on a simple model of human physiological regulatory response. ASHRAE Transactions, 77, p. 247-62.

Galvano, G. 1968. A statistical study on the effect of some climatic factors (temperature and humidity) and nutrition on milk production of Friesian Cattle. Animal Breeding Abstracts, 36, 3416.

Gates, D.M. 1980. Biophysical ecology. Springer – Verlag, N.Y., p. 611.

Gerret Rietveld, V.T. 2002. Heat stress in dairy cattle. Ontario Ministry of Agriculture, Food and Rural Affairs. In: WWW. Omaf. Gov. on. Ca.

Givoni, B. 1969. Man, climate and Architecture, Elsevier, Amesterdam.

Govoni, B. 1963. Estimation of the effect of climate on man. Development of a new thermal index. UNESCO Res. Rep., U.S. Dept. Hlth. Educ. Welf., Haifa.

Guerreno Mena, Y., Gomez Caberia, A., J.M., Serradilla Manrique, and C.C. Balch, 1998. Heat stress on milk production and milk protein content in Friesian Cows. Indian Jour. of Animal Sciences, 68, 9, p. 985-87.

Hancock, J. 1954. The direct influence of climate on milk production. Dairy Science Abstracts, 16, 2, p. 89-102.

Hardy, R.N. 1979. Temperature and Animal life. Edward Arnold, London.

Heffernon, G.A., H.J. Hittner and S.L. Kocis, 1953. Effective temperature as a measure of heat effect. U.S. Armed. Forces Med. J., 4, p. 73-80.

Hemmingsen, A.M. 1960. Energy metabolism as related to body size and respiratory surfaces, and its evolution. Rep. Steno. Hosp., 9, 2, p. 1-110.

Herrington, L.P. 1954. Biophysical adaptations of man under climatic stress. Meteorological monograph, 8, 30, Amer. Meteorol. Soc., Boston, Massachusetts.

Hildebrandt., G. 1962. Arch. Phys. Ther. (Leipzig), 14, p. 39.

Hobb, J.E. 1980. Applied Climatology, Dawson Westview, Boulder, Colorado, p. 218.

Irving L. and J. Krog, 1954. Body temperatures of arctic and sub-arctic birds and mammals. J. Appl., 6, p. 667-80.

Jaglou, C.P. and W.E. Milar, 1925. ASHVE Trans., 31, p. 89.

Johnson, H.D. 1965. Response of animals to heat. In: Meteorolograph Monograph, 6, 28, p. 109-119. Ed. P.E. Waggoner et al., 1965. America Meteorological Society, Boston, Mass, 02108.

Johnson, J.E. 1958. The effect of high temperature on milk production. Jour. of Heredity, 49, p. 65-68.

Kabuga, J.D. 1991. Effect of weather on milk production of Holstein-Friesian cows in humid tropics. Agriculture and Forest Meteorology, 57, p. 09-21.

Kadervel, R. 2002. Limited investment but consistent growth. In: Ravi, N. (ed.). The Hindu Survey of Indian Agriculture, 2002, p. 147-50.

Kathilankal, J.C. 2003. Studies on climatic parameters influencing milk production in cattle. M.Sc. Thesis Submitted to CCS Haryana Agril. University, Hisar.

Kerdö, L. 1967. Biometeorology II, Pergamon, Oxford, p. 147.

Kerslake, D.M. 1972. The stress of hot environments. Cambridge Unvi. Press, London, p. 316.

Kleiber, M. 1932. Body size and metabolism. Hilgardia, 6, p. 315-53.

Kleiber, M. 1965. Metabolic body size. Energy Metabolism (K.L. Blaxter Ed.) Academic Press, London, p. 427-35.

Ladell, W.S.S. 1955. Physiological classification of climates with reference to Nigeria. Proc. Intern. West African Congress, Ibadan, 1949, p. 1-18.

Ladell, W.S.S. 1957. The influence of environment in arid regions on the biology of man. Human and Animal Ecology/Ecologic Humanin et, Animals. UNESCO, p. 43-99.

Ladell, W.S.S., J.C. Waterlow and M.F. Hudson, 1944. Desert climate physiological and clinical observations. Lancet, 247, p. 491-7; p. 527-31.

Landsberg, H. 1954. Bioclimatology of housing. Meteorol. Monograph, 8, 81, Amer. Meteorol. Soc., Boston, Massachusetts.

Landsberg, H.E. 1969. Weather and Health – an introduction to Biometeorology Garden City, New York, Doubleday.

Largent, E.J. and W.F. Ashe, 1983. Upper limits to thermal stress for workmen. J Amer. Industr. Hyg. Ass., 19, p. 246-50.

Lee, D.H.K. 1957. Climate and economic development in the tropics. Harper, N.Y.

Lee, D.H.K. 1964. Intern. J. Biometeorol., 8, 1, p. 61.

Lee, D.H.K. and J.A. Vaughan 1964. Temperature equivalent of solar radiation on man. Intern. J. Biometeorol., 8, p. 61-69.

Lee, D.H.K. and A. Henschel, 1963. Evaluation of thermal environment in shelters. Rep. U.S. Dept. Helth Educ. Wefl., P.H.S., Tr-8.

Lowry, W.P. 1969. Weather and life – an introduction to Biometeorology, A.P., New York and London, p. 219-60.

Mahadevan, P. and H.G. Hutchison 1964. Animal Prod., 6, p. 331.

Mandal, D.K., S. Rao, K. Singh and S.P. Singh, 2002. Effect of macroclimatic factors on milk production in a Frieswal herd. Indian Jour. of Dairy Science, 55, 3, p. 166-70.

McArdle 1947. Mem. Med. Res. Counc. (London), Rep. RNP, 47/391.

McDowell, R.E., N.W., Hooren, and J.K. Camoens. 1976. Effect of climate on performance of Holstein first lactation. Jour. of Dairy Science, 59, p. 967-73.

Mclean, J.A. 1974. Loss of heat by evaporation. Heat Loss from Animals and Man (Monteith, J.L. and Mount, L.E. Eds). Butterworths, London, p. 19-31.

Minard, D., H.S. Belding and J.R. Kingston, 1957. Prevention of heat casualities, J. Amer. Med. Ass. 165, p. 1813-18.

Missenard, H. 1948. Chaleur et. Industrie.

Mitchell, D. 1974. Convective heat loss. Heat loss from Animals and Man (Monteith and Mount Eds.), Butterworths, London, p. 60-76.

Monteith, J.L. 1973. Principles of Environmental Physics. Edward Arnold, London, p. 241.

Morrison, P.R. and F.A. Ryser, 1952. Weight and body temperature in mammals. Science, 116, p. 231-32.

Odera, B.A., S.N. Kaushik and B.G. Kalpatal, 1978. Studies of variation in milk production in Gir Cattle. Indian Jour. of Animal Sciences, 48, p. 1-6.

Olgyay, V. 1954. Application of climatic data housing research. Housing and Home Finance Agency, Washington, D.C., p. 152.

Olgyay, V. 1960. Bioclimatic evaluation method for architectural application. In: Proc. of 2nd Intern. Bioclimatological congress held at London, 4-10 Sept., p. 246-62. Pergamon Press, Oxford, London.

Olgyay, V. 1963. Design with climate: Bioclimatic approach to architectural regionalism. Princeton Univ. Press, Princeton, N.J., p. 190.

Oosterlee, C.C. 1959. The influence of climatic factors on the heat balance and milk yield of cow in relation to the design of farm building in the Netharlands. Int. J. Bioclimatol. Niometerol. III, Sect. C.

Ozhan, M., N. Tuzemen and M. Yanae, 2001. Buyukbas Hayvan Yetistirma Ocuncu baski. Afatark Unviersitesi Zeraat Fakultesi offset Tesisi, Erzurum.

Palia, S.K. and C.L. Arora, 1983. Factors affecting production traits in Jersey cattle in temperate climate of Palampur. Indian Jour. of Animal Sciences, 53, 6, p. 642-44.

Panda, P.B. and D.D. Sandhu, 1983. Some genetic and non-genetic factors affecting milk production in crossbreds of Holstein and Jersey with Haryana and desi Bengal cows. Indian Veterinary Jour., 60, p. 99-106.

Primault, B. 1971. Conditions atmospheriques limits pour transport des animaux domestiques, p. 4. Rapport de Travail Inst. Suisse Meteorol., 15.

Primault, B. 1979. Optimum climate for animals. In: Agrmeteorology J. Seemann, Y.I. Chirkov, J. Lomas and B. Primault. (eds.). Springer – Verlag, N.Y., p. 182-89.

Ramachandran, V.S. and N.M. Murali 1992. On the relationship between milk yield and climatic factors, Mausam, 43, 2, p. 195-98.

Renner, C.W. 1976. Effect of climatic factors on performance of dairy cattle. Animal Breeding Abstracts, 46, 4296.

Rivolier, J. 1975. Intern. J. Biometeorol., 19, p. 12.

Robertshaw, D. 1976. Evaporative temperature regulation in ruminants. Progress in Animal Biometeorology (S.W. tromp and J.J. Bouma Eds.) Swets and Zeitlinger, Amesterdam, Vol. I, Part I, p. 132-39.

Robinson, D.E., G.S. Campbell, and J.R. King, 1976. An evaluation of heat exchange in small birds. J. Comp. Physiol., 105, p. 153-66.

Robinson, K.W. and D.H.K. Lee, 1946. Animal behaviour and heat regulation in hot atmospheres. Univ. Queensland papers, Deptt. of Physiology, 1, p. 1-8.

Robinson, K.W. and P.R. Morrison, 1957. The reaction to hot atmospheres of various species of Australian marsupial and placental animals. J. Cell Comp. Physiol., 49, p. 455-78.

Schickele, E. 1947. Environment and fatal heat stroke. Millitary Surgeon, 100, p. 235-55.

Schmdit–Nielsen, K. T.J., Dawson, and E.C. Crawford, 1966. Temperature regulations in the echidna. J. Cell Physiol., 67, p. 63-72.

Schmidt–Nielsen, K. 1969. The neglected interface: the biology of water as aliquid – gas system. Q.J. Biophys, 2, p. 283-304.

Schmidt – Nielsen, K. 1972. How animals work. London Combridge University Press.

Schmidt–Nielson, K., J. Kanwisker, R.C. Lasiewski, J.E., Cohn, and W.L. Bretz, 1969. Temperature regulation and respiration in the Ostrich. Condor, 71, p. 341-52.

Shinde, S. and V.K. Tanja, and Avtar Singh, 1990. Association of climatic variables and production and reproduction traits in crossbreds. Indian Jour. of Animal Sciences, 60, 1, p. 81-85.

Shinde, S. and V.K. Taneja 1986. Effect of physical environment on daily milk yield in crossbreds. Third world congress on genetics applied to livestock production, Lincoln, Nebraska, U.S.A., July 16-22.

Shinde, S.K. 1984. Effect of physical environment on reproduction and production traits in crossbred cows. M.Sc. Thesis submitted to Kuruksheltra University, Kurukshetra.

Shiv Prasad, Ramkaran Singh and G.S. Bist, 1999. Measure of persistency and its relationship with peak yield and lactation milk yield. Indian Jour. of Dairy Science, 5, 5, p. 308-14.

Shkolnik, A. and Schmidt – Nielsen, 1976. Temperature regulation in hedgehogs from temperate and desert environments. Physiol. Zool., 49, p. 56-64.

Singh, D. and A.S. Yadav, 1999. Studies on genetic and nongenetic factors causing fluctuation in progressive milk records of crossbred cattle. Indian Joru. of Dairy Science, 52, 1, p. 31-35.

Siple, P.A. 1960. Climatologist and Military Geographer, Gen. Staff, Deptt. of Army.

Siple, P.A. 1945. Proc. of Amer. Phil. Soc., 89, p. 177.

Siple, P.A. and C.F. Pssel, 1945. Measurements of dry atmospheric cooling in subfreezing temperatures. Proc. Amer. Phi. Soc., 89, p. 177-99.

Smith, C.V. 1964a. Ventilation and associated patterns of air flow. Agril. Meteorol., 1, p. 30-41.

Smith, F.E. 1955. Indices of heat stress. Med. Res. Council, Memorandum No. 29. (McArdle, 1947).

Smith, L.P. 1967. Forecasting annual milk yield. Agricultural Meteorology, 5, p. 209-14.

Smith, S. and V.K. Teneja 1986. Effect of physical environment on daily milk yield in crossbreads. Third World Congress on Genetics Applied to Livestock Production, Lincohn, Nebraska, USA.. July 16-22.

Steadman, R.G. 1971. Indices of windchill of clothed persons. J. Appl. Meteorol., 10, 4, p. 674-83.

Szues, E., Mika, J. J. Nagy, T. Tran-Anh, I. Gyorkjos, and A. Kovacs, 2001. The role of meteorological factors in milk production of Holstein – Friesian cows. Allattenyesztes – es – Taka – rmanyozas, 50, 3, p. 215-28.

Szokolay, S.V. 1985. Thermal comfort and passive design. In: K.W. Boer (ed.) Advances in solar energy -2, N.Y., American Solar Energy Society.

Thom, E.C. 1957. Air conditioning Heat. Ventil., June, 1957.

Thomas, C.K. and R.M. Acharya, 1981. Notes on the effect of physical environment on milk production in Bos indicus x Bostaurus crosses. Indian Jour. of Animal Sciences, 51, 3, p. 351-56.

Thomas, M.K. and D.B. Boyd, 1957. Candian Geograph, 10, p. 29.

U.S. National Weather Service 1976. Livestock hot weather stress. Regional operations manual letter C-31-76, p. 5.

Univ. of Wisconsin Cooperative Extension, 2002. Heat stress based on temperature humidity index.

Vernon, H.M. 1930. J. Physiol., 70, p. 15.

Whittemore, C.T. 1980. Lactation of dairy cow. Longman, London and N.Y.

Wilson, O. 1962. Metabolism, 5, p. 543.

Winn, P.N. and F.F. Godfrey 1967. The effects of humidity on growth and feed conversion of broiler chickens. Int. J. Biometerol., 11, p. 39-50.

Winslow, A.E.A., L.P., Harrington and A.P., George, 1937. Physiological reactions of the human body to varying environmental temperature. American Jour. of Physiology, 120, p. 1-20.

Yaglou C.P. 1926. The thermal index of atmospheric conditions and its application to sedentary and industrial life. J. Industr. Hyg., 8, 15-19.

Yousef, M.K. 1985. Stress physiology in livestock. Vol. II. Ungulates CRC press Inc., Boca. Raton., Florida.

Yousef, M.K. 1972. Physiological adaptation, Desert and Moutain. Academic Press, New York, p. 107.

7

Monsoon Agroclimatology

7.1 Introduction

Regional weather and climate are manifestations of the global climate system involving the atmosphere-oceans-land surface-cryosphere interaction in complex ways over the different range of time scales. It has been difficult to predict the interactive system with the use of global circulation models.

The major weather systems and atmospheric motion over the earth are created by the unequal heating between the equator and the poles (Schneider and Temkin, 1977). Stommel (1979) reported that ocean heat transport over the globe plays important role in atmospheric motions (Fig. 7.1). Tropics intercept a much larger fraction of the incoming solar energy than do the polar zone giving rise to the motions which regulate the climate. Temperature gradients between equator to poles raises the warm air in the tropics which moves polewards to cooler regions until a state of quasi-equilibrium is attained. Excess heat from low to high latitudes takes place in both hemispheres. Due to rotation of the earth, air motions are complicated than in case of a simple convective circulation. Therefore as the air parcel moves polewards, its radial distance from the earth's axis decreases and gains momention due to fast rotational speeds of the equator to create westerly winds in mid-latitudes. Similarly, the return flows in the lower layers of the atmosphere become easterlies which are called trade winds. Due to tilt of earth's axis, the zone of maximum heating shifts north-south during the year creating the summer to winter seasonal cycle. The system is continuously changing to acquire equilibrium at any location. Therefore, weather and climate are the

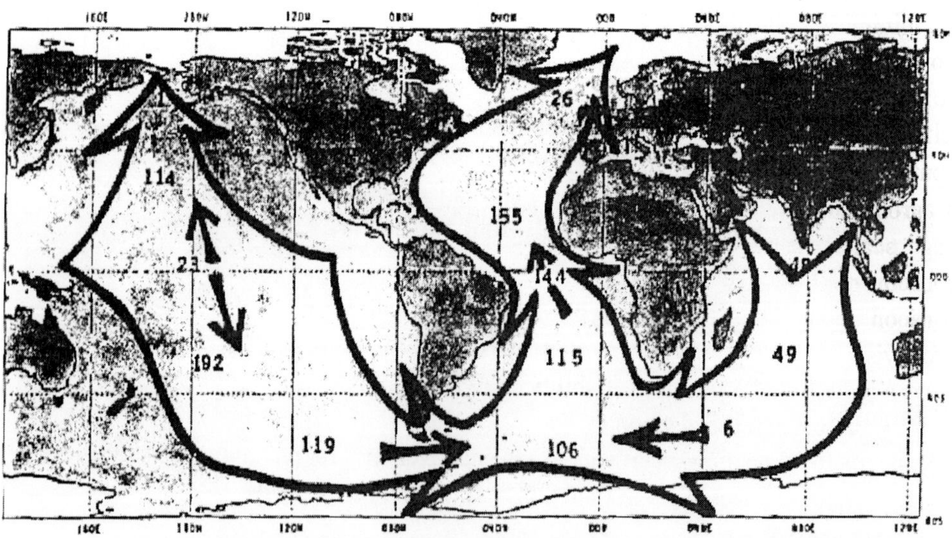

Fig. 7.1: The ocean mean heat transport in 10^{13}W (After: Stommel, 1979).

consequences of differential solar heating, flow instabilities, and interactions between the atmosphere and the underlying ocean land surfaces, vegetation, snow/ice covers. The climatic fluctuations are also influenced by changes in the radiation balance components with the time.

A vast amount of literature is available for understanding, modelling and prediction of the monsoon system during the last 50 years. Several field observational programmes have been undertaken at National and International levels beginning with IIOE (1963-65) to ARMEX (2002, 2003) programmes to provide new insight in the monsoon description. Now it has been conformed that the monsoon circulation is of global scale primarily driven by the changing meridional thermal gradients from winter to summer to winter following the seasonal migration of the sun's position relative to the earth's equator, and is the largest perturbation in the global general circulation during any year. Ocean surface temperature, land-sea temperature differences, position of the ITCZ, interaction with trade wind system and tropical/ mid-latitude feedback are all the parts of the inherent processes which together evolve the characteristics of the monsoon circulation. Adequate experimentally observed data is available to understand the regional/global monsoon system as a complex interacting land-biosphere ocean-atmosphere system over monsoon region.

Agricultural production in the monsoon region is strongly influenced by the monsoon rainfall in summer as well as winter seasons. Monsoon rainfall influenced both the short term (tactical) and long term (strategic) decisions in order to harness the maximum benefit in quality and quantitative natural production. Almost every aspect of agricultural crop production is influenced including operational, cultural, pest-diseases occurrence and their multiplication, soil water availability to plants, proper choice of crop and their genotypes, and minimization of adverse effects of abnormal weather associated features during crop growing seasons (frost, hailstorm, drought/wet spells, etc.). Therefore a through vigilant monitoring of crop weather interaction is required in crop growing seasons in monsoon belt to achieve maximum advantage through the management of monsoonal non-cash inputs. Almost half of the world's population is depending on monsoon's behaviour due to agrobased economy. Therefore weather based forecasts at short, medium and long range scales are of immense importance for the farmers to avail the advantages of monsoon rains as and when they occur and to have a grasp of the characteristics of monsoon features in monsoon belt for their utilization and developing need based agroadvisories.

7.2 Important Concepts/Terminologies

El Nino (EN): A term used by Peruvian Firshermen to describe the appearance of warm water in Peru coast (8°S, 80°W) and near Peru Ecuador Coast (0-10°S, 80°-78°W) (Huskhe, 1959); Trenberth, 1997) regions in late December which is sometimes so intense that temporarily displaced the nutrient rich upwelling waters. Every 2 to 7 years this warming apparently spread to Central and Eastern Pacific (Quinn et al., 1978, 1987).

La Nina(C): The non El Nino phase is called La Nena. During a La Nena phase equatorial Central and Eastern Pacific sea surface temperatures are generally below normal. During this phase large scale Pacific tropical convective zone is more or less in its climatologically normal position over the maritime continent (Wright, 1984).

Southern Oscillation Index: Walker around (1923) characterised southern oscillation as "when pressure is high in Pacific Ocean, it lends to be low in the Indian Ocean from Africa to Australia; these conditions are associated with low temperatures in both these areas and rainfall varies in the opposite direction to pressure." The anomaly of pressure difference between Tahiti (18°S, 149°W) and Darwin (12.30°S, 131°E) is adopted as an index of southern oscillation (Fig. 7.2) represented as ISO or SOI (Trenberth, 1976; Wright, 1975).

Pressure See-Saw: the alteration in the sign of pressure anomaly across the tropical oceans. For example, southern oscillation is a global scale see saw of surface pressure with centres of action around Indonesia, North Australia, and SE Pacific.

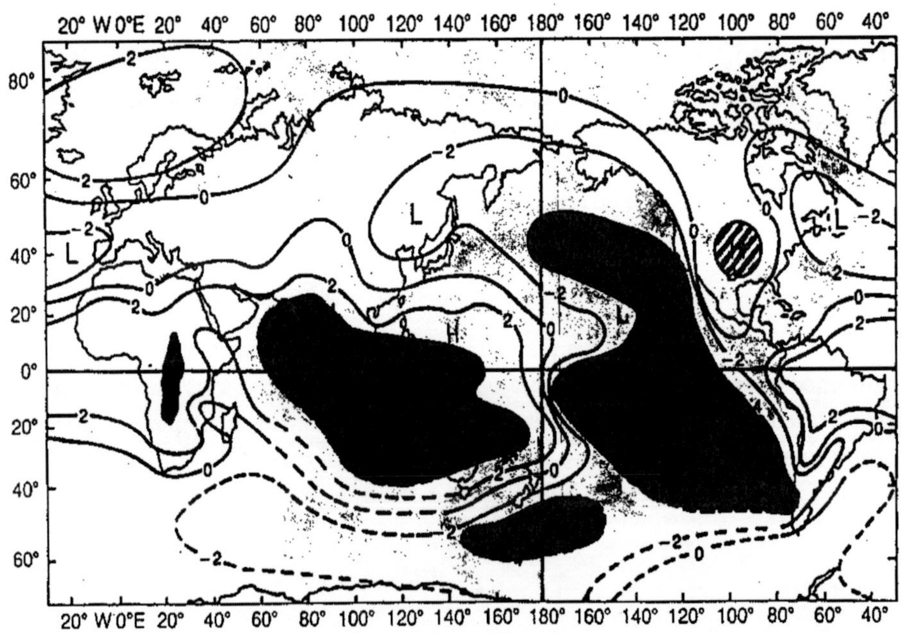

Fig. 7.2: The high positive correlation (X10) of the annual mean sea-level pressues with those for Darwin, Australia and high negative correlations over Tahiti - showing the basis for the choice of the Tahiti - Darwin pressure difference as a Southern Oscillation Index (SOI) (After: Trenberth, 1976).

ENSO: Southern oscillation is linked with El Neno phenomenon to understand global climate system. When SO is below normal, the sea surface temperature (SST) off the coast of South America is substantially above normal and accompanied by above normal SST's over the equatorial, central and eastern Pacific areas indicative of global scale climatic variations. Monthly anomalies are standardised by dividing their long-term standard deviation (Fig. 7.3). (Rasmusson, 1984).

Walker circulation: It is a large scale zonal, west-east, overturning circulation in the equatorial plane. These are large scale tropical convection which acts as a major driving force of the atmospheric circulation through the release of latent heat in mid-troposphere.

Oceanic Kelvin wave: It is an transient response to an abrupt change in oceanic forcing due to atmosphere-ocean interactions in the form of wind induced upwelling and advection, modulation of radiation by cloudiness and heat transfer.

Anticyclone: A high pressure system around which the air moves counter clockwise in the southern hemisphere and clockwise in the northern hemisphere.

Anticyclonic flow: Air motion that turns in a pattern opposite to the earth's rotation clockwise in the northern hemisphere, and counter clockwise in the southern hemisphere.

Cold front: The leading edge of relatively cold air replacing warm air.

Convergence: An atmospheric state that results in the net horizontal inflow of air into an area.

Cyclogenesis: The development of a cyclonic flow pattern in the atmosphere.

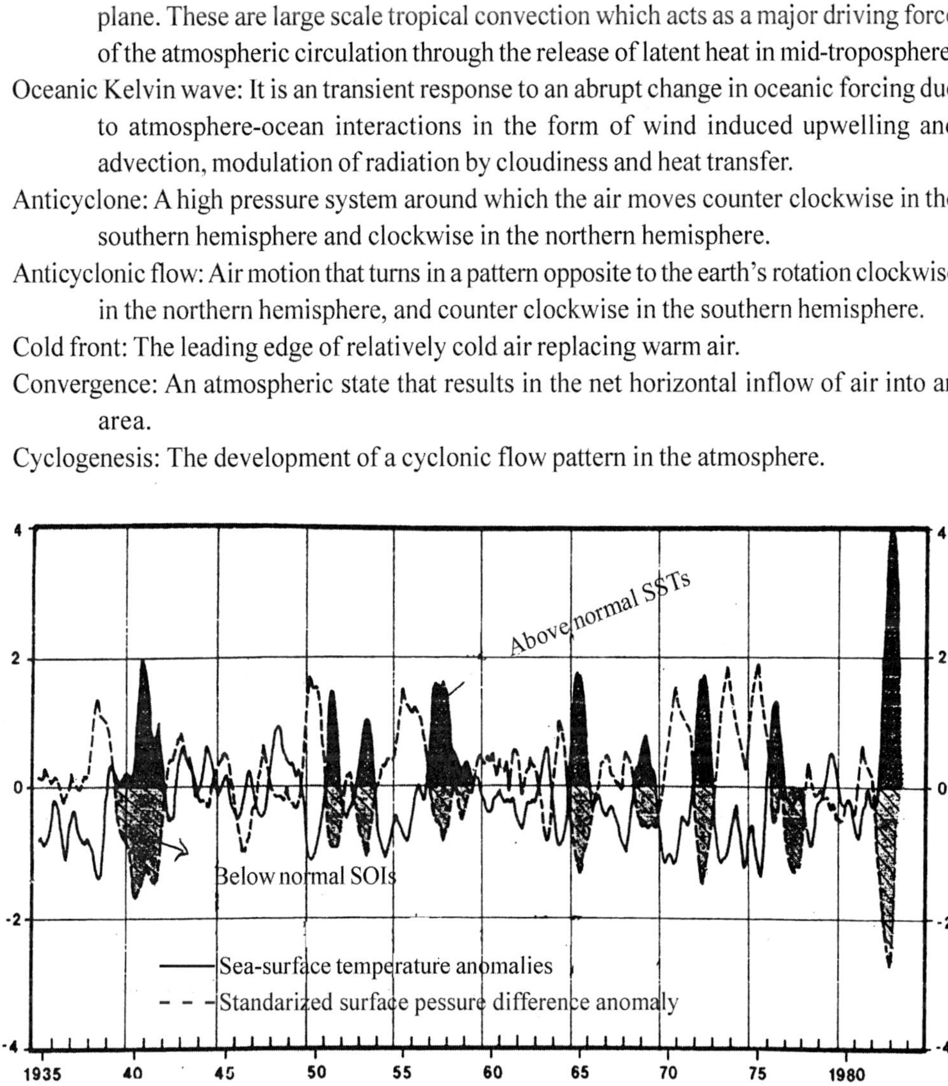

Fig. 7.3: Standardized curve for an index of El Nino-Southern Oscillation (ENSO) (From: Rassmusson, 1984).

Cyclone: A low pressure system having a clockwise rotation in the southern hemisphere and a counter clockwise rotation in the northern hemisphere.

Cyclonic flow: Air motion turning on a pattern having the same direction as the earth's rotation, counter clockwise in the northern hemisphere and clockwise in the southern hemisphere.

Divergence: An atmospheric state that results in the outflow of air from a region.

Easterly waves: A trough in the tropical easterlies that migrates westwards.

Hadley cells: A meridional circulation resembling convective model at low latitudes consisting of ascent near the equator and decent in the sub-tropics, and trade wind at the surface.

Horse latitudes: The oceanic regions near 30-35° latitude characterised by hot, dry conditions with light winds.

Inter tropical convergence zone (ITCZ): A zone at low latitudes seen as a band of cloudiness in satellite photograph caused by the convergence of trade winds from both hemispheres (Fig. 7.4).

Inversion: An increase of environmental temperature with altitude.

Jet stream: A concentration of relatively strong winds within a narrow band in the atmosphere.

Maritime: Air that has developed high humidity at low levels by its residence over a broad water surface.

Meridional: To the north or south, longitudinal.

Wind shear: A local variation of wind in a given direction.

Perturbations: Variations in the assumed stable state.

Warm front: The boundary between cold and warm air masses that moves such that warm air replaces cold air at the surface.

Platenary temperature: The equilibrium temperature controlled by the heat balance of the earth-atmosphere system.

Source region: An extensive area of the earth's surface over which bodies of air remain sufficiently long to acquire characteristic temperature and moisture properties.

Zonal: To east or west latitudinal.

Steering: Movement of a weather system by the atmospheric winds.

Subtropical high: A semi-parmanent cell of high pressure that is strong over the subtropical oceans in summer.

Subtropical jet: The upper-level zone of high winds that develops at 20°-30° latitude to the poleward side of the Hadley cell.

Surge: An abnormally large rise of the sea caused by storm winds.

Trade winds: A system of surface winds that flows equatorward and westward from the subtropics.

Tropical cyclone: A low pressure system with high winds that develops over the tropical oceans.

Tropical depression: A stage of a tropical cyclone with wind up to 27 knots and one or more closed isobars.

Tropical disturbance: A stage of tropical cyclone with upto one closed isobar and a slight surface circulation.

Trough: An elongated region of relatively low pressure.

Turbulence: Motion that has characteristically random, irregular velocities.

Vorticity: The vector measure of the local rotation rate in a flow pattern. The spin in referred as vorticity.

Westerlies: The atmospheric motion in middle latitudes from west to east.

Composite ENSO year: The composite ENSO ("0") year is definsed as beginning in January (0). For a 12 month period from May of the ENSO year to April of the following year is depicted as May (0) to April (+) represents abnormal precipitation for a 12 month period from May of ENSO year to April next.

La Nena Phase: La Nena phase is defined as a period taken for the heat stored in the tropical ocean to be recharged (Wyrtki, 1985).

Warming of sea surface water (w): Presence of maximum (positive anomalies) in the sea surface temperature (SST) in the eastern equatorial Pacific (5°N – 5°S, 180° - 90°W) (Wright, 1984; Angel, 1981).

ENSOW – U: El Neno (EN) during which the SOI is minimum and the equatorial eastern Pacific sea surface temperature (SST) had a maximum (W) in the middle of the calendar year i.e., June to September the period of Indian monsoon season then event ENSOW is unambiguous (U).

ENSOW – A: El Neno years (EN) during which the SOI minima or SST maxima occurred in earlier or later part of the year (not middle of year), then ENSOW events are termed as Ambiguous (A).

Climate variability: IPCC (1996) referred climatic variability to variability observed in climate record in periods when the state of climate system is not showing changes. If the climate state changes, characterised by shift in means, then the frequency of rare events on the side the mean has shifted and might occur more frequently with increasing climatic variability (Salinger, 1994).

7.3 Asian Monsoon as a Global Phenomenon

Monsoon are gigantic convectional system produced by differential seasonal heating of huge Asiatic and Oceanic surfaces. Since seasonal reversal of prevailing winds take place, it is necessary to consider upper air monsoon circulation which is controlled by annual oscillation of temperature. Therefore Flohn (1957) proposed monsoon as a normal seasonal migration of planetary winds following the Sun's trajectory because during northern summer wind belts are displaced towards north, whereas in northern winter shifts take place towards south resulting the mean zonal axis of equatorial winds coinciding with 10°N parallel in July, and 5°S parallel in January-providing two different circulation patterns in the low latitudes according to the land-oceanic interactions.

During winter, intense cooling of huge landmass of Asia forms the high pressure system over the continent with anticyclone centre around 1035 mb over Siberian region along with a secondary anticyclone of 1040 mb level in NW Asian region. Adjoining northern Pacific Ocean is warmer with around 1020 mb level. However, the situation in summer is totally reversed. Thus monsoons are large scale seasonal wind systems blowing over vast areas of the globe in the same direction, and are reversed with the change of season in simple descriptive terms.

Monsoon meteorology has tremendously advanced since 1965s from purely descriptive and empirical approaches to more dynamical and modelling approaches for diagnosing its structure, variability on the synoptic and intraseasonal oscillations (ISO) and interannual scales, and its prediction on meso, short, medium and long range scales using dynamical models. Trade winds flow between equator to both sides-NE winds between equator to 30°N and SE winds between equator to 30°S. The converging place of SE and NE trades from both hemispheres around equatorial belt is called Inter Tropical Convergence Zone (ITCZ) and the low pressure belt winds are calm and variable called equatorial trough. Around 30°N or 30°S there exists high pressure belts with variable winds called subtropical high pressure cells. From these subtropical highs, winds are westerlies towards the poles known as the mid-latitude westerlies with less intensity as compared to the trade winds. But they become stronger polewards. Polar regions are dense because of the low temperatures and are characterised by high pressure. Southern hemispheric westerlies, being oceanic, are stronger and exhibits relatively less seasonal variation in the position of the subtropical highs. High pressure area of winter season over Asian region becomes one of a low pressure region during summer. Therefore winds over the southern portion of the Asian continent completely change their direction from winter to summer reflecting the major monsoonal belt of the world. The general circulation features vary from January to July with attributes to the position of Sun, thermal energy distribution and wind pressure distribution with land-sea interactions (Rama Sastry, 1984).

A comprehensive analysis of global atmospheric research programme experiment on monsoon phenomenon (FGGE) with its regional component MONEX (Monsoon Experiment) in 1979 revealed the Asian Monsoon as a global phenomena. Monsoon system influences the large scale dynamics of entire tropical belt. The monsoon circulation has been described as an equatorially trapped waves in response to localized release of heat (Cadet, 1990).

Fig. 7.4 shows the annual cycle of the planetary scale monsoons with velocity potential at 200 hPa for January and July, 1979—the ECMWF data set to depict the global scale monsoon system circulation in northern hemisphere winter and summer seasons (Krishnamurti and Surgi, 1987). Winter months are characterised by intense Hadley cell, whereas in summer

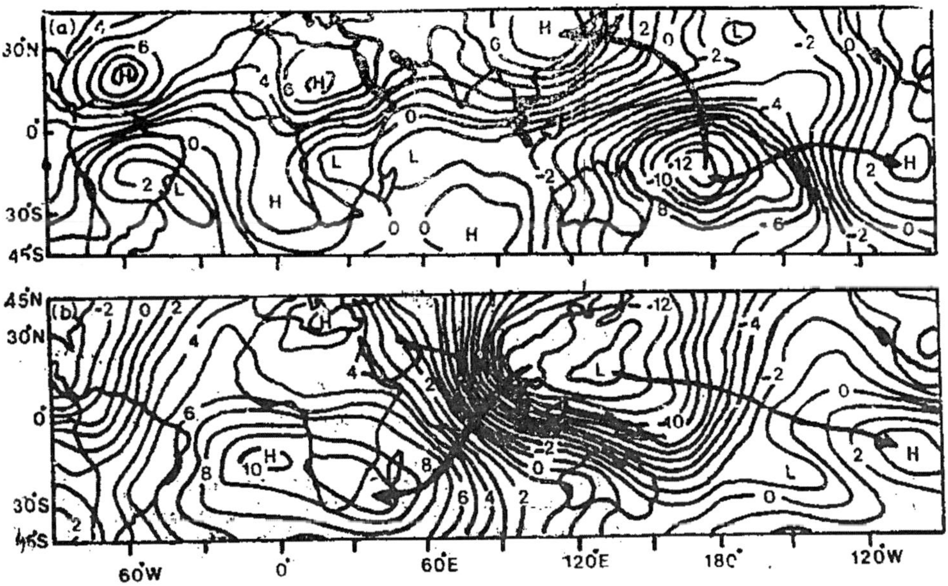

Fig. 7.4: Velocity potentials: (a) at 200 hPa for the month of January 1979 based on European Center for Medium Range Weather Forecasts (ECMWF) data set, (b) at 200 hPa for the month of July 1979. (From: Krishnamurti and Surgi 1987).

months east-west circulations (Walker east-west circulation) are related to the global scale summer monsoon system. Monsoon circulation is considered as a equatorially trapped waves. A prominent signature of the Asian Monsoon is the cross-equatorial wind component directed into the summer hemisphere for the water vapour balance, which is reversed in the winter season. In simple words 'Monsoon' is adopted for wind system where seasonal reversal takes place at least 120° reversal in wind direction with change of season (Ramage, 1971), however Indian Monsoon is an ideal monsoon with immense seasonal wind regimes covering boundaries between 35°N and 25°S and between 30°W and 173°E.

Considering monsoon system as a global phenomenon, a series of International Field Experiment Programmes have been undertaken in the past to understand the genesis of monsoon such as International Indian Ocean Expedition (IIOE) programme during 1963-66, Indo-Soviet monsoon experiment (ISMEX – 73), International Monsoon Experiment (MONSOON-77), first global atmospheric research programme experiment-FGGE and its regional component MONEX-79 during summer 1979. Analysis of this vast experimental data have changed the conceptual understanding of Indian summer monsoon from a regional phenomenon to a global phenomenon. Monex-79 results have shown that the monsoon circulation is in the form of equatorially trapped Kelvin and mixed Rossby – planetary waves which provided the theortically framework to understand the monsoon circulation as a global phenomenon (Cadet, 1990). Since then a number of national field programmes have been launched to understand monsoon dynamics such as Indian Ocean Experiment (INDEX-1980), Monsoon Trough Boundary Layer Experiment (MONTBLEX-1989-90), Land Surface Processes Experiment (LASPEX-1995-96), Bay of Bengal Monsoon Experiment (BOBMEX-1999), Arabian Sea Monsoon Experiment (ARMEX-I in 2002 and ARMEX-II in 2003).

(Cadet 1990) reported that the time averaged motion field around the global tropics in the lower and upper troposphere presented the importance of the monsoon system in the global circulation of the atmosphere. The climatological pattern in the lower troposphere continent (at 850 hPa) asymptotes were well defined between November and April over the eastern Indian Ocean characterising the areas with deep convective cloud activity and intense monsoon rainfall. In upper troposphere, the northern hemisphere winter months were characterized by a pair of upper anticyclones located on poleward sides of the near equatorial lower troposphere convergence zone. The northern anticyclone migrated poleward from northern Malaysia during winter months to reach the Tibetan plateau by July with its maximum amplitude during August. At the end of the summer monsoon with withdrawal of rains, this northern anticyclone shifted back towards equator. Fig. 7.4 shows the annual cycle of the planetary scale monsoons depicting the velocity potentials at 200 hPa for January and July, 1979, expressing the role of monsoon system in global circulations of northern hemisphere winter and summer seasons.

Winter months are characterized by intense Hadley cells, whereas summer months with east-west circulations known as Walker east-west circulations over the Pacific Ocean extending even beyond equatorial belt.

The annual cycle of divergent circulation (divergent wind vector V_D and velocity anomaly χ) at 850 mb (Fig. 7.5) and 200 mb (Fig. 7.6) during the northern summer presents heated Asian landmass which induced the upward motion with convergence of air mass from the oceans over south and southeast asia towards the Asian continent. Air mass at 200 mb diverges away from the Asian continent simultaneously towards the oceans forming a return flow clearly inducing an annual cycle of divergent circulation over Asian Monsoon system (Chen, 1990).

The northward divergent wind of annual cycle component at 850 mb induced the coriolis force which maintains the low level Indian Monsoon Westerlies extending from the east coast of Africa to Himalaya. Since the low level Indian Monsoon Westerlies belong to the part of the Indian Monsoon circulation, this annual cycle of the 850 mb divergent circulation maintains the low level Indian Monsoon. Similarly the southwest divergent wind of annual cycle component at 200 mb induced the coriolis force which supported the tropical easterly jet (TEJ) located between two upper-level monsoon anticyclones to the north and south (Krishnamurti, 1979; Chen and van Loon, 1987). It is worthwhile to mention the three important features of the summer monsoon as:

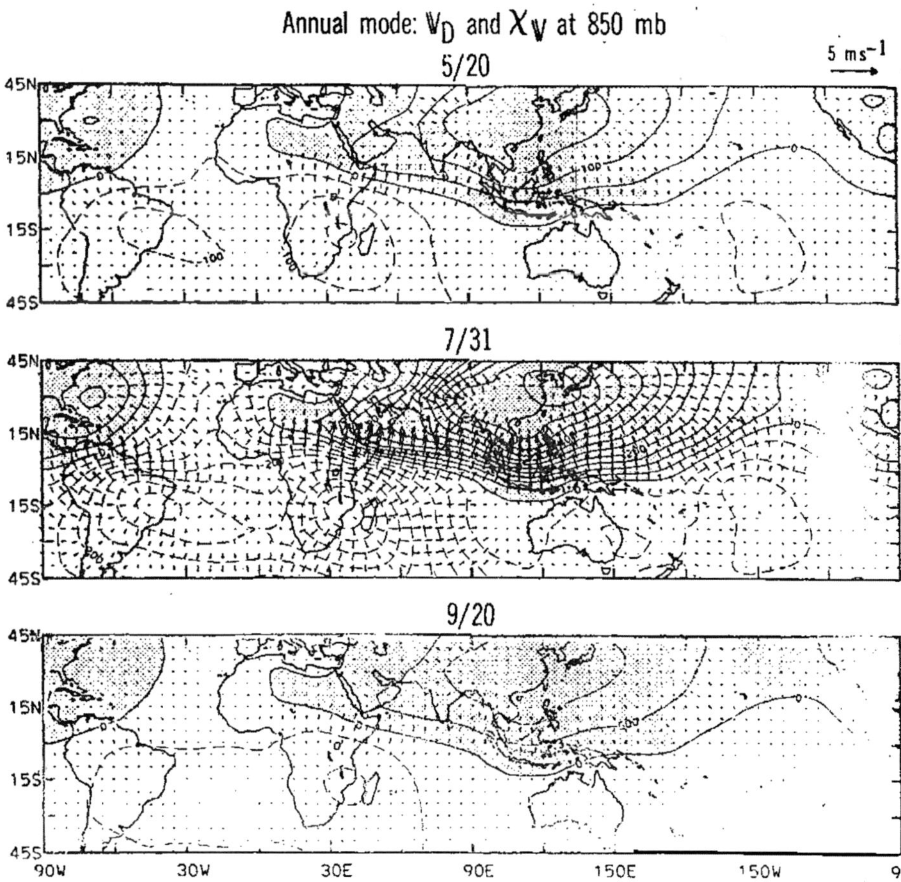

Fig. 7.5: The annual cycle of the divergent circulation (V_D, χ) at 850mb obtained by multiplying the annual cycle harmonic of eigencoefficient time series and eigenvector of the first eigenmode. The contours are the annual cycle anomaly field. The positive values of the χ anomaly field are shaded. The coutour interval is 5×10^6 m^2s^{-1}. (After: Chen, 1990).

(i) The low level flow characterized by intense cross equatorial dynamics of air and moisture flux over the western Indian Ocean by balancing forces with the planetary boundary layer.

(ii) Tibetan plateau which acted as an barrier for monsoon flow as well as an elevated heat source generating temperature gradient in summer as large as land-ocean contrast at the surface which generated the upper level anticyclone over the domain during the summer. Luo and Yanai (1984) analysing this apparent heat source and moisture sink over the plateau reported that sensible heat source over western part of plateau develops dry convections, whereas large values of apparent heat source

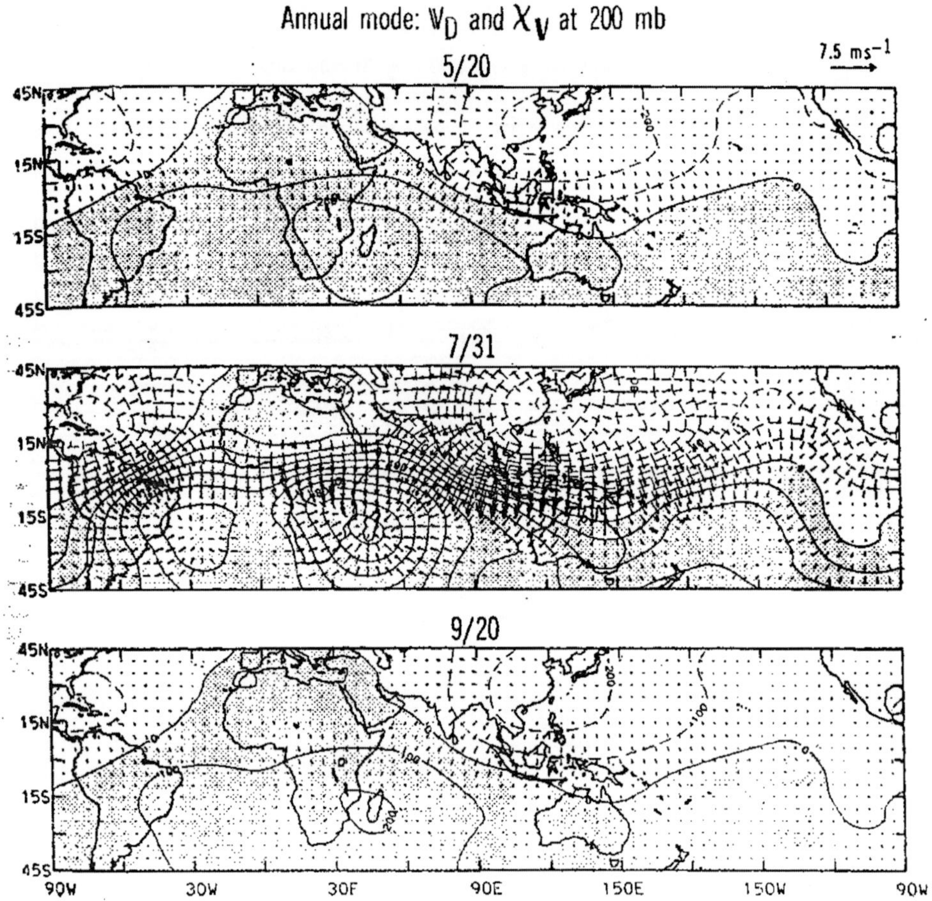

Fig. 7.6: Same as Fig. 7.5 for 200 mb (After: Chen, 1990)

and moisture sink over eastern plateau induces regional monsoon circulation i.e., the development of a low pressure system and upper level Tibetan anticyclone. The heated plateau induces a convergent motion towards the plateau in the middle and lower layers with a compensating divergent flow outward from plateau in the upper layer.

(iii) East-west circulations which derived the planetary scale tropical motions in the northern summer is the major component of the total varience velocity potential (divergent circulation) as well as covariance of the vertical velocity and temperature on the planetary scale since the transfer of available potential energy takes place from the synoptic scale perturbations to the planetary scale systems (Cadet, 1990). When east-west circulations are intense, there is increase of activity leading to tropical depressions, storms and easterly waves of the order to few thousand kilometres. These synoptic scale perturbations takes place in tropics over longitudes along the planetary waves with maximum synoptic scale activity over the monsoon domain between 35°N and 25°S and between 30°W and 173°E. The available potential energy at planetary scale maintained the East-West and Hadley circulations which give rise to tropical Easterly Jet (TEJ)—a major characteristic of the northern summer system.

Krishnamurti (1971) stressed the importance of the global setting of monsoon with planetary scale organization of the summer monsoon circulation system in the upper troposphere with wave number two structures (two centres of divergence over continental regions-Tibetan high and the Maxican high and corresponding centres of convergence over the mid-oceanic region – mid-Pacific and mid-Atlantic troughs).

7.4 Wave Nature of Asian Summer Monsoon

Madan and Julian (1971) reported the frequency oscillation over western Pacific Ocean, but with MONEX-79 experimental observations data strengthened its existence over the Indian ocean due to the break and active oscillation cycle of Indian summer as characterized by an out of phase relationship between convection over central India and equatorial latitudes (Yasunari, 1979; Sikka and Gadgil, 1980). Using FGGE/MONEX data Krishnamurti and Subrahmanyam (1982) found the relationship between the phase of the global low-frequency mode in the tropics and monsoon activity (Fig. 7.7) by ideantifying the motion systems propagating meridionally from the equatorial latitudes to the Himalayas. The active and break cycles of the monsoon were closely coupled to the passage of troughs and ridges of these systems over India. Chen (1990) substantiated the coincidence of these troughs (T)/

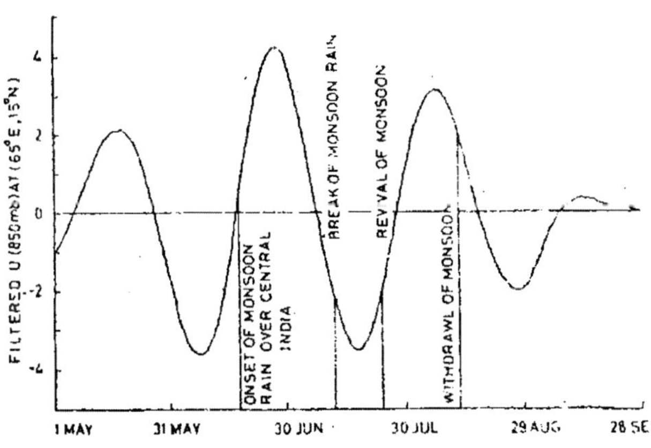

Fig. 7.7: The Observed 30-50 day filtered zonal wind at (65°E, 15°N) for a 120 day period from 28 April to 25 August, 1979. (After: Krishnamurti and Subrahmanyam, 1982).

ridges (R) with the upward/downward branches of 30 – 50 day transient Hadley circulations with y – t diagram of 30 – 50 day outgoing longwave radiation (OLR) (Sikka and Gadgil, 1980 proedure) superimposed with locations of troughs/ridges along 75°E (Fig. 7.8). The negative values of 30 – 50 day filtered OLR indicated deep cumulus convection. Krishnamurti and Subramanyam (1982) clearly depicted that the troughs are associated with cumulus convection which is maintained by the upward motion of the 30 – 50 day local Hadley oscillation. These systems over Asia represent a facet of a global phenomenon which may differ with different frequency modes in different regions. Indian monsoon occurs with 30 – 50 day mode during northern summer (Lorenc, 1984). Kasture and Keshvamurty (1987) also studied 30 – 50 days oscillation and found that it extends throughout the troposphere with slow northward movement and interannual variability.

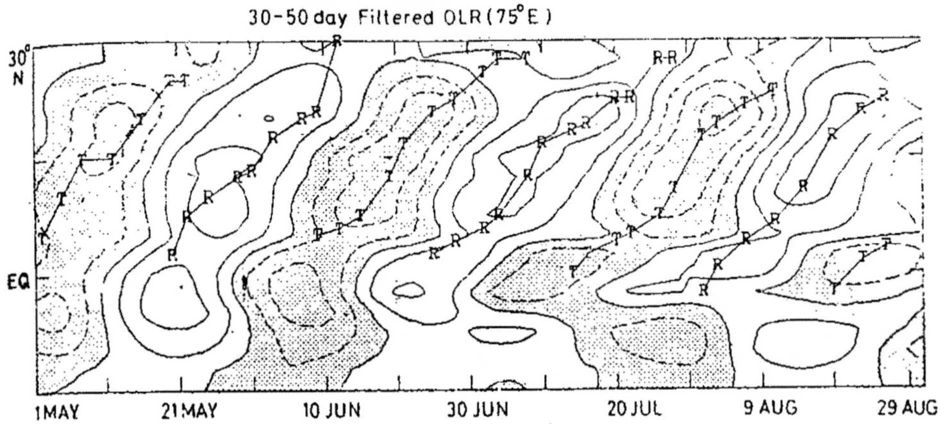

Fig. 7.8: The latitude-time diagram of the 30-50 day filtered OLR superimposed with locations of the KS troughs (T) and ridges (R) along 75°E. The contour inteval is 8 Wm^{-2} and the negative values are shaded. (After: Chen, 1990).

Chen (1990) used the intensity variation of Somali jet over the 1979 Indian summer monsoon as an index. The time series of second order Butterworth band pass filter (Murakami, 1979) 850 mb zonal wind (u) at (65°E, 15°N), the core of somali jet, to extract the 30 – 50 day oscillation of Somali jet (Fig. 7.8). The phase relationship between the annual cycle and 30 – 50 day mode suggested that Indian Monsoon life results from the modulation of the annual cycle during northern summer by 30-50 day mode. The modulation process was accomplished through the coupling between the planetary scale 30 – 50 day mode and 30 – 50 day troughs and ridges induced by the local Hadley circulation by the former mode. Fig. 7.8 clearly presents the eastward propagation of the 30 – 50 day divergent mode. The divergence or convergence centres appeared over the western equatorial Indian ocean and then they propagate towards western Pacific (around 20°N) in around 20 days. The planetary scale 30 – 50 day divergent mode at the upper level is synoptically related to the low level 30 – 50 day mode over the Indian Ocean. These trough and ridge lines (Fig. 7.8) also matched well with the areas of 30 – 50 day filtered 200 mb divergence and convergence respectively indicating the coupling with the planetary scale 30-50 day divergent mode. However, at low level there is direct link between the low level troughs and ridges and the planetary scale divergent mode can be established through the construction of the transient local Hadley circulation. The troughs and ridges are associated with the upward and downward branches of 30 – 50 day transient local Hadley circulations. The deep cumulus convection can be maintained and suppressed respectively by the upward and downward branches of the 30 – 50 day local Hadley circulation which is also supported by plot of 30 – 50 day filtered OLR data (Sikka and Godgil, 1980).

Therefore the coherent propagation relationship between low frequency mode at 200 mb and troughs/ridges at 850 mb (Fig. 7.8) over the Indian Ocean and Indian subcontinent presents that the northward propagation of the latter is steered by the former. These northward travelling 30 – 50 day troughs/ridges interact with the Indian Monsoon trough situated over central India to cause its deepening and filling (Fig. 7.9). This modulation results in the

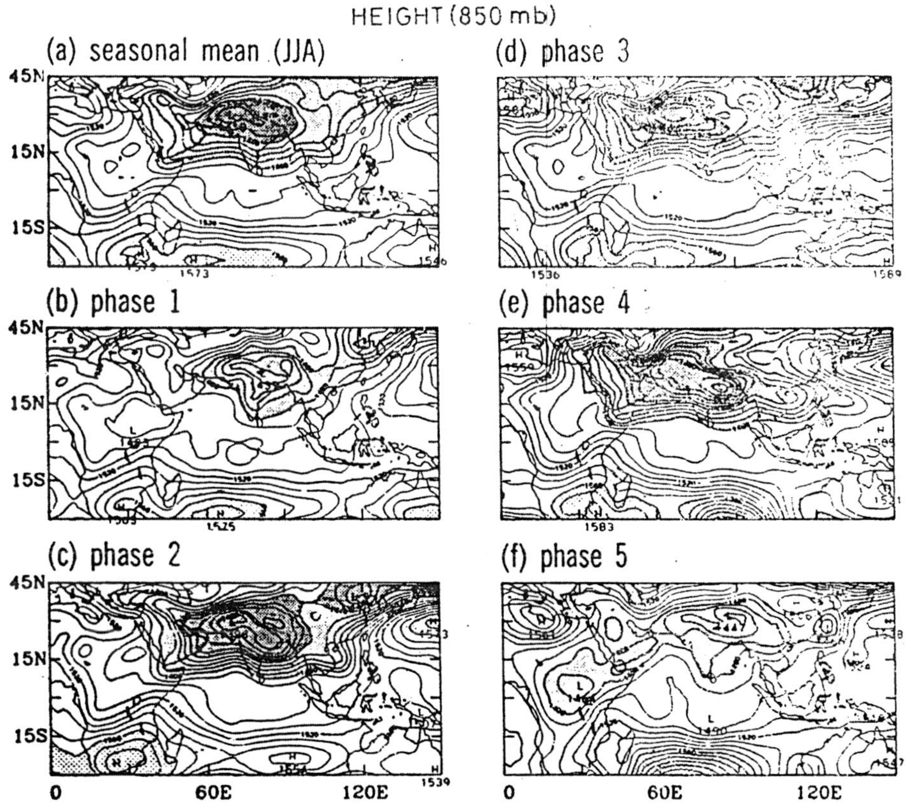

Fig. 7.9: The 850 mb height fields for seasonal mean (a), five phases (b)-(f): 3-13 June, 19-20, 8-18 July, 27 July - 8 August and 17-27 August (After: Chen, 1990).

intensification and weakening of the Indian Monsoon and establish the life cycle (onset – active – break – revival – retreat) of the monsoon (Fig. 7.10) (Chen, 1990).

Keshvamurty et al. (1986, 1988, 1990) studied these aspects with a adiabatic global spectra model with a weak symmetric equatorial heat source to establish low frequency oscillation of 30 – 50 days with northward movement in the India region. Pant (1983) used MONEX data to examine the changes in phases of monsoon in 1979 season within a physical framework. Sikka and Gadgil (1980), Yasunari (1981) emphasized the intra-seasonal oscillation (ISO) of organized convection on 30 – 40 day cycle in which organized convection migrates from near equatorial warm waters towards the northern end of the east Arabian sea, the Bay of Bengal and adjoining continental south Asian region in 3 or 4 episodes in each monsoon season. During active phase of ISO overlapping monsoon low pressure areas are formed and distribute rains over south Asia (Sikka et al., 1986), ISO dominated the monsoon variability (Krishnamurti, et al., 2004; Krinamurthy and Shukla, 2000).

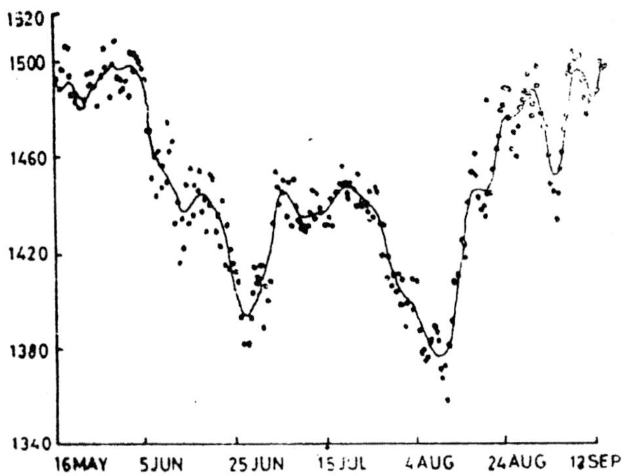

Fig. 7.10: Time series of height at 72.5 E, 22.5°N from 16 May to 15 September 1979. (After, Chen, 1990).

Webster et al., 1999; Saji et al., 1999; Behra and Yamaguta, 2002 have reported the regional scale development of ocean-atmosphere coupled mode between the eastern and the western equatorial Indian Ocean known as Indian Ocean Dipole Mode (IOD) which appears during the monsoon season and peaks in the post-monsoon season to create monsoon fluctuations.

Analysis of satellite based OLR data set on pentad mean basis revealed the interesting features of annual cycle of large scale organized convection over southern Asia to reveal annual and intra-seasonal cycles (Sikka, 2005). During January to end of March the equatorial convection activity lies in the near equatorial South Indian Ocean region (5 to 15°S and 55 to 120°E), and the subtropical ridge lies close to equator along 5°N to 10°N. As the upper troposphere subtropical westerlies dominate most of South Asia, the convective cloudiness belt shifts close to the equator between 5°S to 5°N by April in a disorganized manner by middle of May the OLR minimum shifts from the Indonesian region northwest ward and get organized over the Andaman Sea in Bay of Bengal with advancement of monsoon into Andaman Sea. Upper tropospheric easterlies are also strengthened simultaneously over the region and the subtropical ridge begins to shift near 15°N. With onset of low level strong cross-equatorial flow in the western Arabian Sea in end May/early June and its extension to the central Arabian Sea and Lakshdweep Sea (SE Arabian Sea) an explosive growth of low OLR band is observed which strikes the Kerela coast by the end of May early June. This is beginning of intra-seasonal cycle of monsoon with onset of monsoon. TEJ is also fully established rather abruptly over south Bay of Bengal and south India as the circulation responds to new organization of convections. For the next 25 – 30 pentads (June to mid October) the slow sub-seasonal organization of the OLR field is noticed over south Asia and adjoining west Pacific Ocean. By mid-June, the low OLR band (about 100° longitude over south Bay of Bengal/Arabian Sea) at the time of monsoon onset, shifts eastwards to cover South China Sea and the Philippines Sea (70°E to 130°E along 20°N to 15°N) where 2 to 3 low level convergent circulation centres spaced at 200 – 300 km apart which sparks the spectacular organization of the Asia Pacific Monsoon system on gigantic regional scale over the Asia-Pacific belt. In this process there is first northward moving episode of ISO in the beginning of the monsoon season during mid June to end June. The system oscillates in activity between beginning of July and mid August with relative increase/decrease in OLR in association with the trough-ridge organization of two low frequency modes of the monsoon (10 – 20 day and 30 – 50 day). In some episodes the two modes may annul or enhance each other which may influence the intra-seasonal monsoon rainfall. The subtropical ridge (STR) at the upper troposphere level lies along 30 – 32°N over the Tibetan Plateau region. Towards the end of August, convection is weak due to the period of climatological singularity (Ramaswamy, 1973). However, convection builds up again from near equatorial region which migrates to north Bay of Bengal, passing through central India, it is located over western India by 10th of September. Finally low OLR centre begins to withdraw from western and central India between the mid September to first week of October. The low OLR centre shifts to the central Bay of Bengal by mid October as upper troposphere STR shifts southward to 15°N and weak low level easterlies begin to appear over north Bay which is a signal for the end of the monsoon season and onset of NE monsoon over the northern part of north Indian Ocean. There is considerable inter-annual variability in this slow phase of low OLR over the Asia Pacific region.

BOBMEX-1999 data also showed that the net heating cycle of the Bay of Bengal surface waters is modulated by monsoon disturbance on synoptic and ISO scales with reduction of solar radiation by cloudiness, higher transfer of latent and sensible heat leading to cooling under active monsoon spells and reverse occurs (5 – 20 days) in the weak spells. Overlapping formation of monsoon low pressure systems of super-synoptic scale are the crucial elements in ocean – atmosphere coupling of the monsoon (Sanjeeva Rao and Sikka, 2005).

7.5 Monsoon Trough and Inter Tropical Convergence Zone (ITCZ)

Fig. 7.13 shows that in the surface layers the ITCZ from its mean winter position close to the equator rapidly shifts to about 20° to 25°N with the establishment of monsoon in northern summer (July). Monsoon trough plays a predominant role over India as an anchor of the monsoon rainy period, with more than normal rains when its position lies near its normal or somewhat south of normal position (active monsoon), and the lack of rains when the trough shifted northward close to the foothills of Himalayas (Break monsoon) (Blanford, 1986). ITCZ over Indian and north west Pacific Ocean have been characterized as a narrow shear zone with maximum cyclonic vorticily, convergence, colder temperatures, high moisture content and a organized convective weather (Sikka et al., 1981) based on experimental observed data during MONEX-79 and ISMEX-73. The normal position of axis of monsoon trough over India shows a southward inclination with height up to 6 km with its position from western Rajasthan to Gangetic Plains and northern Bay of Bengal. In its normal position, about 500 km on eitherside of trough axis gets well distributed rainfall due to low level convergence of the trough. The trough is never dormant. There are several mesoscale cyclonic low level circulations located by the wind changes having tendency to move in west northwesterly direction within the trough (Fig. 7.8). Even during the monsoon rainy season, the position of ITCZ line varied with the advancement of summer monsoon season towards north and then with the withdrawal of summer monsoon it again shifts back to the equatorial region. With the onset of monsoon dry convection begins to change over to deep moist convection at the eastern end of the trough. Under fully developed monsoon, land locked boundary layer processes in the monsoon trough region present variations in respect of (i) deep moist convections in eastern end of trough, (ii) dry convections with shallow clouds over western end of trough, and (iii) alternate deep moist and dry convection over east to western region.

The large scale rainfall over the monsoonal region is associated with ITCZ's intensity and variations in its location. Variation in ITCZ is seen to be a succession of events involving genesis, an active spell and demise. Sometimes there are two favourable location of ITCZ – one over seasonal trough on the heated continent, and other over the warm waters of the equatorial Indian Ocean. Therefore the latitudinal distribution of the occurrence of ITCZ is bimodal with the zone of frequent occurrence over the continent and a secondary zone over the equatorial ocean. The most prominent feature of the intra-seasonal variation in the poleward propagation of the ITCZ from the equatorial Indian Ocean to the heated subcontinent throughout the summer monsoon season over a 20° in latitude coherently over a region larger than 30° in longitudinal extent. Such a propagation during the onset phase leads to the establishment of the continental ITCZ and subsequent propagations at intervals of about 40 days leads to its revival from weak monsoon phase. The poleward propagation of the cloud band and rain belt during the onset phase reflects the transition from the spring the summer pattern. ITCZ is seldom uniformly intense throughout its zonal extent as the synoptic scale disturbances are embedded in it (Sikka and Gadgil, 1980). Similarly McBride (1983) showed considerable fluctuations in the ITCZ over Australian longitudes in southern hemispheric summer, but there was no bias in meridional propagation toward poleward as observed over the Indian longitudes in summer monsoon. Gadgil (1990) reported intra-seasonal variations of the ITCZ over the Indian longitudes in summer season with poleward propagation emanating from oceanic regions and traversing both ocean and land to culminate in the seasonal trough were simulated from the modified version of Webster model (1983). The mechanism underlying the propagations in the model involved excess convective heating to the poleward side of the convergence zone relative to the equatorward zone, and can operate on land and sea surface. Joseph (1990) also reported the variability of monsoon rainfall in relation to convective activity of the equatorial trough over Indian Ocean and the Pacific Ocean. Cyclogenesis (tropical cyclones) near the west Pacific equatorial trough is closely related to this variability through a see-saw in convection between this ocean basin and North Indian Ocean, with a period in the range 30 – 50 days. Sea surface temperature anomalies over north Indian Ocean and west Pacific Ocean can cause this variability in the onset of monsoon and rainfall amount over India through a 30 – 50 day mode.

Atmosphere Boundary Layer (ABL) processes across the monsoon trough over India whose eastern end is the seat of frequent convective activity and western end is anchored in the dry sensible heating zone with deep convection in association with the movement of an occasional monsoon disturbance across the Gangetic Plains. MONTBLEX (1989-90) focussed attention on the fluctuations of ABL during the different phases of the monsoon and interaction between eastern and western end of Indian trough. Narsimha et al., 1997; Seetaramaya and Master (1984) studies confirmed the results of the MONEX period. Rao and Mathew (1988) showed that synoptic variability in cloudiness, wind, air atmosphere and SST modulate the air – sea fluxes such that north Bay of Bengal looses net heat during the cloudy and strong wind episodes of this disturbed or active monsoon phase and gains energy during the cloud free and low wind regime of the weak monsoon phase. LASPEX experiment (1995-96) revealed about the land surface interactions with ABL over western Indian region – a region known for sharp contrast in convective organization over annual and intra-seasonal monsoon cycles of monsoon activity, the role played by land-vegetation-atmospheric interactions on intra-seasonal basis.

ARMEX-I (2002) experimental data showed the formation of off shore trough near the surface off the west coast of India and possible existence of meso-scale vortices within the trough. The presence of off shore trough and embedded meso-scale vortices occurred with four episodes in ARMEX-I phase-two active episodes during 26 – 28 June and 6 – 10 August, 2002 (Bhaskar Rao et al., 2003; Mohanty et al., 2003)which resulted in more realistic simulation of the heavy rainfall associated with them along the west coast (Shyamala, 2003). Every fresh surge of monsoon giving heavy rainfall along the west coast is accompanied by the trough development and its movement northwards and a similar situation occured when the monsoon is reviving after a break (Rama Sastry, 1984).

7.6 Onset of Summer Monsoon

John Eliot (1901) was first to report the gigantic cross-equatorial clockwise wind gyre development at the time of monsoon onset in West Arabian Sea linking with south Indian Ocean. Fig. 7.11 gives the mean onset date of the summer monsoon with northward movement of rainfall over India and eastern Asian belt (Tao and Chen, 1987). It is clearly depicted that monsoon advances in a stepwise fashion in eastern Asia, whereas over Indian region it occurs in an abrupt fashion within less than a week. Monsoon onset over India takes place with the establishment of low level southwest monsoon circulation over the India Ocean with its northward migration of the westerly flow with simultaneous replacement with the Tropical Easterly Jet (TEJ) in the upper atmosphere.

A signification feature wind flow in southern Asia during the summer is the low level flow of the monsoon circulation—a special feature emerged from MONEX-79 data. This prominent feature of Asian Monsoon is the cross-equatorial wind component directed into the hemisphere during summer. It is strongest along the east coast of Africa called Somali Jet (Young, 1987) modifying the tropical layer at the crossing of equator to develop eddy circulations. With this strong cross-equatorial atmospheric flow there is water vapour transport

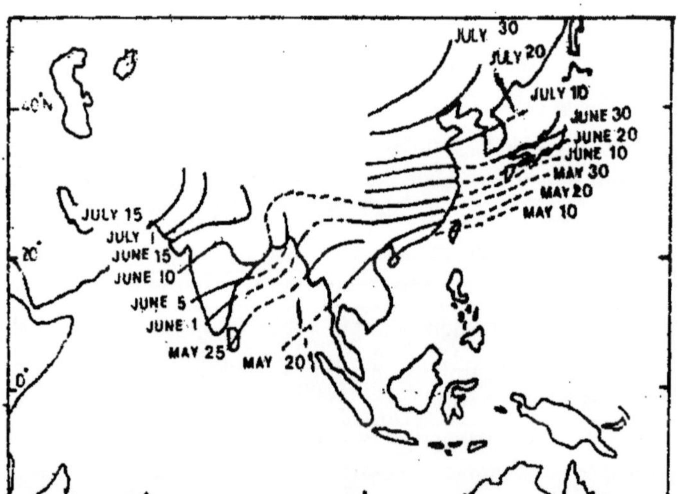

Fig. 7.11: Mean onset date of the summer monsoon (From: Tao and Chen, 1987)

to maintain balance over the region. This bulk water vapour transport by the Somali current from the southern hemisphere generates a major source of water vapour (Cadet and Greco 1987a and b). Moisture flux convergence takes place in the northern Indian Ocean across the equator was closely related to permanence of monsoon over India (Saha and Bavadeker, 1977). Yasunari (1980) and Sikha and Gadgil (1980) showed he east-west oriented cloud land over the Indian longitudes has a tendency to move northward with the first northward movement at the time of onset of monsoon. The break/active cycle in cloudiness change takes place concurrently over the entire Asia including the Indian Ocean and the Western Pacific Ocean. Murakami et al. (1986) argued that onset and withdrawal of the monsoon can be determined by the phase change of the annual cycle and 30 – 50 day low frequency mode established by the annual cycle of the atmospheric circulation over the monsoon region modulated with 30 – 50 day low frequency mode in tropical region. Fig. 7.7 shows the various phases of 1979 Indian Summer Monsoon with weak monsoon conditions along 75°E (10 June, 20 July and 30 August).

Joseph and Pillai (1988) used weekly rainfall index ($\overline{R}$) series to obtain a mean picture of the temporal changes in the rainfall over the east-west belt 8°N to 13°N before the monsoon onset over Kerela coast. The week containing the monsoon onset was considered as zero week and mean of $\overline{R}$ over 25 years (1960 – 84) for 13 weeks period showed rainfall peaks corresponding to the monsoon onset over Kerela and pre-monsoon rainfall peak (Fig. 7.12) about 6 weeks prior to monsoon onset week.

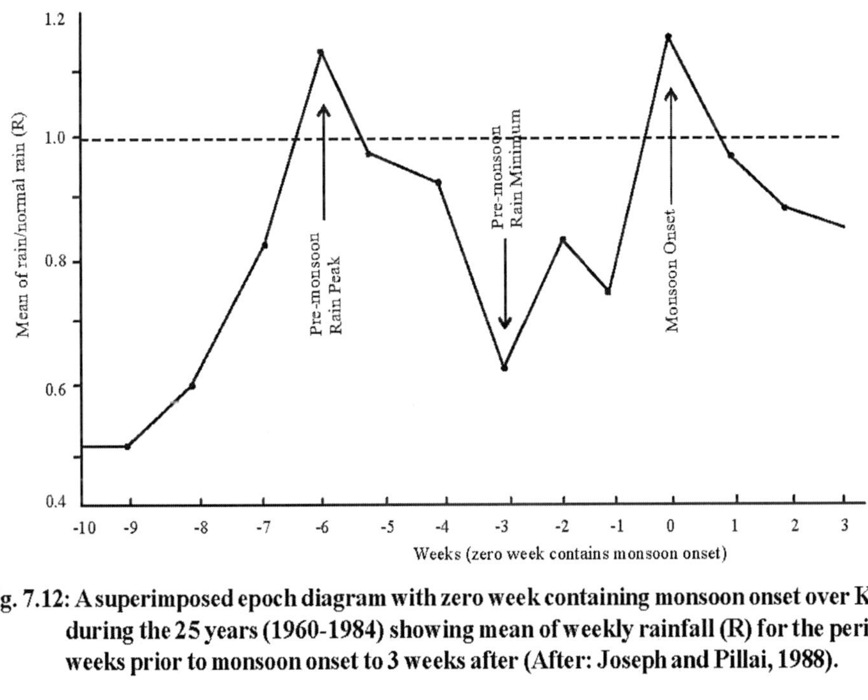

Fig. 7.12: A superimposed epoch diagram with zero week containing monsoon onset over Kerala during the 25 years (1960-1984) showing mean of weekly rainfall (R) for the period 10 weeks prior to monsoon onset to 3 weeks after (After: Joseph and Pillai, 1988).

Sikka (1980a) linked the monsoon onset and its fluctuations with the passage of baroclinic perturbations as extra-tropical storms across the Mozambique Channel in the south Indian Ocean using Monex-79 data. Sikka and Gray (1981), and Saha and Saha (2000) showed that the mid-latitude baroclinic disturbances passing across the Mozambique Channel modulated the northward movement of low level cross-equatorial flow strengthening the equatorial westerlies under favourable conditions and intensified the ITCZ activity over the Indian region resulting in formation of monsoon depressions.

The onset and northward advance of the monsoon over Indian and north west Pacific Ocean takes place with the sudden strengthening of the equatorial westerlies in the lower middle troposphere. The strengthening begins in the Indian sector and takes about a week to 10 days for the westerlies to penetrate up to 140° - 150°E (Paul et al. 1982) as revealed with the 700 mb meridional section along 70°E during the onset phase of the monsoon in 1979. Eastward strengthening of westerlies leads to cyclogenesis in the ITCZ along the preferred zones. The disturbances formed move west/northwestward. The sequence of events suggests that although the group is eastward, but phase velocity is in the apposite direction. Trade winds (easterly flow) extend from the Pacific region towards the Indian region during the weak/break phase. Sikka et al. (1984) reported the phase interlocking of the modes over the Indian region appears to result in active/break cycle of the monsoon. A 30 to 40 day eastward propagation and 10 to 20 days westward propagation with low frequency modes of tropical circulation supports these results.

Awade et al. (1981, 1982 and 1984) related the performance of monsoon with planetary scale energetics in the troposphere (700 and 300 mb levels) and in the lower stratosphere (50 and 30 mb) to reveal that the transport of momentum in middle troposphere is northwards and it is larger in years of good monsoon rains (1969, 1973, 1977) than in years of deficient monsoon rainfall. Transport of sensible heat is northward in deficit monsoon years, whereas in southwards during good monsoon years with a sharp contrast at 30°N.Further it was also indicated that in good monsoon years long waves acts as a source of kinetic energy to the zonal flow. Similarly Raja Rao et al. (1984) also reported in lower stratosphere the transport of momentum and sensible heat having annual variations with the winter maximum and summer minimum.

Flohn (1957), Koleswaram (1958a, b, c) emphasized the role of mid-tropospheric sensible and latent heating over the Tibetan Plateau in the formation of upper tropospheric Tibetan high and the outflow from it leading to the establishment of the tropical easterly jet stream (TEJ). Therefore monsoon trough, Tibetan high, TEJ and the low level strong cross-equatorial flow over the Arabian Sea were important features of Indian monsoon flow. Das (1962) examined the vertical wind circulation over the monsoon trough and showed that moist ascent and convective clouds dominate over eastern India, whereas subsidence prevails over the western India with mostly non-raining low level clouds so called east-west circulation of the regional south Asian Monsoon system. Raman and Ramanathan (1964) discussed the interactions between the lower and upper tropospheric circulation and role of latent heat release in maintenance of the TEJ.

Colon (1964) revealed about the structure of monsoon over the Indian Seas with moist layer in the monsoon air is shallow across the west Arabian Sea due to the presence of capping inversion in the lower atmosphere. The inversion is gradually lifted upward as the air travels over the central Arabian Sea and finally disappears east of 65° - 68°E as the air approaches the convergence zone of the west coast of India and the organized convective clouds are built up (Desai, 1968, 70). Saha and Bavdekar (1973) also reported the bulk of moisture transported across the equator by cross equatorial flow.

Miller and Keshvamurty (1968) established the existence of a new rain bearing transient disturbance called Mid-Troposphere Cyclone (MTC) to reveal the complex dynamical structure after analysing the conventional and drop-sonde data. An important understanding of air-sea interactions during the onset of monsoon resulted from the work of Ghosh et al. (1978) and Rao et al. (1981) with dramatic cooling of SST over the Arabian Sea within a few days after the onset of monsoon and with a similar investigation over the Bay of Bengal (Rao and Mathew, 1988) with a net loss of heat energy during active monsoon spells under higher winds and overcast skies, whereas ocean surface gained energy during weak monsoon conditions under clear skies and weaker wind conditions.

Krishnamurti et al., 1977; Koteswaram et al., 1987; Rao et al., 1987 examined the structure and vertical motion field associated with monsoon depression- the primary rain producing system of monsoons. Jadhav (2002), emphasized on the contribution of monsoon low pressure areas (LPA) and number of LPA days for activity of monsoon rains on inter-annual scale. Misra and Salvekar (1980) studied the dynamical instability of monsoon and its growth as result of barotropic - baroclinic CISK processes. Krishnamurti et al. 1981; George and Mishra, 1993 studied the barotropic – baroclinic energetic and structure of monsoon onset vortex of 1979 using MONEX data.

Pearce and Mohanty (1984) established the built up of the moist layer over the Arabian Sea about a fortnight prior to the monsoon onset as precursor to the establishment of the monsoon season. MOBMEX-1999 experiment showed atmospheric moisture modulation in active-break cycle with moisture extending up to middle upper troposphere under active monsoon conditions and moisture restricted within the lower troposphere in weak monsoon conditions (Bhat et al., 2002). Seetaramayya and Master (1984) using MONEX data showed the build up of the mini-warm pool over SE Arabian Sea, whereas Rao and Shiv kumar (1999) provided occurrence of mini-warm pool over SE Arabian Sea through advection of the low salinity waters from the Bay of Bengal during November to January period enhancing near surface stratification and net surface heating. They also showed the coincidence of monsoon onset variation with this mini-warm pool whose intensity varied year to year. However the monsoon onset vortex does not form every year as it depends on the development of atmospheric instability in spite of the necessary conditions of the environment in terms of the warm ocean surface. ARMEX-II (2003) confirmed the evidence for the existence of low salinity waters over the SE Arabian Sea during December to March, the existence of inversion layers in the 30 – 80 m. depth below the surface waters and their westward propagation during January to March (Shankar et al., 2004 and Sanjeeva Rao and Sikka, 2005). Durrand et al. (2004) even dynamically simulated the inversion layers in a modelling study that these inversion layers could contribute preferential warming of sea surface over SE Arabian Sea during March – April period. However, since the strength of the inversion is rather weak (0.5° to 0.7°C) its predominance over net surface heating under clear skies and low wind conditions could be debatable. ARMEX-II data showed that the warming began by mid-March, and peak SST values exceeding 30.5°C were observed over the area during April to mid May, 2003. In spite of such high SST convection remained suppressed as no atmospheric disturbance formed or moved through the SE Arabian region.

Monsoon onset vortex also did not form and the onset was delayed by about 10 days (Sanjeeva Rao and Sikka, 2005). ARMEX results highlighted that the atmosphere over the Arabian Sea remained decoupled from SST warm zone during March to mid May unless atmospheric instability is initiated. ITCZ begins to appear in equatorial Arabian Sea after mid-May, the atmosphere gets coupled with the sea and the SSTs begin to decrease though still around 29°C above the threshold for convection. By the end of May-early June, with onset of monsoon over SE Arabian Sea, SSTs fall to 28.5°C over the region, while continue to rise above 29.5°C over the NE Arabian Sea. The setting up of this N – S gradient of SST with cooler temperature beneath the active cloudiness regime over the SE Arabian Sea and warming continues over the NE Arabian Sea, may promote northward advance of monsoon along the west coast. Formation of synoptic scale disturbance in the form of off shore trough/MTC/onset vortex, depending on the dynamic instability mechanism operating in the atmosphere, heralds the onset and advance of monsoon along the west coast of India. The warm surface of the north Indian Ocean regulates the monsoonal fluctuations by providing energy for feeding the disturbances. Thus air-sea coupling processes play crucial role in regulating the monsoon on the ISO and synoptic scales rather than initiating atmospheric instability changes and bringing in active/weak/suppressed convective episodes on different temporal and spatial scales. ISOs display their own life cycle of formation, intensification, peaking and decay. Prasada Rao and Krishna Kumar (2005) reported the onset of monsoon over Kerala coast with normal onset dates of June 1 ± 7 days with 28 per cent chances for early or late onset chances. Delay/early onset occurred preferably in the years of El Neno/ La Neno year.

7.7 Withdrawal of Summer Monsoon

With the advent of winter, very large areas of Central Asia become so cold that the air subsides over them creating an intense high pressure cell. The associated wind circulation opposes the monsoonal flow and as a result the withdrawal of monsoon takes place. The withdrawal is preceded with a change in upper level easterlies over to westerlies as an indication. Matsumoto (1988, 1990), using time evolution of large scale circulation over South Asia during the withdrawal phase of Indian summer monsoon from August to October for the period 1979 to 1984 using FGGE level III-b and ECMWF (European Centre for Medium Range Weather Forecasting) operational analysis data, recognized three distinct stages of abrupt transition to establish retreat of summer monsoon (Fig. 7.13). These were

(i) Southward shift of 200 mb subtropical jet stream round the Tibetan Plateau, weakening of 850 mb westerlies around the Indian subcontinent, and decrease of rainfall along the west coast of India, and/or sharp drop of 300 mb zonal mean temperature over the Eurasian continent occurring almost simultaneously associated with these transitions.

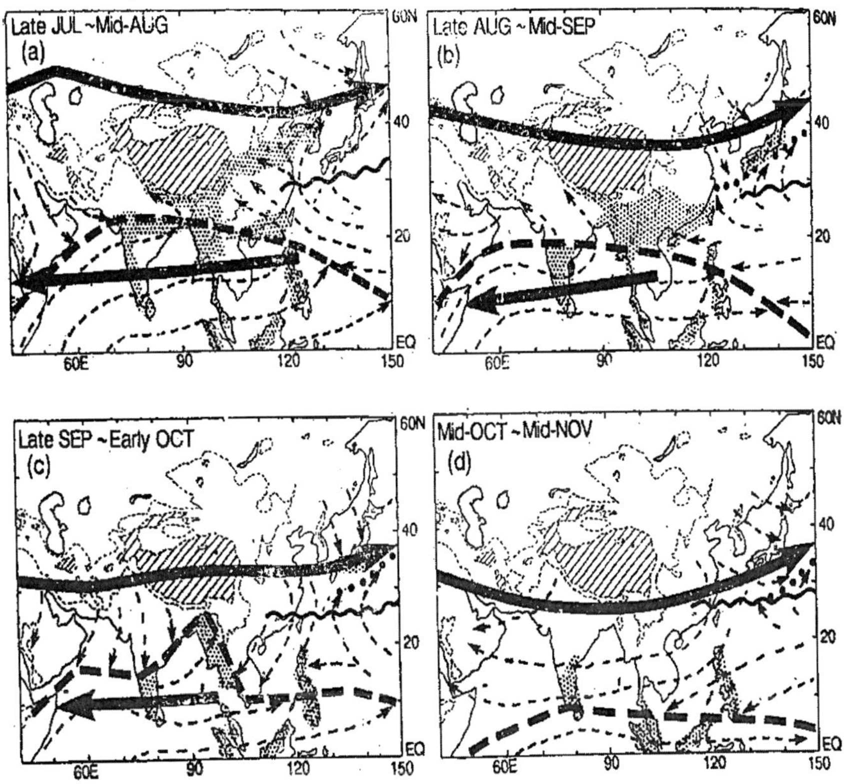

Fig. 7.13: Schematic illustrations of the large scale circulation over south and east Asia from the late August to November 1979. Thick arrows indicate axes of the jet stream at 200 mb. Thin dashed lines and dotted lines indicate locations of the ITCZ and the frontal zones at 850 mb respectively. The wave lines are the ridge lines of the subtropical anticyclone in the western Pacific. The shaded areas denote the heavy rain area. Topographic regions higher than 3,000 m are hatched. (After: Matsumoto, 1990).

Fig. 7.13c shows that remarkable changes occur around 10 October when 850 mb westerlies over the wide regions from 60°E to 100°E at 15°N are abruptly replaced by the easterlies. The prevailing wind over the Arabian Sea is southesterlies for the first 10 days after this change. After 20th October, they are replaced by the northeasterlies establishing the beginning of the NE Monsoon season. The changes of 200 mb wind systems also occur around 10th October. Esterlies are replaced by the westerlies over wide areas from 30°E to 90°E with a sudden southward shift of the subtropical jet stream from 35°N to 28°N at 82.5°E (Fig. 7.13d). After this transition the subtropical jet stream flows around the southern periphery of the Tibetan plateau. In late September there is abrupt decrease of rainfall simultaneously reflecting the end of the monsoon season from west coast of India. Murakami et al. (1986) determined the withdrawal of the Indian summer monsoon in 1979 around 27th September based on harmonic analysis of the seasonal cycles of the outgoing longwave radiation (OLR) data. The time series of zonal mean 300 mb temperature averaged over 30°E to 120°E at several latitudes from 15°N to 75°N showed sharp drop around 22nd September between 45°N to 75°N and around 2nd October between 15°N to 37.5°N in 1979 due to the sudden changes of wind filed and rainfall amounts (Ding et al., 1983).

(ii) In late September when rainfall along the west coast of India, which increases in mid September, decreases drastically. A weakening of monsoon westerlies takes place with a southward shift of 200 mb subtropical jet stream from 40°N to 35°N, and a sharp drop in 300 mb zonal mean temperature takes place at this second transition stage.

(iii) The third transition corresponds to the beginning of northeast monsoon circulation in mid or late October when 200 mb subtropical jet stream begins to flow around the southern periphery of the Tibetan Plateau.

Matsumoto (1988) similarly divided four stages in the seasonal transition from late summer to autumn over east Asia.

The first transition takes place in late August accompanied by a sudden decease in rainfall along the west coast of India and decrease in 300 mb zonal mean temperature averaged between 0°E and 150°E. About 10 days after this transition, the retreat of summer monsoon from northwest India begins. These transitions are closely related to the intra-seasonal low frequency oscillation of Indian summer monsoon and passages of troughs in mid-latitude westerlies over the Tibet- Himalayan region. Duration of monsoon season is shorter in northwest region as compared to southern and eastern region.

7.8 Northeast Monsoon

With the retract of SW monsoon from south India and adjoining Bay or Bengal, the NE monsoon generally sets in with a spectacular change between the two monsoons of the change of wind direction from southwest to northeast, and build up of low pressure over Bay of Bengal, Indian peninsula and Arabian Sea. Tamil Nadu is a major beneficiary with intense rainfall. NE monsoon is also a season of cyclonic storms in the Bay of Bengal which sheds heavy rainfall and are destructive in nature due to strong winds. During this period well marked seasonal trough of low pressure areas also exists in Bay of Bengal which move westwards towards the Indian coast. Generally their latitude of formation shifts from 1°N to about 8°N with the advancement of the season associated with a southward migration of the season near the equatorial trough. The associated upper air cyclonic circulation is hardly seen beyond 850 mb. Some of these shallow systems may be identified as waves in low level easterlies. Since they are weaker than low pressure areas, they provide rains over narrow coastal belt till their wind energy is lost. Seasonal trough extends from surface to about 600 mb sloping east to west across the Bay of Bengal to Arabian Sea.

7.9 Winter Monsoon

There is close relation between planetary scale waves and the winter depressions also known as cold surges. Existence of a 12 – 20 days low frequency oscillation in winter monsoon was found (Murakami and Sumi, 1981; Qin and Wang, 1988) in the cross equatorial air flow at the equator of east Asia. Ding and Krishnamurty (1987) discussed the heat budget of Siberian high with strong radiative cooling and large scale decending motion (a large scale mass convergence over the upper and middle troposphere and divergence over the lower troposphere) which contributed in its rapid build up. Heating in the upper troposphere with heat transfer maintains the mass convergence in upper tropospheric trough accompanied by the enhancement of the warm upper tropospheric auticyclone lying over the Siberian high. In the preformative stage the ridges at 500 mb over Atlantic and Pacific strengthen and expand over Arctic area. The polar vortex then breaks into two parts-one moves southwards over the European and Asia continent, and second one over southeastwards over Russia and China. With the movement of polar vortex, the arctic cold air arrives in the key eastern Siberian region (45°N to 65°N and 70°E to 90°E) which is key region for winter monsoon in east Asia. The cold air stays in the region and enhances its intensity during the formative stage which are divided into tracks of outbreak in different regions of southern, south-eastern and eastern Asia. The strong northerly and northwesterly tracks winds influence the Arabian Sea, the Bay of Bengal and the western Pacific. The belts in south China Sea and western Pacific spreads towards equator to form the strong cross equatorial air flow (Liu and Chen, 1983) over 105°E and 130°E in winter which also influence the summer monsoon in the southern hemisphere. Cold air is strengthened in the region until the breakdown of ridges in its path. At the outbreak stage, the blocking high breaks down and moves forward in eastward. The trough ahead of the high deepens in its intensity and moves eastward or southeast wards with the establishment of trough in east Asian region. If there exists a Bengal trough, the warm and moist southwest air flow will strengthen the cold font causing the winter monsoon to move southward. The dynamical effect of Tibetan Plateau plays an important role on the movement of cold surge/winter monsoons.

A special feature of winter atmospheric disturbances is that they are small in size, less intense slow movement due to the thermally induced shallow monsoonal current. These characteristics are developed due to the upper convergence zone. The winter depressions move parallel along the foothill of Himalaya providing snow and rains in the northern parts of India and keeping lower temperature due to the winds coming from colder areas.

7.10 El Neno-Southern Oscillation Phenomenon

Planetary scale air-sea interactions were studied in terms of El Nino-Southern Oscillation (ENSO) phenomenon (Bjerkenes, 1969) as a result of the world wide weather and regional climatic anomalies observed during the El Nino of 1957. El Nino has been recognised as the global climatic anomalies triggered by the complex air-sea coupling in the equatorial Pacific. To understand and predict these global and seasonal anomalies on intra-seasonal to inter-annual scales, a decade long (1985-1994) Tropical Ocean-Global Atmosphere (TOGA) programme was launched. Sikka (1980b), Shukla and Paolino (1983), Mooley and Paolino, (1989). Krishnamurthy and Goswami (2000) linked the connection of monsoon droughts with the warming phase of El Nino/falling phase of southern oscillation phenomenon in the equatorial central and east Pacific regions. Since only 50 to 60 per cent of monsoon droughts were linked with ENSO signals, Webster and Yang (1952) and Kirtman and Shukla (2000) reported that the ENSO – Monsoon connections were very complex. Wang (1995) and Krishna kumar et al. (1999) suggested that ENSO – Monsoon connections are weaked due to the secular variations in ENSO events. Sikka (2003), linking monsoon drought of 2002 to ENSO build up, expressed that there is no single regional or global parameter which could explain 55 to 60 per cent of monsoon droughts, but with build up of ENSO the probability of excess rains are ruled out during the monsoon season. Raman et al. (1990) ascribed the monsoon drought of 1972 and 1979 as a result of interactions with mid-latitude circulations since drought of 1972 was ENSO related, whereas that of 1979 was not ENSO related. Gadgil et al. (2003) have shown the occurrence of drought over India to be associated with higher cloudiness in eastern equatorial Indian Ocean during the monsoon season. Therefore very large scale teleconnections, operating through the displacement of walker circulations due to changes in the thermal forcing in equatorial region, takes place on a planetary scale. Indian Ocean-West Pacific Monsoon circulation regimes on the super synoptic scale, i.e., the ENSO – Asia – Pacific Monsoon system on planetary scale and IOD on regional scale. Biswas et al. (1998), Krishna Kumar et al. (1999) studied the relationships of these processes with atmospheric circulation features, large scale build up of cloudiness in near equatorial region, SST on regional and extensive scale sub-tropical ridge in mid-tropospheric and mid-latitude circulation etc. ISO-scale variability is prominent in the season and convective clouds move from near equatorial warm ocean toward northern part of the Indian shores. SST fluctuations on ISO scale is 2 – 3°C, which is higher than on the inter-annual scale (Sengupta and Ravichandran, 2001). SST changes on the synoptic scale are usually of the order of 0.5 to 1°C. Convection over the central and north Bay of Bengal fluctuates simultaneously as evident from OLR data analysis on the ISO scale, whereas it fluctuates over the south Bay of Bengal in opposite phase. Warming of the North Bay of Bengal in weak monsoon phase and corresponding cooling over the cloudy south Bay leads to the build up of SST gradients in north-south direction over the area. Thermodynamic coupling of the ocean-atmospheric which cools (warms) the ocean beneath the region of active convective cloudiness (clear area) and warms over north (south) Bay of Bengal during active (weak) monsoon spell, plays a crucial role in regulating the intra-seasonal fluctuations in the monsoon. Convection develops over the south Bay and weakness over north Bay in the weak monsoon spell. Warming of SST in North Bay in weak phase of the ISO cycle could be responsible for the northward movement of convective cloudiness band resulting in re-establishment of another active monsoon phase over north Bay of Bengal in the mid-season (Sanjeeva Rao and Sikka, 2005).

Bjerknes (1969) developed the concept of walker circulation (large-scale zonal, west-east, overturning in the equatorial plane) after studing the El Neno of 1957. Wyrtki (1982) represented the average global tropical circulation characterized by three major convective areas of rising motion, clouds and rain over southeast Asian and western Pacific (area of warm ocean water), tropical south America (Amazonia) and Africa (Congo region) (Fig. 7.14a). Air rises to the upper troposphere and sinks in the descending branches of the planetary scale East-West overturning cells over the Atlantic and central and eastern Pacific. The strongest branch over Pacific is related to SSTs which is warm when air is rising and cool when air is sinking (Fig. 7.14b). Large scale tropical convection acts as a major driving force of the atmospheric circulation which varies with the development of ENSO through the release of latent heat in the mid-tropospheric (Fig. 7.15). Comparing this above normal feature (Fig. 7.14) with the 1982-83 El Neno event (Fig. 7.15) reflecting the shift of Pacific ascending branch to the east to date line and suppressed the convection over rest of the tropics. Southern oscillation index started to fall sharply in June, 1982 and reached a record low values of by the end of December 1982; and then rose to normal values by June, 1983. The SST anomalies appeared in central Pacific first (in June, 1982), and by December entire equatorial Pacific was 1 to 5°C warmer than normal and sea-surface departures reached a record high values (>5°C) in the equatorial eastern Pacific. Fig. 7.15 shows the corresponding changes in the east-west overturning walker cells during 1982-83 ENSO. Comparing with the normal features (Fig. 7.14) there is shift in the convective regions from the Australasia region to the central and eastern Pacific with subsidence and dry weather over Africa, the Indian Ocean and Australasia. There is distruption of rainfall regime around the equatorial belt with floods around SW Ecuador and NW Peru. A shift in the tropical

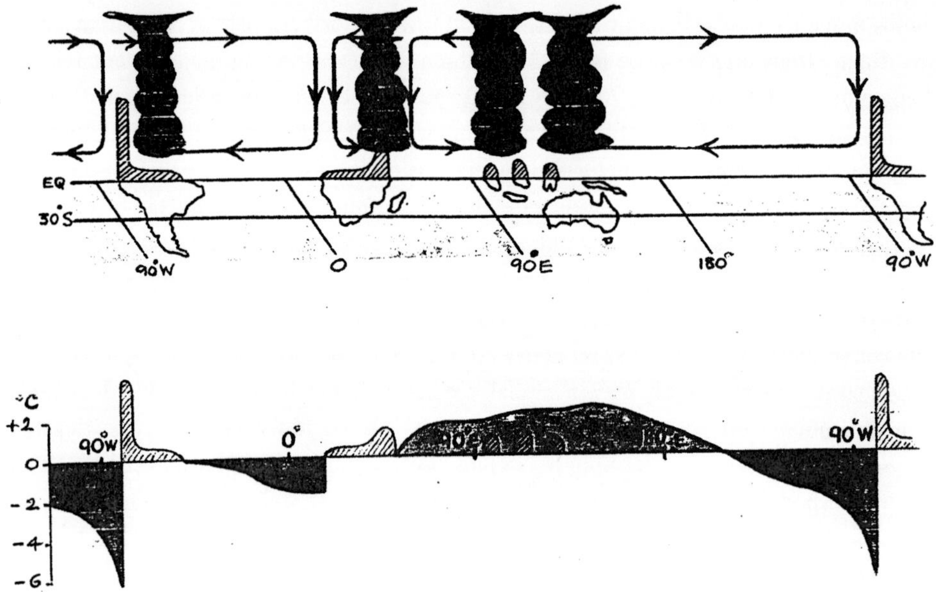

Fig. 7.14: (a) Schematic of the mean Walker circulation along the equator, resulting in rising air and heavy rains over Indonesia and the western Pacific, Africa and the Amazon area in Latin America.

(b) Departure of the sea surface temperature from its equatorial zonal mean is depicted in the lower part of the diagram. (From K. Wyrtki, 1982).

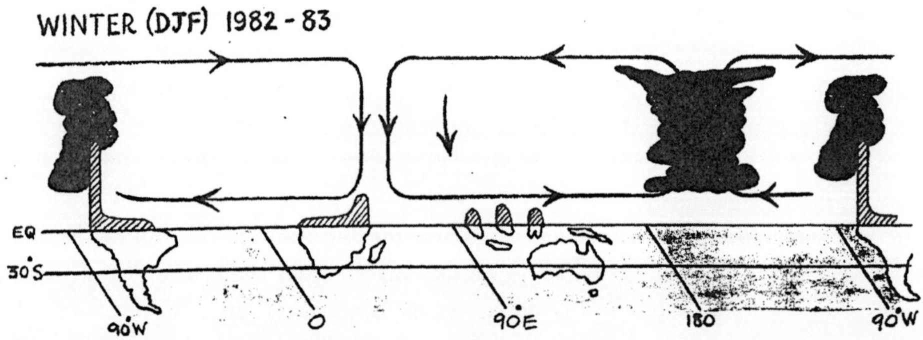

Fig. 7.15: Schematic of the Walker Circulation based on computations of upper and lower troposheric divergent winds with ENSO Winter 1982-83 pattern. (After: Tourre and Rasmusson, 1984).

circulation also created widespread drought in the eastern and southern Africa, Sri Lanka, Phillipines, Indonesia and Australia. The 1982-83 El Nino, as expressed in terms of SSTs, ended by about December, 1983. (Fig. 7.16) as the atmospheric readjustment to a normal state took time even up to 1984 and 1985 called 'Non El Neno' or 'La Nena' phase indicated by normal sea level pressures at Darwin and Tahiti, and the southern oscillation index. El Nino of 1997-98 was throughly studied under climatic variability and predictability project of World Climate Research Programme (1982-2002) (WCRP). Ocean-atmosphere interactions were particularly focussed so as to response atmospheric anomalies, interactions with sea surface temperatures, sea ice, soil moisture, vegetations, and even troposphere-lower stratosphere interactions which could result climatic variability on seasonal to inter-annual time scales. Use of multi-model ensembles has greatly enhanced overall capabilities for climate variability over a few months or seasonal predictions (Busalacchi et al. 2005), El Nino of 1997-98 initiated in February, 1977 and continuously strengthened up to early 1998, severely influenced rainfall in Australia.

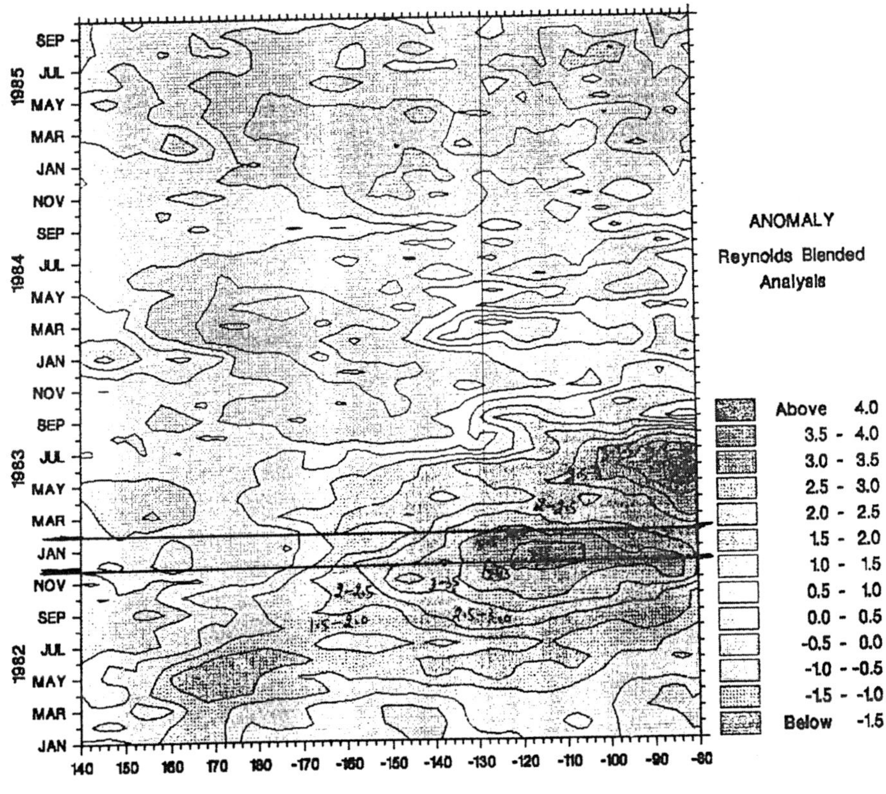

Fig. 7.16: Monthly mean SSTs anomaly for 5°N - 5°S for period January 1982-October 1985. (From CAC/WMC-Washington D.C., WMO, 1987).

The large ocean also has a direct impact on the atmospheric climate through changes in SSTs and the partial pressure of CO_2. In future sea level changes are critical parameters for climatic variability perspective. Southern ocean is very important from global climate point of view since it connects the upper and lower limbs of global overturning circulation, stores about 60 per cent of the total anthropogenic carbon, and may have an annual mode linking to ENSO and sensitive to climate variations globally (Busalacchi et al., 2005). Recent numerical modelling studies (Hurlburt et al. 1976; O'Brien et al. 1981; Webster, 1981) and massive ENSO of 1982-83, Global Climate System Review (WMO, 1987) supported the hypothesis that the anomalous heating of the tropical atmosphere altered the global climate by modifying the tropical train of E – W overturning Walker Circulation Cells. The mid latitude circulations were attributed to the generation of a forced wave-train induced by the tropical diabatic heating and thus creating an obstacle in the basic flow pattern. The general conditions preceding the ENSO events were (WMO, 1987) advocated as

(i) weakening of the pacific trade wind system with westerlies or westerly anomalies appearing the western Pacific,
(ii) rise in surface pressure of Darwin area and fall in surface pressure around central and eastern Pacific (Tahiti) thereby increasing the southern oscillation,
(iii) occurrence of above normal sea surface temperatures in the eastern Pacific persisting even beyond he normal winter time to establish the warming,
(iv) warming of ocean surface from western to central equatorial Pacific leading to a connective activity, and
(v) fall of sea level in western Pacific Ocean with rise of sea level in the eastern Pacific with changes in heat content in the mixed layer depths.

Barnett (1984, 1985) studied the surface wind fields, sea surface temperature, precipitation and sea level pressure in the tropical band, extending $\pm$ 30° of equator from Africa to south America, and evolved the hypothesis that

(i) the Southern Oscillation, El Nino and the monsoon system are all the part of a global scale phenomenon on an interannual time scale,
(ii) the phenomenon appears to have a strong propagating component with initiation in northern Indian Ocean moving towards eastern Pacific Ocean for sea level pressure fields, and
(iii) in the surface equatorial wind field, the large scale signal appears to take the form of a forced atmospheric Kelvin Wave getting energy from the latent heat released with precipitation anomalies since both propagate in phase together (Fig. 7.17).

7.11 ENSO and Regional Precipitation

WMO (1984, 1987) reviewed the relationship between ENSO and regional precipitation during the two major ENSO events of 1972-73 and 1982-83 to quantify relationship between

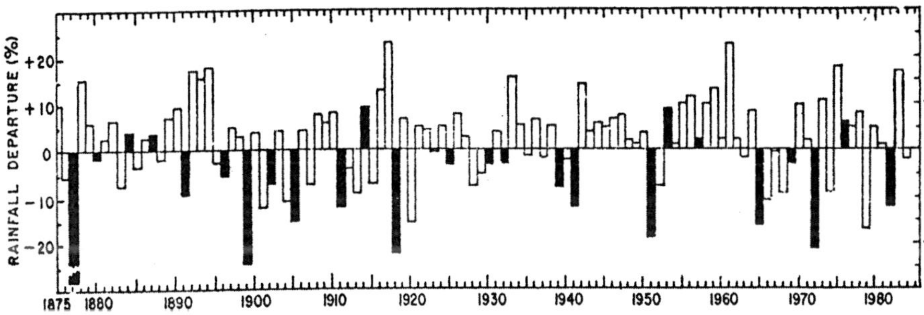

Fig. 7.17: Monsoon rainfall over India expressed in percentage departure, with El Nino events indicated by black shaded pillars. (After: Choudhury, 1985).

them. Pacific Ocean precipitations patterns were directly related to ENSO. Ropelewski and Halpert (1986) identified ENSO related responses specifically four regions in Australia, two regions in each of North America, South America, the Indian subcontinent and Africa, and one region in Central America (Table 7.1). In these regions the season of ENSO related response was found to be in phase with the normal annual precipitation cycle. In each region monthly precipitation data was expressed in percentile ranks which reflect anomalies with different mean and variance thus reflecting the precipitation patterns in regions that have large station to station differences in these parameters. Normally, the ENSO composites of percentile ranked precipitation are formed for 24 months period starting for July preceding the ENSO episode, designated July (–), and continuing through June following the episode, designated June (+). About 25 ENSO episodes were included which occurred during the period 1975-1985.

7.12 ENSO and Precipitation over India and Australia

Since interactive system of ENSO and rainfall patterns over India, southeast Asia and northern Australia are strongly interlinked, therefore it would be useful to examine the characteristics of each episode over the past long records of available data in the region. Kane (1998a, b; 2004) examined the summer monsoon rainfall over India and Australia in relation to El Nino (EN), Southern Oscillation minima (SO) and equatorial eastern sea surface temperature (W or C) in depth to quantify their association (Table 7.2) over the period 1870-1996. Though several workers have reported links between ENSO phenomenon and rainfall extremes in India (Mooley and Paolino, 1989; Ropelewski and Halpret, 1989; Kriplani and Kulkarni, 1997; Tapeshwar Singh, 2000, Medha Khole, 2005), and in Australia (McBride and Nicholls, 1983; Wright, 1988a, b; Ropelawski and Halpret (1989), Nicholls and Wang, 1990; Evans and Allen, 1992). Webster and Yang, 1992 showed that monsoon and ENSO are selectively interactive systems i.e., effects are probabilistic in some areas only (McBride and Nicholls, 1983 for Australia, and Ropelewski and Halpert, 1989) for global scale reflecting

Table 7.1

ENSO related precipitation for different parts of world (1875-1985)

Region	Season	Period	Coherence	Number of Episodes		
				Wet	Dry	Total
Central Pacific	May (0) – April (+)	1932-79	0.98	7	1	8
South central Pacific	July (0) – June (+)	1932-79	0.88	8	0	8
Indonesia/New Guinea	June (0) – Nov. (0)	1879-1984	0.82	5	20	25
Fizi/New Caledonia	Oct. (0) – Mar. (+)	1879-1984	0.95	2	9	11
Micronesia/W. Pacific	Oct. (0) – May (+)	1932-1979	0.91	1	12	13
Hawaii	Nov. (0) – May (+)	1879-1984	0.88	2	9	11
Northern Australia	Sep. (0) – Mar. (+)	1885-1984	0.95	4	22	26
Eastern Australia	Sep. (0) – Feb. (+)	1895-1984	0.94	6	20	26
S. Australia/Tasmania	May (0) – Oct (0)	1895-1984	0.94	6	18	24
Central Australia	Mar. (0) – Feb. (+)	1895-1984	0.86	7	19	26
India	Jun. (0) – Sep. (0)	1875-1984	0.86	5	21	26
Minicoy/Sri Lanka	Oct. (0) – Dec. (0)	1875-1984	0.92	21	5	26
Equatorial east Africa	Oct. (0) – Apr. (+)	1875-1984	0.93	11	2	13
South-east Africa	Nov. (0) – May (+)	1875-1984	0.90	6	22	28
North-eastern South America	Jul. (0) – Mar. (+)	1911-1982	0.91	1	16	17
South-eastern South America	Nov. (0) – Feb. (+)	1911-1982	0.82	18	1	19
Central America/Carribean	Jul. (0) – Oct. (0)	1900-1980	0.77	5	14	19
Great basin of North America	Apr. (0) – Oct. (0)	1879-1984	0.80	9	2	11
Gulf and North Mexico	Oct. (0) – Mar. (0)	1879-1984	0.93	18	4	22

(After: Ropelewski and Halpert, 1986)

that not all El Ninos are equally effective. Therefore Kane (1997, a, b; 2004) attempted a finer classification with use of long range data, and presented a general conclusion that warm water episodes along Peru-Ecuador coast (El Ninos) in conjunction with SO minima and equatorial eastern Pacific SST maxima (warm events), with SO and W in the middle of the calendar year (June to September), seem to lead to deficit rainfall in Indian summer monsoon; while exclusively cold events C (equatorial eastern Pacific SST minima) tend to lead to excess rainfall. In other types of events such as ENSOW-A, ENSO, SOW, SOC, ENW or ENC mixed rainfall features are observed linking to mild droughts or rainfall around normal. Single El Nino or SO minima or equatorial eastern Pacific warm or cold SST events by themselves do not guarantee rainfall deficits or excess. Other factors such as Eurasian winter snow cover, equatorial stratospheric wind QBO (Quasi Biennial Oscillation) also influence the rainfall behaviour (Bhalme et al. 1987; Dey and Bhanukmar, 1983). Kane (1998b) indicated that for Indian monsoon the inclusion of the onset time over the Kerela coast and behaviour of seasonal monsoon rainfall may react differently to various category of events to influence rainfall behaviour since mean onset date around June 1 varied considerably from 7 May to 22 June (Joseph et al. 1994) due to delays in the seasonal transition of equatorial convective cloudiness maxima and El Nino. Also Kane (2004) confirmed that El Ninos which were

Table 7.2
El Neno events, their charcteristics and associated significant features (Kane, 1988a, b; 2004)

Events	El Neno			Southern Oscillation Minima			Sea surface Temperature				Seasonal Rainfall Deviation (%)						Inferences/Conclusion
	S	M	W	S	M	W	Warm S	Warm M	Cold	Non-events or Normal	Extr. Drought (>-15)	Drought	Floods Extr. (>15)	Floods M(9 to 15)	Normal 0 to 9	Normal 0 to -9	
16 ENSOW-U	8	8	-	16	-	-	16	-	-	-	7	4	-	-	1	4	7 extreme droughts, rainfall below normal in rainy season.
14 ENSOW-A	6	5	3	9(4e), (1*l*)	5(1e)	-	10(6e, 2*l*)	4(2e)	-	-	-	-	-	2	8	4	ENSO outside the rainy period season, rainfall deviations are low on both sides
6 ENSO	4	1	1	5	1	-	-	-	-	-	-	1	-	-	1	4	Rainfall deviations were low but rerely positive
10 ENW or C	3	7	-	-	-	-	-	-	4	4	-	-	1	1	5	3	Rainfall deviations were low with no drought, but normal tendency.
46 EL-Neno events	21	21	4	30	6	-	27	4	4	4	7	5	1	3	15	15	
9 SOW	-	-	-	3(e)	3	3	9(2e, 3*l*)	-	-	-	2	3	-	-	2	2	A warm equatorial eastern Pacific does not guarantee a deficit rainfall year, but mixed result.
7 SOC	-	-	-	1	6	-	-	-	4	3	-	1	-	1	3	2	Rainfall deviations are low, but mixed results.
16 Southern Oscillation Events	-	-	-	4	9	3	9	-	4	3	2	4	1	5	4		
Cold Events La-Nena (C)	-	-	-	-	-	-	-	-	36(3*l*)	-	-	-	2	9	3	2	Rainfall above normal is assured.
2 Non-events	-	-	-	-	-	-	-	-	-	2	1	1	-	-	-	-	

S - Strong; M - Moderate; W - Weak, Extr. - Extreme

- SOI minima withfall of pressure difference over 3 or more seasons commencing either inthe same season, earlier seaosn (e) or late season (*l*). strong as 1 to 3 mb, moderate as (0.5 to 1 mb) season weak as (0 to 0.5 mb).
- El Nino events with SST anomaly (Quinn et al. 1978, 1987) strong as >3.0°C moderate as 2 to 3°C, weak as 1 to 2°C, non-event as <1°C (Quinn et al., 1978, 1987).
- Warming (W) and Colling (°C) phase in eastern equatorial Pacific expressed with SST changes over three seasons as (Mooley and Paolino, 1989; Kiladis, and Diaz, 1989) as

Warming	Cooling
0 to +1°C	0.0 to -1.0°C
-0.4 to 2.1°C	-0.4 to -2.1
-2.5 to 0.2°C	+2.5 to 0.1°C

Calender year in which highest (lowest) SST anomaly occurs is taken as warm (cool) year. However when warming (cooling) covers strong when 6 or more consentive seasons then apart from highest (lowest) SST anomaly, the other year is also considered as warming (cooling) year, if for two consecutive season is that year the mean SST anomaly of ESE Pacific is≥1.0 for warm, ≤-1.0 for cold, +1 to -1.0 as normal.

* Normal Rain as Mean ± 0.5 SD, mild flood or drought as Mean ± .5 to 1.0 SD., Severe flood or drought Mean ± >1.0 SD.

active during main rainy season were better associated with droughts, but in years when no El Nino existed were also droughts in India and Australia. Thus El Nino/rainfall connection seems to be loose. Phasing of El Neno duration relation to the main rainy season did matter in favourability for 52 per cent in India, 50 per cent in Queensland, 46 per cent around Sydney/ Queensland, and between 33 to 38 per cent over remaining part of Australia. Relationship of La Nina with floods found better than El Nino relationship with droughts. In Australia some floods were associated with ENSOW also, but more with ENSOW-A than with ENSOW-U probably because some ENSOW-A were short lived (active in early part of year) and are followed by C event in latter part of the same year. El Neno of 1997-98, which initiated in Feb. 1997 and strengthened and continued strong up to early 1998, did not influence much rainfall behaviour over India, but severely affected the Victoria and Tasmania region, and heavy rainfall in August – September, 1998. On the other hand El Neno of 1982-83 started in October, 1982 and strengthened and continued up to August, 1983 resulted in severe droughts over India. Therefore several workers used unrelated parameters to ENSO phenomenon such as Himalayan and Eurasian snow covers as related with inverse relationship with Indian monsoon rainfall (Hahn and Shukla, 1976), but snow depth over Russia better was related to Indian monsoon (Kripalani et al., 1996). Monsoon rainfall relationships with SST and wind speed over the central Arabian Sea (Shukla and Misra, 1977), with stratospheric wind QBO (Bhalme et al. 1986), and with latitudinal location of axis of 500 hPa ridge along 75°E (Krishna Kumar et al., 1992), upper air flow over India, heat low development over southern Asia, meridional pressure gradient and cross equatorial flow over Indian Ocean (Thapliyal and Kulshrestha, 1992).

7.13 Interannual Variability of Monsoon Rainfall

Verma et al. (1984) studied the fluctuations in monsoon rainfall over India and northern hemisphere temperature series for the period (1901-80) to revel that when monsoon rainfall was less than normal is characterised with cooler northern hemisphere, whereas with above average monsoon rainfall period is linked with warmer northern hemisphere. Therefore inter-annual variability of monsoon rainfall depends on the frequency of formation of low pressure system in the monsoon trough than on the monsoon depressions (Sikka, 1980b)

Parthasarthy and Pant (1984) examined 100 years of summer monsoon rainfall with different Southern Oscillation Indices (SOI) to confirm the earlier findings of Walker that rainfall is significantly correlated with the SOI for the preceding, concurrent and succeeding seasons. Mooley and Parthasarthy (1984) developed a negative relationship between the Indian summer monsoon rainfall and the equatorial eastern Pacific Ocean sea temperature (0 to 10°S and 180 – 90° W) anomaly. Sikka (1980b) suggested relationship between the El Neno events over eastern equatorial Pacific Ocean and monsoon failures during 22 El Nino years during (1871-1978) period (Mooley and Parthasarthy, 1983) particularly at locations to west of 80°E and north of 12°N.

Interannual variability of monsoon have been also studied using Atmospheric Global Circulation Models (AGCM) through the diagnosis of sensitivity of the monsoon to various processes like orography, SST, soil moisture, snow cover and snow accumulation (Sud and Smith, 1985; Dirmeyer and Shukla, 1996; Ferranti et al., 1999; Sperber et al., 2001; Gadgil, 2003) reviewed these AGCM models for simulation of south Asian monsoon rainfall and monsoon rainfall simulation in extreme years like warm El Nino year of 1987 and cold El Nino year of 1988 were not satisfactory, as well as another example of abnormal Indian summer monsoon season of 2000 (Krishnan et al., 2003) and the drought of 2002 (Gadgil et al., 2003) could not be simulated using the models. Kang et al. (2002) compared AGCM simulated anomalies with the monsoon of 1997-98 El Nino which was not associated with monsoon drought over India.

The interannual variability of Indian monsoon rainfall has been linked to southern oscillation index (Shukla and Paolino, 1983) and El Neno (Chowdhury 1985) Fig. 7.17 shows the most of the severe droughts years over India were also ENSO years. During (1875-1984) 110 years there were 26 ENSO episodes of which 20 were drought and only 6 wet years with above normal rainfall. The weak monsoon years are generally associated with the above normal Darwin pressure and consequently negative SOI, and vice versa. Chowdhury (1985) compared the monsoon strength (average rainfall over continental India) with the onset date of monsoon over the southern tip of the sub-continent (Kerela) to show that the good monsoon years (normalized rainfall index value greater than 1.4) were associated with early onset date (31 May or earlier) and severely deficit monsoon years (normalized rainfall less than 1.4) were linked with late onsets (7th June or later).

Similarly interannual variability over southeastern Africa over a period of 1875-1884 showed that there were 29 ENSO events of which 22 were accompanied with below normal rainfall linked with ENSO events linking with rainfall variability.

Sikka (1980b), Rasmusson and Carpenter (1983), Ropelawski and Halpert (1996), Mooley and Paolino (1989), studies reported changes in sea surface temperatures over equatorial Pacific Ocean to cause global climate variability. Similar to Pacific Ocean, warmer oceanic temperatures and anomalous convection patterns also occur in central India Ocean (Chandrashekhar and Kitoh, 1998; Webster et al., 1999; Yu and Rienecker, 1999) which caused differences in major precipitation areas through release of latent heat and associated atmospheric heating (Anderson, 1999; Meehl, 1987). Webster et al. (1999), also expressed this phenomenon which occurred in severe El Neno of 1997 event where the SST anomalies developed in Indian Ocean (cold in the east and warm in the west) called Indian Ocean Dipole (IOD) and related the ocean temperature distribution to increasing rainfall over east Africa and reduced rainfall over Indonesia and to the associated shift in atmospheric heating patterns. Saji et al. (1999) used dipole mode in the equatorial Indian Ocean to explain 12 per cent of SST variability. Similarly Singh (1998) reported positive SST anomaly over south Arabian Sea (0 – 10°N) and negative SST anomaly over north Arabian Sea (10° - 25°N).

Singh and Sarkar (2003) reported increasing SST trend in all seasons in coastal region and a decreasing SST trend in winter season in the Island regions.

Medha Khole (2005) examined the interannual and inter-seasonal variability of SST over equatorial Indian Ocean (5°N – 5°S and 50°E – 100°E), and over north Indian Ocean (5°N – 20°N and 50°E – 100°E) during 1961-98 and showed an increasing trend prominently after mid 1970s with rapid global warming. The seasonal SST anomaly for all the four seasons (winter, Pre-monsoon, SW monsoon and Post-monsoon) during 1961-75 was negative; whereas during 1976-98, it was positive in the post global warming period indicating warmer than normal SST values.

7.14 Influence of Oroganphic Features

Numerical simulation of the role of orography by an atmospheric global circulation model was performed by Hahn and Manabe (1975). The orographic features introduced disturbances in the impinging airflow on the different horizontal scales depending on the vertical temperature stratification of the air-stream and the dimensions of the barrier. On smaller scale, the mountain waves are set up on the be side of orographic barrier causing the air turbulence. Sarkar (1965) studied these mountain waves set up by the Western Ghats of the order of 25 km length or more. Their amplitude and magnitude of maximum vertical velocity increased with the wavelength. Similarly De (1971) showed the mountain waves setup during the winter season by the Assam – Burma Hills which were also of 25 km length with maximum vertical velocity of about 2 m sec^{-1} agreed with the cloud bands on satellite photographs. The orographic rainfall increased along the slope of Ghats and reached maximum before the crest of mountain, and after which it decreased sharply (Sarkar, 1967). Gadgil and Sikka (1976) showed that orographic features also caused perturbations in the large scale wind flow by setting up Rossby-type quasi-stationary trough/ridge system depending upon the characteristics of the seasonal zonal flow over Western Ghats. It was envisaged that Western Ghats forced the quasi-stationary trough during the monsoon season in the Bay of Bengal (Bavadekar and Khaladkar, 1984).

Tibetan Plateau affects the atmosphere not only as barrier, but acts as a strong source of heat. Yin (1949), Flown (1957) reported that the subtropical jet stream disappears from the southern side of the Plateau in the beginning of June and shifts to the north of Himalaya and Tibet at 40°N with a slowing down of westerlies over Eurasia with summer heating of Plateau forming a warm core anticyclone in troposphere at 500 mb level called the anticyclogenosis. However, on the southern side of this upper air anticyclone, the direction of air flow is from east to west known as tropical easterly jet (Koteswaram, 1958c). These upper air easterlies descend in the permanent high pressure area formed over south India intensifying the 'High' already present in the area. In October, the conditions are reversed. The middle and upper tropospheric anticyclone over Tibetan Plateau disintegrates. The tropical eastern jet stream is practically non-existent and then subtropical westerly jet stream reestablishes over the northern India with the result the summer monsoon retreats toward the southeast.

7.15 Long Range Forecasting of Monsoon Season Rainfall

Blanford (1884) introduced the long range forecasting by linking winter snowfall in the Himalayas to predict summer monsoon. However, Walker (1923) developed multiple regression models with use of physically linked monsoon predictors. Several multiple regressions were developed using possible predictors of monsoon (Table 7.3) with forecast capability on 65 percent occasions (Sikka, 1980b; Pant and Parathasarthy, 1981; Shukla and

Table 7.3
Predictors of SW monsoon rainfall over India for long range forecast in multiple regression models.

Predictors	Notation
Himalayan snow accumulation at the end of May	P_1
South American pressure (5 March to April + 5 May)	P_2
Mauritius pressure in May	P_3
Zanzibar district rain (April + May)	P_4
Ceylon rainfall in May	P_5
500 hPa Indian April ridge position along 75°E longitude	P_6
Indian east coast temperature (March mean temperature of Calcutta, Visakhapatnam, Chennai and Masulipatnam)	P_7
Darwin pressure (March + April + May)	P_8
Peru coast El Neno category (previous year)	P_9
North Indian temperature (March minimum temperature over Jaisalmer, Jaipur and Calcutta)	P_{10}
East-west extension of 50 hPa trough ridge system over the Northern Hemisphere (January + February)	P_{11}
Central India (Nagpur, Akola) temperature in May	P_{12}
Darwin m.s.l. perssue anomaly (January - April)	P_{13}
10 hPa zonal wind over Balboa in January	P_{14}
Northern hemisphere pressure anomaly over 5° latitude belt (35°N to 40°N) from 0 to 155°E longitude	P_{15}
Northern hemispheric temperature anomaly (January + February)	P_{16}
Eurasian snow cover (previous December)	P_{17}
Argentina pressure (Buenos Aires and Cordoba) in April	P_{18}
Tahiti – Darwin pressure (spring)	P_{19}
Peru coast temperature (previous August)	P_{20}
Peru coast temperature (January - March)	P_{21}
Equatorial pressure (Jakarta, Jan. – April; Port Darwin, Mar. – May; Seychelles, Feb. - March)	P_{22}
Himalayan snow cover (January - March)	P_{23}

Paolino, 1983; Thapliyal, 1984; Verma et al., 1985; Parathasarthy et al., 1988; and Verma et al., 1984).

Thapliyal (1984) developed the following stable monsoon rainfall (R) relationship over India as

$$\hat{R} = 3.2\hat{P}_6 + 7.6\hat{P}_7 - 2.6\hat{P}_8 + 1.1\hat{P}_9 - 0.3\hat{P}_{10} - 1.3 \qquad \text{...(7.1)}$$

where, ^ indicates the departure of the factor from its normal value. Thapliyal (1986) reviewed these large number of predictors, but found P_8, P_6, P_{13}, P_{16}, P_{14} and P_{15} significant parameters to predict Indian monsoon rainfall as a multiple regression expression of the form based on 1958-83 data explaining 85 percent of variations.

$$\hat{R} = 4.31 + 19.65\hat{P}_{12} - 10.73\hat{P}_{13} + 13.51\hat{P}_6 + 31.01\hat{P}_7 + 0.91\hat{P}_{14} - 1..01\hat{P}_{15} + 0.06\hat{P}_{16} \qquad \text{...(7.2)}$$

Keeping in view of large number of predictors, Gowarikar et al. (1989) introduced a new long range forecasting (LRF) technique comprising parameteric and power regression models using 16 antecedent parameters which are significant among the regional and global forcing for the subsequent monsoon. Based on this model, IMD issued LRF for India as whole during 1988-90. Analysis of signals from combinations of a few parameters was performed in the power regression models (Table 7.4).

The sixteen parameters were looked as favourable (positive coefficient of correlation) and unfavourable (negative coefficient of correlation) in different combinations of 2, 3 and 4 parameters. A combination of El Neno (same year) and Darwin pressure were connected with the ENSO phenomenon. Similarly a combination of 3 wind parameters – 500 hPa ridge, 50 hPa wind pattern and 10 hPa wind provided favourable signals in most occasions. A combination of four surface temperatures indicated favourable signals as well as unfavourable signals. Therefore, a combination of 16 parameters predicted better the quantification of Indian Monsoon due to interactive nature of these global and regional forcings governing the monsoonal current of southwest monsoon over India.

(i) **Power Regression Model**: The hypothesis of power regression model stems from the observation that even in the midst of adverse parameteric conditions, Indian monsoon can still turn out to be normal. Therefore by assigning due weightages to signals from different parameters to take into account their interactive behaviour of non-linear characters of different forcings with India monsoon system. The power regression model was expressed as (Gowarikar et al. 1989).

$$R = C_0 + \sum_{i=1}^{13} C_i X_i^{P_i} \qquad \text{...(7.3)}$$

where, R is SW monsoon season rainfall of India as a whole, C and p are model's constants. Data of 1958-88 was used to quantify these constants by stepwise iteration method with an increment of 0.1. The final form of the power regression model was expressed as (Gowarikar et al. 1991).

Table 7.4
Sixteen parameters of power regression model (Gowarikar et al., 1989)

Signals	Parameters in 1988	Notation	Parameters Revised in 2000 Model
Snow cover	Himalayan snow cover (January to March)	X_{15}	-
	Eurasian snow cover (Previous December)	X_2	-
Pressure	Northern hemisphere surface pressure anomaly (Jan. to April)	X_9	-
	Southern oscillation index (SOI) (spring, March - May)	X_{10}	
	Darwin pressure (spring)	X_{16}	Darwin Pressure Tendency (Jan. - Apr.)
	Argentina pressure (April)	X_8	-
	Indian Ocean equatorial pressure (January to May)	X_{13}	-
Temperature	El Nino (same year)	X_{14}	
	El Nino (previous year)	X_{11}	
	North India temperature (March)	X_{12}	Northern hemisphere temperature (Jan. – Feb.)
	East coast of India temperature (March)	X_6	-
	Central India temperature (May)	X_4	-
	Northern hemisphere temperature (January & February)	X_9	-
Wind	500 hPa ridge (April)	X_3	South Indian Ocean SST (Feb. – March)
	50 hPa wind pattern (east wind in January – February)	X_1	-
	10 hPa zonal wind pattern (January)	X_5	Arabian sea surface temperature (Jan. – Feb.)

$$\frac{R+\alpha_0}{\beta_0} = C_0 + \sum_{i=1}^{16} C_i \left(\frac{X_i + \alpha_i}{\beta_i}\right)^{p_i} \qquad \ldots (7.4)$$

where, R is monsoon rainfall over India as a percentage of long period average, X_i are different parameters. α, β, c and p are constants (Table 7.5).

Table 7.5
Constants for power regression model in their decreasing correlation coefficient

i	α_1	β_i	P_i	C_i
0	0	1	-	13.03086
1	50	10	1.4	-0.05824688
2	0	10	-2.0	30.26653
3	0	10	-4.0	-1.733019
4	0	10	4.0	0.6958737
5	50	10	-0.4	-19.56036
6	0	10	4.0	0.4374519
7	0	10	4.0	0.004687574
8	0	100	-3.9	-166292.8
9	50	10	-4.0	-46176.12
10	50	10	4.0	0.05844691
11	50	10	4.0	-0.004456572
12	0	10	4.0	-0.1497441
13	0	1000	3.3	-1183.124
14	50	10	4.0	-0.01344925
15	50	10	-4.0	19403.98
16	0	1000	3.3	1257.985

(Gowarikar et al., 1991)

Multiple Power Regression (MPR) model provided encouraging results in predicting SW monsoon seasonal rainfall, however the best result was given by using the combination of 18 parameters (P_6 to P_{23} of Table 7.1). Thapliyal (1990) observed that MPR models which use more than 14 parameters do not improve upon the forecast and therefore using 14 best parameters with their decreasing coefficient of correlation was developed using data of 1958 to 83

$$R = 4254.4 - 0.322\, P_{11}^{2} + 145.07\, P_{6}^{-0.8} - 22962\, P^{-4}_{12} - 41.665\, P^{1.4}_{13} - 23.332\, P^{0.8}_{14} + 0.015\, P^{4}_{15} + 0.57044\, P^{4}_{7} + 1046.2\, P^{-4}_{17} - 18871000\, P^{-4}_{18} - 0.47151\, P^{4}_{16} - 147810\, P^{-4}_{19} - 0.31496\, P^{4}_{20} - 5318.8\, P^{-4}_{21} - 88.72 P^{-4}_{10} \quad \ldots (7.5)$$

where, P values are used after suitable transformations.

Example 1:

Using the data of monsoon year 1988 assess the SW monsoon season rainfall with multiple regression model's (Thapliyal, 1990) expression as

$$R(\text{mm}) = -22.5097 - 30.1852\, P_6 + 59.6442\, P_7 + 0.1725\, P_{10} + 10.2264\, P_9 - 11.5642\, P_{19},\ R = 0.76$$

given that P_6 = 14.5°N, P_7 = 24.2°C, P_8 = 1009.8 hPa, P_9 = 4.0, P_{10} = 18.5°C

Solution:

Using the above expression with substitution of different predictors we get

$$R = -22.5097 - 30.1852 \times 14.5 + 59.6442 \times 24.2 + 0.1725 \times 18.5 + 10.2264 \times 4 - 11.5642 \times 2.8$$

$$= -22.5097 - 437.6854 + 1443.38964 + 3.19125 + 40.9056 - 32.37976$$

$$R = +994.9 \text{ mm.}$$

Actual rainfall in 1988 monsoon season all India value = 1018 mm

= 119% of normal

Forecasted rainfall = 116.6% of normal.

Thapliyal (1990) proposed a hypothesis that the resultant association (a_r) between monsoon rainfall and the predictor is made up of two types of association – climatic (a_c) and perturbed (a_p) related as

$$a_r = a_c + a_p \quad \ldots (7.6)$$

It is speculated that the perturbed climatic association represents the resultant association between the monsoon rainfall and climate forcings having oscillation period upto 10 years, whereas the basic climatic association between the monsoon rainfall and the mean of the climate forcing over a long period of time of the order of 50 years where the ocean-atmosphere interaction appears to be the most important forcing. Normally the predictors are related to troposphere, stratosphere, ocean surface and ocean-atmosphere coupling. A number of significant coupling are 500 hPa subtropical ridge over India (Banerjee et al. 1978), global parameters of 50 hPa east-west extension of ridge-trough system (Thapliyal, 1984), surface temperature over Northern Hemisphere durig winter (Verma et al., 1984). Thapliyal (1990) improved the LRF by using both the climatic (long-term) and perturbed (short-term) forcings in the LRF model. Another approach was also adopted of using both climatic and perturbed associations in developing LRF models such as power regression, parameteric, auto regressive integrated moving average and dynamic stochastic transfer etc.

(ii) **Parametric Model**: Here a large number of predictors (parameters), well distributed in time and space, are collectively used; and the resultant signal from a group of predictors improved the forecast accuracy (Thapliyal, 1991). A group of 16 predictors of global and regional atmosphere-land-ocean parameters, covering antecedent period from December to May were selected (Gowarikar et al. 1989; 1991) (Table 7.4 from X_1 to X_{16}).

Positive correlation of monsoon rainfall shows favourable (unfavourable) signal for wet monsoon (when rainfall ≥ 90 per cent of long period average) if its anomaly is +ive (-ive). Similarly negative correlation of monsoon rainfall shows favourable (unfavourable) signal for wet monsoon if its anomaly is –ive (+ive). And when monsoon rainfall is normal (within ±10 per cent of long period average) or excess (>110 per cent of long period average), the monsoon was termed wet. On the other hand, when the rainfall is less than 90 per cent of long period average, the monsoon is termed as deficit. Long period average of monsoon rainfall over the country as a whole is about 881 mm.

In parametric approach equal weightage is given to each of the 16 parameters. Percentage number of favourable parameters out of 16 are determined each year of data period and are compared with the performance of the subsequent monsoon rainfall over the country. Thapliyal and Kulshrestha (1992) using 1951-1987 data reported that whenever 70 per cent or more parameters (out of 16) are favourable for wet monsoon (normal or excess), the subsequent monsoons were not only normal but the rainfall was on positive side of the normal on all occasions; and whenever 20 per cent or less parameters were favourable, the monsoons were deficient (rainfall less than 90 per cent of long period average) on 66 per cent occasions only. Similarly, wet monsoons were followed on all occasions when 60 per cent or more parameters were favourable, while deficient monsoons were followed only on 75 per cent occasions when 40 per cent or less parameter were favourable. Therefore parametric model is purely qualitative to indicate whether monsoon would be wet or deficient.

Example 2:

Verify the performance of power regression model using the following data of 16 parameters in a wet and normal monsoon rainfall year.

Parameter	Wet Year	Normal Year
X_1	60.0 °Longitude E	135.0 °Longitude E
X_2	25.7 million sq.km	27.1 million sq.km
X_3	14.5 °N	15.0 °N
X_4	29.1 °C	28.7 °C
X_5	9.0 ms^{-1}	-34.8 ms^{-1}
X_6	24.2 °C	25.3 °C
X_7	50.0 hPa	49.5 hPa
X_8	991.1 hPa	988.1 hPa
X_9	0.58 °C	0.77 °C
X_{10}	2.8 hPa	0.4 hPa
X_{11}	4.0	1.0
X_{12}	18.5 °C	19.4 °C
X_{13}	1010.8 hPa	1011.9 hPa
X_{14}	1.0	2.0
X_{15}	1.19	3.1
X_{16}	1009.8 hPa	1012.4 hPa

Solution:

Using power regression model (Gowariker et al. 1991).

$$R= 13.03086 - 0.05824688 \left(\frac{X_1+50}{10}\right)^{1.4} + 30.26653 \left(\frac{X_2}{10}\right)^{-2} - 1.733019 \left(\frac{X_3}{10}\right)^{-4}$$

$$+ 0.6958737 \left(\frac{X_4}{10}\right)^{4} - 19.56036\left(\frac{X_5+50}{10}\right)^{-0.4} +0.4374519\left(\frac{X_6}{10}\right)^{4} +0.004687574\left(\frac{X_7}{10}\right)^{4}$$

$$- 166292.8 \left(\frac{X_8}{100}\right)^{-3.9} - 46176.12 \left(\frac{X_9+50}{10}\right)^{-4} + 0.05844691 \left(\frac{X_{10}+50}{10}\right)^{4}$$

$$- 0.004456572 \left(\frac{X_{11}+50}{10}\right)^{4} - 0.1497441 \left(\frac{X_{12}}{10}\right)^{4} - 1183.124 \left(\frac{X_{13}}{1000}\right)^{3.3}$$

$$- 0.01344925 \left(\frac{X_{14}+50}{10}\right)^{4} + 19403.98 \left(\frac{X_{15}+50}{10}\right)^{-4} + 1257.985\left(\frac{X_{16}}{1000}\right)^{3.3}$$

Using data of wet year, wet get

$$R = 13.03086 - 0.05824688\,(11)^{1.4} + 30.26653\,(2.57)^{-2} - 1.733019\,(1.45)^{-4}$$
$$+ 0.6958737\,(2.91)^{4} - 19.56036\,(5.28)^{-0.4} + 0.4374519\,(2.42)^{4} + 0.004687574\,(5)^{4}$$
$$- 166292.8\,(9.911)^{-3.9} - 46176.12\,(5.058)^{-4} + 0.05844691\,(5.28)^{4} - 0.004456572\,(5.4)^{4}$$
$$- 0.1497441\,(1.85)^{4} - 1183.124\,(1.0108)^{3.3} - 0.01344925\,(5.1)^{4} + 19403.98\,(5.119)^{-4}$$
$$+ 1257.985\,(1.0098)^{3.3}$$

$$R = 13.0386 - 1.67195 + 4.58244 - 0.39204 + 49.90021 - 9.61692 + 15.00347$$
$$+ 2.92973 - 21.67780 - 70.55083 + 45.42524 - 3.78945 - 1.75403 - 1225.81670$$
$$- 9.09869 + 28.25858 + 1299.12868$$

$$R = 1458.25921 - 1344.36841$$

R = 113.89% of long period average rainfall

Using data of normal year rainfall we get

$$R = 13.03086 - 0.05824688\,(18.5)^{1.4} + 30.26653\,(2.71)^{-2} - 1.733019\,(1.5)^{-4}$$
$$+ 0.6958737\,(2.87)^{4} - 19.56036\,(1.52)^{-0.4} + 0.4374519\,(2.53)^{4} + 0.004687574\,(4.95)^{4}$$
$$- 166292.8\,(9.881)^{-3.9} - 46176.12\,(5.077)^{-4} + 0.05844691\,(5.04)^{4} - 0.004456572\,(5.1)^{4}$$
$$- 0.1497441\,(1.94)^{4} - 1183.124\,(1.0119)^{3.3} - 0.01344925\,(5.2)^{4} + 19403.98\,(5.31)^{-4}$$
$$+ 1257.985\,(1.0124)^{3.3}$$

$$R = 13.03086 - 3.46189 + 4.12120 - 0.34232 + 47.21261 - 16.54396$$
$$+ 17.92306 + 2.81429$$
$$- 21.93562 - 69.50064 + 37.71236 - 3.01496 - 2.12108 - 1230.22438 - 9.83358$$
$$+ 24.40691 + 1310.19975$$

$$R = 1457.42104 - 1356.97843$$

$R \simeq$ 100 per cent of long term average rainfall.

(iii) **Dynamic Stochastic Transfer Model**: Dynamic Stochastic Transfer (DST) model was introduced with the hypothesis that atmosphere is a complex dynamic system where it transforms various inputs into numerious output such as monsoon rainfall of India. In SW monsoon system the transfer dynamic relationship between the continuous input and output series were expressed by a differential equation, and when the series are discrete, the relationship was expressed as

$$R_t = \frac{w(B)}{\delta(B)} Z_{t-b} \quad \ldots (7.7)$$

where, R_t is output, Z_t is input; and constant are of the form

$w(B) = w_0 - w_1B - w_2B^2 - \ldots\ldots - w_6B^6$

$\delta(B) = 1 - \delta_1B - \delta_2B^2 - \ldots\ldots - \delta_6B^6$

B is backward shift operator and w and δ are unknown constant. Thapliyal (1990) developed DST models by considering the April wind patterns at a height of 6 km over India as an input to the atmosphere and subsequent monsoon rainfall as the output of the atmosphere.

The general form of DST model (Thapliyal, 1990) is of the form as given below.

$$\frac{R_t}{R_{t-1}} = \left[\prod_{i=1}^{6} \left(\frac{R_{t-i}}{R_{t-i-1}} \right)^{\delta_i} \exp\left\{ w_0(Z_t - Z_{t-1}) - \sum_{i=1}^{6} w_i(Z_{t-i} - Z_{t-i-1}) \right\} \right]$$

$$\times \left[\exp(-1) \sum_{i=1}^{12} \theta_t - a_{t-i} \right] \quad \ldots (7.8)$$

where, R is seasonal monsoon rainfall for India as a percentage of long period average, Z is location of 500 hPa April ridge line along 75°E longitude over India, w's, δ's, θ's are model constants, and a's are climate forcings.

Therefore at the end of April when wind parameter is available, monsoon rainfall is forecasted using this expression as a function of the previous year rainfall. Model performed accurately comparable with power regression model.

Long range monsoon forecasts are issued in two stages – once before the mid April- a tentative forecast based on available signals from the clinched parameters as well as the estimated signals from remaining unclenched parameters, and then final forecast in last week of May based on Parametric, Power Regression (PR), Dynamic Stochastic Transfer (DST) and updated multiple regression models. But special emphasis is given for PR and DST model outputs due to their encouraging performances over the last two decades.

7.15.1 Pre-monsoon rainfall forecast

Pre-monsoon contributes about 12 per cent of annual rainfall over northern India due to convective currents and thunderstorm activity. Though during this period the evaporation rates are high due to higher temperatures and prevailing dry winds, the conservation of pre-monsoon showers are very useful because during the last decade intense showers have occurred in May – June months delaying the onset of summer monsoon probably due to impact of global warming and seasonal climatic changes. Very few studies are available to predict the pre-season monsoon rainfall. Though the pre-monsoon rainfall is low and erratic, but very effective in decreasing the summer heat load.

Sanjeev Kumar (2003) predicted the summer rainfall amount with rainfall events more than 10 mm, less than 10 mm and all events during April 1 to June 30 period using 1969 to 2000 dada at Hisar (Lat. 29°10'N and Long. 75°46'E) using significant weather variables of the period before the occurrence of rains. The multiple regressions developed for prediction of summer rainfall (R in mm) with amount (less than 10 mm per event) was of the form

$$\begin{aligned} R = & -6.65 + 0.143\,\Delta RH + 0.102\,\Sigma T_{max40} + 0.0383\,I - 0.00164\,\Sigma HU_{20} \\ & (-2.62) \quad (1.07)^{**} \quad (2.29)^{**} \quad (3.39)^{**} \quad (-0.42) \\ & -0.443\,N_w + 2.062\,N_h + 0875\,C_h \\ & (-2.81)^{**} \quad (4.24)^{**} \quad (0.71)^{**} \qquad (R^2 = 0.46, F = 13.64) \end{aligned} \qquad \ldots(7.9)$$

Rainfall events with amount more than 10 mm per event was predicted with multiple regression of the form as

$$\begin{aligned} R = & -18.401 - 0.0294\,\Delta RH + 0.309\,\Sigma T_{max40} + 0.0218\,I + \\ & (-2.8) \quad (-0.116) \quad (2.9)^{**} \\ & 6.886\,N_h - 0.134\,C_h + 0.00757\,\Sigma HU_{20} \\ & (4.8)^{*} \quad (-0.04) \quad (1.04)^{**} \qquad (R^2 = 0.69, F = 12.82) \end{aligned} \qquad \ldots(7.10)$$

where, R is rainfall in mm on nth day, * significance at 0.05 level, ** significance at 0.01 level, ΔRH is increase in relative humidity in percent during last days of rainfall events i.e, during (n – 3) to (n – 1) period, ΣT_{max40} is cumulative maximum temperature values above 40°C during (n – 1) number of days, I is cumulative value of Thornthwaite's heat index during (n – 1) number of days, ΣHU_{20} is cumulative heat units during (n – 1) number of days with base temperature of 20°C, N_w is total number of south and westerly wind directions during (n – 1) number of days using both morning and evening hours of observations, N_h is the amount of lowest cloud during the last days of rainfall events i.e., during (n – 3) to (n – 1) period, C_h is high clouds form during the last days of rainfall events i.e., (n – 3) to (n – 1) to period and E_p is average open pan evaporation rate during (n – 1) period.

Similarly the prediction of summer month's rainfall amount of all events was expressed as

$$\begin{aligned} R = & -5.39 - 0.104\,\Delta RH + 0.0517\,\Sigma T_{max40} + 0.01167\,I \\ & (-2.64) \quad (-2.95) \quad (1.305)^{**} \quad (1.85)^{**} \\ & + 0.164\,E_p + 1.996\,N_h + 8.894\,C_h \\ & (2.78)^{**} \quad (4.405)^{**} \quad (5.79)^{**} \qquad (R^2 = 0.53, F = 26.74) \end{aligned} \qquad \ldots(7.11)$$

Most of the rainfall events during hot summer months (April to June) at Hisar occurred on a single days, but there was a maximum continuity of 4 days rainy spells and such events can be taken care through synoptic charts of short range weather forecasts.

7.16 Impact of Monsoon Variability on Agriculture

7.16.1 Food production

The role of variation of monsoon activity on the variability on the food production over India needs no emphasis. El Neno-Southern Oscillation index, sea surface temperature changes events influence the monsoon rainfall variability during June to September to cause food production variability. Food production in the monsoon belt is strongly influenced by the pattern of monsoon rainfall in the crop growing season i.e., its amount and distribution during different phenophase of crops grown during the rainy season. National level food security strategies are framed based on the monsoon activity—its onset, withdrawal, growing season duration, amount of rainfall and its distribution in growing season according to crop water requirements. Central Research Institute for Dryland Agriculture (CRIDA, 2001) developed a model for assessing the foodgrain production (P) based on monthly rainfall indices (RI) for the monsoon months (June to September) and Southern Oscillation Index (SOI) expressed as

$$P = PI \times T/100 \qquad \ldots (7.12)$$

where, PI is production index and T is technology trend expressed as

$$T = \exp\{(0.0256264 \times \text{year})\ 1.18\ E - 20\} \quad R = 0.94 \qquad \ldots (7.13)$$

$$PI = 72.88 + 0.039\ RI_6 + 0.031\ RI_7 + 0.027\ RI_8 + 0.034\ RI_9 + 0.141\ SOI,\ R = 0.83 \qquad \ldots (7.14)$$

$$RI_i = \Sigma A_i R_i / \Sigma A_i \qquad \ldots (7.15)$$

where, subscripts are months, A_i and R_i are area and rainfall of selected four meteorological subdivision of India.

A_6 represented with Bihar Plateau, west Madhya Pradesh, Konkan and Goa, and Telangana

A_7 represented with east and west Uttar Pradesh, Haryana, Rayalaseema, and Tamil Nadu,

A_8 represented with east Rajasthan, east Madhya Pradesh, and Gujrat, and

A_9 represented with east Uttar Pradesh, east and west Madhya Pradesh, Saurashtra and Kutch, Konkan Goa, Madhya Maharastra, and coastal Karnataka.

Model predicted foodgrains production reasonably will within statistical error limits.

Parthasarthy et al. (1992) forecasted rainy season (Kharif) Foodgrain Production Index (FPI) as a percentage of technological trend and monsoon season rainfall index (MRI) expressed as

$$FPI = 0.374 + 0.667\ MRI,\ (R^2 = 0.77,\ F = 70.7) \qquad \ldots (7.16)$$

where, MRI is percentage of rainfall deviation of particular year from the mean value, and FPI is represented by an exponential curve fitted for actual production called technological trend curve expressed as

$$T_i = 55.4243(1.0205)^{Ni} \qquad \ldots (7.17)$$

where, N_i is the number 1, 2, 3 representing 1966-67 1967-68, . . . respectively. The FPI_i value for any ith year is given by

$$FPI_i = \frac{F_i - T_i}{T_i} \times 100 \qquad \ldots (7.18)$$

where, F_i is rainy season foodgrain production (tonnes) for the ith year. Since both indices are expressed as percentage quantities based on similar transformations they are comparable. Mean and standard deviation for MRI and FPI were 0 and 1, 0.37 and 8.52. Fig. 7.18 can be used to predict the Kharif foodgrain production with use of rainfall index in deficient, normal and excess rainfall years. Long range forecasting models for monsoon activity will assess the most likely rainfall amount as an input for determining the rainfall index to predict the foodgrain production levels.

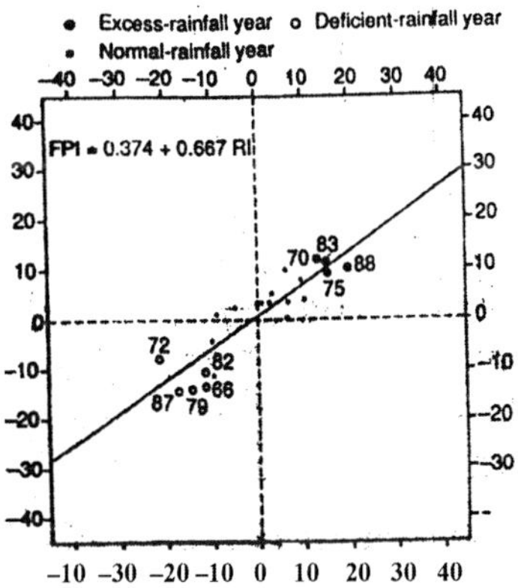

Fig. 7.18: Scatter diagram with fitted regression line showing the relationship between MRI and FPI for 1966-88. (After: Parthasarthy et al., 1992)

7.16.2 Location specific monsoon characteristics for planning crop strategies

In order to harness the benefit of monsoon season rainfall for agriculture it is essential to understand the predictive behaviour of seasonal rainfall for specific crops at a location. It is desired to have the knowledge of pre- plant and planting time decisions as these are influenced by rainfall amount, duration and intensity. The choice of crops and their varieties, monoculture or intercropping, plant population, use of inputs are regulated with the availability of soil moisture in the root zone through seasonal rainfall. Date of onset of monsoon (an effective rainfall shower of 25 mm or a continuous spell of rainy days which meets soil evaporation for at least 5 days for crop germination establishment) at a location or crop growing season as it decides the length of the growing season and predictive behaviour of monsoon rainfall. Under rainfed agriculture it is worthwhile to derive the monsoon season rainfall characteristics at a location in relation to rainfed crop yield variations to envisage the role of conservation farming.

Fig. 7.19 presents the rainfall occurrences at Hisar (Lat. 29°10'N, Long. 75°46'E) in a scatter diagram using long records of rainfall data (past 100 years) to present its behaviour with respect to the date of onset (envelope ABCD) for Pearl Millet cropping (Bishnoi, 1988). Date of onset for Pearl Millet is based on dailywater budgeting process when a soil moisture build up of 25 mm in surface soil is attained with the onset of rainy season for ensuring germination and crop establishment. Depending on the rainfall pattern and soil evaporation losses between the rains at the location, the amount of storage of 25 mm may be attained in a day or in few days. A decreasing linear relation existed between seasonal rainfall amount (R) and onset of growing season of the form.

$$R(mm) = 533.03 - 4.72\, N_1,\ r = -0.56 \quad \ldots (7.19)$$

$$R(mm) = 506.29 - 7.56\, N_2,\ r = -0.79$$

where, N_1 and N_2 are the number of days from June 1 and June 15.

Rainfall variability was quite high at this semiarid location varying from 144.8 to 727.5 mm. The average date of onset of pearl millet season was 8th July with variation from as early as 4th June to as late as 20th August. Normally onset dates are divided into three stages characterized as early, normal and late onset dates for pearl millet production. Scatter diagram clearly shows that the early started season falls in upper part of the rainfall range, whereas late started onset season years were in the lower part of rainfall range as reflected from the above decreasing linear relation. Crop is harvested around October first week (Oct. 5th), and therefore the number of days from onset of rains varied from 47 to 127 days. Growing season availability is normally the period from the date of onset to the date when soil moisture depleted and reduced to permanent willing point in the root zone depth. The growing season duration (GS) is also influenced by the date of onset of monsoon rains expressed as

$$GS(days) = 116.18 - 0.88\, N_i,\ r = -0.62 \quad \ldots (7.20)$$

Fig. 7.19 also presents three sloping lines labelled as 6, 4 and 2 mm day^{-1} representing the average daily rainfall amount (intensity) from onset date to 5th October reflecting that the later onset seasons tend not only to have less rainfall, but also less rainfall per day. This creates soil moisture stress in the crop as evaporative demands sharply increased during 15th August to 15th September, but rainfall amounts are inadequate. Rainfall probabilities of seasonal rainfall of $\geq$ 400 mm, between 200 to 400 mm and less than 200 mm were of the order to 29, 49 and 22 percent respectively presenting the agricultural potentiality of the location. A study on nutrient uptake and water use efficiency of pearl millet crop shows optimum uptake with maximum water use efficiency at a soil moisture consumption level of 284 mm under rainfed and 357 mm under irrigated conditions. Fig. 7.19 also presents that 400 and 200 mm of seasonal rainfall levels corresponding to the maximum and lowest economic yields is expected. Therefore, two horizontal lines corresponding to 400 mm of

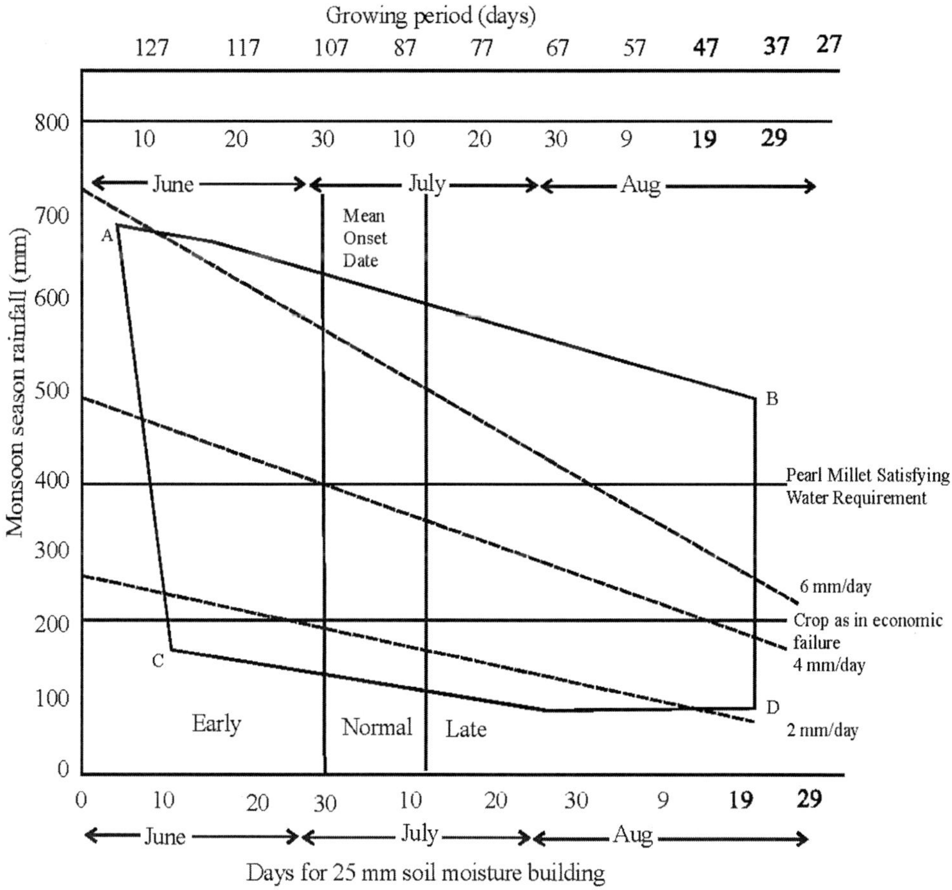

Fig. 7.19: Rainfall pattern, growing period, onset of growing season for pearl millet production

seasonal rainfall at which pearl millet water requirements are satisfied, and 200 mm level where crop behaves as a economic failure. Generally a maximum crop yield is achieved when about 250 to 300 mm of soil moisture (SW) is utilized in successful completion of different phenophases under rainfed conditions as represented by an expression of the form

$$\text{Yield (q/ha)} \;\; 36.3127 \; + \; \underset{(3.265)}{0.2469\,SW} - \underset{(-2.665)}{0.0002279\,SW^2}, \quad (R^2 = 0.76, F = 9.55) \qquad \ldots (7.21)$$

About 29 percent probability of getting more than 400 mm of seasonal rainfall exists with early onset chances and can be preferred for double or intercropping systems under rainfed conditions. With delayed onset, the probability of failure increased to 43 percent (Table 7.6) approaching once in two years at the location.

Pearl millet grain yield are strongly influenced by the nutrient uptake with availability of soil moisture expressed as (Anonymous, 2004).

Table 7.6
Rainfall probabilities and growing season characteristics at a location (Hisar) based on rainfall analysis.

Onset Period	Average Rainfall	Probability (%)	Probabilities for Rainfall Amount				Growing Season Characteristics for Pearl Millet		
	mm		>400mm	335-400mm	200-335mm	<200mm	Onset	Withdrawal	Duration (days)
Early (Before June 30)	445	32	42	17	33	8	June 20 ± 6	Sep. 4 ± 18	76 ± 15
Normal (July 1 – 15)	375	32	25	12	50	13	July 7 ± 5	Sep. 15 ± 16	72 ± 16
Late (July 16 – Aug. 26)	255	36	19	7	31	43	July 28 ± 9	Sep. 8 ± 14	42 ± 18
Season as whole (June 1 to Aug. 20)	335	100	29	12	37	22	July 8 ± 15	Sep. 10 ± 17	65 ± 21

Grain yield (kg ha^{-1}) = – 849.77 + 439.388 log NU_r (R^2 = 0.93, F = 131.37) . . . (7.22)

Grain yield (kg ha^{-1}) = – 1344.3 + 580.082 log NU_i (R^2 = 0.90, F = 94.3)

where, NU is nutrient uptake by pearl millet in kg ha^{-1} with subscripts r and i for rainfed and irrigated conditions. The efficiency of nutrient uptake utilization contributes in the grain yield increase. It is higher at lower nutrient doses and decreases with higher doses of nitrogen depending on the availability of soil water to the plants expressed as (Anonymous, 2004).

NU_r (kg ha^{-1}) = 20.4644 exp{3.188 WUE}, (R^2 = 0.83, F = 48.88) . . . (7.23)

NU_i (kg ha^{-1}) = 16.3529 exp{4.96 WUE}, (R^2 = 0.88, F = 76.2)

where, WUE values are in qha^{-1} cm^{-1} under rainfed and irrigated conditions.

Therefore a detailed analysis of past long record of rainfall data of each location in the monsoon belt needs to be analysed in relation to water requirement of different crops, grain yield achieved with utilization of soil moisture and nutrient uptake analysis for a comprehensive analysis to develop crop growing strategies. Rainfall probabilities (simple, conditional, cumulative) using different statistical techniques (frequencies, Markov chain probabilities, incomplete gamma distribution) can be worked for the location to develop crop planning strategies (Sastry, 2006), but analysis of rainfall data in relation to crop water requirements or application of beta distribution to productive environmental variables (water use, nutrient uptake, water balance parameters) are useful to develop crop production strategies (Yao, 1974). Asimple rainfall analysis pattern for getting monthly rainfall >300, 200-300, 100-200, 50-100 mm for each month of crop growing season (3 to 4 months) was done by National Commission on Agriculture to develop rainfall pattern in relation to the cropping pattern at a location. The criteria for choice of 300, 200-300, 100-200 and 50 to 100 mm monthly rainfall was choosen in order to meet the monthly water requirement of paddy, maize/arhar, sorghum/ pearl millet and mungbean/cowpea/grasses (NCA, 1976). Though arbitrary categorization, but found to have practical utility in the region concerned. Such analysis are useful in relation to crops and be substantiated with computation of water balance parameters in the growing season. Length of the growing season is quantified with rainfall analysis in relation to water requirement of crops or moisture availability periods. Moisture adequacy index (Actual evapotranspiration (AET) as a fraction of potential evapotranspiration (PET) is used for choice of suitable crops depending on moisture availability as

Length of the Growing Period (days)	Cropping System
<75	Grasses, shrubs, silviculture; monocropping of short duration grain crops
75 – 120	Monocropping
120 – 150	Intercropping
150 – 200	Double cropping
>200	Multiple cropping

There are several similar productive environment indices which have been preferred to develop crop management strategies and monitoring of agricultural field operations.

(i) Moisture adequacy index (AET/PET) satisfying the βeta distributionprobabilities (Bishnoi, 1979).
(ii) Moisture availability indices – ratio of 80 to 50 per cent probability amount of rainfall in relation PET for length of growing season of crops (Sarkar and Biswas, 1980).
(iii) Moisture availability in the root zone in relation to water requirement at different phenophase stages of crops (Baier, 1969).
(iv) Crop water stress index (CWSI) as (1 – AET/PET) in terms of growth phases of crops (Jackson et al., 1977).
(v) Water requirement satisfaction index (WRSI) computed with use of water balance procedure in order to monitor the moisture stress in rainfed crops (FAO, 1980).

These productive environment parameters of crops needs to be evaluated with the onset time of monsoon rains at each location to assess the location specific crop management strategies under rainfed conditions.

7.16.3 Impact of ENSO and SST variations on Crop Yield

Climatic variability measured in interannual and interseasonal scales is known as one of the main constraint for agricultural productivity all over the world and there is need for reducing the associated risks to increase productivity. A strong relationship exists between the El Neno/Southern Oscillation (ENSO) and seasonal/annual precipitation over the vast area of Africa and South America (Ropelewski and Halpert, 1987; Kiladis and Diaz, 1989). Sea surface temperatures and pressures in the Atlantic (Nicholson and Entekhabi, 1987) and the Indian Ocean (Jury et al. 1996) are correlated with precipitation patterns over Africa. Crop simulation models have been used to study impact of ENSO on agricultural production in Australia (Ramington and Nicholls, 1993; Hammer et al. 1996; Meinke and Hammer, 1997); in Asian monsoon region and tropical sub-sahara Africa (Manton et al. 1997; Sivakumar, 1997). Phillips et al. (1998, Cane et al. (1994), using the CERES Maize model, with incorporation of ENSO prediction based temperature and rainfall variability during El Nino, normal and La Nena years, simulated the maize yield variability in different agroecological zones of Zimbabwe. It was revealed that a more precise prediction was required for valuable forecasts in crop management decisions as simulated yield results predicted favourable cropping seasons as opposed to the poor ones with respect to better nitrogen management and planting dates for marginal sites for maize management. Two strategies were possible to diminish climatic variability – one by controlling the limiting factors (i.e., introducing irrigation in water limited environments), and the other to adjust production management practices keeping in view the expected/forecasted weather conditions for the upcoming growing season. The appropriate actions can ameliorate the effects of regional food shortages. With the recent capability to predict climate anomalies linked with the onset and intensity of a warm or cold event as part of El Nino/Southern Oscillation phenomenon. ENSO is main source of interannual climate variability in many parts of the world. The probabilistic climate forecasts are being issued for likelihood of expected temperature and rainfall scenarios (below normal, normal or above normal) for the upcoming 3 to 6 months. Using these scenarios, the crop simulation models are being used to assess the effect of varying temperature and precipitation in different crop growth stages to identify the most vulnerable periods of the growing season and to determine the climatic thresholds for maximum/optimum economic crop yields. Even the models are linked with the weather generators conditioned to ENSO phases to evaluate crop management practices which would minimize the negative expected weather conditions and to make advantage of expected favourable conditions. Simulation modelling have been found useful to quantify crop management options in areas where climatic variability is high such as Australia, India, southeast Asia, Africa and South America (Meinke et al. 2001; Travasso et al. 1999; Keating and Meinke, 1998; White et al. 1998; Baethgen and Magrin, 2000; Podesta et al. 1999).

These studies revealed the existence of a near symmetry between the impacts of El Neno and La Nina on precipitation and crop productivity. Positive rainfall anomalies prevailed in El Nino years, whereas negative anomalies prevailed in La Nina years in southeastern Latin America. Impact of La Nina was stronger and less variable on both rainfall and crop yields as compared with the impact of El Nino. Therefore, new weather generators conditioned to ENSO or SOI phases were developed (Grondona et al. 1999). Baethgen (1998) found that probability for obtaining low maize yields in Uruguay was about twice as high as in La Nina years than in normal yeas, and using short season maize hybrids and delaying the planting date by 2 months could minimize the yield losses in La Nina years (Fig. 7.20). Baethgen (1998) found that the benefits of tailoring soybean planting dates to forecast the ENSO phases. Margin et al. (1999) also concluded that delay of planting dates in maize and soybean was the best management option in La Nina years, whereas early sowings optimized the yields in El Nino years. It was also reported that nitrogen rates in maize should be higher during El Nino events to maximize expected maize yield in southeastern Latin America. Similar applications in different parts of the ENSO affected areas will be useful to simulate impacts of ENSO on the crop yields of different crops. The trends may not be similar in different regions. Rezaul Mahamood (1997) showed a non-linear relationship between the decreasing air temperature and the length of initial vegetative growth stage and a largely linear increasing relationship with flowering/heading and maturity stages in Boro rice varieties (Table 7.7). Warmer air temperature conditions resulted shorter growing season with reduced total evapotranspiration.

An increase of air temperature provided longer and more stable thermal conditions for Boro rice maturing stage. A 5 per cent increase and 4 per cent decrease in seasonal total evapotranspiration occurred with each 1°C cooler and warmer air temperature conditions respectively in rice genotypes.

Since ENSO events are incepted 6 months in advance and eastablished 2 months before

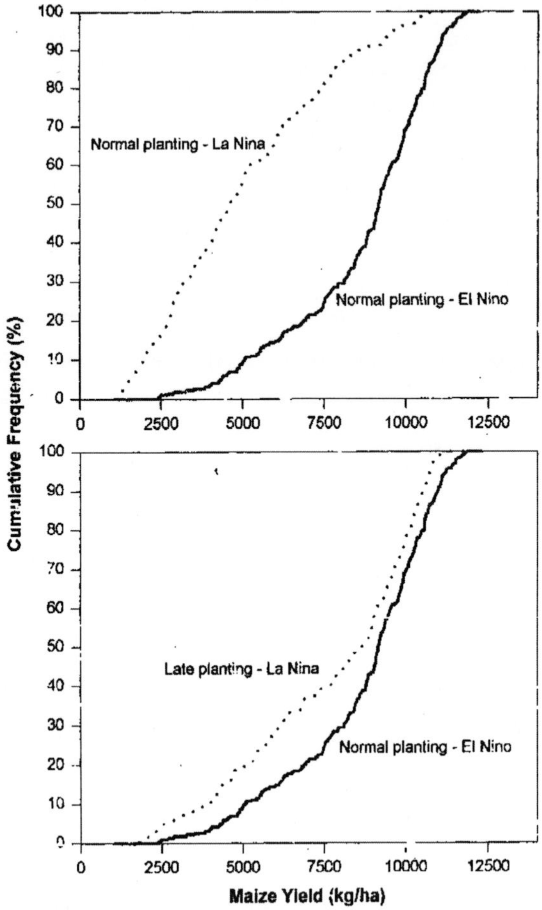

Fig. 7.20: Simulated maize yield distribution for El Nino and La Nina years in Uruguary for normal planting on September 15 with medium duration maize and late planting with short duration maize. (After: Baethgen, 1998).

occurrence of the monsoon rains in India (Table 7.8), it is useful to adopt crop simulation models with historical weather data or use of weather generators data to envisage their potential impacts. Alternative management scenarios can be evaluated with the use of models to identify those scenarios that can reduce the potential impact (Iglesias et al. 1996), and then the optimum scenario can be adopted by farmers to make use of various options for crop management regimes for forthcoming growing season. Gadgil et al. (1999a, b) highlighted the use of peanut model PNUTGRO to link climate, crop models and farmers to draw benefits with early warning.

ENSO leads to short-term climate fluctuation causing severe flooding along SW coast

Table 7.7
Percent time required to complete different growth stages in Boro rice varieties under below normal air temperature

Temperature below Normal (°C)	Initial Growth Stage (%)	Vegetative Stage (%)	Flowering/heading Stage (%)	Maturity Stage (%)
-5	14.00	35.00	32.00	19.00
-4	19.45	33.65	30.40	16.25
-3	21.20	35.00	28.00	15.71
-2	20.00	37.40	27.40	15.20
-1	15.00	42.00	28.00	15.00

(*After*: Rezaul Mahmood, 1997)

of USA, Peru and Ecuador, whereas drought in Australia, India and Indonesia. Trenberth (1998), McPhaden (1999) and Nicholls (2000) described the development of 1997-98 El Nino/Southern Oscillation to severe scale, and predicted the climate impacts with the monitoring southern oscillation index or of east Pacific sea surface temperatures. The applications of El Nino forecasts will be the biggest advance in the use of climate information to mitigate climate damage.

Sinha Ray and De (2003) found a significant decreasing trend in lower stratospheric temperature over equatorial Indian region, but still the lower stratospheric temperature was found to be significantly warmer during ENSO years than non ENSO years. In the upper stratosphere, a significant decreasing trend was noticed only from September to December, whereas in upper mesosphere a decreasing trend in temperature was noticed only during May and June. De and Joshi (1998) recorded the maximum number of cyclonic disturbances occurred during 1941-50 - the decade of fewest ENSO episodes.

The current set of seasonal to interannual forecast models have demonstrated broad skills in predicting global temperatures and precipitation. Use of multimodel ensembles have greatly enhanced overall capabilities predicting climatic variability upto a few months. In the Pacific the intrinsic atmospheric variations due to ENSO and SST variations largely contributed to decadal varibility. The large and vast ocean also influence directly on the climate through changes in SST and the partial pressure of CO_2. Future changes in sea level are a key question from climate change perspective.

Southern Ocean is important from global climate point of view since it connects the upper and lower limbs of global overturning circulation, stores 60 per cent of the total anthropogenic carbon, may have an annual mode linking to ENSO and sensitive to climate variation globally (Busalacchi, et al. 2005). Thus complex atmospheric-sea coupling exists for large scale atmospheric circulations creating climatic variability to influence the resources availability for meeting the requirements of agricultural productivity.

Hill et al. (2000) evaluated three phase method of classifying SOI events – low SOI (El Nino), high SOI (La Nena) and others (Climate Prediction Centre, 1998), and as well as five

Table 7.8
Yearwise El Neno, southern oscillation minima, warming/cooling features over India (Kane, 1998, 2004).

Year	Events	Year	Events
1871	EN (S) SO (M)	1912	EN(S) SO(S)
1872	C(S)	1913	SO(S) W(S)
1873	EN (W) SO (S)	1914	EN(M) SO(S) W(M)
1874	EN (M) C (M)	1916	C(S)
1875	C (S)	1917	C(M) EN(S)
1877	EN (S) SO (S) W (S)	1918	EN(S) SO(S) W(S)
1878	EN (S) SO (S) W (S)	1919	EN(M) SO(M) W(M)
1879	C (S)	1920	W(S)
1880	EN (M) SO (S)	1921	C(S)
1882	C (S)	1922	C(S)
1884	EN (S) SO (W)	1923	EN(M) SO(M) W(M)
1885	SO (M)	1924	C(S)
1886	C (M)	1925	EN(S) SO(S) W(S)
1887	C (M) EN (M)	1926	EN(S) SO(S) W(S)
1988	SO (S) W (S)	1927	C(M)
1889	C (M) EN (M)	1928	C(M)
1890	C (S)	1930	EN(M) SO(S) W(S)
1891	EN (S) SO (S)	1931	EN(M) SO(M) W(S)
1892	C (S)	1932	EN(S)
1893	C (S)	1933	C(S)
1894	C (S)	1934	C(S)
1896	EN (M)SO (S) W (S)	1935	SO(M) C(M)
1897	EN (M)	1936	SO(M) C(M)
1898	C (M)	1938	C(S)
1899	EN (S) SO (S) W (S)	1939	EN(M)
1900	EN (S) SO (S)	1940	EN(S) SO(S) W(S)
1902	EN (M) SO (S) W (S)	1941	EN(S) SO(S) W(S)
1903	C(S)	1942	C(S)
1904	SO(S) W(S)	1943	EN(M)
1905	EN(M) SO(S) W(S)	1944	SO(M) W(M)
1906	C(S)	1946	SO(S) C(M)
1907	EN(M) C(W)	1948	EN(W) SO (M) W (M)
1908	C(S)	1949	SO(M) C(M)
1909	C(S)	1950	C(S)
1910	C(S)		
1911	EN(S) SO(S) W(S)	1951	EN(M) SO(S) W(S)

Year	Events	Year	Events
1953	EN(M) SO(S) W(M)	1972	EN(S) SO(S) W(S)
1954	C(S)	1973	EN(M) C(S)
1955	C(S)	1974	SO(M)
1956	C(S)	1975	C(S)
1957	EN(S) SO(S) W(S)	1976	EN(M) SO(S) W(S)
1958	EN(S) SO(S) W(S)	1977	SO(M) W(S)
1959	SO(M)	1979	SO(M) W(S)
1963	EN(W) SO(S) W(S)	1982	EN(S) SO(S) W(S)
1964	C(S)	1983	EN(S) SO(M) W(S)
1965	EN(M) SO(S) W(S)	1986	W(S)
1967	C(M)	1987	EN(M) SO(S) W(S)
1968	W(S)	1988	C(S)
1969	EN(W) SO(M) W(M)	1991	EN(M) SO(M) W(M)
1970	C(S)	1992	EN(M) SO(M) W(M)
1971	C(S)	1994	SO(M) W(M)

phase method of classifying SOI events – consistently negative, consistently positive, rapidly falling, rapidly rising and consistently zero (Stone and Auliciems, 1992), to assess which method is more useful for wheat producers. It was reported that five phase method is up to 70 times more valuable over the three phase approach in spring and winter wheat production. Five phase method was based on the level and rate of change of SOI over a 2 months period.

Exercise 6

Ex. 1: Using dataset of monsoon year 1989 assess the SW monsoon rainfall of the country using following regression model (Thapliyal, 1990).

$$R = -1.3562 - 28.3105\,X_3 + 67.7058\,X_6 - 3.02592\,X_{12} + 2.1956\,X_{11} + 1.1506\,X_7 - 21.0043\,X_{14} - 12.4341\,X_{10}$$

where X's parameters of

X_3 = 16.5°N, X_6 22.1°C, X_{12} = 18.1°C, X_{11} = 1, X_7 = 53.6 hPa, X_{14} = 1.0, X_{10} = 4.6 hPa and R is total rainfall in mm. [Ans: 954.5 mm]

Ex 2: Verify the performance of power regression model using the following data of 16 parameters to test the predictability of SW monsoon rainfall over India.

Parameter	Year I	Year II
X_1	110.0 °Longitude E	2.0 °Longitude E
X_2	25.6 million sq.km	26.7 million sq.km
X_3	16.8 °N	15.5 °N

X_4	28.3 °C	26.1 °C
X_5	-35.0 ms^{-1}	-11.8 ms^{-1}
X_6	22.0 °C	25.3 °C
X_7	53.8 hPa	52.9 hPa
X_8	9.88.9 hPa	987.0 hPa
X_9	0.46 °C	0.90 °C
X_{10}	4.66 hPa	2.70 hPa
X_{11}	1.0	1.0
X_{12}	18.6 °C	17.4 °C
X_{13}	1010.2 hPa	1011.3 hPa
X_{14}	1.0	1.0
X_{15}	1.2	1.2
X_{16}	1008.8 hPa	1011.0 hPa

Rainfall as percentage of long-term average

[Ans. Year I: 101.66, Year II: 104.24].

Ex. 3: Develop linkage between the El Nino signals (Table 7.8) strength and regional/ location specific monsoon rainfall in different parts of India.
(a) Northwest (b) Northern (c) Southern (d) Central.
leading to drought and flood situations.

Ex. 4: Use regression models to predict the monsoon rainfall in different parts of India (a) Northern (b) Southern (c) Northwest (d) Northeast
using 16 parameter values to verify their applicability at regional levels.

Ex. 5: Analyse past 100 years daily rainfall data of your location in relation to water requirement, nutrient uptake and water use efficiency of important rainfed crops grown at your location to develop crop strategies for efficient utilization of rainfall natural resource.

Ex. 6: Analyse the water availability periods/moisture adequacy index using water balance procedure in relation to monsoon onset date, water requirement/potential evapotranspiration of different crops grown at your location using past long records of rainfall data over weekly periods in summer monsoon season.

References

Anderson, D. 1999. Extremes in Indian Ocean. Nature, 401, (6751), p. 337-339.

Angel, J.K. 1981. Comparison of variation in atmospheric quantities with sea surface temperature variations in the equatorial eastern Pacific. Mon. Wea. Rev., 109, p. 230-43.

Anonymous, 2004. Development of nutrient management strategies for cropping systems in the Indian arid zone: application of simulation models and medium range weather forecast. NATP project report Hisar Centre. CAZRI, Jodhpur.

Awade, S.T., D.R. Sikka, and T.Y. Tyagi 1981. Contrasting features of the large scale dynamics of tropospheric circulation in wave number domain in relation to the performance of monsoon rains over India with special reference to monsoon experiments. Proc. Int. Conf. on early results of FGGE and large scale aspects of monsoon experiment. 12-17 Jan., 1981, Tallahassee, Florida, USA, p. 5-52 to 5-56.

Awade, S.T., M.Y. Tyagi, S.M. Bawiskar and D.R. Sikka 1984. Dynamics of the large scale tropospheric circulation during summer monsoon and tropical droughts. Mausam, 35, p. 315-22.

Awade, S.T., M.Y. Tyagi, S.M. Bawiskar and D.R. Sikka, 1982. Planetary scales abnormalities in 1979, a year of monsoon failure. Presented at the regional scientific conf. on Tropical Meteorology, Truskuba, Japan, Oct., 1982, p. 173.

Baier, W. 1969. Concept of soil moisture availability and their effects on soil moisture estimates from a meteorological budget. Agril. Meteorology, 6, p. 165-78.

Baethgen, W.E. 1998. El Nino and La Nena impacts in SE South America. In: Review, causes and consequences of cold events. A La Nena Summit, M. Glantz (ed.), NCR, Boulder, CO.

Baethgen, W.E. and G.O. Magrin 2000. Applications of climate forecasts in the agricultural sector of SE south America. In: Climate Prediction and Agriculture (ed.), M.V.K. Sivakumar - Proc. Start/WMO International Workshop, Geneva. Intern. Secretariat, Washington, D.C.

Banerjee, A.K., Sen, P.N., and Raman, C.R.V. 1978. On foreshadowing southwest monsoon rainfall over India with mid-tropheric circulations anomaly of April, Indin J. Met. Hydrol. Geophysics, 29, p. 425-31.

Barnett, T.P. 1984. Prediction of El Nino of 1982-83. Mon. Wea. Rev., 112, p. 1403-07.

Barnett, T.P. 1985. Three dimensional structure of low-frequency pressure variations in the tropical atmosphere. J. Atmos. Sci., 42, 24, p. 2798-2803.

Bavadeker, S.N. and R.M. Khaladkar, 1986. Dynamic effects of orography on the large scale motion of the atmosphere (zonal flow and elliptic barrier with maximum height of one km). Mausam, 37, p. 55-61.

Behra, S.K. and T. Yamaguta, 2002. Influence of the Indian Ocean dipole on the southern oscillation. J. Meteor. Soc. Japan, 81, p. 169-77.

Bhalme, H.N., S.K., Jadhav, D.A., Mooley, Bh.V. Ramanamurty 1986. Forecasting of monsoon performance over India. J. Climatol., 6, p. 347-54.

Bhaskar Rao, D.V. and Hari Prasad, D. 2003. The impact of ARMEX-I on the simulation of west coast off-shore trough and heavy precipitation events. Paper presented at ARMEX workshop Chennai, 22-23 December, 2003.

Bhat, C.S., A. Chakraborty, R.S. Nanjundiah and J. Scrinivasam 2002. Vertical thermal structure of the atmosphere during active and weak phases of monsoon over the north Bay of Bengal. Current Science, 83, p. 296-302.

Bhalme, H.N., S.S. Rahalkar and A.B. Sikdar, 1987. Tropical quasi-biennial oscillation of the 10 mb wind and Indian monsoon rainfall-implications for forecasting. J. Climatol., 7, p. 345-58.

Bishnoi, O.P. 1979. The behaviour of moisture adequacy index and its utilization for exploiting the agricultural potential in Punjab and Haryana. Mausam, 31, 1, p. 57-64.

Bishnoi, O.P. 1988. A study of growth and water use simulation submodels in pearl millet. DST Research Report No. ES/92/011/87.

Biswas, N.C., U.S. De and D.R. Sikka 1998. The role of Himalayan massif-Tibetan plateau and the mid-tropospheric tropical ridge over north India during the advance phase of the southwest monsoon. Mausam, 49, 2, p. 28598.

Bjerkenes, J. 1969. Atmospheric teleconnections from the equatorial Pacific. Mon. Weath. Rev., 97, p. 163-72.

Blandford, H.F. 1884. On the connection of the Himalayan snow with dry winds and seasons of droughts in India. Proc. Roy. Soc. London, 37, p. 3-22.

Blanford, H.F. 1886. Rainfall of India. Memoir Ind. Met. Deptt., 3, p. 668.

Busalacchi, A., D. Legler, R. Cattle, T. Palmer and J. Gould 2005. Climate change and variability what can we predict? WMO News Bulletin, 54, p. 51-57.

Cadet, D.L. 1990. Large-scale dynamics of the Indian summer monsoon. Mausam, 41, 2, p. 155-60.

Cadet, D.L. and Greco, S., 1987a. Water vapour transport over the Indian ocean during the 1979 summer monsoon – Part I: water vapour fluxes. Mon. Weath. Rev., 115, p. 653-63.

Cadet, D.L. and Greco, S. 1987b. Water vapour transport over the Indian Ocean during the 1979 summer monsoon – Part II: water vapour budgets. Mon. Weath. Rev., 115, p. 2358-66.

Cane, M.A., G. Eshel, and R.W. Buckland 1994. Forecasting Zimbabwean maize yield using eastern equatorial Pacific sea surface temperature. Nature, 370, p. 204-05.

Chandrasekhar, A. and Kitoh, A. 1998. Impact of localized sea surface temperature anomalies over the equatorial Indian Ocean on the Indian summer monsoon. J. Met. Soc. Japan, 76, 6, p. 841-53.

Chen, T.C. 1990. Effect of the 30 – 50 day oscillation on the life cycle of the Indian monsoon. Mausam, 41, 2, p. 169-76.

Chen, T.C. and H. Van Loon, 1987. Interannual variation of the tropical easterly jet, Mon. Weath. Rev., 115, p. 1739-59.

Chowdhury, A. 1985. The 1985 summer monsoon over South Asia. India Met-Deptt. In: The global climate system CSM R84/86 Report, WMO, Geneva.

Climate Prediction Centre, 1998. El Nino (ENSO) – Previous warm and cold episode years. National oceanographic and atmospheric agency, website http://nic.fb4.noaa.gov.80/products/analysis-monitoring/ensostuff/index. html.

Colon, J.A. 1964. On interaction between the southwest monsoon current and the sea surface over the Arabian Sea. Indian J. Met. Geophys, 18, p. 217-26.

CRIDA, 2001. Verification of foodgrain production model. In: Agricultural Meeorology in India: A Kaleidoscopic View of Activities (ed.) PSN Sastry, DST, New Delhi, p. 113.

Das, P.K. 1962. Mean vertical motion and non-adiabatic heat surges over India during the monsoon. Tellus, 14, p. 212-20.

De, U.S. 1971. Mountain waves over northeast India and the neighbouring region. I.J.M.G., 22, p. 361-64.

De, U.S. and K.S. Joshi 1998. Interannual and interdecadal variability of tropical cyclones over the Indian Seas. The Decean Geographer, 37, 1, p. 5-21.

Desai, B.N. 1968. Interaction of the summer monsoon current in the water surface over the Arabian Sea. Indian J. Met. Geophys, 19, p. 159-66.

Desai, B.N. 1970. Nature of the low-level inversions over the Arabian Sea and the role of Western Ghats in modifying air mass stratification within 500 km of the west coast of the Peninsula. Indian J. Met. Geophys., 21, p. 653-55.

Dey, B. and O.S.R.U. Bhanukmuar 1983. Himalayan winter snow area and summer monsoon rainfall over India. J. Geophys. Res., 88, p. 5471-74.

Ding, Y.H. and T.N. Krishnamurty 1987. Heat budget of the Siberian high and the winter monsoon. Mon. Weath. Rev., 115, p. 2428-49.

Ding, Y.H., T.Iwashita and T. Murakami, 1983. Temperature changes over Eurasia during the late summer of 1979. Sci. Atmos. Sinic, 7, p. 12.

Dirmeyer, P.A. and J. Shukla, 1996. The effect of regional and global climatic expansion of the world's deserts. Q.J.R. Mete. Soc., 122, p. 481-92.

Durrand, F.S.R., S.R. Shetya, J. Vialand, D. Shankar, S.C. Shenoi, C. Ethe and G. Modes. 2004. Impact of temperature inversion as SST evolution during the pre-summer monsoon season. Geophys. Res. Lett., 31, 8.

Eliot, J. 1901. Handbook of cyclonic storms in the Bay of Bengal. GOI printing press, Calcutta experiment. Proc. Indian Acad. Sci., Earth and Planetary Sci., 104, p. 157-87.

Evan, J.L. and R.J. Allen, 1992. El Nino/southern oscillation modification to the structure of the monsoon and tropical cyclone activity in the Australian region. Int. J. Climatol., p. 622-23.

FAO, 1980. FAO report on agroecological zones, Report No. 48, FAO, Rome.

Ferranti, L., Slingo, J.M., Palmer, T.N., and Hoskins, B.J. 1999. The effect of land-surface feedback on the monsoon circulation. Q.J. Roy. Metoro. Soc., 125, p. 1527-50.

Flohn, H. 1957. Large scale aspects of the summer monsoon in south and east Asia. J. Meteorol. Soc. Japan, 75, p. 180-86.

Gadgil, S. 1990. Poleward propagation of the ITCZ: observations and theory. Mausam, 41, 2, p. 285-90.

Gadgil, S. 2003. The Indian monsoon and its variability. Ann. Rev. Earth planet. Sci., 31, p. 429-67.

Gadgil, S. and D.R. Sikka, 1976. The influence of the topography of the peninsular on the low level circulation of the summer monsoon. Proc. Geophys. Fluid Dynamics workshop on monsoon meteorology, I.I. Sc., Bangalore, p. 65-87.

Gadgil, S., P.N. Vinaychandran, and P.A. Francis 2003. Droughts of Indian summer monsoon – role of clouds over the Indian Ocean. Current Science, 85, p. 1713-19.

Gadgil, S., Y.P. Abrol and P.R. Seshagira Rao, 1999a. On growth and fluctuation of Indian food grain production. Curr. Sci., 76, 4, p. 548-56.

Gadgil, S., P.R. Seshagiri Rao, and S. Sridhar 1999b. Modelling impact of climate variability on rainfed groundnut. Curr. Sci., 76, 4, p. 557-69.

George, L. and S.K. Mishra 1993. An observational study on the energetics of the onset monsoon vortex-1979. Q.J.R. Meteorol. Soc., 119, p. 755-70.

Ghosh, S.K., M.C. Pant and B.N. Diwan, 1978. Influence of the Arabian Sea on the Indian summer monsoon. Tellus, 30, p. 117-25.

Grondona, M.O., G.P. Podesta, M. Bidegain, M. Marino and H. Hordig, 1999. A stochastic precipitation generator conditioned on ENSO phase: a case study in SE South America. J. Climate, 13, p. 2973-86.

Gowariker, V., Thapliyal, V., Kulshretha, S.M., Mandal, G.S., Sen Roy, N. and Sikka, D.R., 1991. A power regression model for long range forecast of the southwest monsoon rainfall over India. Mausam, 42, p. 125-30.

Gowariker, V., Thapliyal, V., Sarkar, R.P., Mandal, G.S., and Sikka, D.R. 1989. Parameteric and power regression models – New approach to long range forecasting. Mausam, 40, p. 115-22.

Hahn, D. and J. Shukla 1976. An apparent relationship between Eurasian snow cover and Indian monsoon rainfall. J. Atmos. Sci., 33, p. 2461-62.

Hahn, D.G. and S. Mandabe 1975. The role of mountains on the south Asian monsoon circulation. J. Atm. Sci., 32, p. 1515-41.

Hammer, G.L., D.P., Holzwoth, and R. Stone 1996. The value of skill in seasonal climate forecasting to wheat crop management in a region with high climatic variability. Aust. J. Agric. Res., 47, 5, p. 71-37.

Hill, H. S.J., J. Park, J.W. Mjelde, W.D. Rosenthal, H.A. Love and S.W. Fuller, 2000. Comparing the value of southern oscillation index based climate forecast methods for Canadian and US wheat producers. Agril. and Forest. Meteorology, 100, p. 261-72.

Hulburt, H.E., J.C. Kndle and J.J.O'Brien 1976. A numerical simulation of the onset of El Nino. J. Phys. Oceanogr. 6, p. 621-33.

Hushke, R.E. 1959. Glossary of meteorology. Amer. Met. Sco., Boston, Mass., p. 638.

Iglesias, A., A. Erda, and C. Rosenzweig 1996. Climate change in Asia – a review of the vulnerability and adaptation of crop production. Water Air Soil Pollut., 92, p. 13-27.

I.I.Sc. ,Bangalore 1999. Monsoon anomaly and variability in food grain production. In: Agricultural Meteorology in India (ed.) PSN Sastry 2006. DST New Delhi, p. 111-12.

IPCC, 1996. Climate change 1995: The science of climate change. In: Houghton, J.T., L.G. Meiro Fihlo, B.A. Callandar, N. Harris, A. Kattenberg, and K. Maskell (eds.). Cambridge University Press, Cambridge.

Jackson, R.D., R.J. Reginato, and S.B. Idso 1977. Wheat canopy temperature – a practical tool for evaluating water requirements. Water Res., 13, 3, p. 651-56.

Jadhav, S.K. 2002. Summer monsoon low pressure systems over the Indian region and their relationship with the sub-divisional rainfall. Mausam, 53, p. 177-86.

Joseph, P.V. 1990. Monsoon variability in relation to equatorial trough activity over Indian and west Pacific Ocean – Mausam, 41, 2, p. 291-96.

Joseph, P.V. and P.V., Pillai, 1988. 40 Day mode of equatorial trough for long range forecasting of Indian summer monsoon onset. Current Science, 57, p. 951-54.

Joseph, P.V., J.K. Eischied and R.J. Pyle 1994. Interannual variability of the onset of Indian summer monsoon and its association with atmospheric features El Nino, and sea surface temperature anomalies. J. Climate, 7, p. 81-105.

Jury, R.M., P. Pathack, C.J. De, W. Rautenbach, and J. Vanheerden 1996. Drought over southern Africa and Indian Ocean SST: Statistical and GCM results. Global Atmosphere and Ocean System 4, p. 47-63.

Kane, R.P. 1997a. On the relationship of ENSO with rainfall in different parts of Australia. Aust. Met. Mag., 46, p. 39-49.

Kane, R.P. 1997b. Relationship of El Nino-Southern oscillation and pacific sea surface temperature with rainfall in various regions of the globe. Mon. Wea. Rev., 125, p. 1792-1800.

Kane, R.P. 1998a. El Nino, southern oscillation, equatorial eastern Pacific sea surface temperatures and summer monsoon rainfall in India. Mausam, 49, 1, p. 103-14.

Kane, R.P. 1998b. Extremes of ENSO phenomenon and the Indian summer monsoon rainfall. Int. J. Climatol., 18, p. 775-91.

Kane, R.P. 2004. Relation of El Nino characteristics and timing with rainfall extremes in India and Australia. Mausam, 55, 2, p. 257-68.

Kang, L.S., Jin, K., Lau, K.M., Shukla, J., Krishnamurthy, V., Schubert, S.D., Waliser, D.E., Stren, W.F., Satyan, V., Kitoh, A., Meehl, G.A., Kanamitsu, M., Galin, V. Ya, Sumi, A., Wu, G., Liu, Y., and Kim, J.K. 2002. Intercomparison of atmospheric GCM simulated anomalies associated with the 1997-98 El Nino. J. Climate, 15, p. 2791-2805.

Kasture, S.V. and R.N. Keshvamurty, 1987. Some aspects of the 30-50 day oscillation, Proc. Indian Acad. Sci., Earth Planet. Sci., 96, p. 49-58.

Keating, B.A. and H. Meinke, 1998. Assessing exceptional drought with a cropping system simulator: a case study for grain production in NE Australia. Agril. Syst. 57, p. 315-32.

Keshvamurty, R.N. 1986. Kasture, S.V. nd Krishnakumar, V., 1986. 30-50 days oscillation of the monsoon. A New Theory. Beitr. Phys. Atmos., 49, p. 443-54.

Keshvamurty, R.N., Krishnakumar, V., Kasture, S.V. 1988. Northward propagation of the 30-50 day mode in the Indian monsoon region. Proc. Indian. Acad. Sci., Earth Planet Sci.

Keshvamurty, R.N., S.V. Kasture and V. Krishnakumar, 1990. 30-50 day oscillation of monsoon: Some simple mechanisms. Mausam, 41, 2, p. 189-92.

Kiladis, G.N. and H.F. Diaz 1989. Global climatic anomalies associated with extremes in the southern oscillation. J. Climate, 2, p. 1069-90.

Kirtman, B.P. and J. Shukla, 2000. Influence of the Indian summer monsoon on ENSO. Quar. J. Roy. Meteorol. Soc., 126, p. 213-39.

Koteswaram, P. 1958a. The Asian summer monsoon and general circulation over the tropic. Proc. Symp. "Monsoons of the World", New Delhi, p. 105-10.

Koteswaram, P. and C.A. George 1958b. On the formation of monsoon depression in the Bay of Bengal. Indian J. Met. Geophys., 9, p. 9-22.

Koteswaram, P. 1958c. Easterly jet stream in the tropics. Tellus, 10, p. 43-57.

Koteswaram, P. C. Poornachandra Rao and M. Krishnamurthy 1987. Formation and structures of monsoon revival depression over the Bay of Bengal during a weak monsoon period. Mausam, 38, 1, p. 29-40.

Kriplani, R.H. and A. Kulkarni 1997. Climate impact of El Nino/La Nina on Indian monsoon: A new perspective. Weather, 52, p. 39-46.

Kriplani, R.H., S.V. Singh, A.D. Verneker, and V. Thapliyal 1996. Empirical study on Nimbus-7 snow mass and Indian monsoon rainfall. Int. J. Climatol., 16, p. 23-24.

Krishan kumar, K., K. Rupakumar and G.B. Pant 1992. Pre-monsoon ridge location over Indian and its relation to monsoon rainfall. J. Climate, 5, p. 979-86.

Krishna kumar, K., R. Rajgopalan and M.A. Cane 1999. On the weakening relationship between the Indian monsoon and ENSO. Science, 284, p. 2156-57.

Krishnamurti, T.N. 1971. Tropical east-west circulation during northern summer. J. Atmos. Sci., 45, p. 1304-22.

Krishnamurti, T.N. 1979. Tropical meteorology. In: Compendium of Meteorology, Ft. 4, Vol. II, (ed. E.Wien-Nielsen), WMO – No. 428, Geneva, Switzerland, p. 428.

Krishnamurti, T.N. and D. Subrahmanyam, 1982. The 30-50 day mode at 850 mb during MONEX. J. Atmos. Sci., 39, p. 2088-95.

Krishnamurti, T.N. and N. Surgi, 1987. Observational aspects by summer monsoon. In "Monsoon Meteorology" (ed.), C.P. Chang and T.N. Krishnamurti, Oxford University Press.

Krishnamurti, T.N., J. Motinari, H.L. Pan and V. Wong, 1977. Downstream amplification and formation of monsoon disturbances. Mon. Weath. Rev., 105, p. 1281-97.

Krishnamurti, T.N., P. Ardannuy, Y. Ramanathan, and R. Pasch 1981. On the onset vortex of the summer monsoon. Mon. Weath. Rev., 109, p. 341-63.

Krishnamurti, T.N., T.N. Chakraborthy, N. Cubukca, L. Stefanova and T.S.V. Vijay kumar 2004. A mechanism of the monsoon MJO based on interaction in the frequency domain. Quar. J. Roy. Met. Soc.

Krishnamurty, V. and B.N. Goswami, 2000. Variability and Indian monsoon – ENSO relationship in inter-decadal time scales. J. Climate, 13, p. 579-95.

Krishnamurty, V. and J. Shukla, 2000. Intraseasonal and interannual variationof rainfall over India. J. climate, 13, p. 4366-75.

Krishnan, R., M. Mujumdar, V. Vaidya, K.V. Ramesh and V. Satyan 2003. The abnormal Indian summer monsoon of 2000. J. Climate, 16, p. 1177-91.

Liu, X.L. and L.X. Chen 1983. Influence of the winter monsoon in east Asia on the summer circulation in the southern hemisphere. First Intern. Conf. on Southern Hemisphere Meteorology. Brazil.

Lorenc, A. 1984. The evolution of planetary scale 200 mb divergence during FGGE year, Quart. J. R. Met. Soc., 110, p. 427-44.

Luo, H. and M. Yanai, 1984. The large scale circulation and heat sources over the Tibetan Plateau and surrounding areas during the early summer of 1979 – Part II: Heat and Moisture budgets. Mon. Weath. Rev., 112, p. 966-89.

Madan, R. and P. Julian, 1971. Detection of a 40-50 day oscillation in the zonal wind. J. Atmos. Sci., 28, p., 702-08.

Manton, M., A. Phelan, and H. Virji (eds.) 1997. Workshop on climate variability, agricultural productivity and food security in the Asian Monsoon Region, Report No. 2, START, Washington, D.C.

Margin, G.O., Grondona, M.O., Travasso, M.I., Boullen, D.R., Rodringuez, G.R. and Messina, C.D. 1999. ENSO impact on crop production in Argentina's Pampas region. In: Proc. of 10th Symp. on global change studies, Dallas, Tx, AMS Boston, MA, p. 65-66.

Matsumoto, J. 1988. Large scale features associated with the frontal zones over East Asia from late summer to autumn, J. Met. Soc. Japan, 66, p. 565-79.

Matsumoto, J. 1990. Withdrawal of the Indian summer monsoon and its relation to the seasonal transition from summer to autumn over East Asia. Mausam, 41, 2, p. 196-202.

McBride, J.L. 1983. Tellus, 35A, p. 189-97.

McBridge, J.L. and N. Nicholls 1983. Seasonal relationships between Ausralian rainfall and the southern oscillations. Mon. Wea. Rev., 111, p. 1998-2004.

McPhaden, M. 1999. Genesis and evolution of the 1997-98. El Nino. Science, 283, p. 950-54.

Meehl, G.A. 1987. The annual cycle and its relationship to interannual variability in the tropical Pacific and Indian Ocean regions. Mon. Wea. Rev., 115, p. 27-50.

Medha Khole, 2005. Interannual and decadal variability of sea surface temperature (SST) over Indian Ocean. Mausam, 56, 4, p. 803-10.

Meinke, H. and G.L. Hammer 1997. Forecasting regional crop production using SOI phases: an example for the Australian peanut industry. Aust. J. Agric. Res. 48, p. 789-93.

Meinke, H., W.E. Baethgen, P.S. Carberry, M. Donatelli, G.L. Hammer, R. Selvaraju and C.O. Stokle, 2001. Increasing profits and reducing risks in crop production using simulation approaches. Agril. Syst., 70, p. 493-513.

Miller, F.R. and R.N. Keshvamurty, 1968. Structure of an Asian summer monsoon system. IIOE. Meteorological Monograph, East-West Centre Press, Hawaii, USA, 1983.

Misra, S.K. and P.S. Salvekar, 1980. Role of baroclinic instability in the development of monsoon disturbances. J. Atmos. Sci., 37, p. 384-94.

Mohanty, U.C., N.V. Sam, and S. Dash 2003. A study on the convective structure of the atmosphere over the west coast of India, during ARMEX-I. Paper presented at ARMEX workshop, NIOI, Chennai, 22-23, December, 2003.

Mooley, D.A. and B. Parthasarthy, 1984. Fluctuations in all India summer monsoon rainfall during 1971-78. Clim. Change, 6, p. 287-301.

Mooley, D.A. and B. Parthasarthy, 1983. Indian summer monsoon and El Nino. Pure Appl. Geophys., 121, p. 339-52.

Mooley, D.A. and D.A. Paolino 1989. The response of the Indian monsoon associated with the change in sea surface temperature over the eastern south equatorial Pacific. Mausam, 40, 4, p. 369-80.

Murakami, M. 1979. Large scale aspects of deep convective activity over the GATE area, Mon. Weath. Rev., 107, p. 997-1013.

Murakami, M., Chen, L.X. and Xie, A. 1986. Relationship among seasonal cycles, low frequency oscillations and transient disturbances as revealed from outgoing longwave radiation data. Mon. Weath. Rev., 114, p. 1456-65.

Murakami, T. and A. Sumi, 1981. Large scale aspects of the 1978-1979 winter circulation over the greater WMONEX region. Part II: Long period perturbations. J. Met. Soc., Japan, 58, p. 646-71.

Narsimha, R., D.R. Sikka and A. Prabhu, 1997. The monsoon trough boundary layer. India Acad. Sci., Bangalore, p. 422.

National commission of Agriculture, 1976. NCA Report, Vol. IX, Krishi Bhawan, New Delhi, India.

Nicholls, N. 2000. El-Nino/southern oscillation: after math of the 1997/1998 event and prediction of future events. WMO News Bulletion, 49, 3, p. 26-29.

Nicholls, N. and K.K. Wang 1990. Dependence of rainfall variability on mean rainfall, latitude and the southern oscillations. J. Climate, 3, p. 163-70.

Nicholson, S.E., and D. Entekhabi 1987. Rainfall variability in equatorial and southern Africa: relationship with sea surface temperatures along the southwestern coast of Africa. J. Climate Appl. Meteorol., 26, p. 561-78.

O'Brien, J.J., A. Busalacchi and J. Kindle, 1981. Ocean models of El Nino. Resource Management and Environmental Uncertainty, p. 159-212.

Pant, G.B. and Parthararthy, B. 1981. Some aspects of an association between the southern oscillation and Indian summer monsoon, Arch. Met. Geophys. Biokl., B29, p. 245-52.

Pant, P.S. 1983. A physical basis for changes in the phases of the monsoon over India. Mon. Weath. Rev., 111, p. 487-95.

Parthasarthy, B. and G.B. Pant, 1984. The spatial and temporal relationship between the Indian summer monsoon rainfall and the southern oscillation. Telus, 36A, 3.

Parthrarthy, B., K. Rupakumar and A.A. Munot, 1992. Forecast of rainy season foodgrain production based on monsoon rainfall. Indian Jour. Agric. Sci., 62, 1, p. 1-8.

Parthasarthy, B., Diaz, H.F. and Eischeid, J.K., 1988. Prediction of all India summer monsoon rainfall with regional and large scale parameters, J. Geophys. Res., 93, p. 5341-50.

Paul, D.K., P.V. Puranik, V.R. Deshpande, V.R. Mujumdar and D.R. Sikka. 1982. Characteristics of the near equatorial oceanic ITCZ over north Indian Ocean and the adjoining Pacific during the different phases of monsoon over India during MONEX-79. Proc. Int. Conf. on the Scientific results of the monsoon Ext. Bali. Indonesia, Oct., 81, p. 7-17 to 7-19.

Pearce, R.P. and U.C. Mohanty, 1984. Onset of the Asian summer monsoon 1978-82. J. Atmos. Sci., 41, p. 1620-39.

Phillips, J.G., Cane, M.A. and C. Rosenzweig 1998. ENSO, seasonal rainfall patterns and simulated maize yield variability in Zimbabwe. Agricultural and Forest Meteorology, 90, p. 39-50.

Podesta, G.P., Messina, C.D., Groundona, M.O. and Magrin, G.O. 1999. Associations between grain yield in central and eastern Argentina and El Nino- Southern Oscillation. J. Appl. Meteorol., 38, 10, p. 1488-98.

Prasada Rao, G.S.L.H.V. and K.N. Krishna kumar, 2005. Monsoon onset over Kerela. Jour. of Agril. Meteorology, 7, 2, p. 161-67.

Qin, B. and J.Y. Wang 1988. The simulation effect of low latitude perturbation in the southern hemisphere on the northern cross-equatorial air flow. Acta. Ocean Sinica, 10, p. 253-57.

Quinn, W.H., D.G., Zoff, K.S. Short and R.T.W. Kuo Yang, 1978. Historical trends and statistics of the southern oscillation, El Nino and Indonesian droughts. Fish. Bull., 76, p. 663-78.

Quinn, W.M., V.T. Neal, and S.E. Antunes de Mayolo 1987. El Nino occurrences over the past four and half centuries. J. Geophys. Res., 92, p. 1449-61.

Raja Rao, K.S., S.T. Awade, and M.V.H. Nair, 1984. Dynamics of the stationery eddies in the lower stratospheric circulation in the wave number domain and monsoon activity. Proc. Ind. Acad. Sci., Earth Planet Sci., 93, p. 1-16.

Rama Sastry, A.A. 1984. Weather and Weather Forecasting. Publication Division, Ministry of Information and Broadcasting, GOI, New Delhi, p. 1-203.

Ramage, C.S. 1971. Monsoon Meteorology, Academic Press, p. 288.

Raman C.R.V. and Y. Ramanathan 1964. Interaction between lower and upper tropical troposphere. Nature, 20, p. 31-35.

Raman, C.R.V., Y.P. Rao, and S.M.A. Alvi 1990. The role of interaction with mid latitude circulation and behaviour of southwest monsoon on 1972 and 1979. Current Science, 49, p. 123-29.

Ramaswamy, C. 1973. A normal period of large scale break in the southwest monsoon season over India. Current science, 42, p. 517-23.

Rao, R.R. and B. Methew, 1988. On the observed synoptic variability in the thermal structure of the upper northern Bay of Bengal during MONEX-79. Proc. Ind. Acad., Earth & Planetary Sci. Soc.

Rao, R.R. and R. Shiv kumar 1999. On the possible mechanism and evolution of a nimi-warm pool during the pre-summer monsoon season and onset vortex in the southeastern Arabian Sea. Quar. J. Roy. Met. Sco., 125, p. 787-809.

Rao, R.R., S.V.S. Somanatham, S.S.V.S. Ramakrishna, and R. Ramanadham, 1987. A case study on the genesis of monsoon depression in the northern. Bay of Bengal during MONSOON-77 experiment Mausam, 38, 4, p. 387.

Rao, R.R., U.V. Gopikrishna, and S.V. Babu 1981. A case study of the influence of surface thermal structure and ocean mixed layer depth in relation to surface energy exchange processes during Monsoon-77. Mausam, 32, 1, p. 85-92

Rasmusson, E.M. 1984. El Nino: The ocean/atmosphere connection. Oceanus, 27, 2, p. 5-13.

Rasmusson, E.M. and T.H. Carpenter 1983. Mon. Wea. Rev., 111, p. 517-28.

Remmington, G.M. and N. Nicholls 1993. Forecasting wheat yields in Australia with the southern oscillation index. Aust. J. Agric. Res., 44, p. 625-32.

Rezaul Mahmood 1997. Impacts of air temperature variations on the boro rice phenology in Bangladesh: implications for irrigation requirements. Agricultural and forest meteorology, 84, p. 233-47.

Ropelewski, C.F. and M.S. Halpert 1986. Global and regional scale precipitation patterns associated with the El Nino/Southern Oscillation. Mon. Wea. Rev., 114.

Ropelewski, C.F. and M.S. Halpert 1987. Global and regional scale precipitation patterns associated with the El Nino/Southern Oscillation. Monthly Weather Rev., 115, p. 1606-26.

Rupelewski, C.F. and M.S. Halpert 1989. Precipitation patterns associated with the high index phase of the southern oscillation. J. Climate, 2, p. 268-89.

Rupelewski, C.F. and M.S. Halpert 1996. Quantifying southern oscillation precipitation relationships. J. Climate, 9, p. 1043-59.

Saha, K.R. and S.N. Bavadeker, 1973. Water vapour budget and precipitation over the Arabian Sea during the northern summer, Quart. J. Roy. Met. Soc., 99, p. 273-78.

Salinger, M.J. 1994. Climate variability, agriculture and forests. WMO Tech Note 196, WMO, Geneva.

Sanjeev Kumar 2003. Assessment of summer rainfall events using weather variables under Hisar conditions. M.Sc. Thesis in Agricultural Meteorology, CCS HAU Hisar, p. 1-60.

Sanjeeva Rao, P. and D.R. Sikka, 2005. Intra-seasonal variability of the summer monsoon over north Indian Ocean as revealed by the BOBMEX and ARMEX field programmes. Pure. and Appl. Geophys, 162.

Sarkar, R.P. 1965. A theoretical study of mountain waves over Western Ghats. I.J.M.G., 16, p. 565-84.

Sarkar, R.P. 1967. Some modification in the dynamical model of orographic rainfall. Mon. Wea. Rev., 95, p. 673-84.

Sarkar, R.P. and B.C. Biswas, 1980. Agroclimatic classification for assessment of crop potential and its application to dry farming tracts of India. In: climatic classification ICRISAT – a consultant meeting 14-16 April, 1980. ICRISAT, Patancheric, A.P. p. 89-100.

Sastry, P.S.N. 2006. Agricultural meteorology in India – a kaleidoscopic view activities. Department of science technology, GOI, New Delhi.

Sazi, N.H., B.N. Goswami, P.N. Vinaya chandran, and T. Yamagata 1999. A dipole mode in the tropical Indian Ocean. Nature, 401 (6751), p. 360-63.

Schneider, S.H. and R.L. Temkin, 1977. Water supply and the future climate. Climate, climatic change and water supply. Studies in Geophysics, p. 25-33.

Seetaramaya, P. and A. Master 1984. Observed air-sea interface conditions and a monsoon depression during MONEX-1979. Arch. Met. Geophy. Bimet. Ser A, 33, p. 61-67.

Sengupta, D. and Ravichandran, 2001. Oscillation of the Bay of Bengal Sea surface temperature during the 1998 summer monsoon. Geophys. Res. Lett., 28, p. 2033-36.

Shah, K.R. and S. Saha 2000. On the onset of summer monsoon over India in relation to interaction of monsoon stationary wave with transient baroclinic wave leading to monsoon cyclogenesis. Mausam, 51, 1, p. 1-16.

Shankar, D., V.V. Gopalkrishana, S.S.C. Shenoi, F. Durrand, S.R. Shetye, C.K. Rajan, Z. Jonhson, Araligidad and G.S. Michael, 2004. Observational evidence for westward propagation of temperature inversions with southeastern Arabian Sea. Geophy. Res. Lett., 31, 8.

Shukla, J. and B.M. Misra 1977. Relationship between sea surface temperature and wind speed over the central Arabian Sea and monsoon rainfall over India. Mon. Wea. Rev., 105, p. 998-1002.

Shukla, J. and Paolino, Jr. D.A. 1983. The southern oscillation and long range forecasting of summr monsoon rainfall over India. Mon. Weath. Rev., 111, p. 1830-37.

Shyamla, B. 2003. Observational study on off-shore vortices during ARMEX-2002. Paper presented at ARMEX workshop, Chennai, 22-23, December, 2003.

Sikka, D.R. 1980a. Southern hemispheric influences and the onset of the southwest monsoon of 1979. Results of summer MONEX Field Phase Research (Part B) WMO – GARP – FGGE Operations Report, 9, p. 23-26.

Sikka, D.R. 1980b. Some aspects of large scale fluctuations of summer monsoon rainfall over India in relationship to fluctuations in the planetary and regional scale circulation parameters. Proc. Indian Acad. Sci., Earth and Planetary Sci., 89, p. 179-95.

Sikka, D.R. 2003. Evaluation of monsoon monitoring and forecasting of summer monsoon over India and a review of monsoon drought of 2002. Proc. Indian Nat. Sci. Acad., BA, p. 479-504.

Sikka, D.R. 2005. From the International Indian Ocean Expedition (IIOE) to the Arabian Sea Monsoon Experiment (ARMEX)- four decades of major advances in monsoon meteorology. Mausam, 56, 1, p. 19-36.

Sikka, D.R. and Gadgil, S. 1980. On the maximum cloud zone and the ITCZ over Indian longitudes during the southwest monsoon. Mon. Weath. Rev., 108, p. 1840-53.

Sikka, D.R. and W.M. Gray, 1981. Genesis of monsoon disturbances in north Indian Ocean with the passage of baroclinic waves across the southern Indian Ocean. Proc. Int. Conf. on Scientific Results of the Monsoon Experiment. Bali., Indonesia. WMO, Geneva, 4-29 to 4-34.

Sikka, D.R., D.K., Paul, V.R. Deshpande, V.R. Mujumdar, and P.V. Puranik. 1986. Sub-seasonal scale fluctuations of ITC over the Indo-Pacific region during summer monsoon. Part I: Features over the Indian region. Proc. Ind. Acad. Sci., Earth & Planetary Sci., 94, p. 47-54.

Sikka, D.R., D.K. Paul, and S.V. Singh, 1981. Some aspects of the oceanic ITCZ and its disturbances during the onset and established phases of summer monsoon studied with MONEX-79 Data. Ind. Inst. of Tropical Meteorology, PR 32.

Sikka, D.R., D.K. Paul, V.R. Deshpade, V.R. Mujumdar, and P.V. Puranik, 1984. Study of interactions between the ITCZ over the Indian and the west Pacific regions during the summer monsoon and their modulation by the 30-40 day and 10-20 day modes. IITM Publication.

Singh, O.P. 1998. Sea surface temperature and sea level pressure anomalies over the Arabian Sea during ENSO years. Mausam, 49, 2, p. 264-66.

Singh, O.P. and M.A. Sarkar 2003. Recent sea surface temperature variability in the coastal regions of the north Indian Ocean. Indian Journal of Marine Sciences, 32, 1, p. 7-13.

Sinha Ray, K.C. and U.S. De 2003. Climate change in India as evidenced from instrumental records. WMO News Bulletin, 52, 1, p. 53-59.

Sivakumar, M.V.K. (ed.) 1997. Climate variability prediction, water resources and agricultural productivity: Food security issues in tropical sub-saharan Africa. Report No. 3, START, Washington, D.C.

Sperker, K.R., C. Brankovic, C.S. Deque Frederiksan, R. Graham, A. Kitoh, C. Kobayashi, T. Palmer, K. Puri, W. Tennant and E. Volodin, 2001. Dynamical seasonal predictability of the Asia summer monsoon. Mon. Weath. Rev., 129, p. 2226-48.

Stommel, H. 1979. Oceanic Warming of Western Europe. Proc. Nat. Acad. Sci., USA, 76, p. 2518-21.

Stone, R.C. and A. Auliciems 1992. SOI phase relationships with reinfall in eastern Australia. Intern. J. Climatol., 12,, p. 625-36.

Sud, Y.C. and W.E. Smith, 1985. Influence of local and land surface processes on the Indian monsoon. A numerical study. J. Climatol. & Appl. Meteorol., 24, p. 1015-36.

Tao, S. and Chen, L. 1987. A review of recent research on the East Asian summer monsoon in China. In 'Monsoon Meteorology' (eds.) C.P. Chang and T.N. Krishnamurti, Oxford University Press.

Tapeshwar Singh, 2000. Links between ENSO phenomena and drought episodes in India. Intern. Conf. on Managing Natural Resources for sustainable Agricultural Production in the 21st Century, Vol. 4, WTC, IARI, New Delhi, p. 82-84.

Thapliyal, V. 1990. Long range prediction of summer monsoon rainfall over India: Evolution and development of new modes. Mausam, 41, 2, p. 339-47.

Thapliyal, V. 1984. Prediction of Indian droughts with lower stratospheric winds. Mausam, 35, p. 367-74.

Thapliyal, V. 1986. Long range forecasting of monsoon rainfall in India. WMO/TD/ 87, p. 723-32. WMO, Geneva.

Thapliyal, V. 1991. Important aspects of long range forecast models of southwest monsoon rainfall in India. First Rajkristo Dutt Memorial Award lecture delivered on Jan. 4, 1991 at the 79th Annual session of Indian Science Congress Association at Baroda.

Thapliyal, V. and S.M. Kulshretha 1992. Recent models for long range forecasting of southwest monsoon rainfall in India. Mausam, 43, 3, p. 239-48.

Tourre, Y.M., and E.M. Rasmusson, 1984. The Atlantic region during the 1982-83 Pacific Warm Event. T.O.A.N. 25.

Travasso, M.I., Magrin, G.O. and Grondona, M.O. 1999. Relation beween climatic variability related to ENSO and maize production in Argentina. In: Proc. 10th Symp. on global change studies, Dallas, Tx., AMS Boston, MA, p. 67-68.

Trenberth, K.E. 1976. Spatial and temporal variations of the southern oscillation. Quart J. Roy. Met. Soc. 102, p. 639-53.

Trenberth, K.E. 1997. The definition of El Nino. Bull Amer. Met. Soc. 78, p. 2771-77.

Trenberth, K.E. 1998. Development and forecast of 1997/98 El Nino: CLIVAR scientific issues. CLIVAR Exchanges, 3, p. 4-14.

Verma, R.K., Subramanian, K. and Dugam, S.S., 1984. Interannual and long term variability of the summer monsoon and its possible link with northern hemispheric surface air temperature. Proc. Indian Acad. Sci., Earth and Planet Sci., 94, p. 187-98.

W.M.O. 1987. The global climate system, autumn, 1984 – spring, 1986. Climate System Monitoring, CSM 84/ 86, World Climate Data Programme, W.M.O., Geneva, p. 1-87.

Walker, G.T. 1923. Correlation in seasonal variation of weather VIII: A precliminary study of the world weather (World Weather I), Men. India Met. Dep., 24, p. 75-131.

Wang, B. 1995. Interdecadal changes in El Nino onset in the last four decades. J. Climate, 8, p. 167-285.

Webster, P.J. 1981. Mechanisms determining the atmospheric response to sea surface temperature anomalies. J. Atmos. Sci., 38, p. 554-71.

Webster, P.J. 1983. J. Atmos. Sci., 40, p. 2110-2124.

Webster, P.J. and S. Yang, 1992. Monsoon and ENSO, selectively interactive systems. Quar. Jour. Roy. Meteorol. Soc., 118, p. 877-926.

Webster, P.J., A.M. Moore, J.P. Loschinigg and R.R. Leben 1999. Evidence of coupled oceanic – atmospheric instabilities. Nature, 401, p. 356-60.

Webster, P.J., A.M. Moore, J.P. Loschnigg and R.R. Leben, 1999. Coupled ocean-atmosphere dynamics in the Indian Ocean during 1997-98. Nature, 401 (6750), p. 356-60.

White, D.H., S.M. Howden, J.J. Walcott, and R.M. Cannon 1998. A framework for estimating the extent and severity of drought based on a grazing system in S.E. Australia. Agric. Syst., 57, p. 259-70.

WMO, 1984. The global climate system. A critical review of the climate system during 1982-84. WMO Geneva.

Wright, P.B. 1975. an index of southern oscillation. Climatic research unit CRU RP4, University of East Anglia, Norvich, p. 22.

Wright, P.B. 1984. Relationship between indices of the southern oscillations. Mon. Wea. Rev., 112, p. 1913-19.

Wright, W.J. 1988a. The low latitude influence on winter rainfall in Victoria, southeastern Australia, climatological aspects. J. Climatol., 8, p. 437-62.

Wright, W.J. 1988b. The low latitude influence on winter rainfall in Victoria, southeastern Australia II. Relationship with the southern oscillation and Australian region circulation. J. Climatol., 8, p. 547-76.

Wyrtki, K. 1982. The southen oscillation, Ocean-atmosphere interaction and El Nino. Marine Techn. Soc. J., 16, 1, p. 3-10.

Wyrtki, K. 1985. Water displacements during 1982-83 and genesis of El Nino and the southern oscillation. Intern. Conf. on the TOGA Scientific Programme. WCRP report No. 4, III, p. 1-10.

Yao, A.Y.M. 1974. Agril. Meteorology, 13, p. 405-17.

Yasunari, T. 1979. Cloudiness fluctuations associated with northern hemisphere summer. J. Met. Soc. Japan, 45, p. 227-42.

Yasunari, T. 1980. A quasi-stationary appearance of 30-40 days period in the cloudiness fluctuations during the summer monsoon over India. J. Met. Soc., Japan 58, p. 225-29.

Yasunari, T. 1981. J. Met. Soc. Japan, 59, p. 336-54.

Yin, M.T., 1949. Asymptotic aerological study of onset of the summer monsoon over India and Burma. J. Met. Soc., 6, p. 393-400.

Young, J.A. 1987. Boundary layer dynamics of tropical and monsoonal flow. In: "Monsoon Meteorology" (eds), C.P. Chang and T.N. Krishnamurti, Oxford Univ. Press.

Yu, L. and M.M. Rienecker 1999. Mechanisms for the Indian Ocean warming during the 1997-98 El Nino, Geophys. Res. Lett., 26, 6, p. 735-38.

8

Applied Microclimatology

8.1 Introduction

Microclimate – the climate of smaller regions influenced by the effects of relief, topography and the lower surface features which create disparity between soil and air temperatures, humidity and wind speeds. For example, different types of forests, orchards, cropped fields, lakes, meadows, etc., special type of microclimate is generated in plants called phytoclimate. Sharp differences in microclimatic parameters is generated near the soil-air layers, but will diminish with the air mixing with the height above the surface. Field microclimate can be modified in order to generate optimum climatic variables for maximum growth and rapid development with the advancement of the season. Modification of field microclimate can be altered through several ways with cultural and operational management applications which effects either energy balance components or water balance hydrological components or both through wind turbulence in shelterbelt/wind break erections. A phytoclimate i.e., a modified microclimate is produced among the plants which is controlled by plant height, density of plant cover, leaf area distribution with height, row to row and plant to plant spacings, and adoption of input management practices (soil, water, fertilizer and agrochemicals). There are several cultural practices (shading, row orientation w.r.to the solar trajectory, mulching, intercropping, irrigation, shelterbelts and windbreaks, ridge – furrow sowing, recycling of surface runoff and deep drainage of soil water etc.) which will help in modifying the energy balance or hydrological water balance components so as to modify the microclimate of the surface features to provide the optimum conditions for maximum growth and development. Diurnal microclimate of orchards, forests, hill slopes terrace cropping is quite different than levelled field crops. Therefore, it is worthwhile to understand these different types of microclimates so that exact situation of field microclimate is reflected in the agroadvisory services in specific agroecological regions. A vast literature is available on modification of field microclimate, but here we are interested in their different applications for the benefit of farmers and agroadvisory service.

In each agroecological zone benefit of field microclimate needs to be harnessed for the benefit of farmers to raise productivity both the quantity and quality depending on the availability of seasonal/daily variations in plant's microclimate. The appropriate technology can be adopted to draw benefits of the available agroclimate resources with their efficient utilization. A vast review on these aspects is available in order to use the field microclimate for beneficial effects and should be kept in mind for their applications. (Monteith, 1963, 1973, 1975; WMO, 1972; Geiger, 1965; Slatyer and McIlroy, 1961; Sutton, 1953; Munn, 1966; Rosenberg, et al. 1983; Arya, 1988; Oke, 1987; Waggoner, 1965; Kulik and Sinelsnchikov, 1961; Stigter, 1986; Bishnoi, 2007).

8.2 Modification of Field Microclimate

Fournier d'Albe, 1958; Lowry, (1969) devised three ways of modifying the field microclimate with (i) control of heat balance parameters, (ii) control of water balance parameters, and (iii) control of atmospheric turbulence.

8.2.1 Control of heat balance parameters

The heat balance equation over a soil-vegetative surface is expressed as (Fuch, 1972; Jackson, 1983; Rauner, 1960)

$$R_s(1 - \alpha) + Lw_d - Lw_u = LET + A + G \qquad \ldots(8.1)$$

In order to generate optimum temperature (A) and evapotranspiration rates (LET) of a cropped surface it is desirable to change the heat balance parameters (Fuch, 1972) in temperate winter season through

(i) heat trapping with increase of effective Rs and decrease of albedo of the surface,
(ii) use of white plastic sheets in open space between rows to reflect more light in lower part of crop canopy,
(iii) orientation of row directions so that more radiation is intercepted by the foliage,
(iv) improving the receipt of longwave radiation from the and sky, and
(v) planting of crops on southern sunny slopes in northern hemisphere and vice versa in southern hemisphere to trap more radiation.

Similarly for summer crops in tropical and subtropical arid and semiarid areas the excess available heat load can be shed away through

(i) heat evasion with decrease of effective Rs (shading of plants), increase of albedo of the vegetative surface (Kang, 1978). For example, light shade in tea production (Obaga, 1984), intercropping shading in agroforestry (Stigter, 1984, 1985) with manipulation of foliage cover density,
(ii) decrease of LET through stomatal closer with increase of CO_2 levels and spray of antitransparents, and
(iii) clear sky conditions with rise of the net effective longwave radiations to decrease net radiation levels.

In addition to above tactics to control the heat balance, it is also possible to use cultural practices over small scale areas to retard the sensible and latent heat flux from the soil through

(i) covering the heat sensitive horticultural and vegetable value added crops (flowers, tomatoes, peas, etc.) with plastic enclosures during the night to avoid heat loss through nocturnal cooling, and
(ii) mulching and ploughing to decrease heat loss from soil and to preserve the heat loss and also to decrease transpiration (Russell, 1973). Mulches are useful and effective even to conserve soil water at the reduced rate of evaporation losses. Also albedo of the soil surface is decreased with use of burnt straw mulches and warming the top soil layer.

Ecological differences are adopted for development of natural plant communities in open fields and in shades. Perennials prefer shade for longer periods. In shade, lower net radiation will provide lower soil and air temperatures during daylight hours, whereas during night these may be higher than the open environment. There is reduction in ET. Total radiation and light energy is the main difference for sun and shade species. Shading, mulching and wind protection are concerned with changing of surface exposure without modifying the initial surface properties. But in some cases surface properties are modified to get results such as tillage, ridge-furrow sowing.

8.2.2 Control of water balance parameters

The hydrological water balance equation over a time period at the soil-vegetative surface can be expressed as

$$P + I = PET/ET \pm \Delta SM + SR + DP \qquad \ldots (8.2)$$

where, P and I are inputs of precipitation and irrigation, PET is potential evapotranspiration loss, ΔSM is the change in soil moisture status in the root zone depth of the crop, SR is surface runoff and DP is deep percolation below the root zone depth.

Precipitation is a natural occurring input variable at a location which needs to be conserved most efficiently to avoid surface runoff through bunding of microplots to recharge the soil in arid and semiarid areas. However in high rainfall hilly areas the surface runoff is very large and can be collected in ponds/lakes in low lying valleys for its recycling and better use as per requirements during rainless/drought periods. Irrigation amount is the regulation of the controlled water in order to utilize as per needs to meet evapotranspiration requirements in moisture stress conditions. Deep percolation loss is not utilized by field crops, but is of immense importance for deep rooted horticultural, orchard, and forest vegetations. In order to achieve maximum crop yields, it is desirable to maintain optimum evapotranspiration so that environmental conditions provide maximum growth and rapid development of the crop. Two possible situations may exist-first, the inadequacy of soil water in the root zone depth during anytime of the growing season, and second by reducing the excess evapotranspiration losses leading to the excess vegetative growth. The inadequacy of soil moisture in the root zone depth can be improved through bunding (controlling surface runoff) and strip or terrace cropping, and increasing the soil infiltration rate after every effective rainfall through soil mulching or soil tillage. Straw mulching to conserve rain water is also an effective method is low rainfall areas (Kang, 1981; Midmore, 1984). Water harvesting (interrow or interplot) methods are useful methods to control the surface runoff water to saturate the soil profile with enough soil moisture in order to meet the crop evapotranspiration requirements during the growing season otherwise the soil profile is not adequate to tailor the crop requirements during the growing season.

Reduction of soil evaporation or transpiration from vegetative surfaces can be achieved through several ways by decreasing the capilliary continuity of surface soil layer, decreasing capilliary flow and moisture holding capacity of the surface soil layer, by chemical treatment to control soil evaporation such as hexadecanol, i.e., a long chain alcohol mixed with the soil; spread of thick film forming materials such as latex, waxes, plastics, etc.; stomata closing materials such as glycoryl half ester of decenyl succinic acid; use of reflectants such as kaolinite.

8.2.3 Control of atmospheric turbulence

Wind protection properties are well known with strip cropping practices (Radke and Hagstrom, 1976) in temperate hilly regions, but in tropics as a wind barrier of trees/bushes (Stigter, 1985b; Eimern et al. 1964). Wind speed is the main causative agent for generation of atmospheric turbulence. Wind affects the plants directly in mechanical damage for tall crops and trees, increasing transpiration rate and intake of carbon dioxide. However, seasonal winds in combination with temperature increases its severity. In summer the dry hot winds in combination with temperature generate heat waves desiccating the plants in the intercellular spaces leading to dwarfing of the plants, deformation of shoots. During winter strong winds in rainy periods create lodging in tall crops such as winter maize, wheat, sugarcane etc., whereas in tall trees mechanical damage and even uprooting of trees may take place. Strong cold winds during winter raises the severity of coldness in lowering the effective temperature causing frost damage and causalities of newly born animals. Cyclonic winds near coastal areas spoils the soils and large scale damage due to high wind speeds and deposition of salts from sea water. This will render coastal lands into unproductive unless the excessive salts are leached down.

Wind turbulence though raises the intake of carbon dioxide for higher photosynthesis, but it will be saturated at a certain wind speed depending on the availability of soil water and water vapours in the intercellular spaces. Similarly wind increases transpiration only to a certain point beyond which it becomes constant.

Shelterbelts have been commonly adopted in order to modify the impact of high wind speed on agricultural field to reduce wind speed in order to avoid soil moisture stress of crops and severe mechanical damages. Even the suitable combination of tall crops such as maize, sorghum, sunflower, wheat, oats, sugarcane with smaller height crops like sugarbeet, soybean, chickpea, groundnut, tomato, potato in alternate rows temporally behaves as a wind barrier in arid, semiarid area to harness the benefit of reduction of wind speed in better economic returns. In winter the shelterbelts reduces the severe cold winds to avoid frost damage. The perpendicular direction of the prevailing high wind speeds is to reduce the wind speed to improve the field microclimate in the leeward side to increase production of agricultural crops.

8.2.3.1 Modification of microclimate in leeward side

An effective shelterbelt, with proper height (h) and foliage density and tree spacings in perpendicular direction of prevailing winds speeds at a location in two to ten rows is created depending of dimensions of the agricultural fields to be protected in the leeward side. Higher tall trees are preferred when the area to be protected is higher. A dense barrier impact have been observed up to 10 to 15 h, whereas with 50 percent porosity barrier may modify microclimate up to 20 to 25 h. Wind breaks are porous near the ground and winds can penetrate to larger distances. Even on the windward side the microclimate is improved up to

6 to 8 h distance. A number of reviews are available to show that modified microclimate in the leeward side of shelterbelt is much moderate than the open unsheltered climate (Hagen and Skidmore, 1974; Rosenberg, 1967). Microclimatic and plant physiological parameters on leeward side upto 20 to 25 h distances (in terms of height of shelterbelt) and upto 6h in windward side are improved as compared to the open field climate (Table 8.1).

Shelterbelt reduces the vertical gradient of wind speed by mixing the eddies generated and also simultaneously results in variation of temperature gradients. Increase in temperature during daytime may extend up to 5 to 10h leeward as a shadow effect of the shelterbelt, but after 10h reduction in wind speed is less and slowly approaches towards an open environment. Hagan and Skidmore (1974) reported that air temperature in the leeward side can be assessed from the increase or decrease of evapotranspiration as latent and sensible heats will share the available net energy. Soil temperatures in the daytime increases due to barrier effect of shelter and reverse effect is generated during nighttime with fast nocturnal cooling in warmer locations.

Relative humidity in the leeward side varied considerably due to variations of soil moisture, plant height, plant vigour, evapotranspiration rate and variability of CO_2 levels due to

Table 8.1
Modified microclimatic and plant physiologic parameters in the sheltered environment

Microclimatic/Plant Physiologic Parameters	Sheltered Environment on Leeward (20-20h) and Windward (6h) Distances
Wind speed	Maximum reduction at 4h, significantly reduced all over
Solar radiation	Average daily values are significantly reduced
Net radiation	Average daily value are significantly reduced
Daytime temperature	Greater
Nighttime temperature	Lowered
Daytime humidity	Increased
Daytime vapour pressure gradient	Increased
Nighttime humidity	Increased above daytime levels
Soil temperature	Higher during daytime and lower during nighttime
CO_2 concentration	Decreased relatively
Leaf turgidity	Greater
Stomatal aperture	Wider
Stomatal resistance	Lowered
Evaporation or drying rates	Reduced depending on the reduction in wind speed levels
Evapotranspiration	Lowered, but occasionally increased due to reduced stomatal resistance
Transpiration	Increased due to better leaf area, plant height and roughness of plant size, but with ample water supply difference decreases
Aerial resistance for transport of water vapours	Greater

differences of wind speed reduction and plant growth variability at different distances in leeward side-maximum being at 4h distance in leeward side. CO_2 concentration varies due to decreased turbulence at low wind speeds. In sheltered areas, the reduction in evaporation/ evapotranspiration is less than the corresponding reduction in wind speed in open environment so that available soil moisture in the root zone lasts longer as compared to soil water available in open environment, and thus affecting the plant physiological characters regulating the moisture control in plants. This leads to greater turgidity, lower stomatal resistance and wider stomal aperture in sheltered plant leaves – highest values being at 4h directly depending on available soil moisture status in the root zone depth. The moisture savings due to sheltered environment is beneficial for plant growth, however the rate of water use may increase with benefits in leaf area, plant height and plant growth. Eventually the water use in sheltered area is higher as compared to open environment.

Plant growth and yield attributes are greater in sheltered environment, maximum being at 4h (Davis, et al. 1975a and b). The higher net assimilation rate may be due to higher CO_2 concentration, lower nocturnal respiration and better moisture status in plants as compared to the plants in the open environment. Sheltered environment also improves water use efficiency. In arid and semiarid areas most pronounced results have been reported all over, and farmers in such region should harness the benefit of reduction of wind turbulence with erection of shelterbelts in N – S direction to counteract westerly dry hot winds. Similarly in cold areas NW cold winds are very severe several times which can be counteracted with shelterbelts in NE direction with tall tree's vegetation species having adequate foliage density variations with height.

8.3 Modification of Soil Temperature

Soil temperature variations depends on soil moisture, soil properties and soil aeration. Soil climate can be altered in order to generate optimum conditions through irrigation, shading, soil mulching, shelterbelts so that favourable root zone environment is available for growth of seedlings and roots. Soil moisture is most important variable influencing soil temperature. Soil climate is characterized with diurnal and annual variation of soil temperature, heat capacity of the dry mineral soils, thermal conductivity of soil, thermal diffusivity of soil. The depth related dynamics of the soil temperature of seasonal or annual periods are of great practical significance in agriculture and in installing water pipes below the freezing depths. Soil topography also influences considerably on soil temperature through influx of sunlight. In northern hemisphere, slopes with southern component receive more heat than northern side due to the solar trajectory. Similarly daytime heating and night time cooling values are higher for valley type relief and lowest for hilltop shapes due to intensity of air interchange which is lowest in valleys and highest at hilltop. Open soils are more intensely heated as compared to vegetatively covered soils. A temperature difference of 9 to 10°C can be created in this manner. Even in a ridge-furrow system a difference of 1 to 2°C exists due to exposure of land surface to sun rays. Even shading with a plant cover, the soil surface shows lower temperature in summer and warmer in winter as compared to the bare soil exposed surface by 3 to 5°C. Forest canopy shows mean annual soil temperature is colder in forest at one metre depth by 1°C, but in summer even at 30 cm depth is 5 to 6°C colder than in timber free plot (Chirkov, 1979). Even snow surface (of the order of 30 cm) has a heat insulating effect on the soil as its thermal conductivity is low which will protect the young winter crops as well as fruit trees. In winter it is advisable to increase the temperature of the soil for germinating and rooting of seedling, and speedy growth and development. Conversely in summer the excess heat will overburden the plant and it is necessary to devise ways too decrease the soil temperature particularly of top soil layer. Soil mulching is simplest method of rapid exchange of heat within the soil layer which reduces the temperature of under soil layers by 2 to 3°C in summer, whereas soil compacting will increase the soil temperature by 1 to 2°C. Similarly spreading of mulches will reduce the soil temperature during the summer rainy season period. However, spreading of mulches such as dry straw, coal dust, plastics in furrows, will increase the soil temperature by protecting the release of longwave radiations from the earth's surface (Davies, 1975a). Even charcoal spray in furrows will increase the soil temperature at surface due to relatively higher absorption of incoming solar radiation and decrease of albedo component over the surface.

Irrigation strongly influence the decrease of soil temperature as compared to unirrigated areas in summer even in lower depths due to percolation of soil water. However, in winter an increase in soil temperature takes place but to a much lesser extent as compared to the lowering of soil temperature in summer. Shelterbelts and windbreaks are also useful in increasing the soil temperature in leeward side during daytime – a useful feature in winter, but decrease of minimum temperature during night time is an adverse feature, however it is counteracted with the improvement of accumulated soil moisture. But due to reduction of wind speed in leeward side the lowering of air temperature is relatively reduced.

Germination of crop is directly linked with availability of soil water for the seed to imbibe water and radical emerge from seed coat with soil water above 80 per cent of field capacity and optimum soil temperature above the threshold temperature for the completion of germination process. Table 8.2 shows the optimum soil temperature at 5 cm of soil depth assuming soil water is not a constraint for germination.

8.4 Microclimate of Complex Agrosystems

Microclimates of field crops have been studied and vast literature is available on this subject, however microclimate of specialized cropping systems still deserve attention such as pastures and rangelands, grasses, meadows, orchards, forest vegetations of different types. Light, heat and moisture variations are essential parameters for growth, development and yield of all plants, but in these specialized systems the requirement of specific vegetations needs to be met with the agroclimatic conditions of the area.

Table 8.2
Optimum and threshold temperatures and accumulated heat units required for germination process in different field crops

Crop	Threshold Temperature	Optimum Temperature °C	Accumulated Heat Units (°C days)
Wheat	2.6	20-25	77.9
Barley	2.6	20-25	79.3
Oats	2.2	22-24	91.1
Maize	9.8	28-34	60.8
Sorghum	10.6	30-32	47.9
Pearl millet	11.8	33-34	39.5
Finger millet	13.5	35	40.1
Field pea	1.4	20	110.3
Lentil	1.4	20	90.1
Soybean	9.9	30	70.5
Peanut	13.3	34	76.6
Cowpea	11.0	32-33	49.6
Mungbean	10.8	32-34	49.6
Cotton	14.4	35	58.2
Pigeon pea	12.8	32-35	32.4
Sugarcane (Sprouting sets)	12.8	30	-
Guar	12.8	32-35	32.4
Rape	2.6	22-24	79.0
Safflower	7.4	25-30	69.7
Sunflower	7.9	30	69.7
Linseed	1.9	22-25	88.8
Castor	12.5	30-35	95.4
Buckwheat	11.1	25-30	36.7
Amaranthus	11.7	30	31.9
Sesame	15.9	32-35	63.6
Iapanese millet	7.9	25-30	21.3
Rice (winter temperate)	10.0	20-24	63.6
Rice (summer)	10.5	30	65.8
Chickpea	5.0	20	95.4

(After: Angus et al., 1980/1981)

8.4.1 Microclimate of pastures, rangeland and meadows

It is different ecosystem where aerial and edaphic microclimates influence growth and development of green vegetations, activity of consumer and decomposed elements (Whiteman, 1969). Climate and soil conditions both parameters quantify the propagation of grasses in these specialized microclimates (Ripley and Redmann, 1976). Based on heat and moisture requirements three classes of grasses have been defined as hygrophytes (saturated soil water soils for marshes, meadows flooded areas), mesophytes (meadow phytocenoses grown in forests), and xerophytes (grass species having tolerance for dry soil and air situations in arid, semiarid, and steppe regions).

Temperature and availability of soil water determine the growth and accumulated biomass. Drought situations with temperatures above 35° to 40°C adversely affects their growth. Available moisture scarcity in top surface soil layer, temperatures above 18°C and saturation deficit of 10 mm of Hg are agrometeorological indicators of the commencement of the extinction process for these grasses. Magnitude of soil heat and humidity determine the productivity of grasses in natural regions (Chirkov, 1979). Irrigation during summer months becomes necessary to sustain drought conditions. In arid to semiarid areas with use of irrigation it is possible to produce 5 to 6 grasses/fodder crops per year. There are large annual and diurnal variations of temperature ranges, precipitation is low during summer, evaporation rate much higher, solar radiation and wind speeds are relatively higher in these areas. The dense mat of dead leaves at the surface is a distinctive feature and with moisture it decomposes enriching the soil, but favours for many insects near the ground. It also helps in surface-atmosphere energy exchange process as it functions as mulch conserving moisture in rainy season. Therefore the green and deed leaves functions in different manner due to difference in their role in microclimatic modification (Waterhouse, 1955). Metabolism of soil organisms generates carbon dioxide which diffuses upwards for assimilation by green vegetations synthesizing with plant material. Interseasonal variability in biomass yield exists due to continuity of pasture life cycle where different phases experience different microclimatic conditions influencing growth and development (Kalma, 1970).

8.4.2 Microclimate of tree

In this system crop surface conditions, shape of trees, height variations of vegetative surfaces, and external climatic conditions will very the microclimates. Geometric shape of trees influence the radiational climate directly to influence evapotranspiration, growth and development (Gerber et al. 1974.)

Shape of tree is most important to generate microclimate variations with height as well as inside the foliage. A tree may be of a spherical or vertical cylindrical vegetative shape with a short or long vertical cylindrical stem of very small diameter. Radiational environment is uniform in spherical shapes as incoming (direct and diffuse components) radiations are received uniformly from all sides. However in vertical cyclindrical trees radiation environment will vary with height due to variations of diameter of trees with height. In conifer forest, the shorter height of stem loses its importance in comparison with conical shape of upper foliage shape. The crown of the tree may be compared with a upright cone with a large base. Here the radiation exchange between the base of conical shape of tree and earth may be negligible as compared with direct and diffuse radiations from conical surfaces. In more complex shapes of trees a combination of spherical conical or cylindrical shapes at different heights may represent the shape.

Surrounding atmospheric environment has immediate impact on the development of the tree plant. Radiation penetration through outer crown periphery of the tree provides energy for conversion of CO_2 into photosynthates and other organic components. Longwave radiation emitted from the soil surface also provides thermal energy. Temperature of leaves exposed to Sun have higher temperature than those inside which are partly shaded. Temperature variations influence he activity of cells and movement of cells juice, and tree activity. Crown periphery is useful for intercepting direct, diffuse and longwave radiations, and simultaneously also influence the movement of air inside the tree to establish equilibrium with the outer environment. Due to transpiration and respiration of leaves relative humidity inside the crown of trees are usually higher than open environment. Precipitation is largely intercepted by the outer surface of the tree and deflected outwards of the tree, and therefore periphery of crown is relatively better watered than the base of the tree.

Soil temperature and available soil moisture in the root zone depth of the tree are equally important variable to absorb water and minerals. Activity of roots depends on the soil temperature to maintain equilibrium of root growth with aerial parts of the plant maintaining harmony with each other. Availability of soil water maintains balance through transpiration in exploiting aerial environment to produce dry matter accumulation and its distribution among different plant organs.

8.4.3 Microclimate of orchards

Orchard, being a systematic arrangement of trees in rows or in a quadrilateral fashion, there is exchange of radiated energy between the evergreen leaves and heat resistant qualities of trees in addition to energy received by an isolated tree (Primault, 1979). However, if the branches of one touches the other tree in a particular direction, then the total cylindrical shape can be considered as a single unit. In a rectangular or quadrilateral arrangement wind energy will pass through the passage between the trees and will strike the trees of second row or third row so that wind speed is reduced. If trees are closer barrier will be more effective. In a linear arrangement of trees wind will penetrate inside the orchard to supply energy for evaporation from soil as well as transpiration. If the tree barrier is perpendicular to wind speed, then it will act as shelterbelt to modify microclimate otherwise soil evaporation between the rows will be large. Cassin et al. (1969) reported minimum temperature of 12.5 to 13°C, optimum 23 to 24°C and maximum temperature of 37 to 39°C for citrus growth. Net radiation load on vegetative surface is most important factor for growth and yield (Newman et al. 1967). Temperature, humidity, rainfall variations. There may not be any practical difference between an isolated tree and an orchard tree. In an orchard of a systematic regular arrangement the trees behave independently w.r.to. meteorological variables except wind. Planting distance strongly influence the radiations environment of the orchard (Cooper et al. 1963). Average leaf area index for citrus orchard is 6.5 where radiation is fully intercepted (Kalma, 1970). Orchard, with tree arrangement in rows, behaves as a single unit when twings and branches touch each other. Radiation environment is uniform along the rows when tree heights and vegetative growth is uniform. Temperature and humidity gradients will be through the tops and sides of the each row creating microclimate difference between rows and inside the crown of trees. Wind effects will depend on the direction of wind speed w.t.to the arrangement of the trees. If wind in perpendicular to the tree row arrangement, there will reduction of wind speed inside the orchard rather a turbulent flow with eddies between the rows and an accelerated flow above the orchard will be experienced.

Radiation climate in orchards can be controlled by manipulation of cover density and soil reflectivity as these parameters interact each other (Monteith, 1975).

8.4.4 Microclimate of forests

Forest system behaves as a compact, aggregate vegetative mass over the topographic relief features. It behaves as a irregular terrin relief. If forest tree-tops are uniform, then it will behave as vegetative surface similar to the ground surface (Primault, 1979). Radiation penetrates through the foliage to the ground surface depending on the exposure and slope of ground w.r.to the solar trajectory at the location. The interception of radiation by crowns of trees decreases the temperature gradient inside the forest during daytime with enough transpirational cooling, but during night the longwave radiation from ground are intercepted by the leaves and returned back to the soil surface thus giving a warming effect inside the forest. Wind speed will flow uniformly if tree tops are of same height, but with difference of tree tops height will create turbulence and may not be able to penetrate inside the forest. Therefore, forest acts a whole to generate its own microclimate and with decrease of wind there will be reduction in evapotranspiration and increase in humidity of the system. Microclimate varies with distribution of leaf area, light interception and photosynthesis as a function of incoming radiation load.

The edge of the forest will behave similar to the first row of an orchard arranged in rows and will be influenced by the prevailing environment. The inner side of forest trees will behave similar to trees inside the orchard area. Therefore the trees at the edge of forest are asymmetric due to development of side crowns with receipt of the diffuse and reflected radiations and direct impact of wind speed for carbon dioxide uptake and transpiration. However, inside the forest due to shade there is less penetration of radiation reducing the plant growth.

Treetops and peripherical crowns will experience direct intense global radiations similar to bare ground surface and are intensely heated to produce longwave radiations with increase of evaporation and decrease of relative humidity. Wind perturbations are high when treetops are irregular creating eddies or wind gust. The outer surface of the forest becomes warmer than the surrounding air during daytime and colder by night.

Inside the forest microclimatic conditions are quite different. Only limited radiation reaches the inside forest depending on the crown foliage interception. Only transmitted radiations reaches the inside of forest changing the morphology of plant shade leaves. Crown of trees acts as a shelter thereby decreasing the thermal fluctuations inside the forest. Therefore heating by day and cooling by night are slowed down due to transpiration of plant and evaporation from soil surface. Consequently relative humidity is higher inside forest than outside. In the forest, the crowns of trees forms a uniform vegetative layer where the rainfall in the initial stages of shower are absorbed by leaves and twigs (4 to 5 mm), but then the rain drops will fall to the ground to saturate surface soil as if it is bare soil. Therefore exposure of trees and topography of vegetative or soil surface plays important role in generating specific microclimatic conditions of the forest vegetative system.

8.5 Controlled Climate

Man has not only adapted to existing microclimatic conditions, but also modified the agricultural lands by irrigation, improving soil health, clearing unsuitable species to replace with favourable crops, controlling weeds and development of high yielding fertilizer responsive crop varieties suitable in different agroecologic conditions. Scientific technology simultaneously have also advanced to utilize the better plant responses with controlled optimum condition of environment for maximum growth of the crop species (Table 4.5 and 5.1) to meet the ever increasing demand for food with green houses, glasshouses, plant growth chambers and phytotron controlled climates at different phenophases as per requirements of different crop species (Downs and Hellmers, 1976). Though these are costly propositions over large scale, but have been adapted for value added cash crops, floriculture and horticultural crops to tailor them at a time when these commodities are scarce and costly in open market.

Light intensity, temperature and humidity levels are controlled in plant growth chambers as per requirements of the crop species at different phenophases. However specific designs are also available for specific requirement. For example, seed germinators where temperature control is more important from 5 to 40°C at saturation humidity levels. Specific chambers are designed for plant disease investigations, for photosynthesis studies, for air pollutants–vegetation interactions, for tall crops like sugarcane and forest species etc. The concept of commercial growing rooms have come up in advanced countries for production of seedlings, production of flowers, vegetable and fruits. Phytotrons are complex form controlled environment system to provide different types of plant growth chambers accumulated at one place to study the different combinations of environmental factors simultaneously.

8.5.1 Characteristics of greenhouse climate

Greenhouse climate provides useful features for improving crop production and altering climatic conditions as per requirement of plants. Specific features of greenhouse climate are:

(i) With use of glass or plastic covers the intensity of solar radiation is reduced with absorption of ultraviolet radiations. The longwave radiations are absorbed inside, however in plastic longwave radiations are transmitted partially. Since reflection at top of glasshouse is considerable, therefore glass roof is inclined at standard roof slope of 26.5° to make use of incidence of solar radiation throughout the year in east-west location for better light availability in winter.

(ii) In winter poor light intensity restricts growth of the plant, therefore in order to have sufficient CO_2 assimilation and photoperiodic effects artificial light is supplied before daylight or after darkness. With additional light intensity from fluorescent lamps photosynthesis is enhanced and nighttime temperature is again controlled for optimum respiration. Artificial illumination propagates the light during dull light periods. Also certain plants only begin to bloom when they are exposed to a short day (daylight of 10 hrs or less) known as short day plants, whereas long day plants require above 14 hrs or more of daylight during pre-anthesis/vegetative growth periods. Therefore with the aid of artificial daylight illumination it is feasible to determine the time of flowering within given limits in short day and long day plants.

(iii) Heat balance over greenhouse controls its thermal environment inside the greenhouse. During summer due to excess thermal radiation a proper ventilation is essential to maintain desired level of air temperature and humidity. During night, even in colder days, the greenhouse heating will be a source of energy to maintain night temperatures. Soil heating will function as an additional energy for maintaining night temperatures.

(iv) A large portion of greenhouse energy is utilized for potential evapotranspiration, whereas is a part of energy as latent heat liberated by condensation of water vapours on the inside of the glass surface, a part of energy is lost in conduction through soil and glasshouse material, and part by air exchange through airing and wind velocity.

(v) The air temperature inside the greenhouse are warmer during day and night as compared to the open environment depending on the intensity of solar radiation, size of greenhouse, and air volume inside. Vertical temperature gradient in open environment decreases with height, whereas in greenhouse upper air layers are warmer than lower layers. However with proper roof ventilations similar temperature gradients are maintained so as to represent the open environment. In a heated greenhouse the temperature is controlled by temperature control through reduction of net radiations, by increasing air exchange and increasing evaporation with supply of moisture.

(vi) Leaf temperature are higher than the open environment, but will be lower than the air temperature during cloudless sky at night.

(vii) Temperature and humidity are controlled in the greenhouse by regulating air temperature and ventilation rate. Dry or moist air can be maintained through regulation of evaporation form soil with control of air temperature.

(viii) With sufficient water supply to the soil in greenhouse evapotranspiration can be maintained from actual to potential levels.

8.5.2 Applications of controlled climate studies

Controlled climate studies are desired to understand the plant growth physiological interactions to quantify

(i) Microclimatic parameters affecting plant development at different growth stages.
(ii) Maximum yield potential of economic crops at a location.
(iii) Support field research insights to tailor new ideal genotypes with suitable crop geometry suiting to a particular environment.
(iv) Introduction of new crops or crop species to replace with unsuitable crops/genotypes for better economical returns.
(v) The impact of environmental stress factors on different crops at different growth stages.
(vi) The impact on biochemicals, fertilizers, pest and disease infestations, growth regulators.
(vii) The growth potential of plants for commercial applications. Evaluation of the best crop in shortest time when critical values of temperature, day length light intensity and quality, moisture stress and nutrition are known. Particularly through manipulation of temperature and photoperiod the time from sowing to anthesis can be reduced to 50 per cent while the time of anthesis can be spread over several weeks (Hellmers and Burton, 1972).
(viii) The advancement of basic knowledge of plant physiological behaviour such as the role of phytocrome in photoperiodism and photomorphogenesis (Hendricks and Borthwick, 1965). Photoperiod control has now become a production tool for many greenhouse crops as a basic research technique in plant breeding and forest species research.
(ix) Simulation modelling for different complex crops such as soybean, cotton, sugarcane and cropping systems such as EPIC model have been developed to understand in depth crop-soil-atmosphere-water system.

Farmers have been practising environmental control in the fields through irrigation, land levelling, shelterbelts, intercropping, shading, removal of weeds, fertilization, frost control, moisture conservation, recycling of runoff water, water harvesting, use of plastic covers to protect young seedlings from frosts in winter, etc. Green house applications, where higher nighttime temperature, higher CO_2 levels and light intensity are controlled, are common research tools to optimise the water and nutrient needs, and to study the impact of soil-water pollutants.

In spite of these efforts it is very difficult to understand the complexity of interactions.

The combination of climatic variables required to produce the morphological, physiological and biological characteristics of field crops in controlled environment are unknown limiting their transfer into field production.

8.6 Improvement of Crop Production

It has always been a constant effort to improve the crop production, quantitative as well as qualitatively, to meet ever increasing needs of human and animals population al over the world. Microclimate, being one of the most interlinked component of productivity system, it has been exploited during the last century to meet the demands of productivity through different scientific approaches. A vast literature is available over this subject, but we will list only significant microclimatic processures and methods evolved which farmers can utilize for their benefits. The important microclimatic procedures to improve the production is through the efficient utilization of microclimatic resources-radiation interception, thermal energy utilization, soil water use, shelter from wind etc.

(i) Efficient utilization of radiation resource through radiation interception in relation to crop geometry for maximizing the photosynthetic productivity of crops and animal species.

(ii) Efficient utilization of heat energy for rapid development of crops and animal species in combination with daylength changes.

(iii) Efficient utilization of soil water to meet evapotranspirational needs of crops and animals in order to minimize the environmental stresses induced in their interaction.

(iv) Shelter from the excess wind speed to minimize mechanical damage and to improve microclimate in the leeward side of shelter to increase productivity of crops and animals. Seasonal adverse impacts of wind energy in combination with seasonal extreme temperatures are most crucial both for the crops and animals, and should be minimised through third interlinking source of irrigation.

(v) Shade from excess incoming solar radiation in tropical and subtropical areas as well as the needs of certain crop species even in temperate environments such as for tea and spices cultivation.

(vi) Crop or soil cover through mulches, plastics and glass houses are useful tactics to protect the plants from frost and to supply warmth to provide higher beneficial temperatures for rapid growth and development. Similarly better adaptive animal housing with proper radiant, thermal energy and ventilation are useful microclimatic procedures for rapid growth, development and raising productivity levels.

(vii) Improvement of soil climate with maintenance of proper soil temperature and soil moisture in the crop root zone depth for rapid efficient growth of roots to exploit soil fertility resource for higher productivity. Water use efficiency should be maintained according to the crop's requirement at different growth stages with efficient utilization of soil water and nutrients resource.

(viii) Microclimate of complex agrosystems- there is competition of exploitation of input resources of radiation and light energy, soil water and fertilizer among the components of agrosystems. Components may be two crops or a crop-weed combination or two systems of cultivation flat-ridge furrow sowing, a short-tall crops combination, vertical cultivation of coconut-spices. Efficiency of the system as a whole needs to be improved with efficient utilization of resources in such a way that components of the systems are complementary to each other.

(ix) Phototropism character in certain plants helps to exploit microclimatic sunlight variations for higher productivity. For example sunflower turn towards the movement of solar trajectory provides preferential use of sunlight for the development of seeds and better quality of oil contents. Similarly biotic events in cashew tree, i.e., number of flushes per sq.m., number of panicles per sq.m. and number of fruits per sq.m. were influenced by sunlight with a 7 to 10 days delay in initiation of these biotic events of cashew towards north as compared to the southern side. Yield pattern of cashew were influenced by sunlight with higher values towards south as compared to other directions.

Exercise 7

Q. 1. Develop microclimatic profiles of radiation, wind, temperature, humidity, vapour pressure, carbon dioxide for field crops at your location, and develop microclimatic model for the foliage canopy.

Q. 2. Develop microclimatic profiles of radiation, wind, temperature, humidity, vapour pressure, carbon dioxide for complex agrosystems: intercropping, mixed cropping, trees, orchards, forests; and analyze the competition of use of radiation, soil water, nutrients among the components of agrosystems.

Q. 3. Quantify the benefits of sheltered environment in the Leewardside for crops grown in your region.

References

Angus, J.F., D.H. Mackenzie, R. Morton and C.A. Schafer 1981. Phase development in field crops II. Thermal and photoperiodic responses of spring wheat. Field Crop Research, 4, p. 269-83.

Angus, J.F., R.B. Cunningham, M.W. Moncur, and D.H. Mackenzie 1980-81. Phasic development in field crops I. Thermal response in the seedling phase. Field Crop Research, 3, p. 365-78.

Arya, S.P. 1988. Introductionto Micrometeorology, Academic Press, San Diego, California, USA.

Bishnoi, O.P. 2007. Principles of Agricultural Meteorology. Oxford Book Company, Jaipur, p. 213-224.

Cassin, J., J. Bourdeaut, A. Fougue, V. Furon, J.D. Gaillard, J. LeBourdelles, g. Montagnt, and C. Moreuil 1969. Proc. 1st Intern. Citrus Symp., 1, p. 315-23.

Chirkov, Y.I. 1979. The climate ofmeadows and pastures. Ch. 6, p. 142-44. In: Agrometeorology by Seeman et. al. (eds.) Spring - Verlag, Berlin/N.Y.

Cooper, W.C., A. Peynado, J.R. Furr, R.H. Hilgeman, G.A. Cahoon, and S.B. Boswell 1963. Proc. Am. Soc. Hort. Sci., 82, p. 180-92.

Davies, J.W. 1975a. The control of soil climate by mulching. In: LP Smith (ed.), The affect of weather and climate on plants. Progress in Biometeorology, Vol. I, Swets and Zeitlinger, B.V., Amsterdam, The Netherlands.

Davies, J.W., 1975b. Mulching effects on plant climate and yield. WMO, No. 388, Tech. Note No. 136, Geneva, Switzerland.

Downs, R.J. and H. Hellmers, 1976. Controlled climate and plant research, WMO – No. 436, Tech. Note No. 148, WMO, Geneva, Switzerland, p. 1-60.

Eimern, J. Van et al. 1964. Windbreaks and shelterbelts. WMO, No. 147, Tech. Note No. 59, Geneva, Switzerland.

Fournier d'Abde, E.M., 1958. This modificationof microclimates. In: Climatology, reviews of research. Aridzone Research Series, XI, UNESCO, Paris, France.

Fuch, M. 1972. The control of the radiation climate of plant communities. In: D. Hillel (ed.), Optimizing the soil physical environment towards greater crop yields. Academic Press, New York, USA.

Geiger, R. 1965. The climate near the ground. Harvard University Press, Cambridge, Massachusetts.

Gerber, J.F. and J.D. Marlsolf and J.F. Bartholic 1974. The climate of trees and vines. Hort. Science, 9, 6, p. 569-72.

Hagan, L.J. and E.L. Skidmore 1974. Reducing turbulent transfer to increase water use efficiency. Agril. Meteorol., 14, p. 153-68.

Hellmers, H. and G.W., Burton 1972. Photoperiod and temperature manipulation induces early anthesis in pearl millet. Crop. Sci., 12, p. 198-200.

Hendricks, S.B. and H.A. Borthwick 1965. The physiological functions of phytocrome, p. 405-36. In: Chemistry and Biochemistry of plant pigments. (T.W, Godwin, Ed.) Academic Press, London.

Jackson, J.E. 1983. Light climate and crop-tree mixtures. In: P.A. Huxley (ed.) plant Research and Agroforestry, ICRAF, Nairob, Kenya.

Kalma, J.D. 1970. Some aspects of the water balance of an irrigated orange plantation, Ph.D. Thesis. Volcani Institute of Agril. Res., Bet Dagan, Isreal.

Kang, B.T. 1978. Light transmission of various crop mixtures to determine their efficiency in controlling weed growth by their shade. Annual report, IIA, Box 5320, Ibadan, Nigesia.

Kang, B.T. 1981. Soil moisture under different live mulches grown with crops to prevent soil erosion to control weed and to maintain productivity under conventional and notillage conditions. Annual report, IITA, Box 5320, Ibadan, Nigeria.

Kulik, M.S. and Y.Y. Sinelsnchikov 1961. Meteorology and Agrometeorology. Izbramyee Sochineniya (selected works) English translation, Gidrometeoizat, USSR. Isreal Programme of Scientific Translation with National Science Foundation and Dept. of Agri., U.S.A.

Lowry, W.P. 1969. Weather and Life an introduction to biometeorology, Academic press, New York and London, p. 260.

Midmore, D.J. 1984. The use of traditional mulches for introducing growth of potatoes in the hot tropics and the use non-traditional mulches to assit in the interpretation of the beneficial effects of traditional one (soil temperature, reflectivity of radiation and light measurements). Intern. Potato Centre, Lima, Peru.

Monteith, J.L. 1963. Micrometeorology in relation to plant and animal life. Proc. Linnean Soc., London, 171, 1, p. 71-82.

Monteith, J.L. 1973. Principles of Environmental Physics. Edward Arnold, London.

Monteith, J.L. 1975. Vegetation and Atmosphere. Vol. I, Academic Press, New York.

Munn, R.E. 1966. Descriptive Micrometeorology, Academic Press, New York.

Newman, J.E., W.C. Cooper, W. Reuther, G.A. Cahoon, and A. Peynodo, 1967. In: Ground Level Climatology by R.H. Shaw (ed.), p. 127-47. AAAS Washington, D.C.

Obaga, S.O. 1984. Shade trees in tea – review. Tea, 5, 1, p. 39-47.

Oke, T.R. 1987. Boundary layer climates. Halsted, New York, p. 435.

Primault, B. 1979. The climate of orchards and forests, p. 152-162. In: Agrometeorology by J. Seeman, Y.I. Chirkov, J. Lomas and B. Primault. Springer - Verlag, Berlin Heidelberg/N.Y.

Radke, J.A. and R.T. Hagstrom, 1976. Strip intercropping for wind protection. In: R.A. Papendick et al. (eds.) Multiple cropping. ASA, Madison, WIS, USA.

Rauner, Y.L. 1960. Formation mechanism of the heat balance and microclimate in arid zones. Academy of Sciences. Moscow, USSR.

Ripley, E.A. and R.E. Redmann 1976. Grassland. In: Vegetation and Atmosphere by J.L. Monteith (ed.), Academic Press, London, p. 351-96.

Rosenberg, N.J. 1967. The influence and implication of windbreaks on agriculture in dry regions. In: Ground level climatology, AAS, USA.

Rosenberg, N.J., B.L. Blad, S.B. Verma 1983. Microclimate – The Biological Environment. Willey (Interscience), New York.

Russell, E.W. 1973. Soil conditions and plant growth. ELBS and Longman, London, England.

Slatyer, R.O. and I.C. McIlroy, 1961. Practical Micrometeorology. UNESCO, Paris, p. 214.

Stigter, C.J. 1984. Forecasting and weather services in Africa. Int. J. Environment Studies, 19, p. 269-71.

Stigter, C.J. 1985a. Research methodology to quantify and understand the modification of energy flux at soil/ atmosphere interfaces. Proc. of Intern. Colloquium on energy flux at the soil/atmosphere interface, ICTP, Trieste, Italy, p. 16.

Stigter, C.J. 1985b. Agricultural physics: The practice of a preacher. In: C.J. Stigter (ed.), Ninth Annual Report, Univ. of Dar as Salaam, Tanzania.

Stigter, C.J. 1986. Microclimate management and manipulation in traditional farming. Report of working group on microclimate management and manipulation in traditional farming. Report no WCP-1762, WMO, Geneva.

Sutton, O.G. 1953. Micrometeorology, McGraw Hill, New York.

Waggoner, P.E. 1965. Microclimate and plant disease. Ann. Rev. Phytopath., 3, p. 1003-26.

Waterhouse, F.L. 1955. Quar. Jour. Roy. Met. Soc., 81, p. 63-71.

Whiteman, W.C. 1969. In: Grassland Ecosystem – A Preliminary Synthesis by R.L. Dix and R.G. Beidleman (eds.), p. 40-64. Range Sci. Ser. No. 2, Fort Colins, Colorado.

WMO, 1972. The application of micrometeorology to agricultural problems (ed), L.P. Smith, WMO Tech. Note 119, WMO No. 298, WMO, Geneva, p. 74.

Volume-II

Agroclimatological Advisory Services Confronting Climate Change Complexities

9

Agrometeorological Advisory Services

9.1 Introduction

Agriculture is a specific industry dealing with the cultivation of living organism-plants and animals primarily under natural environment. Climate strongly influences the natural selection and pattern of plant and animal kingdom on the earth's surface. Importance of climate and weather to agriculture is very diverse not only from growth, development, yield of crops; but also in terms of field operations and efficient utilization of available resources. Therefore, for an effective use of climate and weather conditions in agriculture, the knowledge of agrometeorology/agroclimatology is essential. In order to understand the weather variability, atmospheric dynamics and forecast methods a basic understanding of Synoptic and Dynamic Meteorology is essential. Since we are interested in studing the climate and weather in relation to agricultural needs, the interactive discipline – Agrometeorology becomes the obvious choice for agrometeorological services to help agricultural community particularly farmers.

Agrometeorological services have been enlarged to meet he urgent needs of agriculture in terms of short range (1 to 2 days), medium range (3 to 10 days) and long range (11 days to monthly or seasonal). In addition timely urgent warnings about dangerous weather phenomena such as frost, strong winds, heavy rains, cloud brust, cyclones, thunderstorm have been monitored through satellite weather photographs to supplement the synoptic weather forecast for accurate description and their movement with time over space, and are very beneficial for protection of agricultural wealth.

For smooth conduct of agricultural activities, two to five days weather information in necessary about the prevailing and anticipated weather conditions and their effect on the agricultural production. Long range forecasts of anticipated weather characteristics for the forthcoming months are also useful in terms of temperature and precipitation departures from normal, spell of hot or cold waves, drought or wet spells etc. Therefore, crop planning is framed with choice of crops, availability of soil water and thermal heat adequacy etc. Fruit crops also call for phenological services during the flowering period for pollination, initiation of fruit buds, sprouting and ripening of fruits.

Agrometeorological forecasts are becoming popular among farmers and agricultural extension specialists in their application to the agricultural constraints in crop production and efficient utilization of available natural resources. Agrometeorological services are not only restricted to operational and raising of crop production, but also includes much wider aspects of production including animal husbandary, forestry, transportation of produce and efficient use of costly inputs i.e., water, nutrients, healthy varietal seeds of foodgrains including pulses, oilseeds, spices, fruits and vegetables, agrochemicals. A large area of Asian region is influenced by the monsoon rains which is highly fluctuating year to year due to worldwide climatic changes and global warming resulting in about 30 to 50 per cent variations in food grain production. Therefore, weather based tactical and strategic decisions are essential to counteract the adverse effects of adverse weather. Weather based agrometeorological services are beneficial in efficient and proper utilization of natural resources, their conservation, and achieving optimum grain yields and minimising the agricultural losses. Therefore agrometeorological information has to cater the needs of farmers, policy makers, extension and research workers both for long term planning and day to day agricultural operational needs for efficient and economic utilization of resources. Department of Science and Technology, Govt. of India, realizing the needs of medium range weather forecasting and the availability of super computer facilities, a National Centre of Medium Range Weather, Forecasting was established in 1988 to develop operational medium range (3 to 10 days advance) weather forecasting capabilities agroadvisory services for 127 agroclimatic zones of India (Fig. 1.1).

Weather and climate strongly impringe upon both strategic and tactical decisions of crop production for agrometeorological services.

(i) Strategical decisions- these are based on long term climatic data over space and time to develop

  (a) Sustainable agricultural weatherwise planning at a location, zonal or national level based on agroclimatic resource analysis for different types of cropping systems.
  (b) Choice for homogeneous agroclimatic or agroecological zones for seed production programmes.
  (c) Choice of land suitability and agricultural enterprise for grain/horticultural/forestry or animal husbandary production.
  (d) Demarcation of potential productivity zones for specific biological entity.
  (e) Assessing climatic resources for favourable crop planning, irrigation resource management, choice of crops and vegetables for different cropping systems.
  (f) Premanagement of costly inputs seed, fertilizer, agrochemicals keeping in view of long range weather forecasting and likely crop yield potentials in view.

(ii) Tactical decisions- these are based on short term (daily to weekly) weather conditions at microlevel (field or zonal level)

  (a) Farm operational crop management decisions-land preparation, choice of crop variety, germination, crop establishment, growth, reproduction, harvesting and storing of agricultural production.
  (b) Phenological development and behaviours of pest and disease organisms during the crop growth cycle, and choice of agrochemical for plant protection.
  (c) Monitoring of track of soil moisture conditions for irrigation planning during crop's life cycle to regulate the impact of dry or wet spell periods on crop productivity.
  (d) Monitoring of adverse weather conditions during crop growth cycle such as frosts, hailstorms, thunderstorm, cyclone, heavy rains, cloudburst, cold or hot

winds, etc. Farmers need to be briefed immediately for remedial measures to avoid heavy losses due to such untimely weather features. There is strong need to schedule weather sensitive activities.

With the advancement of scientific technology on Remote Sensing and GIS techniques and Crop Growth Simulation models, it has been possible to facilitate the mapping of weather and crop information at micro and macro-scales, flooded and drought areas, ground water availability, natural disaster management in cropped areas due to passage of cyclones, hailstorm, thunderstorm, intense untimely rains etc. Crop simulation models are also important tools for projecting the impact of climate change and climatic variability scenarios. Inter-Governmental Panel on Climate Change (IPCC) of United Nations has confirmed the global warming trend and alarming future projections (rise of 1.4 to 5.8°C). However, over India, the warming is observed in post-monsoon and winter seasons with little variation during the monsoon period. It has been projected that over India increase in temperature is likely to be less in summer than in winter season. Winter rains may have higher variability, whereas summer and monsoon seasonal rainfall may increase by about 10 per cent. This would alter agricultural activities considerably through its direct and indirect affect on crops, cropping systems, livestock pest and diseases, soil water availability, crop yield variations, and crop management strategies.

Also there is need to develop the risk management strategies in each of the agroclimatic zones based on weather to interpret the crop weather relations and operational management strategies to harness the benefit of crop insurance schemes.

There is need to issue special forecasts at specific events of weather phenomenon damaging the crops (drought, intense heavy rains, frost, hail etc.), distribution and extent of damage by pests and diseases (locust, rusts, etc.), distribution of soil moisture storage with onset and withdrawal of monsoon (for sowing, crop phenological schedules for specific field operations), yield forecast assessment for management of food stocks will certainly help the farmers to draw the advantages of scientific technology of new emerging interacting disciplines.

With advancement of agricultural technology, precision farming practices and new regulatory laws for food quality and environmental protection has led to cautious use of weather/climate information and forecasts. Efforts are always focussed to sustain agricultural system with optimum use of inputs, environmentally conserving resources in order to generate a profitable agricultural economy. Now the efforts are to use climatic information and weather outlook forecasts for agribusiness decisions (Lamb et al., 1985; Sonka et al., 1992) in the areas of seed production, crop consulting, chemical and fertilizer manufacturing and marketing, weather insurance, and agricultural cooperatives (Changnon et al., 1990). Reports exist about under utilization of forecasts for operational decisions (Seelay et al., 1990), and even some government agencies are attempting to tailor weather forecasts to meet the specific needs of agricultural clientele including livestock in U.K. and USA (Seelay, 1994; Perry,

1993). United States Department of Agriculture (USDA) have launched a national policy to meet climatic anomalies such as drought, disaster declarations. U.K. is providing agrometeorological services in addition to daily forecasts (Table 9.1) for benefits of farmers. Similarly in U.S.A. National Oceanic and Atmospheric Administration (NOAA) weather radio offers agrometeorological services for operational decision in rural agricultural areas, and even risk based marketing strategies are adopted for weather sensitive vegetables/ fruits/floriculture commodities. However increasing environmental regulations are viewed as a serious threat for agricultural production (Miller, 1992) though the communication technology have reduced the gap for operational agrometeorological services (Rijks, 1992). The agrometeorological services with be a continued effort to serve the farmers with improved and needy services from time to time.

Table 9.1
Fee-based Agrometeorological Services in United Kingdom

Service	Description
Return periods	For extreme events (heavy rains, maximum temperatures)
Accumulated temperature or thermal time	Above or below any threshold base temperature (Ontario heat units)
Frost	Frequency analysis
Soil moisture estimation	For drainage and evaporation calculations
IRRIPLAN and IRRIGUIDE	Strategic and tactical irrigation planning policy
Odour plume analysis	For livestock housing
Glasshouse heating	For greenhouse cultivation of vegetables and flower production
Spraying hours	Climatic probabilities for suitable conditions for spray of agrochemicals
Shadow length	Shading calculations for structures
Pest emergence	Forecast for the emergence of significant crop pests
Sugar level	Estimations for first cut silage
Climatic analysis	For marginal or alternate crops
Physiological ageing	For storage of seed potatoes
Windbreaks/shelterbelts	To express optimum siting
Windchill forecasts	For livestock managers (Sheep)
Weather risk analysis	Site specific frequency distribution of significant thresholds
Local weather monitoring	On-site temporary monitoring to determine microclimate and landscape effects
Land classification	Climatic indices for land classification

(*After*: Seelay, 1994).

9.2 Weather Forecast Services

The success of agrometeorological services to farmers depends on the quality of weather forecast. The weather forecast involves a systematic scientific approach based on data assimilation, a deep understanding of processes involved in the weather system, and use of computing skills to predict the future state of weather conditions adopting the present state of boundary conditions. Weather forecast skills have advanced with the availability of fast super computers, large data observations through radar and satellite's continuous monitoring, telecommunication systems penetration into rural areas, advancement in numerical weather prediction global models to tackle complex weather processes. The improvements in all these interlinking sub-systems have facilitated the weather forecasters to predict weather features closer to reality on different time scales such as

(i) **Now-casting**: Weather forecast from zero to 12 hours are based on the intensive observations over sea and land surfaces, upper air data commonly used for aviation purpose. Radar and satellite observations are useful for cloud pattern analysis for rainfall prediction. Similar analysis are useful for hazardous weather phenomenon for its monitoring at short time intervals. Therefore, it is essential to quantify the time rate of change of phenomenon such as severe convection, intense cyclones, cloud brusts, deep depressions etc., to derive conclusions for user's interest. However, weather events such as thunderstorm, hailstorm, tornadoes etc., may have some limitations about specific location with time.

(ii) **Short range forecasts**: Weather forecast for next 12 to 24 hours period is commonly developed on synoptic scale based on atmospheric numerical weather prediction approach or synoptic technology. Numerical weather prediction models represent the atmosphere on a three dimensional grid with typical operational systems. Powerful super computers are useful for these forecast in order to process the vast data.

(iii) **Medium range forecast**: Medium range forecasts are issued for 3 to 10 days in advance for agricultural field operation and crop management. Though quality of prediction decreases with increasing forecast period, however using numerical prediction models forecast of one month averages of sea surface temperature anomalies were significant. National Centre for Medium Range Weather Forecasting (NCMRWF) issues these forecasts twice a week for different agroclimatic zones. (Table 9.2).

(iv) **Long range forecasting**: Weather forecast of more than 10 days to a season are useful for seasonal strategic planning of farmlands. Also, to a limited extent, the temperature and precipitation anomalies have been predicted over seasonal basis over a large spatial scale to monitor the interactions between atmosphere, ocean and land surface.

9.2.1 Meteorological data representation

Meteorological observations are recorded at 3 to 12 hourly intervals with initiation at 00 hours GMT i.e., 5.30 hours IST all over the world. Internationally agreed numerical codes consisting of five digit figures were developed for quick transformation and timely forecast

Table 9.2

Application of weather forecast

Type of Weather Forecast	Forecast Features	Applications
Now-casting	Cloud pattern, rainfall and abnormal weather features arising instantaneously, special forecasts over space and time with the track of abnormal weather behavior	Instantaneous weather hazards impact or damage of natural wealth and minimization of causalities
Short range	Extreme temperature and pressure, wind speed and direction, sunshine, precipitation records, relative humidity, visual observation features.	Frost liability, irrigation, spray and efficiency of chemicals, pest and disease impacts, extreme temperature protection of plants and animals, evapotranspiration needs, day to day field cultural activities.
Medium range	Continuity of weather system, rainy days and rainfall amount continuity, variation in temperature, humidity, evaporation, sunshine/cloudiness, wind speed and direction w.r.to. the normal features; moisture scarcity and dry spells.	Field preparation and sowing of crops, field operation decisions for irrigation, fertilization, spray applications; phenological forecasts, soil moisture status for amount of irrigation needs, crop water requirement determination; crop yield assessments; field tactical decisions.
Long range	Seasonal rainfall forecasts, abnormal weather behaviour in temperature and rainfall over normal pattern.	Strategic planning for crop and animal's requirements, management of climatic resources for improving efficiency in view of prevailing weather, weather modification possibilities, recycling of stored water.

through microwave links and satellites through exclusive meteorological telecommunication channels specially designed by WMO under their World Weather Watch programme. Meteorological observations are plotted symbolically on a weather chart. Plot of a station data is represented as follow as an example.

Station Plotting Model

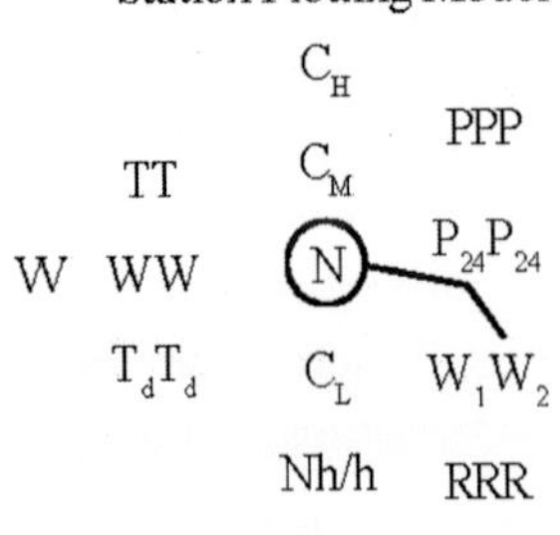

Plot of Surface Observations

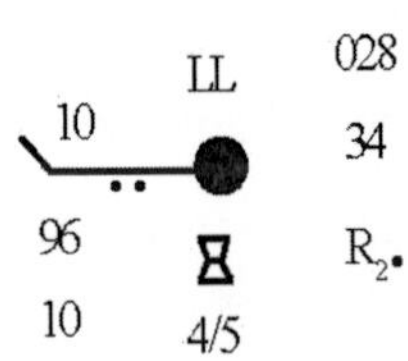

where,

C_H – Form of high clouds – No clouds seen

C_L – Form of low clouds – above half sky covered

C_M – Form of medium clouds

h – Height above ground of the base of lowest cloud seen at about 800 m

N – Fraction of sky covered by cloud of all types – sky is fully covered

N_h – Fraction of sky covered by low clouds reported for C_L. When no C_L cloud exists N_h will refer to amount of cloud for C_M

PPP – Pressure of mean sea level in tenths of mb. omitting the thousand's digit i.e., 1002.8 mb

RRR – Precipitation since 0300 GMT in second group between 20 to 30 mm

TT – Air temperature in tenths of °C

T_dT_d – Dew point temperature in tenths of °C about 9.9°C

VV – Horizontal visibility of 4000 m at which objects are seen

w_1, w_2 – Past weather conditions – R.

ww – Present weather

$P_{24}P_{24}$ – Change of station level pressure in last 24 hours in tenths of mb i.e., 3.4 mb fall in pressure

Wind – Wind is approaching the station from west with wind speed of 9 knots.

9.2.2 Analysis of synoptic charts

The meteorological elements observed at various station at surface and upper air are plotted on a location charts. These charts are analysed to assess the present state of the atmosphere at the observed time from surface to the troposphere in order to predict the subsequent weather. Important useful weather charts for analysis are

(i) Isobaric analysis of sea level and upper air charts.
(ii) Wind field analysis.
(iii) Stream line – Isotach analysis.
(iv) Vertical time section charts.
(v) Vertical cross-section charts.

These charts are supplemented with tephigrams to help in understanding the thermodynamical features and stability/instability over the stations, and satellite and radar cloud pictures over oceanic and data sparse regions to provide useful informations of present state of atmospheric conditions.

(i) **Isobaric analysis of sea level and upper air charts**: The main meteorological elements plotted are mean sea level pressure, surface wind direction and speed, visibility, dry bulb and dew point temperature, present and past weather, types and amount of cloud, upper air pressures at different heights. These charts are analysed by drawing isobars to delineate the closed lows and high pressure systems, troughs and ridges, and cloud distribution.

(ii) **Analysis of wind field**: Wind observations are recorded at discrete points, but its vector field is continuous. Analysis of both direction and speed of wind field is carried out in three ways by

(a) drawing lines of equal wind speed called isotachs to describe wind field completely,
(b) analysing the wind field in terms of zonal and meridional components, and
(c) analysing the wind field with the help of stream lines and isotachs. Stream line is a line parallel everywhere to the instantaneous wind direction at any instant of time. For example, northerly 15 knots is represented as 360/15.

(iii) **Streamline – isotach analysis**: Stream lines analysis can be carried out in two ways by

(a) Isogon method: Numerical values of wind direction and speed are plotted at the stations. Isogons are drawn by scalar interpretation at the intervals of 30°, whereas across each isogon a number of short segments are drawn parallel to the wind direction corresponding to the isogon which can be utilized. The line segments are connected to tagential curve to obtain streamlines.
(b) Freehand method: Streamlines are drawn everywhere parallel to the wind direction which are continuous lines to meet in the region of confluence. However, when the system is weak we can draw less number of streamlines. In surface friction layer, cyclonic in draft and anticyclonic out draft can be depicted. At higher levels with the availability of good data, inflow and outflow coverage can be drawn. However with the less data, the cyclones and anticyclones can be drawn as closed vortices. Normally anticyclones (A) are marked in blue, whereas cyclones (C) are marked in red. Trough lines are denoted as dashed line, whereas ridge line as zigzag line.

Streamline analysis will provide useful features on

(a) Cyclonic or anticyclonic circulations vortices represented by regions of cyclonic or anticyclonic flow in the analysis pattern.
(b) Asymptotes are lines along which the two neighbouring streamlines tend to converge or diverge.
(c) Singular points as the points into which more than one streamline can be drawn or about which the stream lines form a closed curve.
(d) Neutral points as the points at which two asymptotes – one representing confluence and the other diffluence intersect each other.
(e) Waves can be depicted as sinusoidal configuration of parallel stremlines.

(iv) **Vertical time section charts**: Usually the geopotential height of the pressure level, 24 hours height changes, wind direction and speed, mixing ratio values, dry bulb temperature are plotted at various standard levels. The meteorological element over a period of time at a location is plotted. Supporting informations on tropopause, maximum wind speed, and freezing level are also plotted at appropriate levels. The surface observations are plotted at the appropriate date and time at the bottom. Isotherms, isohygric lines and isotachs are also drawn on these maps. Such a map will depict the time sequence of the meteorological variable with time at the location. Movement of waves, vortices are also delineated on the charts. Availability of moisture, fluctuations in tropopause conditions, jet streams, freezing level can also be identified.

(v) **Vertical cross-section charts**: In these charts, the ordinate is height or pressure along a vertical plane, and the horizontal space scale is taken as abscissa. The wind direction and speed, geopotential height of the pressure level, dry bulb temperature are plotted at standard levels. The surface observations, freezing level, tropopause and maximum wind data are also plotted at the appropriate time. Such a chart presents the distribution of various meteorological elements at a number of selected stations at a given point of time. Daily vertical space cross-sections along heavily travelled air routes are useful in aviation services. The vertical time sections and vertical cross-section charts provide an effective link between the thermodynamic surrounding diagrams and synoptic charts.

9.2.2.1 Supplimented weather informations:

(i) **Tephigram ($T-\phi$)**: Tephigram is a thermodynamic diagram where the coordinates are temperature and entrophy. It is an energy diagram where the area is proportional to energy and represents the graphical display of the processes to which the air may be subjected such as isothermal dry adiabatic, isobaric and pseudo adiabatic etc. Fig. 9.1 illustrates the $T-\phi$ diagram. Using tephigrams we can compute he geopotential height of pressure levels, various atmospheric thermodynamic parameters and moisture variables, static stability and various stability parameters.

(ii) **Geopotential height between two pressure levels**: The height difference in gpm can be estimated by two methods-isothermal and adiabatic methods. In isothermal method the geopotential height between two pressure levels P_1 and P_2 with corresponding Z_1 and Z_2 as geopotential height respectively above sea level can be expressed as

$$z_2 - z_1 = \frac{-R}{g}\int_{P_1}^{P_2} T^* d(\text{Ln } P) \qquad \ldots (9.1)$$

where, T* is the virtual temperature, R is specific gas constant. If we plot T* against pressure P, the $\frac{R}{g}\int_{P_1}^{P_2} T^* d(\text{Ln } P)$ is proportional to the area between P_1 and P_2 to the left of the virtual temperature curve. On tepigram, a mean virtual temperature $\overline{T}^*$ can be determined by selecting a vertical isotherm which is cooler than the surrounding temperature for part of the pressure increment and warmer for the remainder. Therefore after selecting $\overline{T}^*$, the height difference is assessed as

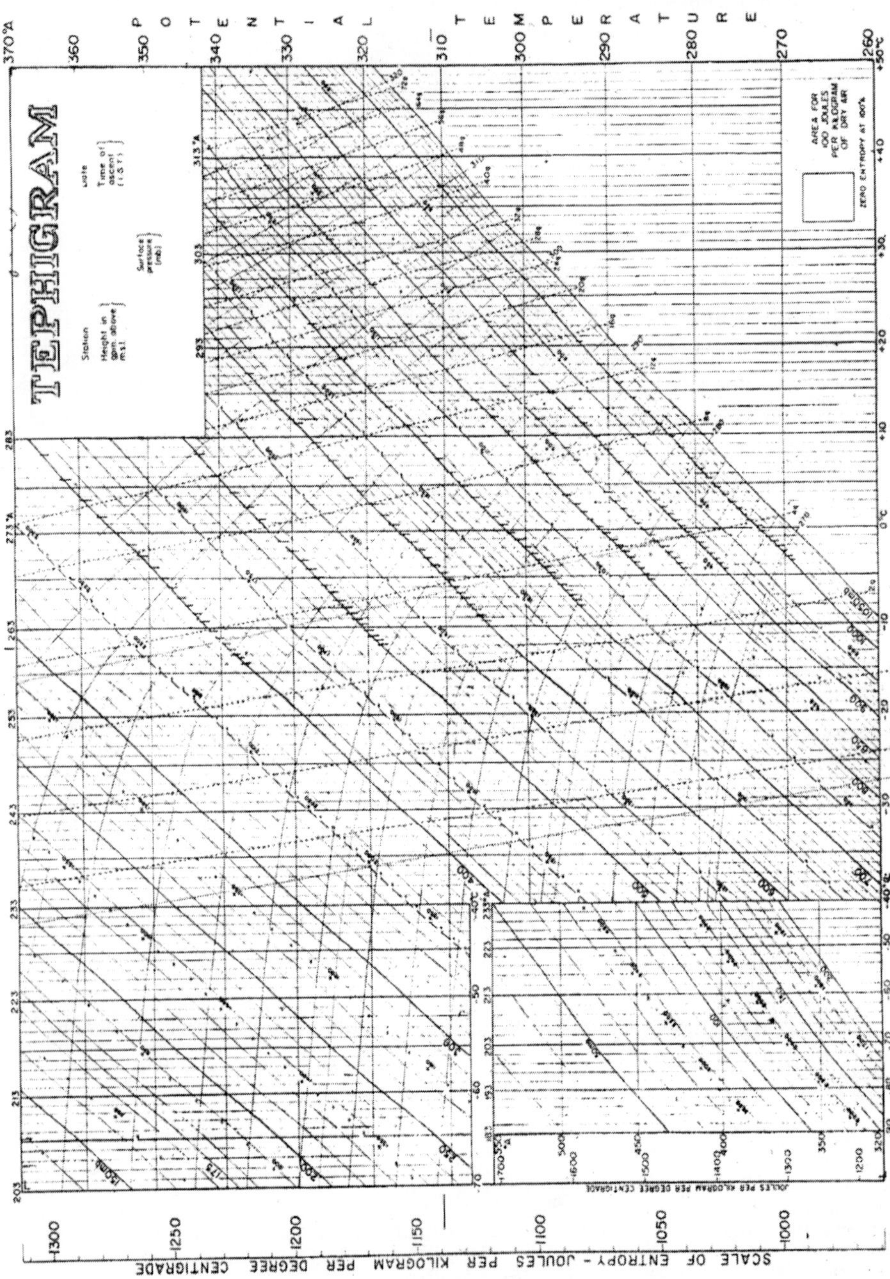

Fig. 9.1(a): T − φ diagram [*After:* Anantakrishnan and Yagnanarayanan, 1989]

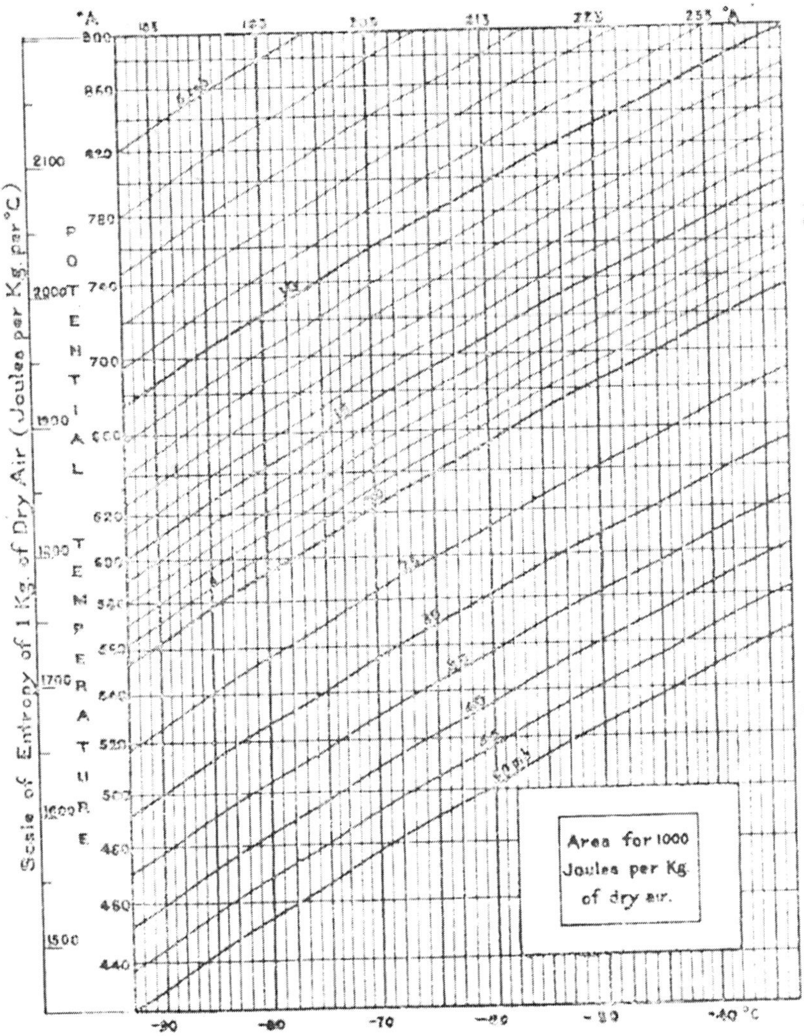

Fig. 9.1(b)

LEGANDS

1. Vertical lines are of equal temperature, marked in degrees Centigrade at the bottom and in degrees Absolute at the top.
2. Horizontal lines are of equal potential temperature, or dry adiabatics, marked in degrees Absolute. The entropy seals, which is actually a logarithmic scale of potential temperature is given alongside, in joules per kilogram per degree Centigrade.
3. Sloping lines, rising towards the right and nearly straight, are isobars marked with pressure, in millibars.
4. Curved lines, rising towards the left, are saturated adiabatics. They refer to saturation over ice for temperatures less than 0°C. For four values, viz., 300°, 290°, 280°, 270° A, lines for saturation over liquid water have been shown, dotted, in addition.
5. Dotted lines, running from top to bottom inclined to isotherms, are dew-point lines (humidity-mixing ratio lines). They describe the pressure and temperature conditions in which given amounts of water vapour will just saturate 1 kilogram of dry air. The lables which 2g, 4g, 8g, etc., marked on the dew-point lines, show 1000 times the humidity-mixing ratio for saturated air. The dew-point lines refer to saturation over ice for temperatures less than 0°C. For two values, viz., 0.5 and 4g/kg., lines for saturation over liquid water have been shown, dotted, in addition.
6. Thickness of layers between standard isobaric levels are given in geopotential metres, along the mid-level isobar.
7. The intervals between consecutive offset marks along 1000, 900, 850, 800, 700, 600, 500, and 400 millibar isobars give the approximate virtual temperature correction for saturated air for the range of temperature between consecutive offset marks.

$$z_2 - z_1 = \frac{R}{g}\overline{T}^{*}\text{Ln}\frac{P_2}{P_1} \quad \ldots(9.2)$$

$$z_2 - z_1 = -287.04\frac{R}{g}\overline{T}^{*}\text{Ln}\frac{P_2}{P_1} \text{ in gpm.} \quad \ldots(9.3)$$

Therefore, geopotential height corresponding to standard pressure levels are indicated in the tephigram.

In adiabatic method one mean dry adiabat is selected by equal area method which cuts the isobars P_1 and P_2 at temperature T_1 and T_2 respectively, then the height is estimated as

$$Z_2 - Z_1 = C_p (T_1 - T_2) \quad \ldots(9.4)$$

$$= 100.44(T_1 - T_2) \text{ in gpm} \quad \ldots(9.5)$$

(iii) Atmospheric thermodynamic parameters:

(a) **Potential temperature (θ)**: It is temperature of a sample of dry air that would have if it is compressed or expanded dry adiabtically from a given pressure P and temperature T to 1000 mb. Using Poisson's equation, (θ) can be expressed as

$$(\theta) = T\left(\frac{1000}{p}\right)^{R/C_p} \quad \ldots(9.6)$$

where C_p is specific heat at constant pressure.

In tephigram, adopting the dry adiabat through the dry temperature at the given pressure (P) level upto 1000 mb and then the corresponding temperature at 1000 mb is the geopotential temperature at the given pressure level.

(b) **Mixing ratio (w)**: It is the ratio of mass of water vapour to the mass of dry air. However when expressed in parts per thousand (gm kg^{-1}) it becomes 1000w gm kg^{-1}. In tephigram, the mixing ratio at a given pressure, we record the saturation mixing ratio line that corresponds to T_d (dew point temperature) at that temperature.

(c) **Specific humidity (q)**: It is the ratio of mass of water vapour present in unit mass of moist air. However when expressed in parts per thousand it is expressed as

$$q = \frac{1000w}{1000 + w} \quad \ldots(9.7)$$

After determining the mixing ratio w as above, we can determine q.

(d) **Saturation mixing ratio (w_s)**: It is the mixing ratio of a sample of air that would have if it is saturated. Therefore at a given pressure, we can record the saturation mixing ratio line passing through T(dry bulb temperature) curve at that pressure.

(e) **Relative humidity (RH)**: It is ratio of actual mixing ratio of a sample of air at a given temperature and pressure to the saturation mixing ratio of the air at the same temperature and pressure expressed in percentage as $w/w_s \times 100$.

(f) **Vapour pressure (e):** It is the partial pressure exerted by water vapours in the atmosphere. Using tephigram from the T_d curve at the given pressure follow he isotherm to 622 mb. The value of the saturation mixing ratio line through thus point at 622 mb gives the vapour pressure in mb at a given pressure.

(g) **Saturated vapour pressure (e_s):** It is partial pressure exerted by water vapour when the air is saturated. In tephigram from the T-curve at a given pressure, follow the isotherm to 622mb. The value of the saturation mixing ratio line through this point at 622 mb level gives the saturation vapour pressure in mb at the given pressure.

(h) **Wet bulb temperature (T_w):** It is the temperature to which air must be cooled by evaporating water into it at constant pressure until it is saturated. It is assumed that the heat required for evaporation is taken from air itself.

In tephigram the wet bulb temperature at any level is determined as

1. First the T_d curve at a given pressure draw a line upward along saturation mixing ratio line.
2. Then from T-curve at a given pressure draw a line upward along dry adiabat, until it intersects the line drawn in step above. Then the point of intersects is called Normand's point.
3. From Normand's point follow pseudoadiabatic back to the given pressure. The corresponding temperature is equal to the wet bulb temperature.

(i) **Dew point temperature (T_d):** It is temperature to which moist air must be cooled isobarically and isohygrically to make it just saturated with respect to water.

In tephigram we follow the dry adiabat through the drybulb temperature (T) at a given pressure, pseudo adiabat through the wet bulb temperature (T_w) at that pressure to intersect at the Normand's point. Then from the Normand's point follow saturation mixing ratio line back to the given pressure. The corresponding temperature is equal to the dew point temperature.

(j) **Wet bulb potential temperature (θ_w):** It is wet bulb temperature that a sample of air would have if it were brought pseudo adiabatically to a pressure of 1000 mb.

In tephigram we follow pseudoadiabat from the wet bulb temperature to 1000 mb isobar, the corresponding temperature is equal to wet bulb potential temperature at the given pressure level.

(k) **Equivalent temperature (T_{eq}):** It is the temperature that a sample of air would have if all its moisture were condensed out by a pseudoadiabatic process and then the sample is brought adiabatically to the original pressure.

In tephigram we can follow pseudoadiabat, through the Normand's point, upward to a pressure where both pseudo and dry adiabat become parallel. From this point we can follow the dry adiabat back to the original pressure. The corresponding temperature at this point will give equivalent temperature. Approximately the T_{eq} is about T + 2.5W.

(l) **Equivalent potential temperature (θ_{eq})**: It is the temperature that a sample of air would have if all its moisture were condensed out by a pseudo adiabatic process and the sample is then brought back adiabatically to 1000 mb level.
In tephigram we follow the pseudo adiabatic through the Normand's point upward to a pressure where both dry and pseudoadiabat become parallel. From this point then we follow the dry adiabat back to the 1000 mb pressure level to provide the corresponding temperature at this point gives the equivalent potential temperature. In approximation θ_{eq} is about θ + 3W.

(m) **Virtual temperature (T*)**: It is the temperature that a sample of dry air would have if its specific volume is same as that of the moist air sample at the given pressure. In approximation T* is T + w/6.
In tephigram hatch marks are provided to calculate the virtual temperature. The temperature difference between two hatch marks gives the virtual temperature correction for saturated air for the range of temperature occurring between the hatch marks.

(n) **Lifting condensation level (LCL)**: It is the pressure level at which a parcel of air is to be lifted dry adiabatically to make it saturated. In tephigram, the LCL is the Normand's point for the lowest pressure level.

(o) **Convective condensation level (CCL)**: It is the height to which a parcel of air, if heated sufficiently from below, will rise adiabatically until it is just saturated. In tephigram we can proceed upward along the saturation mixing ratio line through surface dew point temperature until this line intersects the drybulb curve. The CCL is at the height of this intersection.

(p) **Convective temperature (T_c)**: It is the surface temperature that must be reached to start the formation of convective clouds by solar heating of the surface air layer.
In tephigram, after determining CCL point on the drybulb curve we can proceed along the dry adiabat to the surface pressure isobar. The temperature at the intersection is the convective temperature.

(q) **Level of free convection (LFC)**: It is the height up to which a parcel of air is lifted dry adiabatically until saturation and pseudoadiabatically, thereafter would first become warmer than the surrounding air. The parcel would then continue to rise freely above this level until it becomes colder than the surrounding.
In tephigram LFC is located at the height where the pseudoadiabat through the wet bulb temperature intersects the environmental curve.

(r) **Equalibrium level (Eq.L)**: It is the height at which the temperature of a buoyantly rising parcel again becomes equal to the temperature of the environment. In tephigram Eq.L is determined where the pseudoadiabat through LFC intersects the drybulb curve.

(s) **Satibility parameters:**
   (a) Real latent instability: In tephigram +ive area > –ive area
   (b) Pseudo latent instability: In tephigram –ive area > +ive area
   (c) Layer of latent instability: In tephigram plot the dry bulb and wet bulb curves. Draw the pseudoadiabat passing through the lowest value of the dry bulb curve. The layers below the level which is cuts wet bulb curve is called the layer of the latent instability.
   (d) Environment of instability: In tephigram plot the dry bulb and wet bulb curves. Draw the pseudoadiabat passing through the highest value of the wet bulb curve. The layer where it cuts the environmental curve is called the layer of environmental instability.
   (e) Potential instability: In tephigram draw the wet bulb curve. The layer where the lapse rate of the wetbulb curve is greater than psedoadiabat curve that layer is called he convective or potential instability.

9.2.3 Analysis of radar and satellite informations

Satellite informations are playing an increasing role in supporting and improving weather analysis and forecasting. The satellite interpretation from basic cloud identification to detailed cloud processes for assessment of precipitation, wind and temperature variations. A number of polar and geostationary meteorological satellites are operating in space for continuous monitoring of atmospheric conditions over the same area in tropical and temperate latitudes through geostationary satellites, whereas monitoring of atmospheric conditions in polar areas by the polar orbiting satellites.

The geostationary satellites are in orbit directly over equator at bout 36000 km above the earth to cover one round in 24 hours. Their speeds are tuned exactly with the rotation of earth so that atmospheric condition over the same spot over the earth are monitored with time. INSAT takes cloud pictures every half an hour over an area of 52°S to 52°N in latitude and 32° to 150°E in longitude. However, polar satellites are at about 1000 km to circle the earth over both the poles to scan the entire globe in one strip over a swath of 2800 km east-west, and several degree of longitude in north-south at a transmission range of 135 to 148 MHz in polar satellites, whereas in geostationary satellites it is about 1691 to 1694 MHz. The orbit period of an earth satellite natural or manmande for a circular orbit is determined as

$$T = 84.4\left(1+\frac{h}{R}\right)^{3/2} \text{ minutes} \qquad \ldots (9.8)$$

where, R is radius of the earth and h is height of satellite above earth surface. the area covered by a satellite (east-west width) is expressed as

$$d = \phi_c \times 111 \text{ km} \qquad \ldots (9.9)$$

where, d is distance from a point just below the satellite to the picture point for viewing angle from the satellite, ϕ_c is the geocentric angle. Meteorological satellites contain two types of sensors in the visible 0.58 to 0.68 mμ and infrared 0.73 to 1.1 mμ range. Visual and infrared imageries are produced from the earth scan radiation simultaneously – visible by silicon detectors, whereas infrared by metal detectors which are digitized and transmitted to the earth. Visible satellite photographs looks like a standard black and white photograph. Clouds in the visible are white, land masses are grey, whereas water bodies appears very dark as a result of reflected sunlight.

Infrared imagery is just a picture portraying different temperature as black, white or some shades of grey. High clouds are very cold so they appears white, whereas mid clouds are somewhat warmer so they appears grey, and low clouds are still more warmer so that they are of darker shade of grey. Land and water contrast on infrared imagery depends on temperature contrast. Normally infrared imagery is taken during midday in summer when water and land temperature differences are the maximum.

Several useful inferences can be drawn from these cloud pictures in framing the agroadvisory services for farmers about future rainfall 3 to 4 days ahead. Special photographs are available in meteorological observatories to match with these specific features.

(i) ***Cloud***: Clouds can be classified by assessing their average height above ground and their features as given in Table 9.3. Snow and cloud reflect most of the sunlight and therefore look alike, but can be distinguished with terrin features.

(ii) ***Characteristics of weather systems***: Different kinds of weather systems produce their own characteristics cloud systems (rainfall and wind), and therefore may be considered as "signatures" of weather systems producing them. So from cloud patterns we can have a look of the weather system. Having the knowledge of the behaviour of weather systems in the past, we can anticipate the future behaviour (movement, intensification or weakening) of weather system. Therefore an observed cloud pattern seen in satellite cloud pattern will enable the farmers to anticipate rainfall and strong winds likely to occur in next 3 to 4 days (Table 9.4.)

9.2.4 Adverse weather phenomenon affecting agricultural resources

There are several weather phenomenon which adversely influence or damage the crops such as agricultural droughts and floods, dry hot and cold winds, dust storms, hails, frost, wet spells with cloud brust or intense rainfalls cyclonic storms and depressions, excessive or defective insolation etc. Therefore, there is need to develop insurances to cover weather risks, efficient networking and quick weather services. Monitoring of adverse feature's impact on farm lands through satellite will assess the economic losses and will devise ways and means to minimise losses and causalities. A special forecast needs to be issues for farmers for their timely control.

Table 9.3
Cloud features and types

Average height above mean sea level (m.)	Thickness (m.)	Cloud's Observed Features	Cloud Type
10,000	500 in vertical extent	Daytime white colour of feathery type, sometimes varying form with straight streak or lines, plumes or curved threads, and these may merge as a thin sheet, Ice crystals compounds.	Cirrus (High clouds)
3000-6000	2 km in vertical and few hundred km in horizontal	Form of layer or sheet composed or rounded masses with some regular structure	Altocumulus (Medium clouds)
	2 km in vertical extent and few hundred km or more in horizontal	Striated or fibrous veil without any structure appearing as thick dark grey sheet, often light to moderate rainfall, they move horizontal extent and last for few days.	Altostratus (Medium clouds)
1500 base	500	Non-rainy fair weather clouds with flat base and heaps like cauliflower tops.	Fair Weather Cumulus
1500 base	500	Sky is covered with same base and same thickness and appear like steps of a staircase. Not raining even in rainy season.	Stratocumulus clouds
300 to 400 base	100	Falling rain evaporates and saturates the air below the cloud system and layered cloud appear	Stratus Low clouds

500	Deep clouds of vertical extent up to 10 km to 15 km, cloud at base 20 to 30 km and top 5-7 km across.	During summer afternoon and rate night over sea, clouds of vertical extent develop with strong winds and carry forward ice crystals giving appearance of high cirrus at top. At such height temperature increases with height flattening the cloud top and blown forward by the wind to appear like a blacksmith's anvil. Such clouds carry electric changes producing lightening and thunder with moderate to heavy rain lasting for an hour or so.	High cirrus cloud at the top side Thunder clouds or Cumulonimbus

Table 9.4

Observed cloud features in relation to weather system

S.No.	Characteristics of Cloud Pattern	Weather System
(a) Cyclone Features		
1.	Inward spiralling cloud bands with small cloud free area at centre (eye).	Cyclone over sea surface
2.	Magnitude of cloud area where the spiral arms have joined to form a continuous cloud deck.	Severity of winds in cyclone
3.	If eye is not present	Depression not a cyclone
4.	Average sized spiral banded cloud system	Wind speed of 100 kmph
5.	When size of dense cloud area is larger than normal in spiral banded cloud system	Wind speed of 150-200 kmph
6.	Cyclone after crossing into the land area weaken into depressions	Wind speed continuously reduces to 40 kmph or less
7.	Horizontal displacement of cyclone over sea or land surface is about 12 kmph	Position of eye is known along the track for monitoring
8.	Monitoring of track direction with movement of cyclone/ depression from the usual west northwest ward to north or northeast ward direction	Change of direction will reflect decrease in displacement of centre in next 24 hrs
9.	Cyclone is embedded with deep convective clouds in cloud disc and spiral arms i.e., of large vertical extent with brighter white spots in satellite pictures	Heavy rains from such brighter spots.

10.	When cyclones/depressions move west northwest wards across Gangetic plain or central parts of the India.	Heavy rains (10 to 15 cm or more in a day) will occur in west and west-southwest sector 500 km ahead of the centre of the system to issue warning for heavy rains.
11.	If the cyclones or depression recurves towards north or northeast	Heavy rains will be expected in northern and northeastern sectors.

(b) Cloud pattern of ITCZ

1.	An east-west oriented elongated low pressure area near the equator such that air from northern and southern hemispheres converge around equator forming a broad clouded 2000 – 3000 km long strip along east-west and 300 to 500 km wide in north-south direction. It is not a uniform clouded area, but rather a globular cloud masses irregularly arranged system appears in beginning of May.	The maximum cloudiness band is known as inter tropical convergence zone (ITCZ) cloud system.
2.	ITCZ cloud system move northwards up to latitude of 10°N or so and dissipate by the 3rd or 4th week of May to produce thundershowers in Sri Lanka and Indian coast with moderate to heavy rains and rain squalls over the area.	Onset of SW Monsoon exact date can be monitored by watching ITCZ cloud zone northwards movement from equator around wind velocity of 100 km a day.
3.	The northward movement of ICTZ cloud system slowed down and remains in quasi-stationary state around Bombay latitude for 4 to 7 days due to larger cloud system over Kashmir and the eastern part of ITCZ cloud system gets a larger northward displacement.	Heavy rainfall over Bombay and adjoining areas around June beginning and 2nd to 3rd week of June rains in central and eastern parts of India
4.	Around fourth week of June the ITCZ band resumes its movement northward for about 15 days depending on the strength of cloud system to cover entire India except western Rajasthan by first week of July in normal course.	Onset of rainy season over different parts of country can be monitored with movement of cloud band towards north.
5.	If the strength of cloud band weakens during its northward movement of ITCZ cloud band may dissipate in its path and we will wait for the next ITCZ cloud band to appear in Indian Ocean around equator.	Break in condition of monsoon.
6.	By beginning of July another ICTZ cloud system would appear near equator and recourse its movement northward march as before taking 35 to 40 days to reach foothills of Himalayas.	Strengthening of monsoon activity.
7.	Three to four times during monsoon months, the eastern end of ITCZ cloud band brightens up as a spiral structure in the cloud band.	Formation of monsoon depressions
8.	Watching carefully northward movement of ICTZ cloud system and eastward movement of Kashmir cloud system in August month, and northeast part, eastern Himalaya will receive heavy rainfall and creating scarcity of rainfall	Break monsoon condition and its possible duration so long as ICTZ cloud system stays in the north-eastern part.

in northwestern and central India. Situation will revive with the arrival of next ITCZ cloud band movement.	
(c) Cloud pattern of western disturbances	
1. By the end of November large cloud systems begins to move from west towards northeast or east across north India north of 30°N. Eastern end of system moves about 800 to 1000 km a day about 10°E longitude per day in the region. Front part of system is thin and fibrous showing high circus cloud patterns giving light rains, whereas moderate rains with strong winds and thunder in the rear part of cloud system.	Western disturbance cloud systems
2. Movement of western disturbance cloud system in satellite pictures	Anticipation of future weather conditions in west and eastwards of the system
(d) Hot summer weather thunderstorms	
1. During the months of April, May and June (northwestern India), there is formation of a vertical convergence zone along the broad belt from Deccan to northern plains of India. Due to summer heating a chain of cumulonimbus, clouds form along this zone and move eastwards to dissipate with tendency to recur along the same chain day after day for two to three days.	In the forenoon cloud free zone, whereas in afternoon cloud patches, and one day's cloud picture will indicate the next day cloud's position to produce thunder showers.
(e) Cloud formation, development and dissipation:	
These features are observed with satellite pictures of earth and atmosphere in the visible and infrared spectra. Picture in the visible region shows cloud formation, their type and movement. The rate of cloud movement will assess wind speed and direction, whereas satellite pictures in the infrared region will assess temperature of earth's surface, temperature of cloud in its different parts, and cloud height.	Temperature of earth surface, cloud temperature and its height, and wind movement.
(f) Abnormal weather events:	
By superimposing the synoptic charts and satellite cloud pictures, weather forecast for occurrence of specific weather phenomenon are predicted such as hailstorm, squalls, flash floods, rainfall amount and time, evapo-transpiration, frost intensity, cloud brusts etc.	Use of surface temperature for specific weather event.
(g) Monitoring of pest and diseases:	
Synoptic charts and satellite pictures are useful in assessing and monitoring of pest infestation and outbreak of diseases. Following examples are illustrated for their use	Confirm with ground truth observations.
1. Assessing rainfall probabilities at seedling stage in groundnut augments the Aspergillus niger – a soil borne pathogen will appear 2 to 3 weeks after	Crown rot disease epidemic in groundnut at seedling stage

	sowing of groundnut due to water logging conditions with continuous rains. (Ramaswamy, 1988; Gadgil et al., 1996).	
2.	A wet spell of 2 weeks duration after pod filling stage in groundnut promoted formation of Cercospora personata sysntoms (Siddaramiah et al, 1983). When the probability of rainfall amount greater than 20 mm per week for two week period is greater than 80 per cent, disease epidemic is alarming.	Incidence of Late Tikka disease
3.	A dry spell during vegetative stage in groundnut with rainfall probability less than 0.4 for a fortnight period is adequate for outbreak of leaf minor causing webbing and dropping of leaves (Gadgil et al. 1996).	Ground leaf minor, Aproacrama modicella Der.
4.	With onset of rainy season testing of water samples for water borne diseases in human beings and animals.	Foot and mouth disease in animals
5.	With onset of monsoon rains after hot summer, the spordic movement of white root grub bettles from soil surface to tree branches.	White root grub beetles.
6.	Rainy season during winter and early spring over southeast Arabia, Iran, Afghanistan and Baluchistan when locust breeding takes place. The swarms produced fly eastwards to lay eggs during SW monsoon in NW India. The next generation swarms return back during autumn to the same area of previous generation.	Desert locust forecasting with advent of rainy seasons.
7.	In early planting of corn, two insect pests – European corn borer (large larve) and crown root worm larvae in the soil emerge in southern part of Michigan stage first and then gradually spreads towards north as the sowing of corn is initiated earlier on southern parts and progressed towards north with favourable weather conditions for corn growth (Gage, 1989).	Spread of European corn borer and crown root worm in corn
	And with normal sowing corn is affected by adults of crown root worm clipping the silks. However under late sown conditions both these pests do not appear.	Adult of crown root worm in corn
8.	Movement of western disturbance from north to northwestern region during winter season with cloudiness, decrease of air temperature in the region by 6 to 10°C, cold wind flow and humidity above 90 per cent deposits the inoculum of mustard aphids in the region influenced by western	Spread and multiplication of mustard aphids, insect pests of Lepidoptera family.

	disturbances. Rapid multiplication with rise of temperature after passing away of western disturbance (Bishnoi, et al. 1992).	
9.	Areal insects of Lepidoptera migrates with favourable weather conditions. Studies of Heliothis sp. indicate that there are large scale movements of moth population (Raulson et al. 1982).	Heliothis have a life cycle of 4 weeks in hot summer to 10 weeks in cooler regions, and needs more sprying in hot conditions.
10.	The trajectory of dispersal of Puccinia graminis tritici uredospore is similar to sequential location of tropical cyclones in India. The deposition of spores occurs exactly below the dissipation of cyclones (Nagarajan and Singh, 1975). Spread of stem rust of wheat from oversummering of Nilgiris to Central India annually takes place due to cyclone movement (Nagarajan and Singh, 1990).	Transport of fungi in relation to wheat rust. Appearance of stem rust can be predicted one month in advance (Nagarajan and Joshi, 1980)

(i) **Agricultural Drought**: Agricultural drought is a complex phenomenon resulted from prolonged moisture scarcity in crop root zone deteriorating soil water availability to plants and heavy reduction of plant productivity. Such features are quite common in arid to dry subhumid areas in tropical to even temperate belts. When rainfall occurrence is not in balance with the requirement of plants, the soil moisture scarcity is developed in the root zone leading to different levels of agricultural drought termed as mild, moderate, severe and extreme categories. These are developed slowly and slowly with cumulative effects of moisture scarcity, and similarly these extremeness of drought conditions vanishes with cumulative build of soil moisture reserve in the crop root zone with the occurrence of rainfall in relation to water need with time. Water deficit can be monitored over short periods (5 to 7 days) to avoid drought through irrigation resources. Thus, drought is a culmination of a set of weather sequences which require extended periods of its development to give a serious drought situation depending on nature and duration of water deficiency. Palmer's concept on drought (1965) is very useful in determining drought's spatial and severity spread, its initiation and termination over region with time. For example using Plamer's concept, the extreme drought developed around Alwar in Feb., 1974 with index value of -4.11, and drought extended to north covering wide areas in Punjab and Haryana by Nov. 1974 touching index value of -5.0, and by December, 1974 Ludhiana and Karnal region acquired drought index of -5.0 in two pockets, but still drought index was less than -4.0 over wide area; drought situation improved due to winter rains by March 1975, but still drought index was less than -3.0, and drought severity further aggravated by June, 1975 with index value touching -5.0, but drought ceased with onset of monsoon in July 1975. (Bishnoi, 1980)

(ii) **Wet spells**: Wet spells occur during rainy season when intense heavy rainfall occurs continuously over a period and there is practically nil evaporation and leaching of soil water leading to heavy surface sunoff. This occurs during the monsoon rainy season when occurrence of heavy downpour due to cloud brust in steep hilly topographic areas or cyclonic storms near the coastal areas. Normally such heavy rainfall occurs instantaneously over a short period to generate a wet spell completely spoiling the crops unless the excess soil water is drained out immediately. With improvement of climatic conditions, potential evaporation of available water at surface will enhance the workability conditions. Depending on the storage of excess moisture different categories wet, moderately wet, severely wet and extremely wet conditions are expressed. Water surplus conditions can be monitored over short periods to control excess water availability in the root zone and to store for recycling during drought periods in the crop growing season. Similar to Palmer's drought concept, wet spells severity and its spread can also be monitored over space and time. For example, extreme wet spells of 1955-56 with sudden development in Sept. 1955 with index value of +4.0 over a wider area in Punjab, in October, 1955 index value touched 10.2 at Ludhiana and slowly diminished to 3.55 in June, 1956. Severe wet spell extended from Pathankot – Amritsar to Ambala – Chandigarh (Bishnoi, 1980) areas.

(iii) **Dry hot winds**: Dry hot winds is a meteorological phenomenon characterised with low humidity (<20%) and high temperature (above 45°) in the near-earth layer, and of intense wind speed common from dry zones. Such dry winds will intensify evapotranspiration rate, creating rapid water scarcity in the root zone damaging plant organs or even mechanically damaging the plant stem. Normally such winds are frequent in arid and semiarid areas during hot summer months in tropical to sub-tropical belts. But in temperate areas dry winds occur as a result of the transformation of air masses (warming) over anticyclone periphery. The degree of dryness of a region determines the agrometeorological criteria of dry winds. In severe dry hot wind conditions evaporation is above 10 to 15 mm/day. Wind speed also intensity the severify of dry wind's impact on crops.

Impact of agricultural droughts and dry wind can be minimised with choice of drought resistant crop species, protective irrigation from stored water or water harvesting during rainy spells to saturate the root zone profile of lower topographic areas or bunding of the farmlands with erection of shelterbelts to reduce to impact of dry winds and evaporation checks. During hot summer months (April to June) the maximum temperatures over north western parts with light sand or sandy soils is above 45°C and more.

(iv) **Cold waves**: During western disturbances cold winds with temperatures below 0°C flow leading to frost, spread of inoculum of pest and diseases in plain areas. A tall shelterbelt perpendicular to these disturbances are useful to control the impact of cold winds as well as lower temperatures.

(v) **Duststorms**: Duststorms are associated with unsuitable modes of cultivation in a climatic region caused by strong winds, loose dry upper soil surface, poor plant cover in vast open land surfaces. These are observed during hot summer season when relative humidity is less than 40 per cent and dry high winds are blowing. Loose light soil particles are carried away along with the wind. Usually the duststorm are followed with light rains due to deposition of water vapours available in atmosphere with transpiration of plants and heavy evaporation from soils of irrigated field during this period with cultivation of cotton and fodder crops in small pockets. To avoid these duststorm it is necessary to carry out programmes with drought resistant bushes and vegetation to stabilize the loose surface soils with moisture conservation measures. During summer hot months over northwest India, there is little moisture in lower atmosphere of 1 to 2 km height above ground, but moisture influx at higher altitude cause a thunderstorm. Raindrops evaporate before they reach to ground, but the cooled air falling strikes the ground, raises dust and rushes forward as a wind squall called "Duststorm". High wind speeds are associated in squalls and occasionally gusty winds are also associated touching 120 kmph.

(vi) **Hailstorm**: Hails develops during the seasonal transition months when thermal convection is very intense which form a massive cumulonimbus form. In these clouds a zone develops at a rate greater than maximum velocity of rising currents where larger drops condense and accumulate to form hail. Lifted by rising air currents to upper portion of clouds where large drops freeze to form hailstone nuclei and grows rapidly as other supercooled drops settle on hails nuclei. The cloud zone where growth of hails occur is known as hail focus. Growth of hailstones occurs rapidly up to 10 to 30 mm size. Higher the velocity of rising currents (>10 m sec^{-1}) longer they rise in upper part of clouds (where cloud peak has a temperature of the order to -25 to -30°C) to form larger the size of hailstones. Hails damage the standing crops creating heavy losses. Hails suppression is possible by retarding the formation process with bombardment of silver iodide or lead iodide. These compounds create a large number of crystallization nuclei onto which the reserves of supercooled water localize to retard hail formation. Hail is a precipitation of solid ice. If a large hailstone is to grow to a diameter of 10 cm then an updraft in excess of 50 m sec^{-1} is required (Ludlam, 1961) with parent air mass be highly unstable. In thunderstorm activity air masses is always associated with a relatively humid layer near the surface and a thick dense layer of cloud exists between cloud base and the freezing level. The terminal velocities of large rain drops are of the order of 8 m sec^{-1}, whereas updrafts in large tropical cumulus clouds are normally three times greater than this. Ludlam (1963) showed that the diameter of large stones can be only slightly diminished by melting. Hailstones of diameter less than 1.5 cm may completely. The frequency of hail and size of the stones have shown a marked correlation with elevation in the tropics. Fawbush and Miller (1953) devised a nomogram to forecast the size of hail from a radiosonde trace adjusted for time of occurrence of storm. The forecast depends on the release of atmosphere's potential energy by taking into account the field of atmospheric motion with assessment of upper air structure with time in future when hailstorm will occur. A rough approximation for diameter (d) in inches of the hailstone was expressed as $(0.008\ bh)^2$, where b and h are base and height of the positive triangle and these are assessed in the nomogram (Fig, 9.2) by adopting following steps.

(a) Evaluate the mean mixing ratio of the moist layer.

(b) Record the temperature at the time of storm.

AEROLOGICAL DIAGRAM

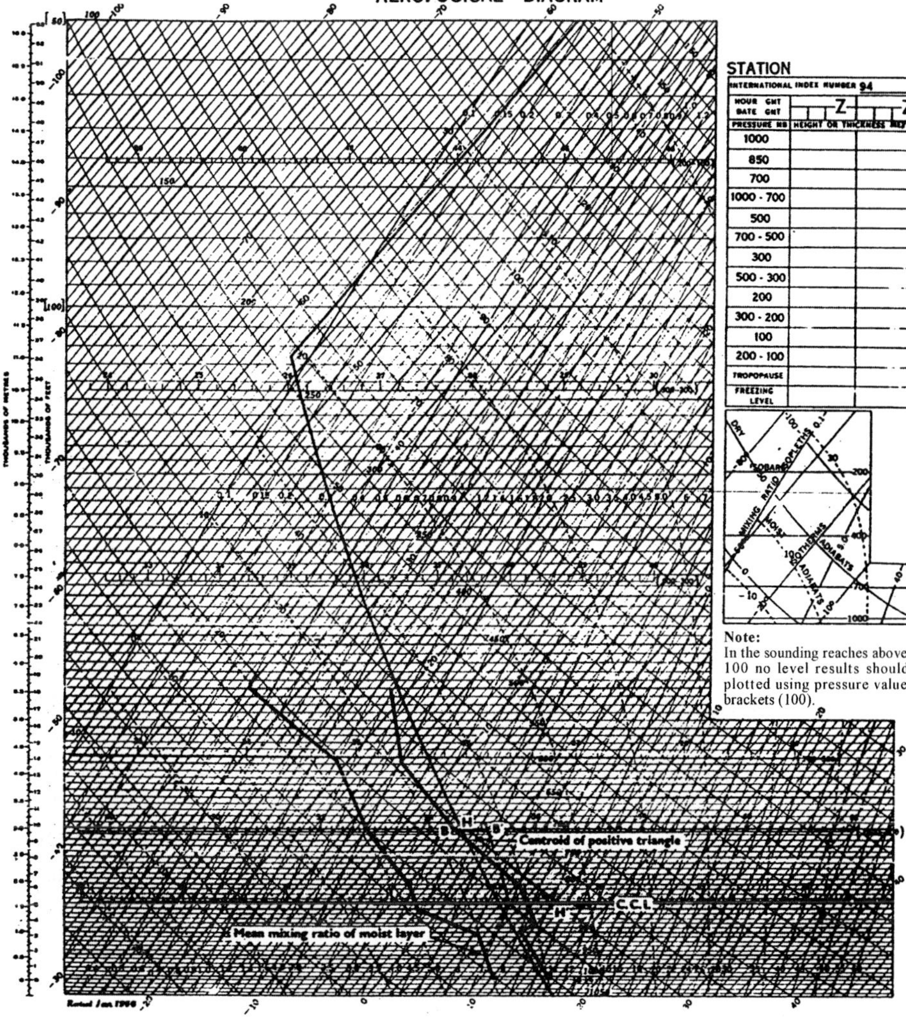

Fig. 9.2: Parameter evaluation for Fawbush-Miller hail-size determination. (After: Fawbush and Miller, 1953)

(c) Using mixing ratio and temperature values determine the convective condensation level (CCL).
(d) Draws the saturated adiabat through CCL through the point where the temperature sounding is -5°C.
(e) The isobar where the temperature is -5° forms the base (b) of the positive triangle.
(f) Calculate the centroid of the positive triangle and draw the dry adiabat through centroid to intersect the -5°C isobar at H and the CCL at H^1. The base (BB^1) and height (HH^1) re measured in °C.
(g) The positive triangle represented the strength of the updraft, whereas the temperature at the top of the triangle will quantify the temperature significant in the formation of hail.

Fawbush and Muller (1953) indicated that the dry air aloft played a critical part in the production of very large hail. The height of the wet bulb freezing level above the terrain is of greatest importance, when this level is below 5000 ft large hail rarely forms, and when it is more than 11000 ft above the terrain hail may form aloft but only small sizes reaching the surface. The optimum height of the wet bulb freezing level for large hail is 8000 ft.

(vii) **Tornadoes**: The conditions favourable for formation of hail are also favourable for the development of tornadoes. The destructive storms-extreme thunderstorm gusts, hail and tornadoes are characterised on the radiosonde trace by pronounced conditional and convective instability through deep layers, irrespective of lack of moisture aloft. Such instability and a powerful lifting mechanism due to vigorous cold front are necessary around the time of maximum heating for tornado formation. There is need of a strong advection of moisture on the low level with dry air aloft. Generally such conditions are available in area 6° latitude N or S of the axis of high or low level jets. Showalter (1953) developed in index value of -6 for tornado occurrence. This index is calculated by obtaining the difference in the potential temperatures between the top of moist layer and the wet bulb freezing level. A zero or negative tornado index is associated with favourable conditions for tornado formation probability. Positive values will show greater stability.

(viii) **Fog**: It is a weather phenomenon when extremely small water droplets remains suspended in the lower atmosphere touching the ground and reducing the horizontal visibility. It occurs when wind is light or calm, relative humidity is above 80 per cent and there is inversion or cold air mass advection in lower air layers touching the ground. Fog develops when air is stable. Fogs are of two types – radiation fog and advection fog. Radiation fog is resulted from rapid night time cooling and water vapours in lower air layer touching the ground condenses on nuclei available suspending in the air such as dust particles, smoke, impurities etc. Radiation fogs are common in winter in low lands valleys where lake/stream is flowing. Advection fogs occur when a warm, moist air flows over a cold snow or water surface. Radiation fogs occur during night or early morning when nocturnal cooling is rapid, whereas advection fog may occur at anytime of the day. Also when a warm moist air is passing over a cold air near the ground, there is condensation of water vapours resulting into a high inversion fog.

Fog restricts horizontal visibility and acts as a cover over and inside the vegetative parts. Fogs are classified as thick fog, fog, moderate fog and thin fog restricting visibility up to 45, 180, 150 and 900 m respectively.

Mist is also a less dense fog where the suspended water droplets restricts visibility double the distance of fog restriction. Here the water vapour particles are small in size and therefore visibility is better than fog. Fog provides a shelter over vegetative surface and is washed away with the warming of lower atmosphere with sunrise and wind motion.

Thick fog is developed with temperature conditions near the ground due to anticyclonic weather conditions. If this situation occurs in industrial areas (Steel or Zinc plants or coal burning sources) will lead to develop a strong air pollution crises influencing the chemical irritation of exposed membranous surfaces causing death due to preexisting cardiorespiratory diseases particularly due to SO_2 with particulates (Heimann, 1961). Within 12 hours of heavy fog and air pollutants accumulation, a large number of people will become ill with respiratory tract symptoms (Fry, 1953).

(ix) **Thunderstorms**: It is an atmospheric disturbance accompanied with thunder, lightening, wind squalls, the towering cloud, sharp rains and sometimes hail. Its characteristics are an immense cumulonimbus cloud followed with rain for short period, sharp drop in temperature and destructive outrushing heavy squall winds before occurrence of rains. Normally thunderstorms occurs all over with decreasing frequency with increase in latitude. Thunderstorms are of two types due to local and cyclonic disturbances.

(a) **Local thunderstorms**: With strong heating of the ground surface, the overlying air is heated to form local convection (Fig. 9.3). The expansion of heated air and its forced ascent results into cooling of air at an adiabatic rate of 10.0°C per km. until dew point is reached (10 to 12 km). Therefore, a cumulus cloud is formed at this level which continues to develop so long as warm air is rising. Cumulonimbus clouds are formed with condensing of upward motion of the humid air. Raindrops are formed by condensation of water vapours in the foreword portion of thunderstorm. In this strong upward motion of relatively warm air in the front and its mixing with cold decending air, a squall or round cloud is formed at lower attitude. Also the falling raindrops are broken up and are electrified positively. The air with its negative charge passes on the other parts of the clouds. Large quantities of positive and negative charges accumulate to develop electric forces to create lightening flash. With lightening air again

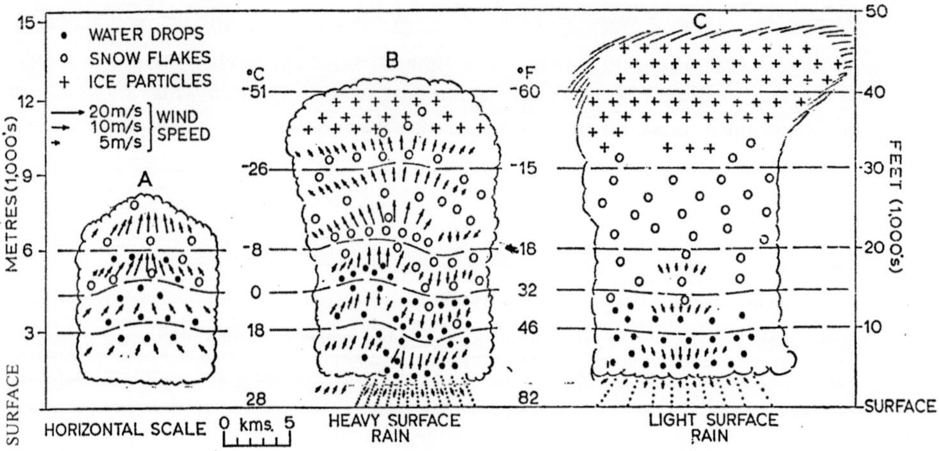

Fig. 9.3: The cycle of thunderstorm development. (a) developing stage, (b) mature stage with updraughts and downdraughts, and (c) dissipating stage dominated by cool downdraughts. (Adapted from Barry and Chorley, 1971)

expands due to warming leading to dissociation of atoms and molecules of air with large thunder noise. Rising warm air mixes with the cold decending air, a squall cloud develops at a lower altitude and moves along the front of thunderstorms. The electric discharges produce oxides of nitrogen which are washed down to earth by rain as nitrogenous fertilizer. Therefore, thunderstorm acts as a valuable friend of farmer. Thunderstorms move in a direction with a speed determined by the mean motion of the atmospheric layer from ground to 15-20 km height. A average thunder cloud is 20-30 km across with average horizontal speed of 20 km ph. In some thunderstorms of vertical exten hailstone is formed which is frozen rain water. A 5 cm diameter hail will have a fall velocity of about 100 kmph. Thunderstorm forecasting is a difficult task and needs the expert skills to predict it. To predict severe thunderstorm activity detailed synoptic observations on expected maximum temperature (above 27°C), trajectories of low level moisture advection (dew point above 10°C), isallobars, proximity of upper trough, proximity of frontal trigger as upmotion agent, trough aloft (cool pool – as severe storms are about 2 to 14° longitude ahead of trough), proximity of low and high level jets if any etc. Similarly radiosonde data is analysed for existing and adjusted traces to investigate past history, absolute stability and instability, conditional instability, lifting condensation level (LCL), convective condensation level (CCL), level of free convection (LFC below 700 mb for tornado formation), height of wet bulb freezing level above locality (8000 ft optimum), relative humidity above 75 per cent from CCL to 500 mb.

Indices have been developed to predict thunderstorm activity resulting from convection, fronts or troughs, non-frontal phenomenon. Galway (1956) determined mean mixing ratio and potential temperature in lower 900 m for the time of maximum heating to locate the mean values of lifting condensation level (LCL) where saturated adiabat through this point intersects 500 mb level. The 500 mb temperature will indicate the updraft temperature within a cloud if developed. Index value is determined by subtracting the updraft temperature from the environmental temperature. High values of Rackliff (1962) index (ΔT) will show instability (>30 critical in non-frontal situations) assessed as

$$\Delta T = \theta_{w900} - T_{500} \qquad \ldots (9.10)$$

where, θ_{w900} is 900 mb wet bulb potential temperature (°C) and T_{500} is 500 mb dry bulb temperature. Boyden (1963) developed index (I) for advective change in upper air (upto 700 mb) as (Z – T – 200), where Z is 1000 – 700 mb thickness (in decametres) and T is 700 mb dry bulb temperature in °C. The threshold values for thunderstorm was given at 94, and therefore isopleths of the index on charts will show its movement with the 700 mb wind.

Jefferson (1963) modified Rackliff's method with use 700 mb dew point depression to allow for dry air in middle levels. The index (T_j) was expressed as

$$T_j = 1.6\,\theta_{w900} - T_{500} - \tfrac{1}{2}T_{d700} - 8 \qquad \ldots (9.11)$$

where, T_{d700} is 700 mb dew point depression in °C. The threshold value was taken as 28 for thunderstorm activity.

Fawbush – Miller (1954) developed method for hot weather air mass thunderstorms with turbulence extending to the surface. The air of downdraft (near freezing level) warms wet adiabatically as it descends through the cloud. Since rains below the cloud base saturates the air, the lowest temperature during the rain (downdraft) at the ground will be almost the potential wet bulb temperature corresponding to zero wet bulb. The level at which the wet bulb temperature is zero is called wet bulb freezing level. The drop in temperature during thunderstorm rain is closely related to strength of a peak gust at the surface. Fawbush and Miller (1954) used this relationship to forecast the strength of thunderstorm gusts using the nomogram. The peak gust at surface was expressed as:

$$\text{Peak gust} = 7 + 3.06t - 0.0073\,t^2 - 0.000284t^3 \qquad \ldots (9.12)$$

where, t is the difference between the surface temperature just prior to the storm and immediately after the first shower. Strong gusts associated with non-frontal thunderstorm were expected at the earth's surface only in areas where the wet bulb freezing level is below 11000 ft. High wind speeds are experienced in squalls associated with thunderstorms which last for a few minutes in localized area. Sometimes gusty winds are also associated with squalls.

(b) **Cyclonic storm**: It is an intense low pressure system with pressure deficiency at centre varying from 20 to 50 mb, and a large revolving mass of air about 5 km thick and a diameter varying from about 200 to 1000 kilmeters encounting heavy winds above 100 kmph to 150 kmph, and occasionally 200 kmph (Fig. 9.4). In northern hemisphere, it has been observed that winds around a low blow with low pressure to the left, the cyclone rotates in anticlockwise direction with rings of maximum winds of 100 km in diameter and wind speed decrease both inwards and outwards from the ring. The system moves as whole covering about 200 to 400 km a day with displacement of centre along the track of movement. The rings of storm or hurricane winds are associated with heavy torrential rains touching even 20 to 30 cm per day. Such intense rains with fierce winds will damage the exposed property, standing crops and recharging the soils with sea water salt contents. Moore and Osgood (1985) assessed the sugarcane crop damage and sugar loss caused by high winds of hurricane in terms of sum of losses due to stalk growth termination, metabolic depletion and reduced assimilation, and reported that estimated loss was within 10 per cent of the apparent losses at the mill. Normally a few hours before the centre of a severe cyclonic storm over the sea crosses the coast, there is rapid rise in sea level and a surge of sea water of 5 to 10 m. due to fierce winds may takes place known as storm surge or tidal wave. This would continue for 2 to 3 hours and thereafter rain would cease with clear sky and winds weakening to almost a calm.

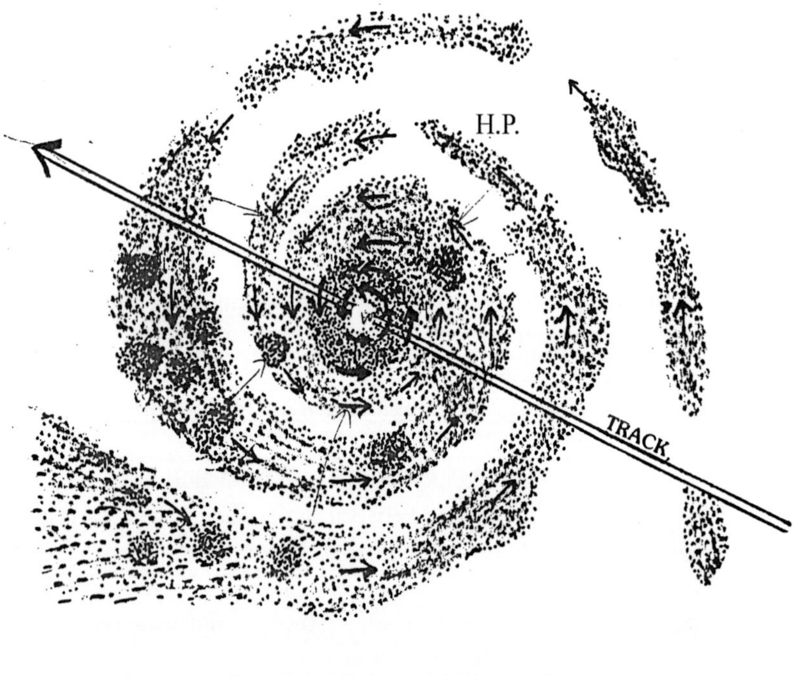

9.4: A horizontal section of tropical cyclone (After: Pisharoty, 1986)

Then all of a sudden the sky would become overcast with commencement of heavy rains and blow of winds from opposite direction due to the other part of the ring of storm or hurricane winds over the location for about 2 to 4 hours. Winds will gradually decrease with decrease of rains turning to a normal situation with passing over of the cyclone. Southern India from Bay of Bengal side experiences a number of cyclones (Fig. 9.5). High wind speeds are encountered in gales associated with

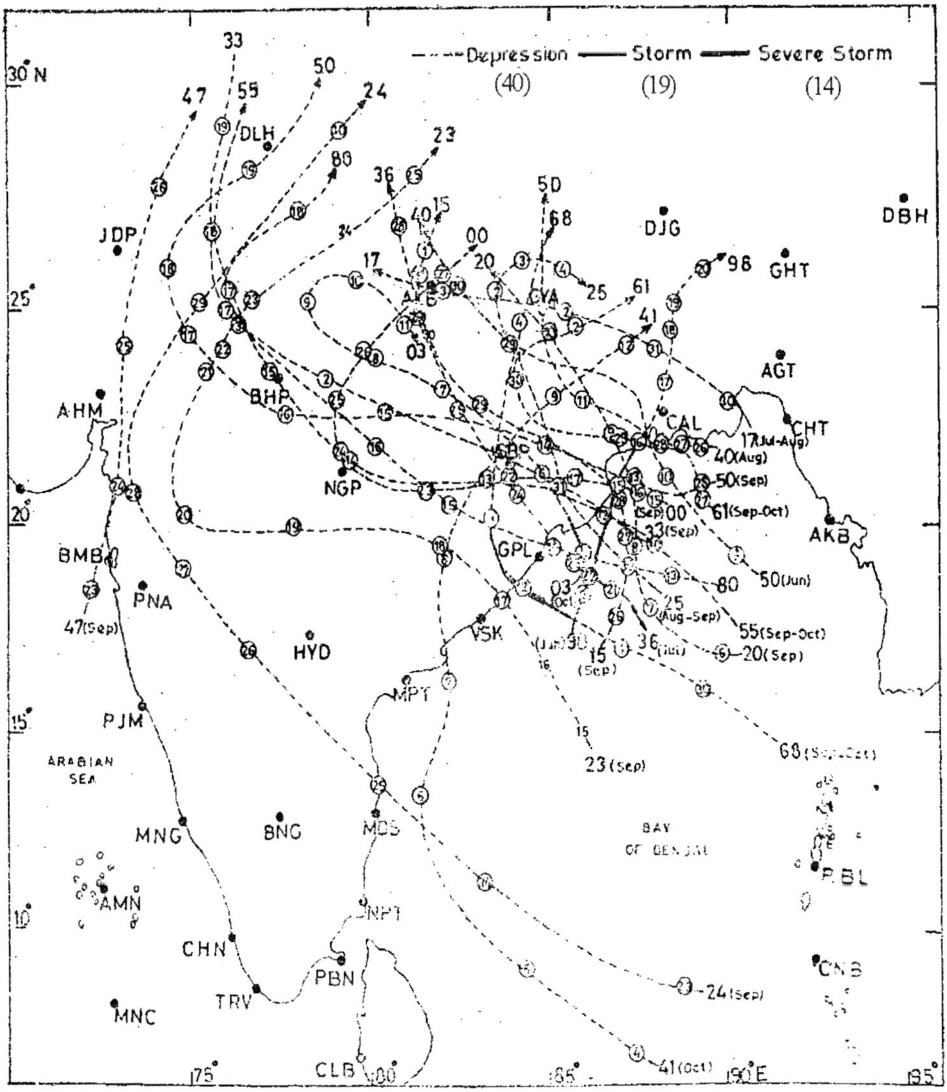

9.5: Storm tracks of recurving disturbances towards north India from Bay of Bengal. (Source: Dhar and Nandargi, 1993)

cyclonic storms and last much longer period (few hours) and are more widespread. Satellite monitoring of their trajectories is essential at hourly intervals to avoid heavy losses.

Such large scale tropical disturbances originate over the warm oceans where abundant of atmospheric moisture, sea surface temperatures above 26/27°C and practically little frictional surface resistances to the wind flow. For forecasting these cyclonic activities observations are recorded for closed circulations, isallobars, pressure regions below 1000 mb, significant increase in wind strength or sea tidal waves, upper level wind flow, jet streams isotherm patterns, movement of system etc. to develop forecasting indicators.

(ix) **Excessive heavy rainfall:** Rainfalls of 40 to 80 cm a day occur at one or two places every year in India due to monsoon depressions or cyclones. Such rainy spells of excessive rains are of continuous spells of 2 to 5 days touching a total of above 100 cms. As an example Prasad et al. (1994) reported a heavy rainfall occurance during 9-10 July, 1990 due to cloud brust over Pauri-Garhwal area with development of synoptic situation causing heavy rainfall (Fig. 9.6). The cyclonic circulation embedded in monsoon trough and extending up to 3.1 km above sea

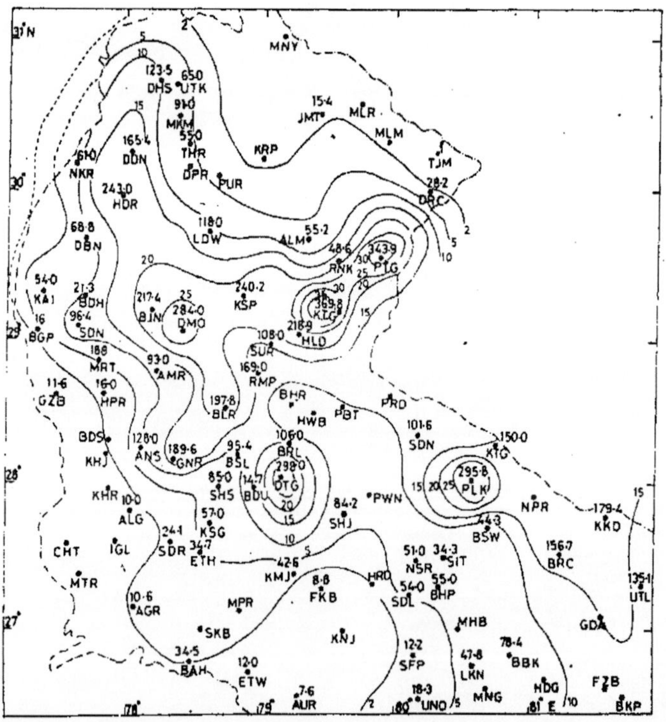

9.6: Isohyetal pattern or rainstorm on 9-10 July 1990 (After: Prasad et al. 1994)

level with a trough aloft. The lower level convergence area was overlain by upper air divergence. This caused heavy landslides and casualities. 21 to 37 cm rainfall occurred in day. Normally a heavy rainfall is preceded or followed by another of nearly same intensity. Such heavy rains occur in short spells of only a few hours duration each. Dhar et al. (1993) studied spatial distribution of 97 severe rainstorms of India during 1980 to 1990 (Fig. 9.5) and reported that most of rainstorms have preferred zone of occurrence corresponding to the orographic boundaries where they occur, and there are cells of heavy rainfall.

(x) **Extremes of temperature**: During the hot weather period April, May and June maximum temperature over different parts of northern India experiences temperatures above 45°C. Pisharoty (1986) reported highest maximum temperatures in different parts of India, whereas during cold weather period January and February minimum temperatures over Northwest India were 5°C or less. Since grass minimum temperatures are about 4 to 4.5°C less than their corresponding minimum temperatures, so even less than 0°C grass minimum temperatures are experienced such that extreme northwest part of India is liable for frost in cold season with 5 to 20 days in each month from Nov. to Feb. months.

(xi) **Excessive or defective insulation**: Excess radiation heat during summer or defective insolation during monsoon rainy and winter seasons needs to be monitored to provide optimal radiational needs of the plant. Excess radiation load in summer would need heavy irrigation and protection by shading of tall tress, whereas less radiation will provide reduced photosynthesis, delays maturity and reduces dry matter and grain yield productivity.

9.2.5 Pre-season long range forecasting

Section 7.15 have been already dealt to forecast the monsoon season rainfall in India using multiple regression models, power regression models, parametric models, and dynamic stochastic transfer model predicting seasonal total rainfall. These models have been used in India predicting seasonal rainfall quite accurately within the variations of rainfall features. Sections 7.10 to 7.12 have also clearly brought out that La Nena phenomenon in Eastern Pacific Ocean is strongly linked with above normal rainfall, whereas El Neno are linked with drought conditions over Asian region completely ruling out the possibilities of excess rainfall in India. El Nino and La Nena signals are very useful in assessing the monsoon variability and should be used as a forecasting tool as they are initiated at last 4 to 6 months in advance. Recent advances in understanding the ocean – atmosphere processes have further improved (GCM models) the accurate seasonal forecasts. For example, ENSO phenomenon is strongly related to the reversal of the easterly trade winds in the Western Pacific, and thereby is associated with extreme weather events such as floods in peru and drought extending over the southwest Pacific, northwest Australia, southeast Asia and Indian sub-continent. The status of ENSO is reported in monthly World Climate Bulletin of WMO and even the sea surface temperatures are modelled by ENSO models upto one year in advance, and can be used for yield forecasting well in advance. Therefore, such advance warning of drought or favourable season would enable us to take useful crop management decisions in different regions such as the inputs required in agriculture in terms of seeds, fertilizer, agrochemicals, amount of irrigation water etc. Further agricultural policy management can be thought of well in advance to

(i) increase or decrease area under different crops in different regions,
(ii) increase or decrease of livestock population keeping fodder production in view,
(iii) choice of drought tolerant or flood tolerant crops and their varieties,
(iv) management of irrigation water over different regions keeping in view of choice of crops and intensity of cropping planned,
(v) to manufecture/distribute greater or smaller quantities of costly inputs in terms of seeds, fertilizers and agrochemicals, and
(vi) to manage in advance about the food shortage or sale of produce.

Therefore strategic long term planning of cropping systems can be developed keeping in view of available agroclimatic resources of moisture and thermal conditions. Long term climatological data is analysed to quantify the crop growing durations to fit in the short, medium or long duration varieties of different crops. Early sown (first) crops and late sown (second) crops are both exposed to greater risks of water deficits than crops grown in the middle of the rainy season. Angus et al. (1987) linked the water balance model with double rice cropping in rainfed environment of Phillipines (Bolton and Zandstra, 1981) and then linked with long term weather data to test the double rice cropping for 58 years. It was reported that model simulated growth of a single crop in all the seasons, but a second crop was possible in a relatively few years. However, when the time of establishment of first crop and the turn around period between crops was varied from 30 to 10 days, the number of simulated second crop increased from 5 to 55 per cent years. The simulated yield also increased in 10 days over the longer 30 days turn around period. Therefore a quick field management is required to reduce the turn around period after the first crop to sow the second crop within 10 days to utilize efficiently the seasonal environmental resources. Section 7.16 shows how farmers can alter nitrogen management strategies with early, medium and late sown cultivars in crops to draw the benefits of El Nino/La Neva years with use of climatic scenarios. Such applications should be developed in different parts of Asia, Australia, Africa and South America regions to develop appropriate management strategies to harness the benefit of natural clamities in monsoon system.

9.3 Agrometeorological Forecast Services

9.3.1 Crop development and phenology

Development is the ontological change in an organism as it progresses through the life cycle i.e., change from seed to vegetative, vegetative to reproductive habit of plants. Similar changes takes place in pest and disease organisms such as changing instars of insects and latent period of fungus organisms. In crops the development process is called phasic development i.e., progress of life cycle through seeding to emergence, emergence to tilling, tillering to vegetative, vegetative to anthesis, anthesis to reproductive. Morphological development in plants refers to initiation and termination of each plant organ from leaf to seed. Phenology refers to the study of these changes of biological forms.

The product of the duration of these inter-phases (D) and the temperature above a threshold level ($T - T_b$) was constant called thermal time or heat units– a characteristics of the species.

$$D(\overline{T} - T_b) = k \text{ (constant)} \qquad \ldots (9.13)$$

This expression can be expressed in terms of daily rate of development (RD) which is inverse of the phase duration i.e.

$$RD = \frac{1}{D} = (\overline{T} - T_b)/k \qquad \ldots (9.14)$$

This linear relation between RD and $\overline{T}$ is useful in quantifying the threshold temperature (T_b) and k determining the thermal time. However, in warm environments the physiology of organism may be a non-linear response such as in rice (Yashida, 1981) in the humid tropics. Also when these is a photoperiod response, the pattern may not be linear due to short day or long day type. In such situations it is useful to describe the combined effect of temperature and photoperiod on phasic development with defining the threshold P_b as

$$RD = \frac{(T - T_b)(\overline{P} - P_b)}{k'} \qquad \ldots (9.15)$$

where k' is photothermal time constant of the phase duration.

Temperature and photoperiod are main weather factors controlling phasic durations of different crop species. There are three significant levels of temperature for growth and development of plants called cardinal points. Similarly photoperiod do have threshold, optimum and inversion levels for photoresponsive crops. For tropical to subtropical crop species the cardinal levels are 8 to 12°C, 25 to 30°C and 45 to 50°C respectively, whereas for temperate and cold regions there levels are 0 to 5°C, 15/20 to 25°C, and 40 to 45°C respectively. Angus et al. (1980/81) reported variations in threshold temperature and total heat units required for seedling emergence phase of different crops, and also highlighted that the threshold temperature steadily increased through the ontogeny of wheat species with values of 2.6 ± 0.2, 3.3 ± 0.7°C, 5.1 ± 2.1, 3.5 ± 0.6°C, and 8.9 ± 2.6°C during sowing to emergence, emergence to floral initiation, floral initiation to anthesis, emergence to anthesis, anthesis to maturity phases respectively. Angus et al. (1981), while studying the phasic development of spring wheat in a multilocation trial between latitudes of 14° to 66° suggested that the development responses to both temperature and photoperiod were non-linear between seedling emergence and anthesis, and similar to threshold temperature there exists a critical photoperiod during pre-anthesis period which varied between 7.95 to 9.16 hours in two contrasting development features of cultivars and varied from 4.77 ± 2.68 hours at emergence to floral initiation period to 9.27 ± 0.45 hours at floral initiation to anthesis period. There are several factors which influence the thermal time or duration of different plant processes in phasic durations such as variations in threshold temperature, threshold photoperiod, vernalization and genotypes of short, medium and long durations particularly hybrids.

Phenological forecasts i.e., the onset dates of the main phases of crop growth in the prevailing environment, each of which has a different climatic requirements. Since agricultural requirements of crops are not identical in different periods of vegetation, it is assumed that moisture availability is not a constraint so that it is easy to assess an overall assessment of the conditions of crop formation with onset of main development phases. It is useful to issue independent phenological forecasts for farmers such as forecasts of periods of crop maturity, flowering period of fruit crops etc. in different thermal belts. In order to quantify the dates of onset of the main development phases of crops (emergence, jointing, heading, flowering, anthesis, soft dough, hard dough) and the duration of interphase periods, two approaches are commonly used.

(i) Accumulation of active temperature: The growth rate of plants and period of onset of growth phases depend mainly on the environmental temperature. Accumulation of temperature is done by various methods. Accumulation of active temperatures involves the summing up of the mean diurnal temperatures commencing from the day when this temperature exceeds that at the commencement of plant growth. However this method does not take into account the temperature on days when it becomes less than the known biological minimum of a given plant. The probable dates of onset of growth stages (D_2) by the method of active temperatures is determined by knowing the date of the onset of the preceding phase (D_1) and the total of active temperatures required for reaching the desired interphase period. The mean diurnal temperature exceeded the lower limit for the growth of a given crop (T_l). The actual mean diurnal temperature until the day of forecasting are added and the anticipated temperatures for the rest of future period are taken from the short and medium weather forecasts. The day when the temperature total adds up to the value required for the interphase period can be regarded as the day of onset of the anticipated phase.

$$D_2 = \sum\left(\frac{T_{max} + T_{min}}{2}\right)\Bigg| T > T_e + D_1 \qquad \ldots (9.16)$$

(ii) Accumulation of Effective Temperature: To determine the duration of interphase periods by summing up of effective temperature, the expression used is

$$D = \frac{A}{(\overline{T} - B)}, = \frac{A}{\overline{T} - T_l} \text{ where } \overline{T} - T_l = 0 \text{ if } T_l > \overline{T} \qquad \ldots (9.17)$$

where D is duration of the interphase period, A is total effective temperatures or thermal time required for the onset of a given stage from the level B which represents the lower temperature limit at which the processes of a given stage commence and $\overline{T}$ is mean diurnal air temperature over the duration period (Robertson, 1968; Wang, 1967).

Therefore the main difference between the totals of effective and active temperatures is that in computing the total of effective temperatures the lower growth limit is deducted from the mean diurnal temperatures, while in computing the total of active temperatures the whole of them are taken and not exclusively on those days when they exceeded the lower growth limit.

The lower limit of development is about 5°C for plants of temperate climate in vegetative period, whereas for plants of tropical and subtropical climates it is taken as 10°C in vegetative period. Thus using effective temperature values the date of occurrence of different phenological stages and their duration can be determined.

Example 1:

Calculate the probable date of hard dough of winter wheat variety WH147 when ear emergence occurred on 10 February and it is known that this variety needs an effective temperature of 500 day °C (with threshold lower temperature limit of 5°C) is required from the stage of ear emergence to hard dough of wheat. Forecast data is available from 15th March for 10 days period. The observatory actual temperature is available before 15th March as given below:

Date	Tmax(°C)	Tmin(°C)	Effective Temperature (°C)	Accumulated Effective Temperature (°C)
Feb. 11	19.6	3.8	6.7	6.7
12	14.1	7.5	5.8	12.5
13	16.3	5.5	5.9	18.4
14	20.1	4.5	7.3	25.7
15	22.1	5.0	8.5	34.2
16	22.6	5.4	9.0	43.2
17	26.0	10.4	13.2	56.2
18	25.4	12.0	13.7	69.9
19	28.3	12.5	15.4	85.3
20	27.3	9.0	13.2	98.5
21	28.0	9.7	13.8	112.3
22	29.0	15.5	17.2	129.5
23	27.0	14.1	15.5	145.0
24	17.0	8.9	8.0	153.0
25	24.9	7.4	11.2	164.2
26	22.5	8.5	10.5	174.7

27	22.0	8.3	10.2	184.9
28	23.1	8.8	11.0	195.9
March 1	25.1	13.1	14.1	210.0
2	23.5	13.7	13.6	223.6
3	20.4	7.5	9.0	232.6
4	23.0	6.5	9.8	242.4
5	24.0	7.1	10.5	252.9
6	25.5	5.6	10.6	263.5
7	27.1	7.1	12.1	275.6
8	28.0	13.6	15.8	291.4
9	27.0	14.0	15.5	306.9
10	28.5	13.5	16.0	322.9
11	27.0	8.6	12.8	335.7
12	25.9	8.6	12.2	347.9
13	27.4	10.1	13.7	361.6
14	28.2	7.9	13.1	374.7
Forecasted data on medium range forecast:				
15	28.3	8.5	13.4	388.1
16	29.5	8.9	14.2	402.3
17	31.4	10.0	15.7	418.0
18	32.0	10.5	16.2	434.2
19	33.1	11.5	17.3	451.5
20	33.2	12.5	17.8	469.3
21	35.4	16.4	20.9	490.2
22	32.3	13.5	17.9	508.1

Solution:

$$\text{Date of Hard dough} = \frac{500}{\sum\left(\frac{(T_{max} + T_{min})}{2} - T_l\right)} + \text{Date of ear emergence}$$

$$= \frac{500}{\sum_{i=1}^{n}(\overline{T}_i - T_l)} + \text{February 10}$$

$$= \text{40th day} + \text{February 10}$$

Date of Hard dough = March 22.

Example 2:

Compare the values of effective and active temperature values for the data given below when T_l is 5°C

Solution:

Day	T_{max}(°C)	T_{min}(°C)	T(°C)	Effective Temperature (°C)	Active Temperature (°C)
1	12.8	3.8	8.3	3.3	8.3
2	12.0	5.6	8.8	3.8	8.8
3	10.6	3.0	6.8	1.8	6.8
4	10.2	-1.0	4.6	0.0	0.0
5	10.4	-2.4	4.0	0.0	0.0
6	11.0	-2.8	4.1	0.0	0.0
7	10.0	-4.0	3.0	0.0	0.0
8	8.0	1.0	4.5	0.0	0.0
9	10.6	2.0	6.3	1.3	6.3
10	12.4	0.4	6.4	1.4	6.4
11	8.7	-2.3	3.2	0.0	0.0
12	9.0	2.0	5.5	0.5	5.5
13	10.0	-2.0	4.0	0.0	0.0
14	12.0	1.2	6.6	1.6	6.6
15	14.0	0.0	7.0	2.0	7.0
Total				**15.7**	**55.7**

Example 3:

Calculate the duration required from ear emergence to grain filling stage in pearl millet with threshold temperature of 10°C and total effective temperature required from ear emergence to grain filling in hybrid pearl millet is around 250 day °C.

Solution:

Day	T_{max}(°C)	T_{min}(°C)	T(°C)	Effective Temperature (°C)	Cumulative Effective Temperature (°C)	Mean Temperature (°C)
1	36.0	25.8	30.9	20.9	20.9	30.9
2	35.5	25.7	30.6	20.6	41.5	30.7
3	34.4	27.6	31.0	21.0	62.5	30.8
4	36.3	27.5	31.9	21.9	84.4	31.1
5	35.3	25.9	30.6	20.6	105.0	31.0
6	37.2	27.4	32.3	22.3	127.3	31.2
7	38.4	27.8	33.1	23.1	150.4	31.4
8	36.2	24.6	30.4	20.4	170.8	31.3
9	31.5	26.5	29.0	19.0	189.8	31.0
10	35.2	26.2	30.6	20.6	210.4	31.0

Medium range forecast for next days						
11	33.7	26.3	30.0	20.0	230.4	30.9
12	35.0	26.8	30.9	20.9	251.3	30.9
13	35.7	24.9	30.3	20.3		
14	35.9	26.5	31.2	21.2		

Duration of interphase between ear emergence to grain filling

$$= \frac{250}{(\overline{T}_i - T_l)} = \frac{250}{(\overline{T} - 10)} = \frac{250}{20.9}$$

$$= 12 \text{ days.}$$

One of the important modification of the degree day concept is the introduction of the photoperiodism factor in the

$$A' = L.\sum_{i=1}^{n}(T_i - T_l) \qquad \ldots (9.18)$$

where $T_i - T_l = 0$ if $T_l > T_i$, and L is the average daylength during the phenological period and A' is called photothermal constant or photothermal time for photosensitive plants and varieties. The photothermal response of plants may be modified by other environmental variables such as soil moisture status, radiant energy available. Robertson (1968) proposed a non-linear model as the plants do not response to temperature, photoperiod, sunshine, radiant energy etc linearly, but possess a lower and a upper critical limits. There exists an optimum value of each variable for growth, development and yield potential called cardinal points which also change during the life cycle of the crop. For cereals five phenological periods are adequate to provide homogenous physiological periods for phenological studies i.e., planting to emergence, emergence to jointing, jointing to heading, heading to soft dough, and soft dough to ripening.

9.3.2 Soil moisture

The available moisture in top soil is particularly useful for farm operations and agricultural planning. The climatological water balance procedure is very useful in quantifying the soil moisture status depending on rainfall pattern, evapotranspiration losses at the location. The soil moisture status during the crop growing period is useful in regulating the inputs and farm operations. Therefore, keeping rhythum with the track of soil moisture condition is essential in supplying nutrients, scheduling of irrigation, growth and development characteristics of the plant. Baier and Robertson (1966) and Baier et al. (1972) developed a versatile soil moisture budget using 6 layers of soil depth with climatological water balance procedure. For operational purposes of crop conditions with soil moisture availability in USSR (Fedoseev, 1961; Kulik and Sinelschikov, 1961) the following regression procedure was developed

$$\Delta SM = AW + BT + CR + D, \qquad \ldots (9.19)$$

where ΔSM is the change in soil moisture (mm) of a 1 m. soil layer over a 10 days period, W is the soil moisture reserves (mm) at the beginning of 10 days period, T is mean temperature (°C) over 10 days period, R is total precipitation (mm) over 10 days period, and A, B, C and D are regression coefficients as given below. Fig. 9.7 shows the perennial productive moisture reserves in one metre layer of soil for different crops.

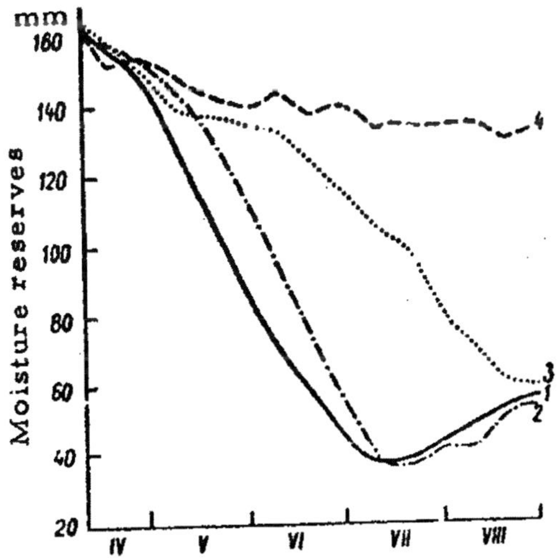

9.7: Average perennial productive moisture reserves in meter layer of the soil (mm) for different crops in the root zone of (1) for winter wheat, (2) for spring wheat, (3) for maize, and (4) for black fallow. (After: Kulikand Sinelsnchikov, 1961)

Soil moisture conditions	A	B	C	D
Greater than 60 per cent of field capacity	-0.36	1.98	1.64	32
Between 60 per cent of field capacity and willing point	-0.22	0.63	0.49	8.2
Less than willing point	-0.15	-0.03	0.49	1.4

Field operations, cultural methods, supply of nutrients and agrochemicals, scheduling of irrigations are strongly dependent on the available soil water in the 1 m. soil profile in a cropped field, therefore monitoring of soil moisture status during the crop growth and development periods is very essential. Over large basis surface soil moisture status can be monitored at regular intervals with microwave radar technology.

Palmer's concept on crop moisture index is also useful in monitoring crop moisture conditions (Fig. 9.8) with the initiation/fall of drought and wet spells with time (section 3.4.4). When used over weekly periods this procedure is very useful in advising farmers about the crop moisture conditions as described below (Table 9.5). Weekly maps are being used in agroadvisories for farmers (Palmer, 1968) in U.S.A. (Bishnoi, 1980).

Table 9.5
Crop moisture indices with crop moisture condition

Palmer's Crop Moisture Index (PCMI)	Crop Moisture Index (CMI)	Notation	Crop Moisture Conditions
>3.0	>4.0	A	Still excessively wet, some fields flooded, high humidity, pest and diseases needs attention immediately.
2.0 to 3.0	3.0 to 4.0	B	More dry weather needed, fields wet in spots, pest and diseases need attention.
1.0 to 2.0	2.0 to 3.0	C	Favourable for crop growth, development and reproduction; prospects good yield potential, optimum soil moisture conditions.
0 to 1.0	1.0 to 2.0	D	Favourable for crop growth and field work, adequate moisture for immediate needs.
0 to -1.0	0.75 to 1.0	E	Top soil moisture inadequate, needs rain, crops can be sown, land preparation feasible.
-1.0 to -2.0	0.50 to 0.75	E’	Too dry, prospects decreases, needs irrigation, drought conditions established, moisture stress features visible.
-2.0 to -3.0	0.25 to 0.50	E”	Too dry, prospects reduced, drought conditions must go to protect heavy losses.
<-3	<0.25	E’’’	To dry, crop yields severely reduced, serious drought situation, irrigation badly needed.

(After: Bishnoi, 1980)

9.3.3 Field operations

Favourable temperature, soil moisture and wind are important meteorological variables for agricultural farm operations as supported with data in Table 9.6.

Table 9.6
Weather variables favourable for different field operations

Specialized field operations	Favourable climatic conditions
Field preparation	Temperature optimum required for crop, soil moisture in top layer about 75 to 80 field capacity. Soil climate plays important role (Chirkov, 1979)

Choice of crops	Favourable climatic conditions over the crop growth duration /season, particularly temperature, precipitation and water requirement of crop, rainfall probabilities over 5 to 7 days period.
Crop planting	Favourable soil temperature at 5 to 10 cm depth of soil surface and 70 to 80 per cent of field capacity soil moisture for at least 4 to 5 days of germination.
Weeding	Crop varieties based on seasonal outlook for rainfall probabilities, after germination of crops the early weeding is most beneficial i.e., after every effective rainfall event.
Fertilizer application	Track of soil moisture conditions during vegetative, flowering and reproductive phenophases; soil and air temperature optimum; low wind speed; rainfall probabilities.
Agrochemical applications	Optimum temperature for crop, low wind speed, no precipitation during next 24 hours.
Irrigation scheduling	To maintain optimum soil moisture conditions above 50 per cent of available water in the root zone profile, weekly water budget during crop growing season is necessary, dry/wet spells durations.
Crop harvesting	Good sunshine and drying rates of crops are needed, precipitation or dew probabilities be low.
Transportation of agricultural	Maximum – minimum temperature range be less, no weather hazards, optimum temperature needed to avoid deterioration of produce and its spoilage.
Control of insect, pests and diseases	Life history of insect, pests and disease organisms and their food habit requirement be matched with phenology of crops, temperature, relative humidity, dew, wind pattern are important variables.
Crop spraying for control of pests,	Spraying be preferred when wind speeds are minimum (below 12 kmph), air temperatures are moderate (10 to 30°C), relative humidity is low and practically lower rainfall probabilities. Wind direction should also be taken care so that only the targeted areas are sprayed with agrochemicals. In winter spraying be done after the evaporation of dew deposited on plant organs.
Wind breaks and shelterbelts erection	Windbreaks/shelterbelts modifies the microclimate in leeward direction favourable for improving soil moisture conditions with reduction of wind speeds for improving water use efficiency and better growth and development of crops.
Thinning of crop	Rainfall probability decreases with advancement of season, soil moisture depleted.

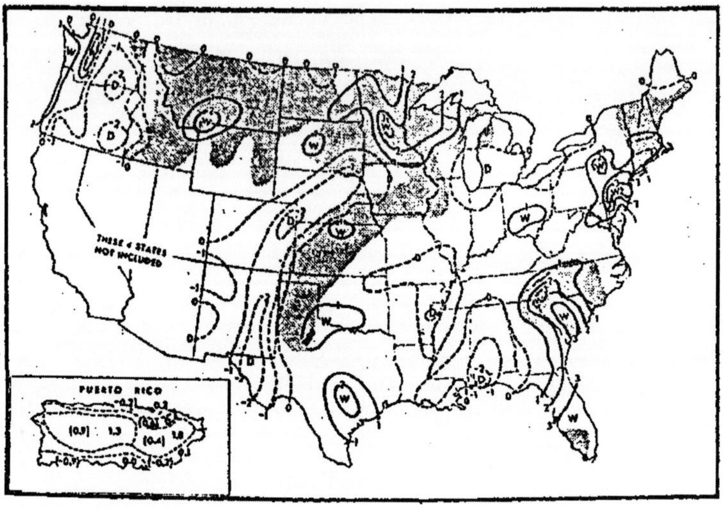

9.8: Palmer crop moisture index variations (After: Palmer, 1968)

9.3.4 Crop water requirement

Lewin (1973), Lomas and Lewin (1979) developed a graphical procedure for assessing crop water requirement of field crops with known ratio of crop coefficient i.e., ratio between crop water loss and pan evaporation or potential evapotrnspiration values for the crop stage in a particular month. Fig. 9.9 illustrates the procedure for crop water needs. Location index in Fig. 9.9 is given below:

Index	Location	Annual PET(mm)	Thornthwaite Climatic Classification Notation
1	Darjeeling	725	$AB_2'ra'$
2	Shillong	800	AB'_3ra'
3	Simla	900	$B_3B'_3wa'$
4	Dibrugarh	1000	AB'_3ra'
5	Aijal	1100	AB'_4ra'
6	Tura	1200	AA'wa'
7	Ranchi	1300	$C_2A'w_2a'$
8	Purulia	1400	$C_1A'sa'$
9	Cuttack	1500	$C_2A'w_2a'$
10	Varanasi	1600	DA'da'
11	Khandwa	1700	DA'da'
12	Indore	1800	DA'da'
13	Bhuj	1900	EA'da'
14	Phalodi	2050	EA'da'
15	Jaisalmer	2100	EA'da'
16	Rajkot	2150	EA'da'

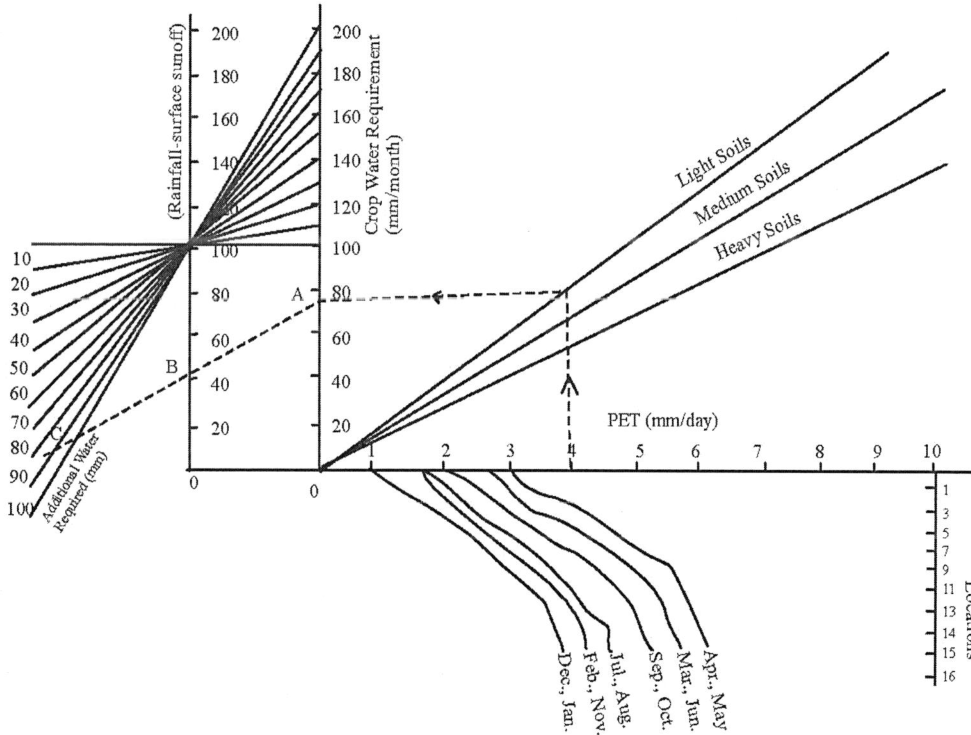

Fig. 9.9: Crop water needs (After: Lomas and Lewin, 1979).

For example, at Ranchi the pan evaporation/PET is 4 mm per day during July-August monsoon months. Assuming light soils 75 mm per month is crop water requirement. If rainfall occurred in July month is 100 mm, then effective rainfall which contribute to meet crop water requirement, after deducting surface runoff and deep percolation, is of the order of 45 mm. Joining AB and comparing the parallel curve w.t.to AB, we get 42 mm. of additional irrigation needed. Depending on the cultivated area we can determine the total volume of water to be supplied with irrigation. However, the surface runoff amount is 55 mm which should be controlled with bunding or soil conservation methods in the soil profile by maintaining its leaching rate with soil mulching after every effective rain. Surface runoff depends on the soil type, intensity of rainfall and percolation rate. Surface runoff was predicted for daily rainfall by using the Soil Conservation Service curve number equation, as (U.S. Deptt. of Agriculture, Soil Conservation Service, 1972)

$$RO = \frac{(R - 0.2S)^2}{R + 0.8S} \text{ for } R > 0.2S \qquad \ldots (9.20)$$

$$RO = 0 \text{ for } R < 0.2S$$

where, RO is runoff and R is rainfall in mm, S is a retention parameter which varies with soil type, landuse, management, stope and with time due to variations in soil water contents. Parameter S is related to curve member (CN) by the equation (U.S. Deptt. of Agriculture, Soil Conservation Service, 1972) as

$$S = 254\left(\frac{100}{CN} - 1\right), \qquad \text{... (9.21)}$$

where, constant 254 gives S in mm in order to express R and RO in mm. Therefore depending on soil moisture conditions curve number can be selected for any area by using the SCS hydrology handbook (U.S. Deptt. of Agriculture, 1972) (Table 9.7).

Assuming that CN_2 is appropriate for a 5 per cent slope and moist conditions. For other slopes CN_2 value can be adjusted using CN_3(wet), CN_1(dry), CN_2(moist) conditions values which are interrelated as (Sharpley and Williams, 1990)

$$CN_1 = CN_2 - \frac{20(100 - CN_2)}{100 - CN_2 + \exp[2.533 - 0.636(100 - CN_2)]} \qquad \text{... (9.22)}$$

$$CN_3 = CN_2 \exp[0.00673\ (100 - CN_2)]$$

9.3.5 Pest and diseases

Crop pests and diseases are biological organisms reducing the agricultural productivity accounting for more than 25 to 30 per cent losses. Weather variables strongly influence their outbreak and multiplication leading to epidemic situation where losses are uncontrollable. In order to develop their forecast it is desirable to understand the population dynamics of pests or disease severity as influenced by various biotic and abiotic factors so that suitable pest/disease management practices are evolved for adoption in the different regions. Biotic factors of diseases are fungi, bacteria, nematodes, viruses, viroids, mycoplasma like organisms, whereas a large number of insect and pest species compete with human beings due to mismanagement of insect's environment and availability of vegetation species throughout the year with increasing intensity of cropping and irrigation facilities. The abiotic factors are climate and weather variables. Actual microclimatic conditions inside the plant's environment influence the outbreak of diseases and development of pests due to congenial environment. Macroclimatic conditions over larger space and time influence the movement of desert locusts with arrival of monsoonal easterly humid winds, movement of mustard aphids in plain areas with the arrival of western disturbances in winter season, movement of wheat rusts along the tracks of cyclonic movement or western disturbances, polycyclic fungal diseases attaining epidemic status such as rusts, mildew, blights, leaf spots, ergots, moulds etc. where a number of disease cycles occur within one host cycle. The losses in polycyclic diseases are manifold as compared to the initial incidence of monocyclic diseases such as foot rot, root rot, wilt, loose smuts etc. Therefore in order to develop weather forecast for pest and diseases, the interaction process between the pest/disease – host plant – weather

Table 9.7
Runoff curve numbers (CN_2) for hydrologic soil cover complexes (assuming antecedent moisture condition II moist and $I_a = 0.25$)

Landuse	Soil Cover: Agronomic crop practice	Soil Cover: Hydrologic surface conditions for infiltration	Hydrologic Soil Group: A	B	C	D
Fallow	Straight row	-	77	86	91	94
Row crops	Straight row cropping	Poor	72	81	88	91
	”	Good	67	78	85	89
	Contoured	Poor	70	79	84	88
	”	Good	65	65	82	86
	Contoured and terraced	Poor	66	74	80	82
	”	Good	62	71	78	81
Small grains	Straight row cropping	Poor	65	76	84	88
	”	Good	63	75	83	87
	Contoured	Poor	63	74	82	85
	”	Good	61	73	81	84
	Contoured and terraced	Poor	61	72	79	82
	”	Good	59	70	78	87
Close seeded legumes	Straight row cropping	Poor	66	77	78	89
	”	Good	58	72	81	85
	Contoured	Poor	64	75	83	85
	”	Good	55	69	78	83
	Contoured and terraced	Poor	63	73	80	83
	”	Good	51	67	76	80
Pasture or Rangeland	Natural	Poor	68	79	86	89
	”	Fair	49	69	79	84
	”	Good	39	61	74	80
	Contoured	Poor	47	67	81	88
	”	Fair	25	59	75	83
	”	Good	6	35	70	79
Meadow	Natural	Good	30	58	71	78
Forest woods		Poor	45	66	77	83
		Fair	36	60	73	79
		Good	25	55	70	77
Farmsteads		-	59	74	82	81
Roads (dirt surface)		-	72	82	87	89
Roads (Hard surface)		-	74	84	90	92

(*After*: SCS Hydrology Handbook, USDA, 1972)

variables are understood. Pests and diseases are spread by air movement over short to longer distances under favourable environmental conditions. Therefore temperature, relative humidity, radiation, rainfall, dew, wetness duration are important meteorological factors for growth and reproduction of organisms in the crop ecosystem. There are several approaches which can be adopted to quantify the epidemic of diseases or outbreak of pests, their multiplication and spread for developing agroadvisory services. Established thumb rules with past experience and empirical relations are commonly used for forecasting the pest and diseases, whereas synoptic features are useful for spatial spread of areal pest and diseases. Crop models are used for linking the crop development phases with the pest/ disease development phases meeting their requirements each other.

(i) ***Thumb Rules:***

For agrometerological advisory services, a set of meteorological and biological conditions are necessary to hasten of the process for pest or disease development, then such thumb rules are very useful for taking a quick decision in adocating the favourable situation for a particular pest or disease event. Here few examples are given to highlight this view.

(a) Weather conditions during pollination is a critical parameter in plant's life during flowering stage for seed production in food crops, vegetables and fruit crops. When number of days with temperature greater than 25°C are available, the amount of honey yield is higher (Smith, 1975). In orchards and forests due to shading effect a light to moderate wind and temperature above 25°C, a bee can take more number of flights for better pollination during flower setting.

(b) Appearance of mustard aphids inoculum with arrival of western disturbance from cold hilly areas to plain mustard grown areas with cloudy, humid cold waves. After passing away of western disturbance, with clear sunshine and rise of temperature by 6 to 8°C the mustard aphids appears within a week after the arrival of western disturbance.

(c) With arrival of SW monsoon rains after dry hot summer there is decrease of soil temperature and increase of soil moisture in top soil. White root grub beetles (Holotrichia sp.) appears at night within a week in larval stage to damage the crops, and can be collected under a tree as they move up in tree branches for food. Soil temperature and soil moisture in root zone influence the insect stages (grass hopper, eggs, larval and pupa) in the soil. The emergence, movement and distribution are governed by weather parameters (Mukharji and Gage, 1978).

(d) Desert locust (Schistocerca greqaur, forsk.) breeds at favourable places in Africa, Middle East, and SW Asia during winter and early spring where good winter rainfall is received. The swarms fly toward east during he summer season and breeds during June to September with arrival of SW monsoon rainfall. The swarms resulting from this breeding return back to the same area during autumn when the winter rainfall is approaching. Therefore desert locusts have stages of eggs, nymph or hopper and winged adults when favourable soil moisture and temperature are available during the rainy season.

(e) Long distance transports of fungi in relation to cereal rust is well established another example of transport of spores by wind currents. Synoptic charts at surface and upper air wind trajectories are useful in determining the arrival and point of source of infection. Nagarajan and Singh (1975) observed the trajectory of dispersal of Puccinia graminis tritici uredospores similar to sequential location of tropical cyclones in India. Spread of stem rust of wheat P. graminis tritici uredospores from the over summering in Nilgiri to central India annually due to cyclones movement (Nagarajan and Singh, 1990). Wind currents spread uridini spores of stem rust from the Nilgiri hills of south India to move in a definite path called 'Puccinia Path' which is also applicable to brown rust. Based on these aspects, a set of upper layer synoptic conditions were advocated known as 'The India stem rust rules' (ISR) (Nagarajan and Singh, 1975). When the set of ISR is satisfied, the disease appears exactly below where the tropical cyclone dissipates. The amount of spores that take off from the Southern Hills depends upon the track of cyclones. Therefore the methodology is used in forecasting the appearance of stem rust (Nagarajan and Singh, 1976) even one month prior to the actual disease appearance in the field (Nagarajan and Joshi 1980). The spread of urediniospores of stem rust and brown rust of wheat from southern parts of India to central and northern parts shows a long distance dispersals (Nagarajan, 1983).

(f) Grass hopper population increased in dry years when compared with following year as the outbreak occurred after two successive years of subnormal rainfall (Smith, 1954).

(g) Late blight of potato and tomato is caused by fungus Phytopthora infestans (Mont.) de Bary. Spores presented in infected potato tubers develop, germinate and reinfect under favourable condition of temperature, humidity and moisture conditions. Potato blight is linked with the abundant dew deposition followed by rainy weather. However, in different parts of the world different thumb rules are adopted.

In Europe this disease is associated with warm or mild weather with the lower temperature infection limit of 10°C. Everdingen (1926) and Beamont's (1947) advocated the outbreak of disease within 7-21 days with critical period of at least 4 hours of dew, T_{min} above 10°C in the night followed by cloudiness of 8/10 in the day, rainfall not below 0.1 mm and relative humidity at 15.00 hours not below 75 per cent.

In USA, Wallin (1951) viewed it as a cold weather disease limited by high temperature and low relative humidity. A critical period of 10 hours of temperature not below

75°F and relative humidity not below 90 per cent are favourable for development of blight.

In South Africa a daily maximum-minimum temperature range between 23.9 to 10°C, RH above 70 per cent at 14.00 hours for at least two days for blight to occur and if for 3 days leads to a potential danger area.

(h) **Wheat rusts**: Wheat rust is present as uredospores and even teleutospores for several days to months, but reinfection and multiplication depends on the host and favourable weather. Depending on weather conditions during the growing season rust will appear in wheat. When a warm weather persists it is conductive for Black rust (stem rust), Punccinia graminis Pers; but when a cool weather persists during the growing season liability for Yellow rust (stripe rust), Punicia striiformis Westend exists; and when intermediate weather persists it is favourable for Brown rust (leaf rust), Punicia recondit Dorms. Lange et al. (1958) expressed Black rust infection units as 9/5(T – 10) D.C., where T is temperature °C, D is dew hours in light and C is a correction factor varying between 0 and 1. Black rust occurs with optimum conditions of bright weather with temperature of 16 to 27°C, excessive dew, little cloudy and high humidity greater than 70 per cent. It is severe in dry and warm weather situations.

Yellow rust is available in host plant grasses in hilly regions. Optimum weather conditions for rapid multiplication are cooler temperatures between 9 to 13°C with relative humidity greater than 70 per cent and partly cloudy conditions for a week's period. Wind factor spreads the rust to distant places. Such weather features are common during the flow of northwestern disturbances along the foothill of Himalayas. A long warm autumn and a short mild winter are favourable environments.

Brown rust inoculum is available in host plant grasses with optimum conditions of 15 to 22°C with overwintering of uredomycelium, relative humidity of above 70 per cent, intermittant clouds and sunshine occurred for a week's period.

(i) **Apple scab**: It is caused by Venturia inequalis. Leaf wetness and temperature determine the critical period for fungus to develop the apple scab called the 'Mills Period'. (Mills, 1944). When the product of leaf wetness (hours) and mean temperature (°C) exceeded 140°C hours, the infection is probable provided the mean temperature is not below 25°C and leaf wetness period is greater than 9 hours for the infection to occur. After an incubation period of 3 to 4 weeks, the fungus breaks through the cuticle as dark brown spots. Fungus spores infection occurs when buds are opening and needs to be controlled at this stage otherwise conidia will spread the disease by infecting young foliage and fruitlets.

(j) Late blight of potato, Phytophethora infestans was predicted based on periods when inflection is likely to oocur with favourable weather situations of (a) 10 consecutive days with cumulative rainfall of more than 30 mm and 5 days average temperature not exceeding 25.5°C, or (b) after accumulation of 18 to 20 severity units (each day is scored on 0 to 4 severity scale according to temperature and hours will relative humidity greater than 90 per cent).

(k) Early blight of tomato, Alternaria solani, occurred with favourable environmental conditions of disease development severity scale (0 – 4) according to periods of wetness and temperature during the forecast period. First spray was applied when at least 35 disease severity values were accumulated and subsequent applications with accumulation of at lest 20 disease severity values.

(l) Peanut leaf spot disease is dependent on the number of hours during which relative humidity remained 95 per cent or more and intensity of lowest temperature 15 to 17°C for more than 10 hours (Valli, 1968). Maximum attack was observed at 20 to 25°C with RH greater than 95 per cent for 12 hours or more and inflection decreased with decrease of temperature.

(ii) ***Empirical statistical relations***

Insect pest and disease forecasting have been done by developing multiple regressions between disease index or outbreak and multiplication of pests with optimum weather variables during the crop growing season. A few examples are given below to forecast the pest and disease multiplication with favourable environmental condition:

(a) Symmons (1959) developed a forecasting expression between rainfall and size of red locusts population as

$$F = 6.518 - 0.16\,P' + 0.425\,PP + 0.192\,P,\ R = 0.93 \qquad \ldots (9.23)$$

where, F is filial population, P' is previous rainy season rainfall in inches, PP is parental population, P is rainfall in inches during the first three months of breeding season that give risk to the filial population. A dry period within a fluctuating climate is also favourable to locust outbreaks.

(b) Gourangakar and Chakravarty (1999) developed expressions for brassica aphids populations in 10 cm of termininal shoot at emergence, establishment and declining phases based on weather variables as

$$Na = 194.63 - 17.4\,\overline{T}_{min} + 0.23\,\overline{RH_E},\ R^2 = 0.8 \text{ (Emergence phase)} \qquad \ldots (9.24a)$$

$$Na = 269.12 - 5.05\,\overline{T}_{max} - 0.4\,\overline{T}_{min} + 0.12\,\overline{RH_E},\ R^2 = 0.8 \text{ (Establishment phase)} \qquad \ldots (9.24b)$$

$$Na = 245 - 32.45\,\overline{T}_{min} + 4.49\,\overline{T}_{max},\ R^2 = 0.69 \text{ (Declining phase)} \qquad \ldots (9.24c)$$

where, Na is number of aphid populations in 10 cm of terminal shoot. In rapeseed the aphid population was expressed as

$$Na = X_1P_{t-1} + X_2T_{max(t-1)} + X_3T_{min(t-1)} + X_4T_{mean(t-1)},\ R^2 = 0.73 \qquad \ldots (9.25)$$

where precipitation (P in mm) and temperature (T in °C) are one week prior (t – 1) to the prediction of population.

(c) Powdery mildew of mango (Oidium mangiferae, Berthet) – a blossom disease severity damaged the fruit setting when relative humidity was greater than 80 per cent and relatively high minimum temperature greater than 10°C for 4 to 5 consecutive days during the flowering season. Sinha et al. (1999) forecasted it in 0 – 10 severity scale expressed as

$$\text{Mildew severity} = 0.0009\, X_h X_t + 0.375,\ R^2 = 0.72 \qquad \ldots (9.26)$$

where, X_h is cumulative hours of humidity (≥80 per cent) for 5 hours at 10°C or above, X_t is minimum temperature.

(d) Solanki et al. (1995) developed a disease index for powdery mildew mustard expressed in percentage as

$$\text{Disease index (\%)} = -196.6 + 1.42\, X_1 + 5.09\, X_2 + 0.11\, X_3,\ R^2 = 0.75 \qquad \ldots (9.27)$$

where X_1 is days after sowing of crop, X_2 is average temperature of previous 10 days, and X_3 is number of conidia trapped.

(e) Karnal bunt disease was accurately predicted with weather variables during seed development stage as disease index in percent

$$DI = 3.961 - 0.141\, T_{max12} - 0.038\, RH_{M12} + 0.047 RH_{E12} - 0.126\, R_{12} + 0.403\, RD_{12} + 0.227\, SS_{12},\ (R^2 = 0.96,\ F = 19.12) \qquad \ldots (9.28)$$

$$DI = -0.065 - 0.39\, T_{max10} + 0.007\, RH_{E10} + 0.045\, T_{max11} + 0.029\, RH_{E11} - 0.048\, T_{max12},\ (R^2 = 0.95,\ F = 22.39) \qquad \ldots (9.29)$$

where, T_{max} is maximum temperature in °C during 10, 11, 12th standard weeks as sub-scripts, RH_M and RH_E in percent are morning and eventing relative humidities during 10, 11, 12 standard weeks as sub-scripts, R_{12} is total rainfall in mm 12th standard week, RD_{12} is number of rainy days during 12th standard week. With arrival of rain during 10 to 12th standard weeks occurrence of Karnal bunt disease can be predicted well in advance.

(f) Ergot infection index in pearl millet was predicted using weather variables as (Gupta et al. 1983)

$$\text{EII (per cent)} = -92.92 + 0.39\, RH + 0.051\, R - 3.9\, T - 3.48\, SS,\quad R^2 = 0.95 \qquad \ldots (9.30)$$

where RH is mean relative humidity in percent, R is total rainfall in mm, SS is sunshine hours and T is mean temperature in °C.

The optimum weather variables were 12 mm of rainfall, 6 hours of sunshine, 20°C as mean temperature and 75 per cent relative humidity from protogeny to early flowering of pearl millet. Temperature and relative humidity variables were contributing to 80 per cent variation of ergot infection. Ergot is mainly spread by the admixture of solerotia with the seeds which reaches the soil and takes about 30 to

45 days to coincide with crop's flowering and receive the airborne ascospores. Under favourable conditions of high humidity, the disease spreads very rapidly through secondary inoculum accomplished through conidia, and are disseminated by raindrops and insects.

(g) Stalk rot of maize developed with high rainfall, humidity and temperature levels (Saxena and Lal, 1986). The disease incidence was predicted as

$$DI = 129.0536 - 2.2591\ T_{max} - 0.9520\ RH_{max} - 0.0193\ R + 9.2595\ SS \quad \ldots (9.31)$$

where, T_{max} is mean maximum temperature in °C, RH_{max} is mean maximum relative humidity in percent, SS is mean sunshine hours for 15 days period after flowering – the critical period for development and spread of the disease.

(h) Ascochyta blight in chickpea was predicted in terms of disease index scale (0 to 1) using standard weekly weather variables as (Jhorar et al., 1997).

$$DI = 4.97 - 0.168\ T_{max,}\ R^2 = 0.85 \quad \ldots (9.32a)$$

$$DI = 1 - 179.4\ \exp(-0.143\ RH_E),\ R^2 = 0.88 \quad \ldots (9.32b)$$

$$DI = -2.47 + 2.77\ HTR - 0.561\ HTR^2,\ R^2 = 0.89 \quad \ldots (9.32c)$$

where, HTR is humidity thermal ratio – a ratio of humidity in per cent and temperature in °C. Standard weeks 9 to 14 were important for disease development. When maximum temperature was less than 20°C the disease was limited by temperature. Favourable conditions for disease development were maximum temperature of 23 to 26°C, afternoon relative humidity 40 to 60 per cent and humidity thermal ratio of 1.3 to 2.5 respectively. Maximum temperature of 29.6°C and lower humidity of 37 per cent limited the disease development.

(i) Blister blight of tea was predicted using weather variables as (Kerr and Rodrigo, 1967a, b)

$$N_b = 33.0 + 0.3451\ X_f - 0.03725\ X_1X_2 \quad \ldots (9.33)$$

where, N_b is number of blisters per 100 shoots above 3 weeks before, X_1 is number of spores caught per m^3 of air, and X_1X_2 is the interaction of X_1 with mean daily sunshine hours (X_2).

(j) Powdery mildew in jujube (Ziziphus mauritiana) infection index (y) were strongly dependent on temperature and relative humidity (Sinha, 2005) with an expression of the form

$$Y = 0.077 + 0.00583t{*}h - 0.000003t^2{*}h^2 - 0.00076t^2 + 0.001h^2 \quad (R^2 = 0.88). \quad \ldots (9.34)$$

where, Y is infection index, t and h are effective temperature and relative humidity hr. respectively. Temperature between 17 to 26°C showed higher infection index with 24-30 hr. relative humidity exposure (above 90 per cent). Maxima infection occurred at 22°C and relative humidity (>90%) exposure above 20 hours duration during fruit development stage.

Disease warning messages can be of two types – strategic (once in growing season) and tactical (several times during growing season). Strategic warnings are w.t.to. a disease with local sources of inoculum such as stripe rust of wheat in the Pacific NW of USA (Coakley and Line, 1981), wheat stem rust with Indian stem rust rules (Nagarajan and Singh, 1975). Tactical disease warnings are based on future infections. Crop phenology based warnings are risk assessment systems (Kelly, 1982) are useful. Further the disease based warning system are not so easy, but can be monitored slow and steady approach such as yellow rust of wheat. However, more explosive diseases can be handled based on weather based system which are easy to monitor regularly such as Downy Mildews. Van Everdingen's (1926) concept of critical period – a system based on few critical weather data only – which can be measured easily rather than to explore the life history of the pathogen. It is also feasible that in different agroecological conditions there may be different solutions for one and the same plant disease. For example Zadoks (1984c) showed that for potato late blight there was no unique solution for all situations and necessitated different approaches in different countries or agroecological areas. Farmers can act on the delivery format as per needs. However a computerised system have been also developed to help farmer such as EPIPRE management in wheat, which encompasses six diseases and all aphid species (Zadoks, 1981; 1984a, 1984b).

9.3.6 Animal diseases

Animal diseases are caused by viruses, bacteria, protozoa, helminths and metabolic disorders (Tromp et al. 1976). Weather variables influence the outbreaks of the diseases (Table 9.8). For example facial eczema of sheep and liver fluke disease of sheep. Weather conditions affect livestock indirectly by the fodder grown in the area which influence the nutrition of the animals causing various metabolic disorders and deficiency diseases. Poisonous plants and toxic fungi are important causes of disease in livestock.

Anhidrosis or non-sweating of horses disease is very severe in temperate zones with high humidity all over the year. There is a progressive loss of ability to sweat even in very hot weather which upsets the heat regulating mechanism and temperature of horse rises very rapidly. Respiration rate is excessively increased to lose surplus heat in the expired air. High temperature with high humidity are crucial factors. Ultimately this disease leads to slow cardiac failure with dyspnoea, weakness and oedematous swelling of tissues of ventral body wall.

Climatic conditions strongly influence the anthropod vectors which transmit several diseases such as tick-borne piroplasmoses, Rift Valley fever trypanosomiasis, rickettsia infections, equine encephalomyelitis, ephemeral fever, of cattle, bluetongue of sheep, horse sickness, pox-diseases etc. Once the anthopod responsible for transmission is detected, temperature and humidity conditions influencing its life cycle are known for the outbreak of the disease (Tromp, 1976a, b).

Table 9.8

Meteorological variables influencing animal diseases

Disease	Symptoms	Causative agents	Animals	Meteorological factors	References
Thermoregulatory diseases	Anhydrosis: Progressive loss to sweat even in warm dry environment imbalancing heat regulation. Temperature of body rises leading to exhaustion and hyperpyrexia	Imbalance of heat regulation	Horses	When high air temperatures combines with high humidity in coastal or continental areas. Temperature upto 43°C may occur and die	Evan et al. (1957)
High altitude diseases	Brisket diseases due to pulmonary hypertension	Serious dyspnoes high altitudes	Cattle	Reduction of oxygen levels	
Bacterial diseases	Braxy: Infection in stomach and intestine walls	Clostridium septicum available in soil	Sheep	With onset of frost sheep grazes very cold grass and stomach is chilled. Due to invasion of bacteria leads to sudden death	
	Haemorrhagic septicaemia	Pasteurella septica	Cattle	Outbreaks with onset of SW monsoon rains with sudden growth of new grass leaves upsetting digestive system.	
Virus diseases	Rift valley fever	Mosquito species	Sheep, goats, cattle and man	In rainy spell periods in low lying areas. Stagnant of water due to wet spells.	
	Foot and mouth disease (Aphtae epizooticae). Highly infectious disease, initiates with high fever followed by	Picoma virus	Cattle, swine, sheep and goat	Damp cold weather favours the survival and spreading of virus. Absence of ultraviolet rays, high humidity, strong downwind,	Smith (1970), Primault (1958)

	vesicles in the mouth. Dripping of mucous on the legs causing vesicles on the feet which becomes swollen and painful, health deteiorates severely, mortality low			rainfall, in night favours the development of epidemics, Excessive sunlight and high temperature do not allow the survival of virus.	
Tickborne diseases	Babesiosis: High fever	Transmitted by a Tick Boophilus microplus	Cattle	Depends on the survival time of female tick to lay eggs etc. which is strongly affected by humidity. Drought conditions kills the tick population. Wet periods favours the transmission in pastures stock.	Wilkerson and Wilson (1959)
Helminth diseases	Male and female worms living in intestinal canal of host to produce eggs which are passed out. Larvae develops into infective stages which are swallowed into body along with food.	Roundworms, tapeworms, and flukes during free living stages outside the animal body	Herbivorous grazing animals, farm animals	Seasonal changes in herbage on which host is feeding. Direct effects of temperature and humidity on hatching of eggs and development of larvae outside host. Microlimate of herbage cover is critical for development of worm parasites.	
	Fascioliasis or liver fluke disease, damaged liver	Fasciola hepatica transmitted by a trematode, Parasite	Sheep	Specific humidity and temperature conditions required	Ollerenshaw and Rowlands, (1959)

		complicates the life cycle in the snail Lymmaea truncatula		for a successful life cycle of parasites.	
Metabolic disorder diseases i.e., indirect meteorotropic diseases	Grass tetany: tetanus convulsions	Hypomagnesaemia	Grazing cattle and sheep	Low minimum temperature, little sunlights, high rainfall, strong winds favours the causative agent	Allcroft (1947)
	Fascial Eczema: Liver disease	Photodynamic substance: Phylloerythrin	Sheep, cattle, horses	Outbreaks after hot dry summer with occurrence of rains, soil temperature and soil moisture	Smith (1970) Filmer (1957) Mitchell et al. (1959)
Myxomatosis	Virus disease	Virus belongs to Poxviridae, genus Leporipoxvirus (L.)	Rabbits	Virus is very sensitive to natural environmental influences particularly high temperature and sunlight	Marshall (1959)
Rinderpest	Severe virus disease – the lesions of the mucous membranes of intestine leads to diarrhoea which results in dehydration and serious disturbance of electrolyte metabolism. There is circulatory breakdown leading to death.	It is caused by a panotropic RNA virus producing destruction of lymphatic tissue.	Cattle, pig, sheep	Ambient temperature influence mortality rate and severity of the symptoms.ecoclimatic conditions are significant for outbreak of disease.	
Sheep Pox and Goat Pox	Virus disease characterized by vesiculo – popular exanthema of skin and mucous membranes.	Capripox virus	Sheep and goat		
Rabbies	A lethal virus disease transmitted to man through all mammals.	This virus belongs to Rhabdoviridae family genus Lyssavirus.	Dog, cat, bats and wild carnivores		

Leptospirosis	Contagious acute chronic disease with fever, abortions, uraemia and icterus.	Leptospira (L.) interrogans	Man, cattle, pig, horse, dog and small ruminants	Body temperature is increased.	
Poultry diseases: Botulism	A few hours after ingestion of toxin, birds are extremely weak, difficulty in swallowing, progressive paralysis of leg, wings and neck muscles ultimately leading to death.	Clostridium botulinum which produces a dangerous toxin.	Birds	Poisonous substances are related to temperature, less toxin is produced at 12°C, but above 20°C much toxin is produced. Disease over during hot summer and in other seasons near power stations causing thermal pollution in soil and water.	Payne (1967)
Marek's disease	A serious neoplastic chicken disease of lymphoid system	Herpes virus in horny layer of skin from infected chicken transmitted by air borne particles	Chicken pheasants	Weather conditions favourable for airborne transportation.	Payne (1967) Tromp et al. (1976)
Keratoconjunctivitis	Initial signs are tearing and tend to keep eyes closed. Blindness may develop with loss of weight	Ammonia fumes over 50 ppm., poorly ventilated spaces	Chicken	Inadequate ventilated poultry houses in winter where chickens are kept on litter or shavings.	Tromp (1980)
Newcastle disease	A virus disease with high mortality. It initiates with coughing and rattling, rapidly spreads in flock, difficulties in coordination, paralysis, coma, egg production drops to nil in few days	Virus which belongs to genus Paramyxoviridae	Chicken, turkey, pheasants	Changes in temperature and humidity favours the disease, excessive sunlight reduces the transmission.	Payne (1967); Tromp (1980)

Rift Valley fever in Africa, the insect transmitted virus disease, by certain species of mosquito in wet seasons in low lying areas in sheep, goat and cattle in epidemic form after 5 years – a period for new population of a susceptible sheep.

Tick-borne diseases in livestock are transmitted by ticks population under favourable conditions of temperature, humidity, exposure to sunlight, life cycle of tick and weather conditions are useful in predicting this disease (Taylor and Muller, 1974).

Farm animals are parasitized by helminths such roundworms, tapeworms and flukes. Meteorological variables influence these helminths so long as they are outside the animal's body. Safer use of pastures is necessary based on meteorological variables.

Metabolic disorder diseases are ketosis, grass tetany, milk fever, pregnancy toxaemia of sheep etc. Grass tetany disease occurs in grazing cattle characterized by sudden development of titanic convulsions causing death due to fall in the magnesium content of the blood. Low level occurs when mean minimum temperature is below 4.5°C with little sunshine, high rainfall, strong winds prevail before the wet period, also there is snow, hail, sleet occurrence (Allcroft, 1947). The magnesium level is the blood is under endocrine control regulating the mechanism influenced by cold, wet and windy weather. Thyroid and adrenal glands regulate the magnesium levels and adverse weather may act as stress factor and precipitation of the disease. Therefore association of weather with occurrence of grass tetany in grazing cattle can be used as a guide to take precautionary measures.

Bacterial diseases multiply only within the body of their host, but epidemic is caused when they pass from one host to another during which weather influence them very strongly. For example, First braxy in sheep caused by infection of stomach and intestinal wall by a sporulating organism Clostridium septicum. This organism is available in the soil and ingested by sheep. But severity of disease and death occurs when frost weather conditions occurs. Another bacterial disease is haemorrhagic septicaemia of cattle caused by Pasteuretica septica. Severe losses occur soon after onset of monsoon rains probably due to sudden flush of young grass upsets the digestive system of cattle weakened by shortage of food in long dry season (Tromp, 1980).

Crawford (1962) distinguished diseases due to non-sweating (thermoregulatory problems), metabolic disorders (nutritional problems), bacterial diseases, virus diseases, diseases transmitted by anthropods vectors (insects – mosquitoes, ticks, flies etc.), and roundworms or nematodes. Smith (1970) also distinguished wind borne diseases (due to bacteria and viruses transmitted by wind currents) and fungal diseases (nutritional diseases)

Nixon-Smith (1972) prepared a nomogram based on physiological reaction of sheep to environmental weather conditions to alert sheep grazers to protect their animals from adverse weather conditions. It is essential that warm blooded animals must maintain body temperature within a relatively very narrow range of temperatures 39 ± 1°C and should get sufficient energy for the operation of vital functions with the help of metabolic processes. Body temperature is controlled with the help of thermoregulatory system which provides the heat balance with the surrounding environment. If the heat drain is so large that animal cannot cope with it, then the body temperature will fall to develop hypothermia. Depending on the intensity of heat loss it may cause even death. Therefore, intense heat losses due to adverse weather conditions in the winter season may lead to sudden losses among the newly born lambs and sheep recently off shears. Even a lesser drain of heat over a longer period may cause deaths in shorn sheep in poor conditions and in young lambs.

In winter season with lowering of temperature, a combination of both wind and rain will aggreviate the environmental conditions to cause loss of body heat by sheep to attain fatal hypothermia even at air temperature of 20°C. Lower temperatures will intensify the cooling caused by the combination of wind and rain. Under frosty conditions, wind and rains are absent, losses of newly born lambs from hypothermia are very less and no alertness on agroadvisories is required even when wind speed is less than 4 knots and no occurrence of rains. Therefore wind speed, rainfall and minimum temperature values are incorporated in nomogram (Fig. 9.10) which indicates the threshold level at which hypothermia may develop in young lambs and newly born sheep for given combination of these weather elements. Fig. illustrates with the use of drawing a line parallel to the abscissa from a point on temperature scale, corresponding to the minimum- a lowest temperature expected during the episode of adverse weather during the forecast period, until it intersects with the slopping line (or interpolated line) corresponding to the highest hourly average wind speed expected during the forecast period or same episode of adverse weather. From this point of intersection, draw a line parallel to the ordinate until it meets the sloping line corresponding to the rainfall total expected during the forecast period. From this point draw another line parallel to the abscissa across to the boxes showing the alerting criteria for (a) newly – born lambs, (b) newly – shorn sheep provided that their coats are wet.

Therefore during short/medium range weather forecast reports of the location, we can issue sheep weather alert warning to the graziers about the adverse weather conditions expected within next 24 to 48 hours in specific areas where losses in (a) new – born lambs or (b) new born lambs and newly – shorn sheep are likely to occur.

Fig. 9.10 shows that considerable rain is required to penetrate and wet the coat of adult unshorn sheep so that losses of such sheep are a relatively rare event.

An episode of adverse weather as short as two hours can lead to some stock losses and in such situations an emergency alertness warning can be issued. The alerting criterian for newly – shorn sheep and young lambs more than four hours old applies only if the fleeces of the animals are wet, so that the full chilling impact of the adverse weather may be felt. This would require about 5 mm of rain during or immediately preceding the episode of adverse weather. The coats of newly born lambs are saturated with foetal fluid and they remain wet for periods of upto 4 hours. However, when the newly shorn sheep are in poor condition due to scarcity of feed (drought stress period), the threshold value may be lowered to that of newly born lambs. In Australia at 40°S latitude a set of critical conditions

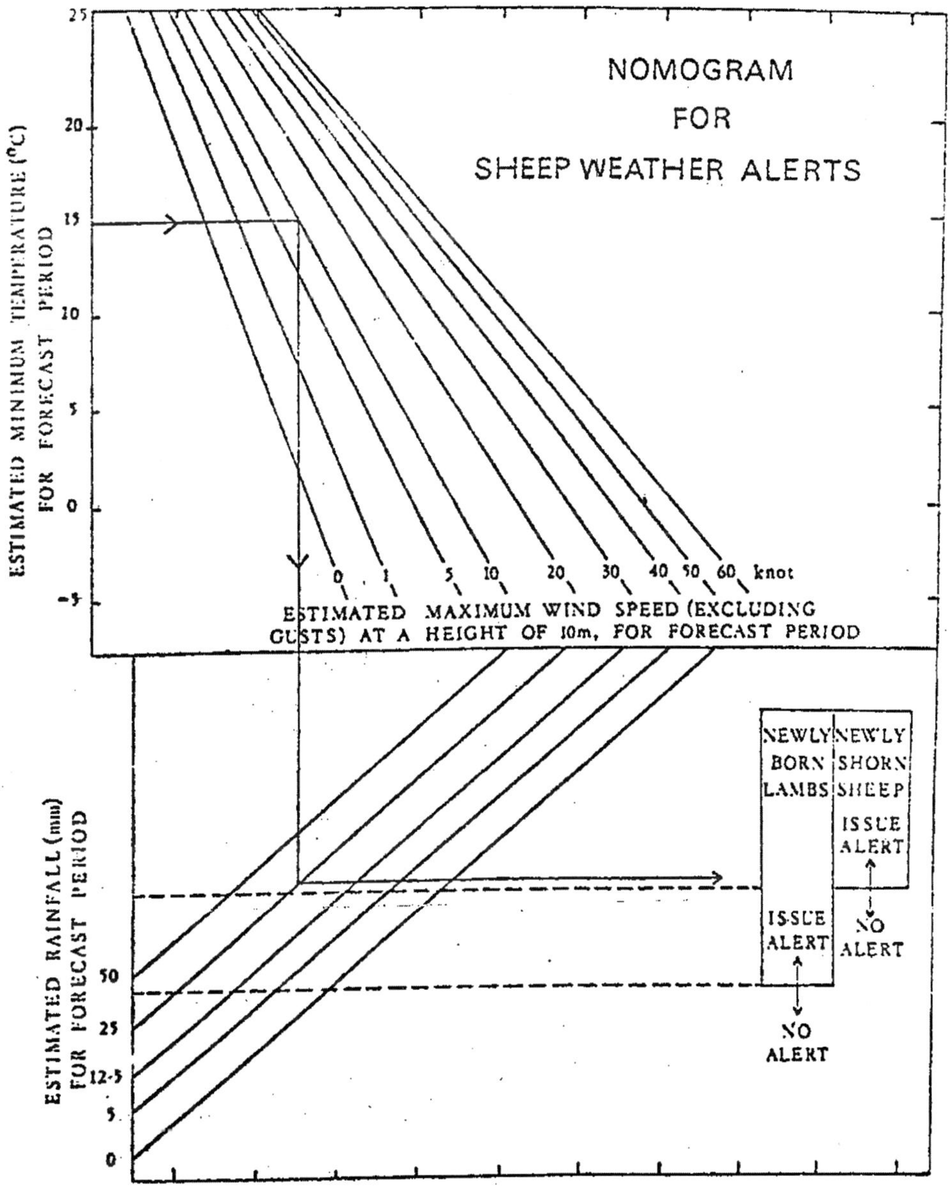

Fig. 9. 10: Nomogram for sheep weather alerts. (After: Nixon–Smith, 1972)

acts as a guide for issuing cold snap warming in winter months as (i) frequent showers and rain, (ii) ambient temperature 10°C or less, (iii) wind speed of 5 to 7 m sec^{-1}. Synoptic maps and satellite photographs can help in issuing early warning for arrivals of cold outbreaks or cold wet windy weather to alert the farmers to protect their animals with detection of first cyclogenesis.

Therefore meteorological data can be used for forecasting of outbreak of diseases in livestock to minimise losses. For example, a serious disease of sheep known as fascioliasis or liver fluke disease is caused by liver damage caused by the trematode parasite Fasciola hepatica. The extent of liver damage depends largely on the number of encysted flukes ingested. Sometimes it causes heavy losses. The bad fluke years were related to weather variables (Ollerenshaw and Rowlands, 1959).

Facial Eczema in sheep, cattle and horses also cause heavy losses. The essential lesion is a cholangistis of the liver which upsets the metabolism of phyllocrythrin which is photodynamic substance causing damage to skin. It has been established that the toxic substance is produced by a fungus sporidesmium bakeri on ryegrass. The disease outbreak is correlated with weather factors particularly the years with warm (hot and dry) summer and autumns over the normal values. Outbreak appears soon after sufficient rains to stimulate new growth of the pasture. The temperature deviation from normal and rainfall occurrence are monitored to issue warning of serious outbreaks (Mitchell et al., 1959). Soil temperature and soil moisture parameters further improved the forecast when soil temperature at 20 cm soil depth reached 17°C or above and soil moisture in soil attained a deficit of 38 mm. The critical conditions were when grass minimum temperatures of 12°C or above (Smith, 1970). Therefore weather relations can be developed for animal disease forecasting as meteorological factors influence the animal diseases by affecting the resistance of host to germs/pathogens as well as resistance and evolution of these germs/pathogens during their biological cycle particularly when part of this cycle occurs outside the host animal.

9.3.7 Human diseases

It is a prognostication of action and consequences of anticipated meteorological conditions and weather types on man. It can be both short and long term prognostication. Daily or weekly medical meteorological forecasts are useful to encounter environmental behaviour. Behavioural alterations and aberration have been correlated with weather types (Buettner, 1957). The incidence of heat illness in hot dry summer in heavy industries should be curtailed with appropriate modification of environmental exposure, moderation of physical work, and adequate environmental protection (Minard et al. 1957). Industrial and domestic pollution in lower atmospheric layer can be controlled by regulating meteorological conditions for dispersal of manmade contaminants (Kleinsasser and Wanta, 1956). Human population shifts to particular climatic region having certain favourable charcterisitics.

Experimentally derived weather indices from physiological studies on healthy young acclimatized men have been developed. Physical environment, clothing and metabolic rate are precisely controlled. These indices are broadly based on meteorological and physiological informations. These have been already discussed under comfort indices (Section 6.5 and

6.6) based on dry bulb temperature, wet bulb temperature, effective temperature, wet bulb globe temperature. Phillips (1897) suggested that a deviation of 8 to 10°F above normal, an average daily dry bulb temperature equal to or greater than normal maximum temperature, and a night temperature in excess of 24°C provided a reliable criteria to predict heat wave to avoid deaths from heat stroke. Haldane (1905) reported that wet bulb temperature above 33.3°C represented the upper limit of tolerance for working man, an effective temperature scale was devised which induced the same sensation of warmth as experienced in open air environment. However Bedford (1946; 1950) substituted temperature of black globe for dry bulb temperature, effect of radiation was incorporated in the index. Yaglou (1926), Yaglou and Minard (1957) used wet bulb globe temperature as a heat index.

In order to study the impact of heat indices on man, physiological indices were preferred. McArdle (1947) predicted sweat loss by an acclimatized man exposed to heat based on globe or dry bulb temperature, wet bulb temperature, air motion and rate of heat production.

The effect of natural chronic cold exposure on the response of human for cold acclimatization was studied with standard exposure of 12-14°C for 2 hours. Shivering, oxygen consumption, rectal temperature and skin temperature were recorded. During winter, shivering was 90 per cent less, cold induced oxygen consumption 50 per cent less, and rectal and skin temperature remained unchanged (Thomas et al., 1960).

In winter there is increased occurrence of difficulty in breathing than during other seasons with younger children this was more marked than with older ones (Menger, 1960). Tromp (1963; 1974) observed that Asthma frequency increased rapidly after a sudden increase in general turbulence of the atmosphere when combined with influx of cold air masses. Periods of steep barometric fall, due to rapidly approaching low pressure areas, have a marked Asthma increasing effect with influx of cold polar or continental air masses followed by one or more active cold fronts. Severe dyspnoea is caused in Asthma patients predominantly by a sudden drop in environmental temperature combined with marked cooling effects resulting from considerable atmospheric turbulence.

Many communicable disease such as malaria, trypanosomiasis, yellow fever, and filariasis depends on insect vector for transmission among humans or between human and animals which act as reservoir for the parasitic disease organism. Malaria caused haemosporozoan Plasmodium is transmitted by mosquitoes. Similarly yellow fever is transmitted by the bite of mosquito Aedes aegypti. Linnaeus (Abushama, 1974) at temperature above 23°C. Similarly house flies transmit pathogenic organisms causing human diseases such as typhoid fever, dysentry, cholera and tuberculosis. Trachoma is another viral disease of eye which is transmitted by house flies after feeding on an infected eye causing high incidence of blind people. Abushama (1974) reported that tsetse fly are blood sucking insects and transmits blood parasites. Tsetse flies feed on the blood of reptiles, birds, and large ungulates indiscriminately. Human diseases are strongly linked to environmental conditions. For example sleeping sickness in Africa or Filariasis in Africa, South America and Southeastern Asia are related to extreme tropical environmental conditions. However there are tropical diseases which occur with the availability of favourable climatic conditions even in subtropical zones such as amebiasis dysentery, the cysts of which is found in the intestines of human beings all over the world in the form of harmless infection of the lumina of the intestines which give rise to dysentery with passing of blood with the stool. The isotherms of warmest month in the northern and southern hemispheres can be used as a boundary lines for this disease to occur. Similarly 10°C as mean annual isotherm represents the northern boundary for the development of phlebotomus in Europe and 13°C mean annual isotherm as northern boundary for the conditions under which Aedes aegypti, the vector of yellow fever and dengue, can develop. In tropics there are no thermal season, whereas in temperate belts there is seasonal rhythm. This seasonal phenomenon can be observed with typhus abdominalis, diphtheria, fever, measles etc. which are evenly distributed over various months of the year in regions belonging to zone without thermal seasons. There are diseases which have their own seasonal rhythm in the tropics such as cerebrospinal meningitis where peak occur at the end of dry season and comes to end with occurrence of rains. In Africa it shows two peaks due to two dry seasons.

The predilection period of an epidemic i.e., period with relatively highest disease cases every year is good indicator of communicable diseases signifying weather based process behaviour of cerebrospinal meningitis in Africa dry season (January to April) where dry dusty winds from Sahara blows.

9.3.9 Frost

(i) *Types of Frost*

Frost prediction depends on their origin and prevailing weather conditions. There are three types of frosts.

(a) **Advective Frost:** These are formed as a result of onset of cold waves during winter months. Synoptic conditions with low temperature, cloudiness widespread, rainfall and wind speed are favourable conditions to cause wide spread frost conditions for few days. Temperature drops to below 0°C over earth surface and surrounding air. With passing away of cold waves over the area climatic conditions with clear sunshine and low wind speed becomes favourable for radiation frost, and therefore advection frost followed by radiation frost intensify the frost down to -8 to -10°C at night. Synoptic maps and satellite pictures are useful in monitoring such frosts as microclimate differences disappear to a large extent. However windward side is more susceptible to frost.

(b) **Radiation frosts:** These are formed on calm, clear nights with fast nocturnal cooling of the earth and earth bound objects. Due to low wind speeds and poor vertical mixing of the ground air layer during radiation frosts, there is high temperature inversion. Diurnal variations of temperature is large. These are location oriented frosts will varying degrees depending on the surface moisture conditions. Dry soil or vegetative surfaces are more liable to frost. Intensity of radiation frosts can be accurately recorded with grass minimum thermometer installed open at the vegetative surface or at a height of 4 to 5 cm above the soil surface as the night radiation from the surface of grass minimum thermometer is closer to the radiation from the vegetative surface. Radiation frosts are morning frosts which are noticed in the night and may stay over vegetative surface for even upto 8 to 10 hours. During clear weather conditions these can be daily feature for a fairly prolonged period damaging newly emerging vegetative organs in plants.

(c) **Advective – radiation frosts**: These are formed as a result of arrival of cold northern wind and its subsequent cooling by night radiation. Here the process of advection and radiations supplement each other. Here temperatures up to -2 to -3°C are attained in the air layers touching the soil surface, but minimum temperature in the Stevenson's screen will show above 0°C. This type of frost is observed before the morning i.e., from mid night to about 3 or 4 A.M. and are mostly continuous for one or two nights. Topography plays important role in duration of frost period. Differences in properties of soil-vegetative surfaces due to varying soil moisture contents, soil and vegetative surfaces colour, dry or moist vegetative contents will provide varying conditions for nocturnal cooling to provide a variety of distribution of frosts in different topographic situations. Warning of advection frost is feasible with inflow of cold waves in the region which takes 3 to 4 days for transformation of air inflow into the local current and also there is rapid change in temperature and cloudiness in the area with arrival and passing away of cold waves. Windward slopes of elevation are more susceptible to frosts.

(d) **White frost**: It is caused by sublimation of ice crystals on objects near the ground surface such as tree branches, metal wires etc. When these objects acquire temperature below freezing as air with a dew point below freezing is brought to saturation by rapid cooling.

(e) **Black frost**: This occurs when vegetation is frozen due to reduction of temperature of air that does not contain sufficient moisture for the formation of white frost.

(f) **Glazed frost**: When falling rain drops freeze on ground and is transparent.

(g) **Hoar frost**: It is deposition in the form of ice crystals having the shapes of feathers, needles, etc. It is white and opaque.

(ii) ***Frost evaluation:***

Protection of plants from low temperature is important in agriculture particularly horticultural and vegetable crops. Vetskevich (1961) presented crop species according to their resistances in terms of temperature tolerance at different growth stages. Normally tropical and subtropical crops are poor resistances to low temperatures. Temperate fruits and vegetables have better tolerance for low temperature and can sustain lower temperatures below 0°C to about to 7 to 9°C at germination, 1 to 3°C at flowering and 2 to 4°C at fruiting. To avoid frost it is necessary to have proper site selection where radiation interception is better over the topography with the solar trajectory in view. Thermal insulation of soil and vegetative surfaces can avoid frost liability which can be achieved through soil manipulation, soil cover with mulches, light to moderate irrigation which will prevent the nocturnal cooling of the surface. Shelterbelts and windbreaks are also useful to reduce the impact of advection frosts. Also other costly methods are available such as air mixing, direct air and plant heating which are not advisable over farm scale. Soil mulching and light irrigation are the simplest methods to avoid the frost liability at the ground surface. Screen's minimum temperature, recorded at 4½ ft height, cannot assess the frost liability as the ground surface is 4 to 5°C cooler than screen's temperature. However, under overcast and partly cloudy conditions difference is less. But under clear sky conditions frost liability is much higher (Fig. 9.11). To have a deeper insight of the frost phenomenon clearly at a location, it is essential to record observations during danger periods from sunset to sunrise on air temperature, cloudiness,

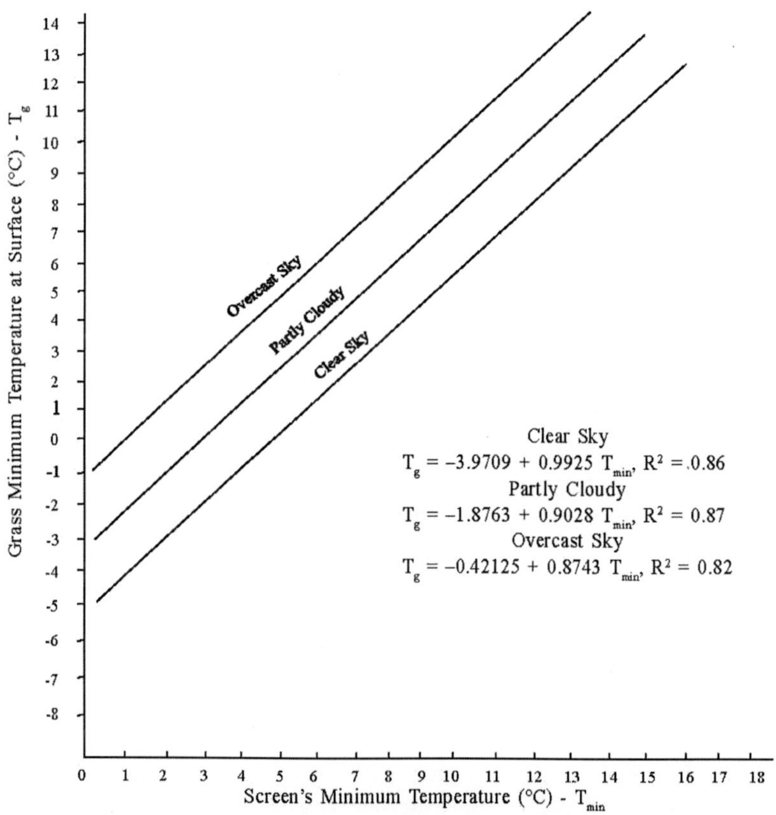

Fig. 9.11: Frost liability under different sky conditions.

wind speed and direction, and humidity at 15/30 minutes time intervals with automatic weather station in the affected crops at the level of more sensitive plant parts. A constant tracking racking in a graphical form will be more useful. A 'danger point' line can be drawn on graph as a horizontal line depending on the frost resistivity of plant organs. Dry bulb and wet bulb temperatures at different height of plant parts, wind speed and direction as well as low and middle clouds are recorded simultaneously. Dew point temperature corresponding to each temperature values are determined. The two curves representing the air temperature and dew point values at different hours from sunset to sunrise will reveal frost liability and frost duration in the night. Fig. 9.12 shows increase in frost danger liability when the air temperature curve dips downwards steeply. Simultaneously dew point curve also indicates the probability of condensation with liberation of latent heat of vaporization which stops the fall in temperature. In may be noted that cooling of lower layer is faster after 2100 hours and reverse is true from sunset to 2100 hours. Therefore, by evaluating both curves we can assess whether

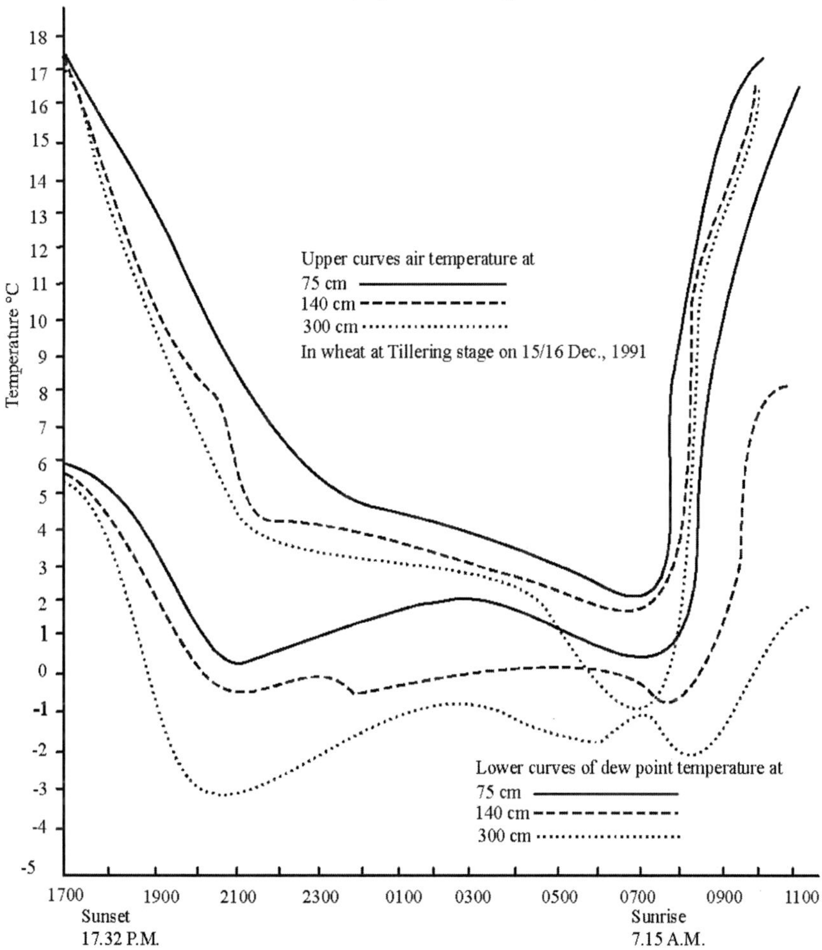

Fig. 9.12: Frost liability in early hours.

condensation is likely to take place or not for frost occurrence during the night hours, and frost duration can also be assessed from the frost point to sunrise. For example if the danger point is -1°C in wheat crop and sunrise is 7.15 A.M. Fig. 9.12 shows temperature and dew point curves to enable forecasts of frost several hours before it strikes to provide time required for taking protective measures. Fig. 9.12 shows that (i) there is steep fall in air temperature as well as dew point temperature from sunset to 2100 hours at the rate of 2.6, 2.88 and 3.23°C per hour at 75, 140 and 300 cm above the ground (ii) after 2100 hours to sunrise there is linear decrease of air temperature at the rate of 0.4, 0.34, 0.32°C per hour at 75, 140 and 300 cm levels (iii) after 2100 hours dew point temperature increases up to one to two hours before sunrise, but sharply decreases just before sunrise at minimum epoch time, and (iv) release of latent heat of vaporization is variable.

Topography of the ground surface plays important role in frost formation. Closed troughs and narrow bending valleys will acquire higher frost liability as compared to crest and upper part of slopes.

The extent of damage caused by frost on agricultural crops/plants depending on the periods of onset, intensity and its duration, phenology growth phase, crop variety, and field agricultural conditions. Plants have been classified into 5 main ecological groups based on tolerance of frost spells (Vetskevich, 1961).

(a) Most stable plants withstanding short spells of frosts to -7 to -10°C or more such as temperate plants.
(b) Stable to frost spells upto -5 to -8°C includes root crops, oil bearing crops and early fibrous crops such as flax and hemp.
(c) Moderate resistant plant withstanding frosts up to -3 to -4°C such as soya, jute etc.
(d) Poorly resistant to frosts such as maize, millet, sorghum, potato, tobacco withstanding frosts up to -2 to -3°C.
(e) Unstable to frosts withstanding upto -0.5 to -1.5°C such as kidney beans, rice, cotton, sesame, groundnut etc. The buds of most plants die at -3 to -4°C. Frosts are particularly damaging to fruit plantations during flowering and ovary formation.

(iii) ***Forecasting of minimum temperature and frosts:***

(a) **Use of past data**: A base guidance map can be prepared for a location using winter season data recorded at 1800 GMT on dry bulb temperature in °F (T_a), wind speed in mph (U), cloud cover in eights of sky (C), number of days since last rainfall (R) (a maximum of 8), and dew point temperature in °F(T_d). Smith (1975) developed a dot diagram with ($T_a + U + C - R$) on Y-axis and T_d on x-axis (Fig. 9.13). A black dot represented frost for nights with a frost and a small dot in a circle for nights with only ground frost, but not air; and a small open circle for nights representing no frost situations. Frost points were closer to the origin and curves were drawn on dot diagram separating in general the different types of night minima. Such maps are useful in characterising the reliable frost forecast assessment for the specific location. The frost duration is also an important parameter for selecting the protection measures.

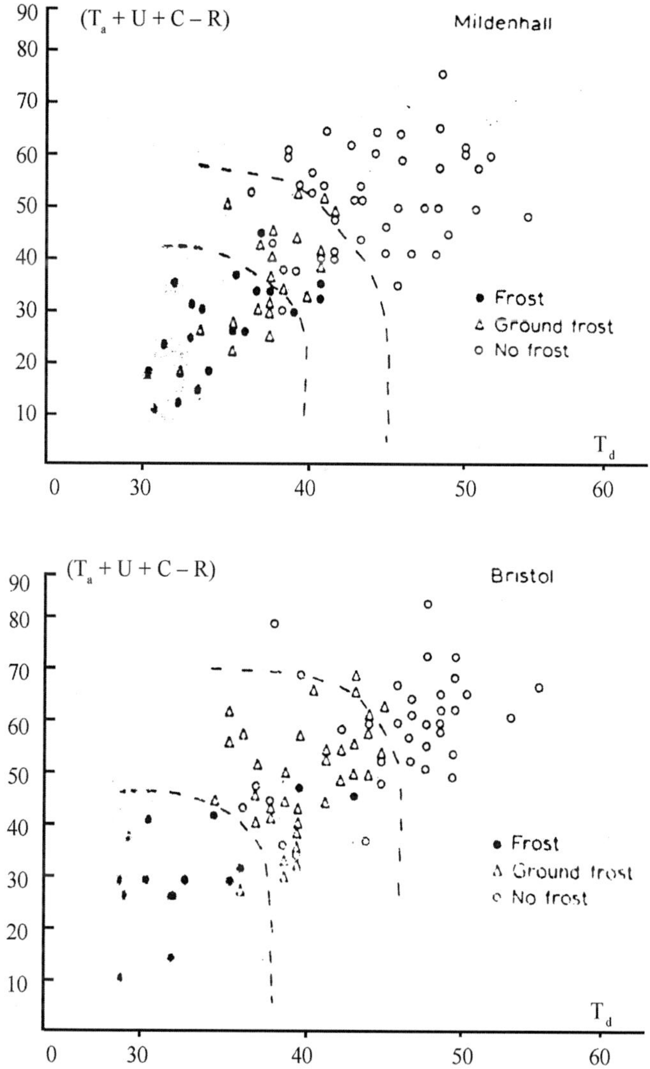

Fig. 9.13: Frost forecasting nomograph for Bristol and Mildenhall. (Smith, 1975)

A fine spray of water is adequate for frost protection upto 5°C. Glass and plastic covers are also effective, but not feasible over large scale.

(b) **Use of synoptic maps**: Synoptic maps are useful for tracing the advancement of cold waves path for frost liability with anticipation of minimum temperatures for frost warning. Mikhalevskii equations (1953) provided good results for frost forecast based on the observations at 13.00 hours and rendered accurate forecast after the observations at 1900 hours as

$$T_{min} = T_w - (T_a - T_w)C \qquad \ldots (9.35)$$

$$T_{gmin} = T_w - (T_a - T_w)2C \quad \ldots (9.36)$$

where, T_a is dry bulb temperature at 1300 hours, T_w is wet bulb thermometer reading at 1300 hours, C is coefficient depending on the humidity of air at 1300 hours. If the T_{min} is less than –2°C, then a frost is expected at -2 to 2°C. At 2100 hours, a correction is made for cloudiness on the basis of observations made at 13 hours. If sky is clear, then value be reduced by 2°C; when cloudiness is 4 to 7 it is not corrected; and when the cloudiness is total, it is increased by 2°C.

$$C = 0.2329 \exp(0.0315\ RH), \quad (R^2 = 0.99) \quad \ldots (9.37)$$

Example: 4

If the air temperature at 1300 hours by dry bulb thermometer is 15°C, wet bulb thermometer at 8.6°C, and relative humidity at 1300 hours is 60 per cent. If sky is clear at 2100 hours, then assess the frost liability.

Solutions:

$$T_{min} = 8.6 - (15 - 8.6) \times 1.5$$
$$= 8.6 - 6.4 \times 1.5$$
$$= 8.6 - 9.6 = -1.0°C$$

Since sky is clear at 2100 hours, anticipated minimum temperature is -3°C. Frost is expected

$$T_{g,min} = 8.6 - 9.6 \times 2$$
$$= 8.6 - 19.2 = -10.6°C$$

Example: 5

If the dry bulb thermometer at 1300 hours is 17.5°C and wet bulb thermometer is 10.5°C with relative humidity being 35 per cent. Sky is clear at 2100 hours, assess the minimum temperature at the soil surface.

Solution:

At 33 per cent RH, C is about 0.8

$$T_{g,min} = 10.5 - (17.5 - 10.5)\ 2 \times 0.8$$
$$= 10.5 - 7 \times 2 \times 0.8$$
$$= 10.5 - 11.2 = -0.7°C$$

with the correction for cloudiness $T_{g,min}$ is -2.7°C and a frost is expected on the ground.

Damage to crops by frost and freezing temperatures can be avoided by improving the thermal balance with microclimatic modifications at the surface. It is possible only if adequate time is available to take such measures at the field level. It is necessary to modify the nocturnal microclimate directly by controlling heat balance, radiation balance and water balance at the surface. Heat is supplied from external sources to raise the surface temperature or nocturnal cooling is cut off to preserve heat energy in maintaining surface temperature.

Sprinkling and irrigation are most common methods to avoid frost, whereas insulating soil surface with mulches, plastic covers, spread of charcoal are feasible in vegetable crops and floriculture. However there are costly methods of supplying heat by burning raw resources fuels and waste chemicals, and wind machines for air mixing, and supplying heat through heaters, wind machines etc which are not economical at farmer's field.

Most forecasting methods are of short term forecast even less than 18 hours for local minimum temperature and frost conditions, and therefore monitoring of synoptic conditions is essential for assessment of probability of occurrence of weather patterns favourable for such conditions. Commonly used synoptic features are

1. dew point temperatures (T_d) < 6°C – the possibility of frost,
2. dew point temperatures <0°C – the probability of frost,
3. maximum day time temperature (T_{max}) ≥ 18°C – no possibility of frost,
4. wind direction between the range SW to SE favours frost,
5. increase/decrease of cloud cover overnight – unfavourable/favourable for frost,
6. Overnight increase in pressure with resultant clearing the sky and decrease of wind speed – favours frost,
7. availability of water vapours even without cloud reduces nocturnal cooling, and
8. expected change in the 24 hour 850 mb temperature cooling due to advection-large change leads to frost conditions.

Thus light or calm winds, low humidity, dry soil or surface conditions, and clear sky favour the nocturnal cooling leading to frost and critical temperatures damaging the young tissues, retarding growth, and crop losses particularly in vegetables, horticultural and fluoricultural species.

(c) **Empirical formulae**: Several empirical expressions have been developed all over the world, but few significant results are being presented to use the available data. Young (1920) forecasted minimum temperature using curvilinear hygrometric expression of the form

$$T_{min} = T_d - \frac{RH - n}{4} + V_d + V_h \qquad \ldots (9.38)$$

where n is 20, 30 and 40 for clear, partly cloudy and cloudy skies respectively, and V_d and V_h are parameters depending on T_d and RH values as $V_d = (T_d + 28)/3$ and $V_h = (RH - 52)/6$ respectively.

However, Allen (1957) provided an alternate version which has been favourably accepted expressed as

$$T_{min} = T_W - \frac{1}{4}(T_d + 16) \qquad \ldots (9.39)$$

where T_d, T_w, and RH are observed at 16.45 hours LST.

Korven (1964) developed an empirical expression for forecasting T_{min} several hours before sunset expressed as

$$T_{min} = T_{17} - \left\{ t + \frac{(T - T_d)_{17}}{2} \right\} \qquad \ldots (9.40)$$

where, T_{17} is the air temperature at 1700 hours LST, t is the number of hours from 1700 hours to sunrise to nearest half hour value, and T is ambient air temperature. Imada and Hirano (1957) assessed T_m as $(T_w + T_d)/3$ posing threat when $T_m < 3°C$, and were divided into 3 groups based on specific humidity. A correction factor of subtracting 3.5°C for lowest specific humidity and to 2.0°C for higher values.

Example: 6

Verify the empirical expressions for prediction of minimum temperature using the following data recorded with automatic weather station over a wheat crop at Stevenson screen's height.

Time (hrs)	Air Temperature (°C) Dry bulb	Air Temperature (°C) Wet bulb	RH(%)	Observed Minimum Temperature (°C)
16.45	22.5	16.3	50	12.7
17.00	20.0	16.0	51	12.7
16.45	12.4	6.0	30	0.1
17.00	11.3	5.2	31	0.1
16.45	16.6	10.4	41	5.6
17.00	16.1	10.6	46	5.6
16.45	18.0	11.2	39	2.9
17.00	17.7	11.1	40	2.9

Solution:

Empirical expressions are location specific and therefore may not hold true at all the places.

	Time	Dew point Temperature (°C)	Wet bulb Temperature (°C)	T_{min} (°C) Using Young (1920)	T_{min} (°C) Using Allen (1957)	T_{min} (°C) Using Imada and Hirano (1957)	T_{min} (°C) Observed
Day 1	16.45	11.4	16.3	11.6	9.5	9.2	12.7
	17.00	11.2	16.0	11.4	9.2	9.1	12.7
Day 2	16.45	-4.9	6.0	2.4	3.2	0.4	0.1
	17.00	-5.7	5.2	1.8	2.6	-0.2	0.1
Day 3	16.45	3.1	10.4	6.8	5.6	4.5	5.6
	17.00	4.4	10.6	7.3	5.5	5.0	5.6
Day 4	16.45	3.5	11.2	7.2	6.3	4.9	2.9
	17.00	3.3	11.1	7.0	6.3	4.8	2.9

(d) **Graphical Approach:** For developing agrometeorological services for farmers simple nomograms or grphical forms are more practically useful to derive quick decision on occurrence or non-occurrence of frost as a function of weather variables at a location.

1. **Jarvi's Approach**: Jarvi (1978) reported that the grass minimum temperature values at the ground surface is useful for frost liability at the ground vegetative surface rather than T_{min} value at screen height. Fig. 9.14 shows the relation between air temperature values at 1400 hours with corresponding dew point temperature values. When the surface temperature on the following night fall below 0°C is marked with (X) otherwise when surface temperature is above 0°C marked as (0). The distribution of (X) and (0) will depict the frost liability. Fig. 9.14 shows the frost liability in south Finland divided into 3 categories of different frost severity levels.

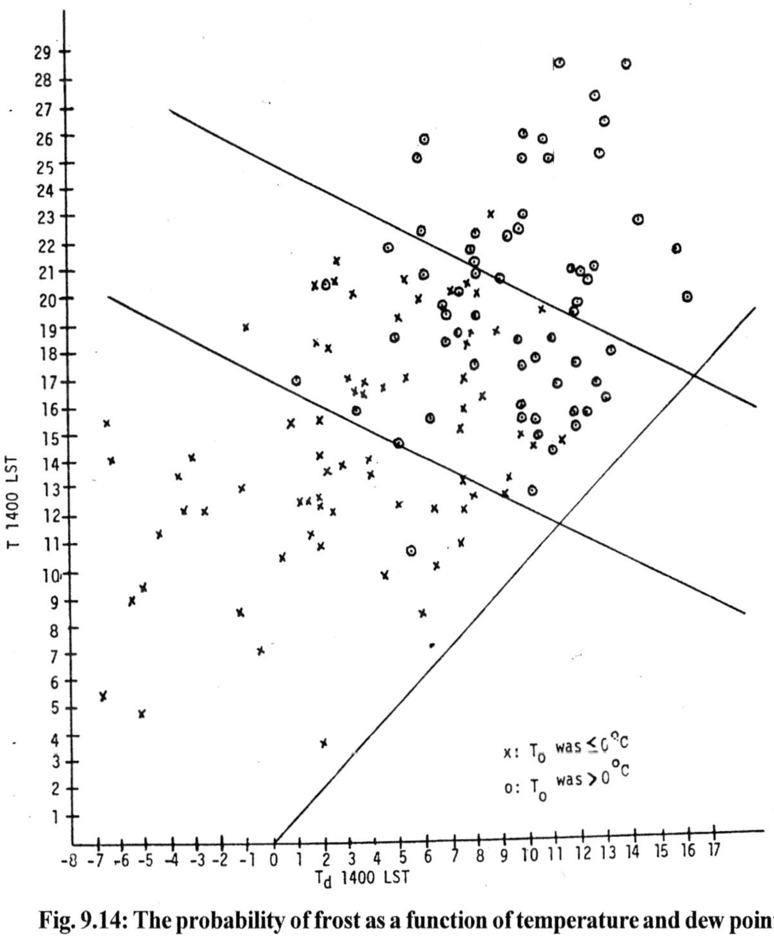

Fig. 9.14: The probability of frost as a function of temperature and dew point. (After: Jarvi, 1978).

2. **Jordian Method**: Jordan Agrometeorological Division for winter crops predicted frost by developing a nomogram graph with plotting of 1200 GMT air temperature vs. 1200 GMT relative humidity with recording simultaneously of grass minimum temperature (T_g). A line is drawn separating the negative values from the positive. A second line is drawn seprating T_g values between 0°C and -1°C from those less than –1°C, and are labelled as areas with no liability, possible occurrence and sure occurrence of frost. Such graphs and drawn for each location and can be quickly used for assessing frost liability after evaluating the synoptic situation, predetermined limits of surface temperatures, 850 mb temperature and height, and 500 mb thickness are satisfied at 1200 GMT at the location concerned (Fig. 9.15) (WMO, 1978).

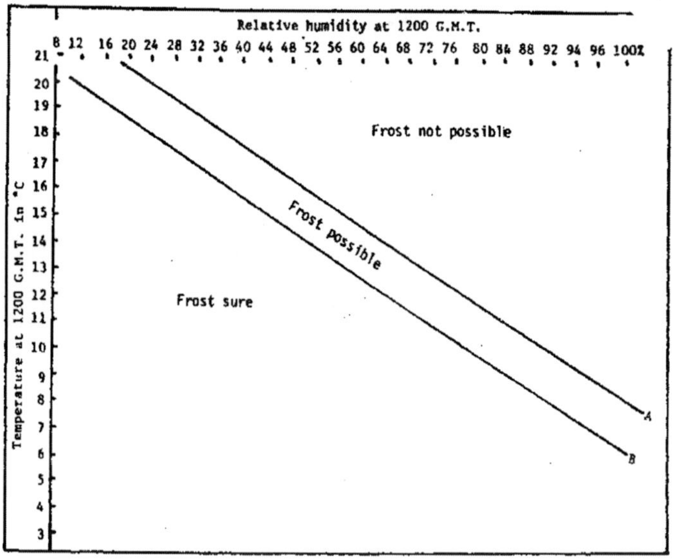

Fig. 9.15: Possibility of frost in the following night, based on combined situations of temperature and humidity at 1200 GMT (After: WMO, 1978).

3. **Soderberg's Approach**: Soderberg (1969) developed a graphical technique using 1600 LST air temperature, dew point, low or middle cloud cover and the expected change in the 24 hour 850 mb temperature (Fig. 9.16) for forecasting of T_{min}. For example, if a 1400 LST air and dew point temperature are 12.5 and -2.5°C respectively, cloud cover of 0.3 and no change of 850 mb temperature is expected, then T_{min} is of the order of 0.5°C under non-advective conditions. However, under advected conditions correction is applied for change in 850 mb temperature and cloud cover conditions.

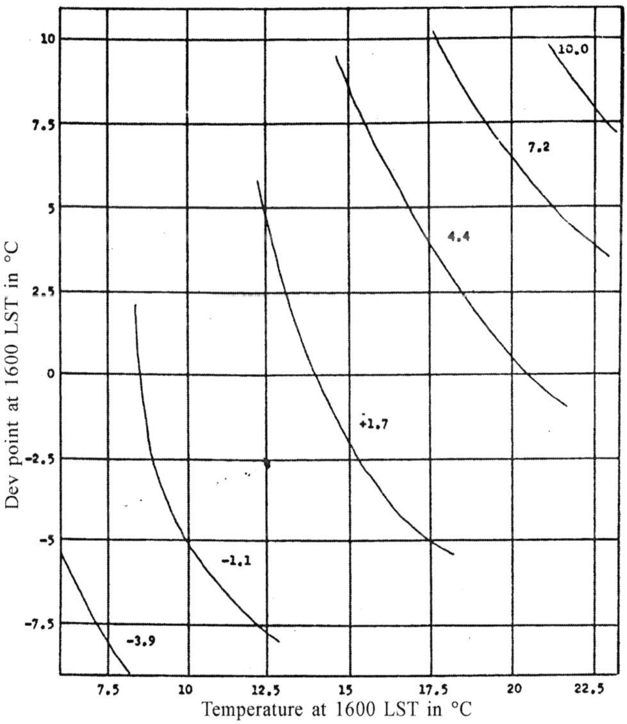

Fig. 9.16: Graph of minimum temperature in °C for non-advective nights when the cloud cover at 1600 LST ranges from 0 through 5-10. (After: Soderberg, 1969)

4. **Brounov's Graph**: Brounov (1957) predictd the probability of the onset of frosts using single factor dry bulb thermometer readings at 1300 and 2100 hours using the graph (Fig. 9.17). Origin of frosts is facilitated by a clear sky, calmness, dry porous soil surface and low humidity of air in the evening hours. On the contrary when it is cloudy there is strong wind, atmospheric and soil moisture is high, then there is usually no frosts at night.
   Also the wet bulb readings at 1300 and 2100 hours were used to obtain minimum temperature. When 2.5 to 3.5°C and 1.5 to 2.4°C were subtracted from 1300 and 2100 hours readings respectively depending on calm, clear sky conditions. Further an additional correction can be introduced to obtain grass minimum temperature at ground surface.

Example: 7

Air temperature recorded at 1 PM and 9 PM were 15°C and 8°C respectively. Use Brounov's graphical method to predict frost liability?

Ans. Temperature at 9 PM is 8°C and difference between 1 PM and 9 PM readings is 7°C. Therefore using Fig. frost probability is 52 percent.

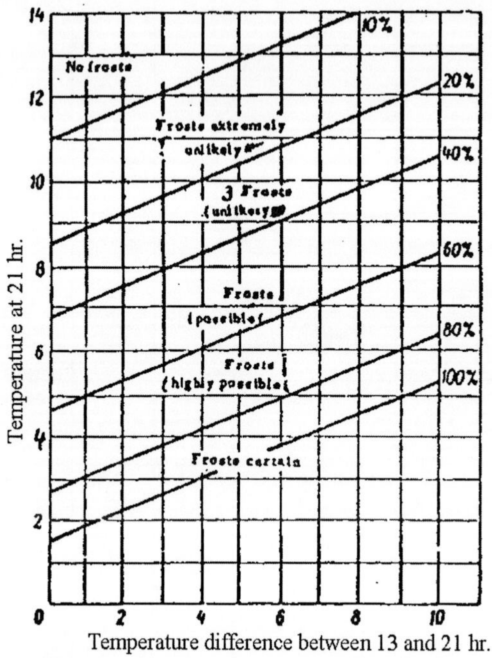

Fig. 9.17: Brounov's graph to determine the probability of frosts. (After: Brounov, 1957)

5. **Surface temperature forecast**: Using the laws of radiative heat transfer from the earth's surface at night, the temperature of the earth's surface is highly correlated with that of the air in the surface layer, and therefore the assessment of minimum temperature of the soil should provide a good forecast of the minimum air temperature which formed the basis of Brunt (1941) formula and its subsequent modifications by Reuter (1951).
   The nocturnal cooling of the earth's surface assessed the air temperature on clear, calm night expressed as

$$\Delta T_0 = \frac{2}{\pi}\frac{\sigma T_0^{\,4}(1-a-b\sqrt{e})}{\rho_s C_s K_s}\sqrt{t} \quad \text{(Brunt's equation)} \qquad \ldots (9.41)$$

$$\Delta T_0 = \frac{2}{\pi}\frac{R_n(0)+\lambda\frac{dT}{dz}+(\Gamma_s+\Gamma)^{CA}}{\sqrt{\rho_s C_s K_s}+C\sqrt{A\rho}} - \sqrt{t} \quad \text{(Reuter's equation)} \qquad \ldots (9.42)$$

where, ΔT_0 is the fall of temperature at ground surface, T_0 is the sunset temperature at the ground surface, t is time in hours from sunset to sunrise, λ is coefficient of heat conductivity of the soil, $\frac{dT}{dz}$ is the change in temperature with depth of the soil,

Γ_s is lapse rate of air temperature at sunset, Γ is lapse rate of air temperature, A is coefficient of eddy conductivity of air, ρ is density of air ρ_s is density of soil, $R_{n(0)}$ is net radiation at sunset over earth's surface, K_s is thermal diffusivity of the soil, C is specific heat of air, C_s is specific heat of soil, A is 0.65 $\overline{U}$ and $\overline{U}$ is mean wind speed in mph, $\sqrt{\rho_s C_s K_s}$ as mean value of 0.29 cal °C^{-1} cm^{-2} min$^{-1/2}$.

Eley (1961) further simplified the Reuter's expression to develop a nomogram for quick assessment of nocturnal cooling to use it for agroadvisory frost services expressed as

$$\Delta T_0 = 2\frac{\sigma T_0^{\,4}(0.56 - 0.08\sqrt{e})}{\sqrt{\rho_s C_s K_s} + C_p\sqrt{A\rho}}\sqrt{t} \quad \text{(Eley's equation)} \qquad \ldots (9.43)$$

$$\Delta T_0 = F.E\sqrt{t}$$

where, E is net outgoing radiation $\sigma T_0^{\,4}(0.56 - 0.08\sqrt{e})$, and F is $2/\left(\sqrt{\rho_s C_s K_s} + C_p\sqrt{A\rho}\right)$. A nomogram was developed to quantify F.E. for various values relative humidity and sunset temperatures (Fig. 9.19), and then using F.E value corresponding ΔT_0 is determined from Fig. 9.18 for the day. However the method can be used only when surface is dry as $\sqrt{\rho_s C_s K_s}$ increases with wet soil

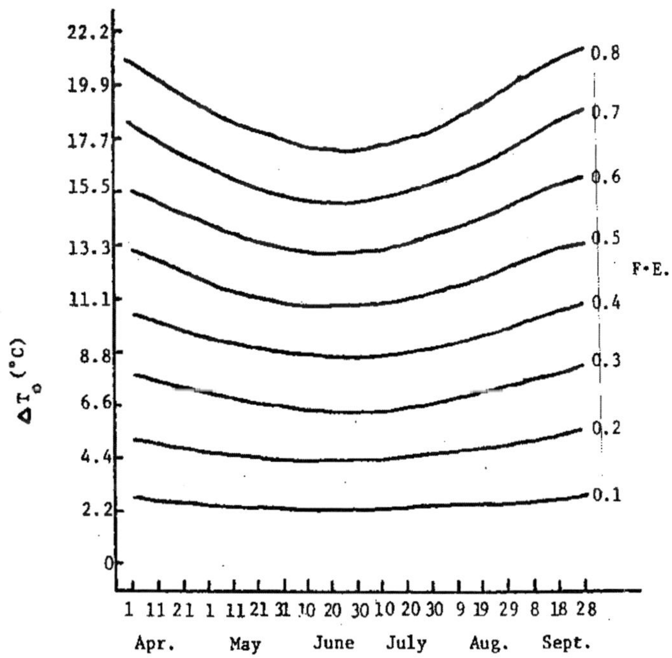

Fig. 9.18: Nocturnal cooling (After: Eley, 1961)

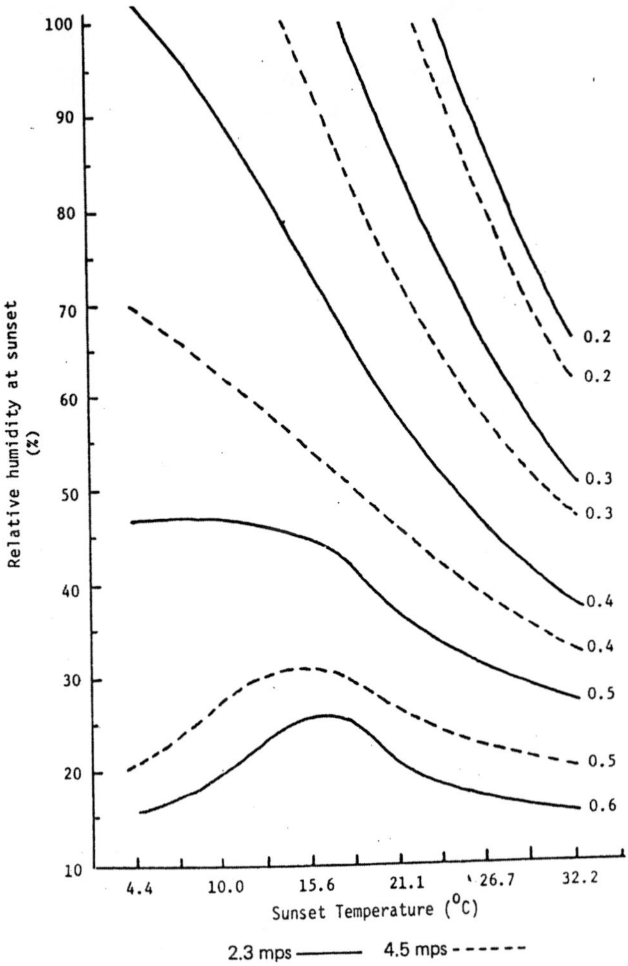

Fig. 9.19: Graphic solution of F . E for wind values (After: Eley, 1961)

and temperature of surface varies. Therefore Kagawa (1968) knowing the soil parameters of a location, simplified the Brunt's expression as

$$\Delta T_0 = CS(0)_n \times 2.03\sqrt{t} \qquad \ldots (9.44)$$

where, $S(0)_n$ is flux of terrestrial radiation with n tenths of cloud expressed as

$$S(0)_n = S(0)_0\,(1 - K_n) \qquad \ldots (9.45)$$

where, K_n is constant depending on cloud type i.e., 0.031 for cirrostratus, 0.063 for altostratus,0.085 for stratus, and 0.099 for nimbostratus. Two nomograms are drawn one for evaluating $S(0)_0$ and other for $2.03\sqrt{t}$ for the day (Fig. 9.20 & 9.21). It is better to mean values of term $\rho_s C_s \sqrt{K_s}$ for good radiation nights at a place or using

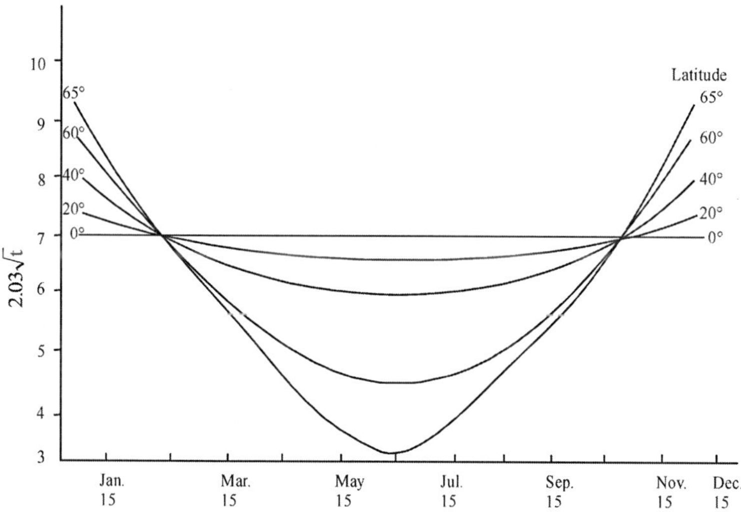

Fig. 9.20: Plot of $2.03\sqrt{t}$ vs. date for different latitudes in N. Hemisphere, t is hours between sunset to sunrise. (After: Kagawa, 1968)

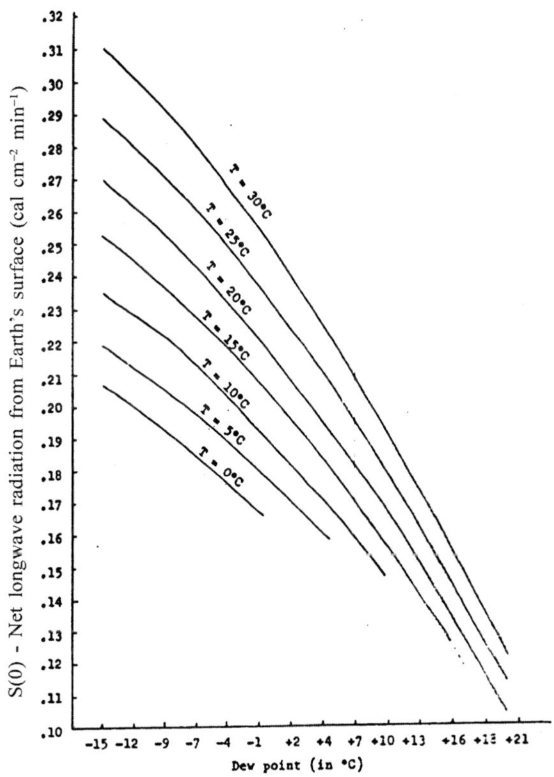

Fig. 9.21: Graph to find S(0), the net longwave radiation from the Earth's surface from temperature and dew point $S(0) = T(1 - .44 - .08\sqrt{e})$ (Brunt's equation) (After: Kagawa, 1968).

soil temperature profiles data a mean value of thermal diffusivity be determined using the expression

$$\frac{\partial T}{\partial t} = K\frac{\partial^2 T}{\partial z^2} \qquad \text{. . . (9.46)}$$

where, $K = K_s/C_s\rho_s$

Davis (1967) provided a graphical solution of the Brunt's equation in the form of

$$\Delta T_0 = \frac{\sigma T_0^{\ 4}(2.26 - 0.32\sqrt{e})}{\rho C\sqrt{K}} \qquad \text{. . . (9.47)}$$

where $\rho C\sqrt{K}$ value of 0.077 was used (Fig. 9.22).

Cellier and Itier (1984) using a one dimensional conservation equation model forecasted the nocturnal cooling of air and soil temperatures of each layer (from one metre deep into the soil up to 200 m in the air above surface) using meteorologial and net radiation data at sunset. In spite of relative roughness of the radiative calculations, the results were in agreement with observed values within 2°C discrepancy both for air and surface temperatures.

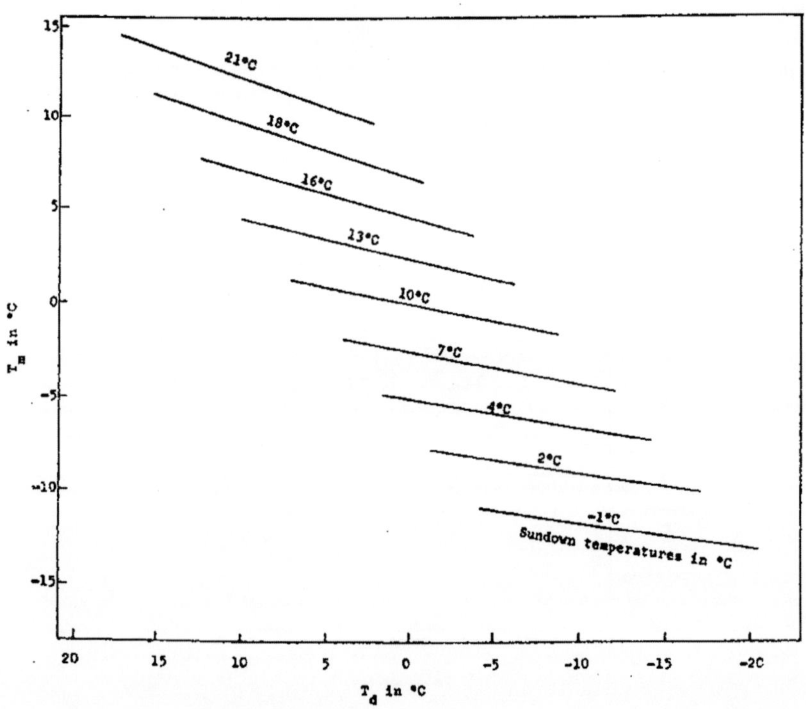

Fig. 9.22: Graphical solution of Brunt's formula (After: Davis, 1967; WMO, 1978).

9.3.10 Forest Fires

Forest fire is a severe problem in dense forest areas caused by lightening, self combustion or manmade affairs. Weather conditions favourable for inducing forest fires are primarily lack of rainfall, low humidity, high temperature and high wind speeds. Frost fire warnings are issued based on combustibility indices (Molga, 1962). Most commonly used Index of combustibility (G) for predicting frost fire danger was proposed by Nestrov (Molga, 1962) as

$$G = \sum_{i=1}^{n} (e_s - e_a)_i T_{ai} \qquad \ldots (9.48)$$

where, T_a is air temperature in °C, e_s and e_a are saturated and actual vapour pressure in mm of Hg, n is number of days since last rainfall.

Index of Combustibility (G)	Fire Danger
<300	No danger
300-500	Slight
500-1000	Modrate
1000-4000	Fir danger
>4000	Extreme fire danger

Meteorological information also helps to control bush fires where a large animal population (cattle and sheep) graze every year. Observations on wind speed and direction, temperature and humidity data will help us to provide useful information to control bush fires more efficiently to save natural pastures for animal grazing. Rizk (1988) reported the use of grassland fire danger meter (Fig. 9.23) for different levels of curing (in per cent) i.e., the risk levels of danger. After setting the meter (Arrow A) at some level against degree of curing (suppose 88 per cent), then set relative humidity in percent (B) value against air temperature in °C (C), finally read of the fire danger index scale (from 0 to 100 scale) and rate of spread fire danger in km per hour against the wind speed value in km per hour (D). The different rotating discs of degree of curving, temperature, humidity, wind speed and fire danger scale are pivoted on a common central axis.

Danger Index Scale	Fire Danger
<2.5	Low
2.5 – 7.5	Maderate
7.5 – 20.0	High
20.0 – 50.0	Very high
50.0 – 100.0	Extreme

Reduction in burnt area will allow grazing additional animals with availability of more pasture lands.

9.3.10 Use of crop weather models

Crop weather models can be put to both strategic and tactical level decisions for agrometeorological services (Baier et al., 1976).

GRASSLAND FIRE DANGER METER

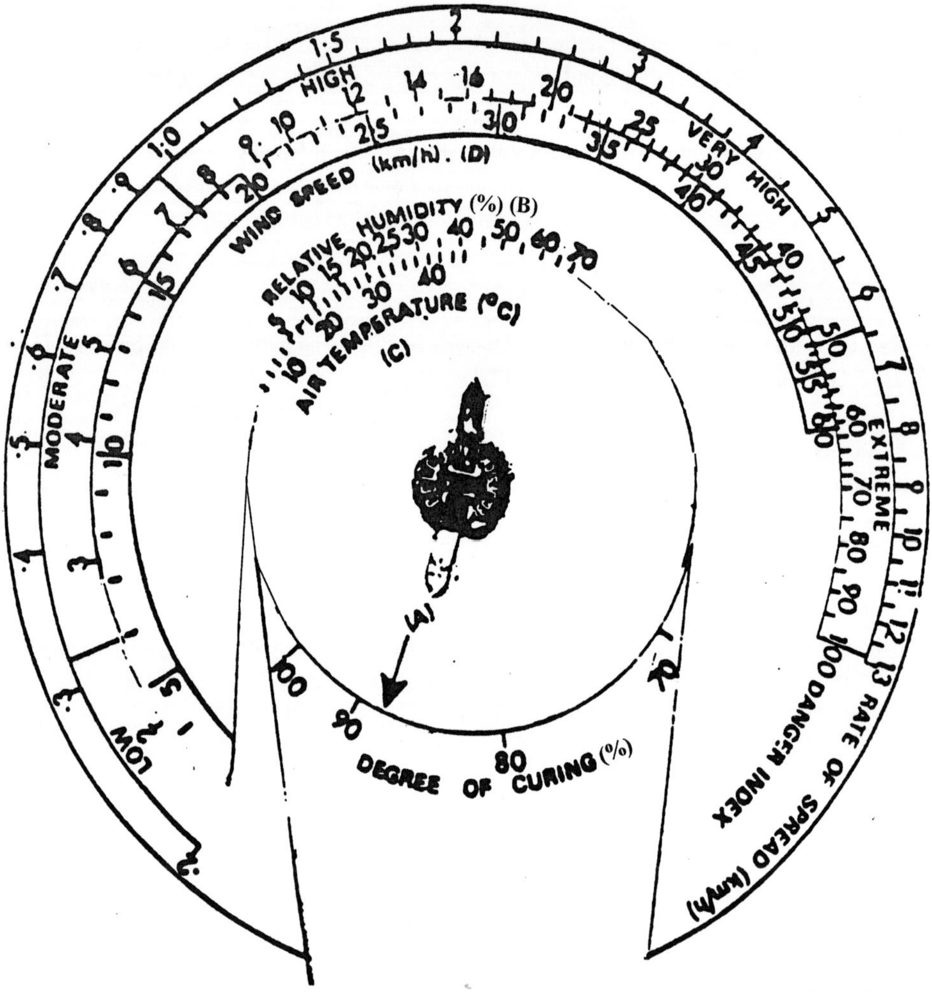

Fig. 9.23

(i) Strategic level:

(a) Crop weather models can be put to fundamental ecophysiological research a long term farm planning, regional resource scheduling, selection of new sites for crops such as Gayule (Nix, 1987) and sunflower (Anderson et al. 1978) in Australia.

(b) With use of optimized parameters- a strategic scheduling of multiple cropping can be envisaged for irrigated or rainfed agriculture. Angus et al. (1974) used CROPEVAL model for alternative grain sorghum cropping system with proper combination of soil type, sowing date, maturity type cultivar and canopy development rate for higher stable yield.

(c) Crop yield forecasting at a location and homogeneous regional levels (Hodges, 1985; Angus, 1984).

(d) Management of national food polices by assessing national, regional and global level food production.

(e) Simulation models can be put to prescribe the optimal time of sowing, maturity type, time of fertilization and agrochemicals based on long term weather conditions such as management options for rice based dryland cropping systems (Jones and O'Toole, 1987).

(f) Long term weather forecasts (monsoon season), pre-advocacy of El Nino-Southern Oscillation (ENSO) offers the rescheduling of regional resources such as irrigation water, fertilizer, yield forecasting, proper choice of agricultural enterprise with likely impact of drought or wet spells in the region.

(ii) Tactical level:

(a) Simulation models can be used for management of individual fields in scheduling irrigation, fertilizer application, use of agrochemicals.

(b) Choice of variety, spraying operations for control of pest and diseases, phenological forecasts with operations of farmers to sow and harvest crops in time.

Crop weather models are based on different basic principles, and therefore can be used for different applications for which their efficiency to quantify the organism's interaction with physical environment is relatively higher.

In empirical statistical approach, one or more agroclimatic resource variables are related to crop responses normally yield. Since these models are dependent on the climate and other local or regional agroclimatic features, they are used in

1. assessing the expected yields over local, regional and large regional scale to manage national food stock.
2. assessing agroclimatic analysis of crop production to illustrate which are significant weather variables for yield variability in the region.
3. evaluation of crop responses to weather elements. Baier's (1973) well known work demonstrated the response of temperature and moisture factors at different phenological stages of wheat.
4. assessment of the impact of natural or man induced variability of climate on crop production. Models can be connected to long term forecast of rainfall with ENSO for regional yield forecasts. Impact of climatic change or weather variability on grain production is much more. It has been suggested that a climate change of the

order of 10 per cent for precipitation or 1°C for temperature might be balanced by the technological progress, whereas 3°C temperature decrease combined with a 30 per cent reduction in precipitation would be expected to reduce yield production by 20 to 30 percent.

In crop weather analysis approach, the product of two or more variable factors, each representing the functional relationship between a particular plant response (normally crop yield) and the variations in selected variables at different plant development phenophases, modify the interactive crop responses the weather and climate variations. Such models are useful in quantifying the

1. daily contribution to final crop yield in response to variations in each of the input variables is quantified,
2. determines the critical threshold values of each of the input variable i.e., the lower and upper threshold cardinal points, and
3. to generate the response pattern of crop yield to daily temperature and moisture parameter at any time of crop development from planting to ripening (Baier et al., 1976). The functional relationship at different development phases was either linear or quadratic for each variable in contributing to final yield.

In crop growth simulation approach, a daily contribution of shysical, chemical and physiological mechanisms between plant and crop growth processes contributes to growth, development and reproduction utilizing the agroclimatic resources surrounding the plant. The daily transformations taking place from sowing of seed to reproduction of seed through different plant development phases is simulated so as to represent the actual behaviour of the plant during its life cycle. Several crop weather models are now available particularly CERES models (IBSNAT programme, 1988); MACROS models (Penning de Vries et al. 1989; Van Keulen et al. 1992) and EPIC model (Sharpley and William, 1984) for various applications in crop yield assessment, potential productivity, genetic improvement, impact of climate change and variability on crop productivity, management of input resources, and for various strategic and tactical decisions for farmer's benefit (Jones, 1990). Simulation models are very useful for tackling practical problems as:

1. when basic plant processes of production and distribution of dry matter and water relations are properly described in the model, entire response of the plant to the environmental conditions are simulated with limiting factors in crop life cycle,
2. to evaluate specific processes such as photosynthesis, transpiration or respiration for accumulation hourly rates over the day and daily rates over the growing season to quantify the total seasonal dry matter production or grain yield,
3. acts as a guide to study in depth behaviour of biological systems to quantify the

relative importance of some factors over the others to explore deeper insight for future experiments, and

4. tactical decisions for management of individual fields in irrigation scheduling and fertilizer application. However at strategic level simulation models have been used to describe the optimum time of sowing and time of fertilization.

Crop simulation models are based on two strong components of growth and development. The crop growth components are modelled based on use of input resources through water balance, intercepted radiation and partitioning coefficients of biomass into different plant organs; whereas development is the progress through an organism's life cycle based on temperature and daylength. The development component is linked with the pest management systems. Therefore, these simulation models have been coupled with pest and disease problems (cotton TEXCIM model, Sterling et al., 1992); sorghum midge model, Baxendale, 1984; sorghum midge SORKAM model, Rosenthal et al. 1989; EPIDEM model for early blight epidemic of potato and tomato, Waggoner and Horsfall, 1969; EPIMAY and EPICORN models for corn leaf blight, Waggoner et al. 1972; Massie, 1973; EPIVEN model for apple scab, Krantz et al. 1973; a flexible plant disease model EPIDEMIC, Shrum, 1975). Even more complex pest and disease management computer model EIPIPRE for wheat have been developed for quick decisions at farmer's field (Zadoks et al. 1984); BLIGHTCAST for late blight of potato (Krause et al. 1975) with spray recommendations based on weather situations; TOMCAST for early blight in tomato (Pitbaldo, 1992) with fungicide schedule based on environmental conditions; PLASMO model for downy mildew in grapevine (Rosa et al. 1993) for simulation of infection development.

Some of these models are being used for forecasting and counselling services on important diseases such as late blight of potato, rice blast, wheat rusts, downy mildew in fruits and vegetables, and apple scab.

Stapper et al. (1986) developed COMAK – a crop management model to help in recording farm activities, use of machinery and manpower, warning on crop management requirements, fertilizer and soil water application, and aiding in financial management. Similarly Poole and Causton (1985) developed AGVISER – a farm management service model to help farmers for field operations. Stapper et al. (1986) developed SIRAGCROP – a crop management model for irrigated wheat and summer crops in Australia to predict soil water contents, schedule of irrigation, and other crop management practices. Even specific models of cropping systems such as nitrogen – tillage – residue management NTRM model (Shaffer and Larson, 1987); cropping system EPIC model (William et al. 1984); ALMANAC model for describing the input resources utilization competition between two crop species or crop-weed (Kiniary et al. 1992) in terms of radiation interception, leaf area development, soil water and nutrient uptake.

9.3.11 Weather watch survey

A regular survey and collection of agrometeorological, phenological observations on pests and diseases in specific agroecological zones is useful to combat insect attacks and disease severity in the crop growing seasons. Multidisciplinary working group comprising experts and extension workers should meet regularly at weekly intervals to develop and disseminate the agricultural consequences of the recent past weather with probable future weather (next 2 to 3 days) and actions to be taken as a result immediately through broadcasting on radio and T.V. channel system. At various times during growing the season, different specific service agencies should be involved such as irrigation water supply, seed multiplication agencies, fertilizer and chemical distribution services, quality control agencies etc. in order to draw maximum benefit at the best opportune time to increase the efficiency or to reduce the quantity of energy need in the complex agrosystem by a judicious choice of the timing of an operation relative to the prevailing agrometeorological conditions. In this way agrometeorlogy can make a significant contribution to the food self sufficiency in the overall economic and social development of agroecological zone and will act as a novel approach for crop protection and its management during the growing season (Mathys, 1987).

With the understanding of various interrelationships between the abiotic variables and the complex of biotic elements including crops, pathogenic organism and other biological agents interfering in the host plant/pest or disease relationship, have led the concept of 'Integrated Pest Management (IPM)'. IPM reconciles the economic interests of the farmer with fundamental ecological and toxicological aspects with a view to take the best advantage of natural elements exerting containment on pests and diseases. Economic threshold limit values are established which delineates the pest population/disease intensity levels below which damage is tolerable and above which specific protection measures are needed. This would require a repeated evaluations of the pest population/disease intensity levels for each major pest or pathogen during the growing season.

Therefore a regular surveillance of crop weather/pest/disease situation during the crop growing season will help us to predict with good accuracy how pest populations/disease intensity levels are going to develop in relation to the prevailing weather and with due consideration of natural enemies exerting containment on pests/diseases. The methods used (Mathys, 1987) in IPM will maintain the physical environment through their mutual interacting effects where use of chemicals is last option when following options are exhausted.

IPM Technology	Methods of Adoption
Biological	Natural enemies – parasitoids, predators, entomopathogens (microorganisms, microbiological agents), biological agents

	attacking weeds. It is intended to create conditions favouring the development of natural enemies.
Biotechnical	Growth harmones, pheromones, autocidal control, attractants-repellants are adopted for growth and development of crop without any side effects.
Resistance breeding	Selection of resistant cultivars to pests and diseases
Cultural methods	Crop fertilization, crop rotation, tillage, intercropping, under sowing, variation in sowing/planting time, pruning, thinning, water management.
Mechanical and physical	Trapping, protection with nets, destruction by hand.
Regulatory	Quarntine, sanitation (pests and pathogen free cultivars eradication programmes).

A development in this direction has been achieved with entomopathogenic bacteria, fungi and viruses some of which i.e., Bacillus thuringiensis have been developed into selective insecticides used in agriculture and vector control of human and animal diseases without disturbing the physical environment. Commercialization of predators and parasitoids with separation of various strains of the same species. A breakthrough in this direction have been achieved with the egg parasitoid Trichogramma for which host specific strains can be isolated, mass – produced, encapsulated and distributed like pesticides. Release of egg parasitoid Encarsia formosa to control while fly Trialeurodes iaporariorum and Phytoseiulus persimilis, a predatory mite attacking tetranychid mites. Phytoseiulus persimilis has been successfully introduced in Africa to control the cassava green mite Mononychelus tanajoa complex. Also synthetic insect pheromones acting as attractants or repellents or causing confusion of insect behaviour such as communication disruption entailing mating suppression (Charmillot and Blaser, 1982) are being used. Another approach is use of insect resistant plants such as Neem tree for playing important role in crop and stored products protection at farmer's level. Solar heating sanitation – a procedure of heating bare soil under plastic sheets kills a large number of pathogens and weed seeds. Israel has optimized citrus production along these lines (Rosen, 1979). Zadoks (1981) computerized management system is used for winter wheat. Federal Republic of Germany and Netherlands have also adopted the integrated programme involving light cultivation, undersowing, intermediate crops, mechanical weed control, minimum fertilizer applications, minimum use of pesticides according to economic threshold of weeds (Steiner, 1979). Such strategies are very useful with climate change scenario for environmentally safe plant protection.

Adequate amount of data on optimum temperature, humidity and other meteorological variables for epidemic of diseases of crops and animals, and outbreak of pest population and their multiplication is available in literature (Mavi, 1986; Bishnoi, 2007; Nagarajan, 1983; Jager, 1999; Gage, 1989) that these variables can be kept in view while monitoring the weather watch survey in the agroclimatic zone for inception of pest and disease problems in the growing season. This will provide strong field in situ input to the weather forecasting station for interacting with surveillance group for feedback to improve the advisory services.

8.4 Food Security in Relation to Weather Advisory Services

Due to climate change and global warming the increasing temperature and associated drought there is increasing climatic variability, and therefore food security all over the world in different regions has attained a grim situation. Reduced rainfall with impact of drought due to limitation of irrigation facilities will further aggrivate the situation of food availability. Sometimes the failure or delay in onset of monsoon rains completely deteriorates the prospects of rainy season crops as well as the winter season crops due to decrease in availability of stored reserve monsoon season rainfall in dams. Import-export of food supplies and agriculture commodity prices are closely linked with the food availability. Therefore there is strong need for a crop weather research on a global scale by the international community which will take into account the ground reality situation in different regions to diversify the sources of exportable surpluses. In many parts of the world, with different environments and cropping systems, the input levels are low to marginal and in such areas tactical advices are more appropriate to manage the non-cash inputs such as operational farm decisions, proper choice and variety of crops, conservation of available soil water, manual handling of weeds, plant populations in relation to crop growing season and water availability periods, use of water harvesting techniques to conserve moisture in low rainfall areas, storing and recycling of stored water in high rainfall belts, etc. will certainly stabilize the grain yield of rainfed crops. Therefore agroclimatic resources should be exploited at every location according to the weather based agricultural planning and agrometeorology will help to seek solutions for better management of natural resources improving far productivity in variable climate conditions. It is necessary to develop in each agroecological zone

1. Specific farmer's requirement for weather information for agrometeorological activities for improving productivity of farming systems including animals, horticultural and forests.
2. Adapt the operational agrometeorological techniques already tested in different agroecological zones to improve the productivity efficiency with efficient utilization of agroclimatic resources (radiant energy, thermal energy, soil water utilization, microclimatic modification).
3. Continuous availability of quality agrometeorological/agroclimatological data to use the quantitative relationships already developed between weather and any component of agricultural production and operational applications for farmer's benefit.
4. Interdisplinary trained expert system is needed to make use of practical operational agrometeorological techniques to improve quantity and quality of agricultural production of farming system prevalent in different agroecological zones. It takes

about 15 to 20 years to develop expert system with a continued effort and maintaining the sound network of agrometeorological data.

5. Use of information technology and mass media to disseminate agrometeorological information until it is available to end users in the desired format so that end users are convinced to apply for their benefit. There is a strong need to link the central agrometeorological advisory unit at national level with agroecological zonal unit where experts of all relevant interdisciplinary disciplines are available to develop, interpret and translate in user's language, and then finally to link with all villages in that zone through telephone answering services and personal computers linking.
6. A stage has now come when operational agrometeorology application must yield economical benefits at farmers and agricultural research institutions.
7. Agrometeorological advices will help the farmers to acquire a less destructive agriculture, preventing soil degradation and promoting the conservation of natural, agricultural and non-renewable resources. Such services will also help to identify underutilization and overutilization of unrenewable resources to modify their use for sustainable exploitation of resources.

Finally the agrometeorological advice will not attain perfection until feedback from farmers becomes so strong to counteract the erroneous statements to improve the advice for its ecological soundness and effective economical benefit of farmer's agricultural enterprise. There are many aspects in agroadvisory forecasting system which cannot be expressed in modelling or synoptic systems where experienced experts can provide reliable advice within the logical framework. Agricultural research service of USDA developed crop management expert system for cotton called COMAX/GOSSYM to advise cotton growers on application of nitrogen, irrigation, and growth regulators (McKinion and Lemon, 1985), and farmers can use these interface COMAX-GOYYSM models to get answers of specific questions. Likewise in agroadvisory service role of expert advice is of immense use in terms of field operational application in different farming system including vegetables, horticulture, agroforestry and animal husbandry components for a balanced requirements of farmer at his own resources.

Beck et al. (1989) developed an expert system SOYBUG to advise Florida farmers on control of four important insect pests of soybean: velvetbean caterpillar, stink bug, corn earworm, and soybean looper with integration of threshold rules based on crop phenology and economics to provide specific recommendations of pesticides and application rates. As a procedural steps, first important phenological stages are identified keeping in view the food habit of the insect pest since velvet bean caterpillar is foliage feeder, whereas stink bug is pod feeder. Then the economic threshold levels are determined at each phenological stage by identifying the levels of infestation with precise available information. For example, Beck et al. (1989) determined the threshold populations for stink bug during early podfill as (0.02 new + 0.02 small + medium large) $\leq$ 0.3 for below threshold, 0.3 to 0.4 for near threshold, ≥0.4 for over threshold; and if medium large ≥ 0.3 then well above threshold. Similarly for velvet cater pillar at early podfill the threshold levels were identified as (medium + 2 large + very large) ≥ 4.5 and per cent damage greater than 16 in terms of defoliation. However if (medium + large + 0.5 very large) ≥5 or (medium + large + 0.5 very large) ≥2 and percent damage greater then 17, then population levels of caterpillar are well above threshold during early podfill stage. Control measures were recommended based on population levels and economic considerations. Expert system also use the generic properties of pesticides (price, residual activity, effectivity) for weed management (Nagarajan, et al. 1987). Knowing the feeding and development behaviour of insects in SOYBUG, Wilkerson (1983) developed the soybean integrated crop management model SICM to simulate the feeding damage on the basis of number of insects in each life stage and insect morphological development based on degree days to assess the damaging population levels. Therefore such expert system should be developed in each zone for delivery of extension information needed for improving the productivity of the farming system.

Exercise 8

Q. 1. Use empirical equations of Young, Allen and graphical methods to test their validity at your locations during winter nights with favourable synoptic situations for frost.

Q. 2. Use graphicl/nomograns to test their validity at your locations and examine which method gives favourable results at your location among the Jarvi, Soderberg, Smith approaches for frost prediction?

Q. 3. Test the validity of Brunt, Reuter, Eley and Davis approaches to quantify the surface temperature during winter months to predict frost liability with use of sunset temperature, sunset relative humidity, sunset prevailing average wind speed using soil parameters of your location.

Q. 4. Determine surface minimum temperature on the following day when sunset temperature is 12°C with sunset relative humidity 70 per cent and prevailing wind speed is 4.5 mps for normal surface conditions. (Ans: Approx. 1.8°C)

Q. 5. Develop relationship between grass minimum temperature and Stevenson screen's minimum temperature recorded at your agromet observatory specifying the soil and vegetative characteristics during the winter season.

Q. 6. Study graphically the frost and minimum temperature impact on wheat at terminal spikelet stage with use of following data recorded with automatic weather station on 17th January, 1992 from sunset (17.46 p.m.) to sunrise (7.15 a.m.).

Time of Day (Hrs)	Air Temperature °C at			Dew Point Temp. °C			Wind Speed (mph)			Wind Direction (deg.)
	75 cm	140 cm	300 cm	75 cm	140 cm	300 cm	75 cm	140 cm	300 cm	
1800	17.5	16.9	16.5	10.6	9.9	6.5	1.89	1.90	4.34	22
1900	15.0	13.4	11.7	11.1	9.7	7.4	3.32	3.7	4.30	22
2000	12.4	10.7	8.5	8.5	6.7	2.5	3.34	3.72	4.33	22

2100	11.1	9.3	6.3	7.6	5.8	1.6	3.41	3.75	4.40	22
2200	10.5	8.3	5.4	7.4	4.9	1.4	3.46	3.81	4.49	23
2300	9.5	7.9	5.2	6.7	4.6	1.7	3.52	3.87	5.54	23
2400	7.6	5.5	3.9	4.7	2.8	1.4	3.56	3.93	4.60	23
0100	6.5	4.1	3.5	4.4	1.8	1.1	3.64	4.00	4.66	24
0200	6.1	3.8	3.1	4.0	1.8	0.8	3.67	4.07	4.73	24
0300	6.1	3.5	2.7	4.4	1.7	0.8	3.71	4.11	4.77	24
0400	6.5	3.1	2.5	4.9	1.3	0.7	3.75	4.15	4.84	24
0500	6.4	3.4	2.4	4.6	1.9	0.3	3.75	4.19	4.89	25
0600	5.5	3.8	2.5	3.8	2.3	1.2	3.80	4.22	4.22	25
0700	4.8	3.6	2.5	3.4	2.3	1.2	3.79	4.23	4.92	25
0800	4.8	3.5	1.9	2.9	1.5	0.6	2.19	2.14	4.03	220
0900	7.1	5.5	4.8	4.5	3.8	2.9	0.15	0.04	0.30	220
1000	10.6	10.7	10.5	7.6	7.4	6.0	0.81	0.42	0.92	227
1100	14.0	15.0	14.9	7.5	8.3	7.9	1.55	1.32	1.80	242
1200	17.2	18.8	18.1	8.7	11.3	10.8	2.19	1.76	2.87	257

Q. 7. Study graphically the minimum temperature impact on wheat crop at elongation stage on 16/17 February, 1992 from sunset (18.10 hrs) to sunrise (7.01 hrs) using the following data recorded with the use of automatic weather station.

Time of Day (Hrs)	Air Temperature °C at			Dew Point Temperature °C			Wind Speed (mph)			Wind Direction (deg.)
	75 cm	140 cm	300 cm	75 cm	140 cm	300 cm	75 cm	140 cm	300 cm	
1800	17.6	17.5	17.2	9.9	9.8	9.4	1.14	1.41	1.79	313
1900	15.1	14.0	13.4	10.4	11.8	8.8	2.61	3.46	4.32	22
2000	12.3	10.4	8.5	8.5	7.5	3.6	2.33	3.74	4.32	22
2100	10.0	8.4	6.5	6.6	6.5	2.0	3.40	3.75	4.33	22
2200	9.5	7.5	5.5	5.8	5.5	1.8	3.45	3.80	4.39	22
2300	9.5	6.7	4.7	6.3	4.1	0.8	3.51	3.87	4.48	23
2400	8.2	6.0	4.2	5.4	3.8	0.6	3.54	3.90	4.54	23
0100	7.4	5.6	4.1	4.9	3.4	1.2	3.58	3.94	4.59	23
0200	6.7	5.0	3.6	4.0	2.6	0.9	3.62	4.0	4.66	24
0300	6.3	4.5	3.6	3.7	2.2	1.0	3.66	4.06	4.71	24
0400	6.1	4.4	3.6	4.0	2.1	1.0	3.60	4.10	4.75	24
0500	5.4	3.9	3.2	3.5	2.1	0.8	3.70	4.11	4.78	24
0600	5.0	3.2	2.4	3.3	1.4	0.3	3.72	4.14	4.82	24
0700	4.8	3.1	2.0	3.1	1.3	0.4	3.66	4.16	4.86	24
0800	4.8	3.2	2.0	2.7	0.6	-1.1	1.48	1.74	2.77	263
0900	8.3	8.4	7.4	4.4	3.8	3.8	0.75	0.43	1.02	318
1000	12.1	13.3	13.1	5.4	6.4	6.2	1.99	1.71	2.80	350
1100	14.9	16.4	16.1	6.0	7.2	5.1	3.03	2.68	4.05	317

Q. 8. (a) Study synoptic weather charts of your agroclimatic zones for advisory services during summer, monsoon and winter seasons.

(b) Study the synoptic weather charts during the specific weather events of cyclones, cloudburst, winter depressions with cold waves or any other weather extreme/ abnormal event occurring in your region.

(c) Support and refine your observations with satellite pictures to prepare the format for weather advisories during hot summer, winter and rainy seasons.

Q. 9. Study T – ϕ diagram with basic data of your location to determine different atmospheric thermodynamic parameters.

Q. 10. Prepare a crop phenological data map of your state for different important economical crops including horticultural and vegetable crops, and develop regressions for predicting different phenological stages.

Q. 11. Using the past records of soil moisture data, precipitation, temperature and evaporation data at your location under bare/cropped soils, develop a soil moisture forecast regression for your location to predict soil moisture at different times of crop sowings/planting.

Q. 12. Develop agrometeorological forecast for field operations at farmer's field using short range/medium range weather forecast report of your zone with use of other supporting informations on crop performance, crop input requirements for soil moisture, fertilizers etc., control of pest, diseases and weeds preferably through IPM controls, impact of prevailing weather on crop's and animal's performance towards yield/yield attributes. Special mention be made for any climatic abnormal feature unfavourable if occurred with its impact on crop/animal's performance such as

(a) Impact of frost and control measures needed.

(b) Prevailing temperatures above/below the normal trend with its impact on crop yield attributes and maturity

(c) Any unfavourable weather event occurred with its impact on crops such as excess rainfall, storm, cyclone, winter depressions with cold waves, heat waves and rainless period of drought severity, lightning and thunderstorm impact of cold waves/wind child indices on human and animal's work performance.

(d) Special mention be made of the benefit of shelterbelts/windbreaks on crop's performance due to its beneficial effects in the region if any farmers have made use in controlling soil moisture, wind speed and as a barrier for cutting of the cold waves, use of cultural/irrigation method to control frost or impact of previous week's agrometeorological forecast report.

(e) Use nomograms, regression models and crop simulation models already developed for assessing the impact of current weather conditions on the crops and animals if any to highlight the food production and management strategies in the zone.

References

Abushama, F.T. 1974. Flies, Mosquitoes and disease in the Sudan. Khartoum Univ. Press, Sudan.

Allcroft, W.M. 1947. Seasonal hypomagnesium of the bovine without clinical symptoms with special reference to the influence of climate on the level of serum magnesium and the incidence of clinical hypomagnesaemia, Vet. J., 103, p. 75-100.

Allen, C.C. 1957. A simplified equation for minimum temperature prediction. Mon. Wea. Rev., 85, p. 119-20.

Anantakrishnan, R. and S. Yagnanarayanan 1989. India meteorological department technical note no. 26-1949, Govt. Photozinco Press, Pune, p. 2-3.

Anderson, W.K., R.C.G. Smith, and J.r. McWilliam 1978. A system approach to the adaptation of sunflower to new environments II. Effects of temperature and radiation on growth and yield. Field Crops Res., 1, p. 153-63.

Angus, J.F. 1984. Rice yield forecasting project progress report 1983. Technical memorandum 84/2. CSIRO Division of Water and Land Resources, Canberra, p. 16.

Angus, J.F., D.H. Mackenzie, R. Morton and C.A. Schafer, 1981. Phasic development in field crops II. Thermal and photoperiodic response of spring wheat. Field crop research, 4, p. 269-83.

Angus, J.F., J.J. Basinski, and H.A. Nix 1974. Weathe analysis and its application to production strategy in areas of climatic instability. Report of IAD/UNDP, Internat. Expet. Consultation on the use of improved technology for food production in rainfed areas of tropical Asia. Food and Agriculture Organization of UN, Rome.

Angus, J.F., M. Damdam, Fazekas de St., C. Growth, H.N.S. Mulyani, M. Sudjadi and R. Wetselaar, 1987. A simulation model of nitrogen response of irrigated rice, p. 137-40. In: Efficiency of nitrogen fertilizers for rice. IRRI, Los Banos.

Angus, J.F., Mackenzie, D.H., R. Morton, and C.A. Schafer 1980/81. Phasic development infield crops I. Thermal response in the seedling phase. Field Crop Research, 3, p. 365-78.

Baier, W., H. Davidson, R.L. Desjardins, C.E., Quellet and G.D.V. Williams 1976. Recent biometeorological applications to crops. Int. J. Biometeorol., 20, 2, p. 108-27.

Baier W. 1973. Crop weather analysis model: review and model development. Jour. of Applied Meteorology, 12, 6, p. 937-47.

Baier, W., D.Z. Chaput, D.A., Russello, and W.R. Sharp 1972. Soil moisture estimator programme system. Tech. Bull., 78, Canada Deptt. of Agriculture, Ottawa, Ont. K1A Oc6.

Baier, W. and G.W. Robertson 1966. A new soil moisture budget. Can. J. Plant Sci., 46, p. 299-315.

Barry, R.G. and R.J. Chorley 1971. Atmosphere, weather and climate. Methuen & Co. Ltd., 11 New Fetter Lane, London EC4, p. 101-02.

Baxendale, F.P., G.L. Teetes 1984. Temperature dependent model for sorghum midge spring emergence. Environ. Entomol., 13, p. 1566-71.

Beaumont A. 1947. Trans. Brit. Mycol. Soc., 31, p. 45-53.

Beck, H.W., P. Jones and J.W. Jones 1989. SOYBUG : an expert system for soybean insect pest management. Agricultural systems, 30, p. 269-86.

Bedford, T. 1946. Environmental warmth and its measurement. Med. Res. Council, War Memorandum No. 17, London.

Bedford, T. 1950. Environmental warmth and human comfort. Brit. J. Appl. Physiology, Feb., 33-38.

Biometeorology by S.W. Tromp 1980. Heyden, London.

Bishnoi, O.P. 1980. Study of climatic water balance parameters in relation to crop planning over Punjab and Haryana. Ph.D. Thesis Submitted to Andhra University, Waltair.

Bishnoi, O.P. 2007. Principles of Agricultural Meteorology, Ch. 10, Oxford Book Co., Jaipur.

Bishnoi, O.P., Harbir Singh and R. Singh 1992. Incidence and multiplication of mustard aphids in relation to meteorological variables. Indian Jour. of Agric. Sci., 62, 10, p. 710-12.

Bolton, F.R. and H.G. Zandstra 1981. A soil moisture based yield model of wetland rainfed rice. IRRI Res. Paper Series, No. 62, p. 10.

Boyden, C.J. 1963. A simple instability index for use as a synoptic parameter. Met. Mag., 92, 1092, p. 198-210.

Brounov, P.I. 1957. Probability of frost determination. In: Agrometeorology (ed.), M.S. Kulik and V.V. Sinelshchikov, Izbramye Sochineniya (Selected Works), Gidrometeoizdat.

Brunt, D. 1941. Physical and Dynamical Meteorology, Cambridge University Press, New York, p. 428.

Buettner, K.J.K. 1957. Present knowledge of correlation between weather changes, surface and air electric space charges, and human health and behaviour. Fed. Proc. 16, p. 631-37.

Cellier, P. and B. Itier, 1984. Un modele de prevision de la temperature minimale nocturne sour abri et a la surface du sol en conditions de gelees radiatives. J. Rech. Atmos., 18, 1, p. 11-22.

Changnon, S.A., S.T. Sonka, and S.L. Hofing 1990. Assessment of uses and values of climate predictions in agribusiness. National climate programme office, grant no NA 89AA – D – CPO35 final report, p. 23.

Charmillot, P.J. and Chr. Blaser 1982. Confusion sexuella contre le carpocapse des prunes Tr. II contribution Bull. Soc Entomol. Suisse, 55, p. 65-76.

Chirkov, Y.I. 1979. Soil climate. In: Agrometeorology by J. Seeman, Y.I. Chirkov, J. Lomas and B. Primault (eds.), Springer – Verlag, Berlin Heidelberg, New York, p. 131-38.

Coakley, S.M. and R.F. Line, 1981. Quantitative relationship between climatic variables and stripe rust epidemics on winter wheat. Phytopathology, 71, p. 61-67.

Davis, G.R. 1967. Some notes on the use of the Brunt formula unpublished research report, National Weather Service, Lakeland, Florida.

Dhar, O.N. and Shobha Nandargi, 1983. Spatial distribution of severe rainstorms over India and their associated areal rain depths. Mausam, 44, 4, p. 373-80.

Eley, J. 1961. Prediction formula for minimum temperatures at Winnipeg. C1R – 3580, TEC – 387, Canada Deptt. of Transport, Meteorol. Branch, Toronto, p. 5.

Evans, C. Lovatt, Smith, D.F.G, Ross, K.A. and Weil-Malherbe, H. 1957. Physiological factors in the conditions of dry coat of horses. Vet. Rec. 69, p. 1-9.

Everdingen, E. Van 1926. Tijdschrift Plantenziekten, 32, p. 129-40.

Fawbush E.J. and R.C. Miller 1953. A method for forecasting hail stone size at the earth's surface. Bull. Amer. Met. Soc., 34, 6.

Fawbush E.J. and R.C. Miller 1954. A basis for forecasting peak wind gusts in non-frontal thunderstorm. Bull. Amer. Met. Soc., 35, 1.

Fedoseev, 1961. Soil moisture content and moisture provision of agricultural crops, water balance of agricultural field. In: Meteorology and Agrometeorology by M.S. Kulik and V.V. Sinelshehikov, Ibramye Sachineniya (Selected Works) English Translation, Gidrometeoizat, USSR. Isreal Programme of Scientific Translation and Deptt. of Agriculture, USA.

Filmer, J.F. 1957. Losses from facial Eczema can be prevented. New Zealand Dept. Agric. Bulletin No. 388.

Fry, J. 1953. Lancet, 1, p. 235.

Gadgil, S., P.R. Sheshagiri Rao, N.V. Joshi, and S. Sridhar 1996. Foecasting rain for groundnut farmers. How good is good enough. In: Climate Variability and Agriculture (eds.) Y.P. Abrol, S. Gadgil and G.B. Pant. Narosa Publ. House, New Delhi, p. 306-19.

Gage, S.H. 1989. Predicting pest occurrence and impact in a dynamical environment. In: Albert & Weiss (eds.), Climate and Agriculture System Approach to Decision Making Charlston, S. Carolina.

Galway, J.G. 1956. The lifted index as predictor of latent instability. Bull. Amer. Met. Soc., 37, p. 528.

Gourangakar, and N.V.K. Chakravarty 1999. Aphid population on Brassica spp. In relation to phenology and dynamic crop simulation modelling for agroadvisory services for farmers. NCMWRF, DST, p. 239-47.

Gupta, G.K., G.V. Subba Rao, and M.B. Saxena 1983. Relationshipp between meteorological factors and the occurrence of ergot disease of pearl millet. Trop. Pest Management, 29, p. 321-24.

Haldane, J.S. 1905. The influence of high air temperature. J. Hyg. Camb., 5, p. 494-513.

Heimann, H. 1961. Effect of air population on human health. In: Air pollution, Columbia Univ. Press, NY, p. 159-221.

Hodges, T. 1985. Using the CERES – Maize model to estimate production for the US cornbelt. Agron. Abst., 77, 12.

IBSNAT Project 1988. Experimental design and data collection procedures (CERES models) for IBSNAT. 3rd Edition, Univ. of Hawaii, Hawaii.

Imada, M. and Y. Hirano 1957. Forecasting of late frost in Tokyshima prepicture, Rep. of Agric. Meteorol., 203, Japan Meteorological Agency, Tokyo, p. 6.

Jager, M.J. 1999. Improved understanding of dispersal in crop pest and disease management: current status and future directions. Agril. and For. Meteorol., 97, p. 331-49.

Jarvi, P. 1978. In: WWO Tech note No. 157, WMO, Geneva, p. 16 and 35.

Jefferson, G.J. 1963. A further development of the instability index. Met. Mag., 92, p. 313.

Jhorar, O.P. S.S. Mathauda, Gurdip Singh, D.R. Butler, and H.S. Mavi, 1997. Relationship between climatic variables and Ascochyta blight of chickpea in Punjab, India. Agri. and For. Meteorology, 87, p. 171-77.

Jones, C.A. 1990. Use of crop simulation models in dryland agriculture. Advances in soil sciences, 13, p. 333-45.

Jones, C.A. and J.C. O'Toole 1987. Application of crop production models in agroecological characterization: simulation models for specific crops. In: A.H. Bunting, ed. Agricultural Environments: Characterization, Specification and Mapping. Commonwealth Agricultural Bureau International, London.

Kagawa, N. 1968. A evaluation of Brunt's cooling formula for minimum temperature prediction at five prairie locations. TEC, Canada Deptt. of Transport, Meteorol. Branch, Toronto, p. 19.

Kelly, J.R. 1982. Rationalizing the need for a fungicide programme on winter wheat using a risk assessment scheme. ECPC Monograph No. 25, p. 77-89.

Kerr, A. and W.R.P. Rodrigo 1967a. Epidemeology of blister blight III spore deposition and disease prediction. Trans. Br. Mycol. Soc., 50 4, p. 609-14.

Kerr, A. and W.R.P. Rodrigo 1967b. Entomology of Tea blister blight IV disease forecasting. Trans. Br. Mycol. Soc., 50 4, p. 609-14.

Kiniry, J.R., J.R. Williams, P.W. Gassman and P. Debacke 1992. A general process oriented model for two competiting plant species. USDA – ARS, Iowa State Univ. and INRA contribution report no ALMANAC 2, p. 1-34, BRC, Temple, Texas.

Kleinsasser T.W. and R.C. Wanta 1956. The development of a forecasting service for use in air pollution control. A.M.A. Arch. Indust. Health, 14, p. 307-18.

Korven, K.M. 1964. Forecasting minimum temperature at R.C.A.F. Station, Penhold, Alberta, CIR-4060, TEC-524, Canada Deptt. of Transport, Meteorolo. Branch Toronto, p. 9.

Krantz, J., M. Mogk and A. Stumpf 1973. Z. Pflanzenkrankh, 80, p. 181-87.

Krause, R.A., L.B. Massie, and R.H. Hyre, 1975. BLITECAST: a computerized forecast of potato late blight. Plant Dis. Rept., 59, p. 95-98.

Kulik, M.S. and Y.Y. Sinelsnchikov, 1961. Meteorology and Agrometeorology. The Isreal Programme of Scientific Translations pursuant to an agreement with the National Science Foundation and Department of Agriculture, U.S.A.

Lamb, P.J., S.T. Sonka and S.A. Changnon, 1985. Use of climatic information by U.S. agribusiness. NOAA Tech. Rep. NCPO 001, p. 67.

Lange, C.T., C.H. Kingsolver, J.E. Mitchell and E. Cherry 1958. Determination of the effects of different temperatures on uredial infection with Puccinia graminis var. tritici. Phytopathology, 48, p. 658-60.

Lewin, J. 1973. A simple soil moisture simulation model for assessing the irrigation requirements of citrus. Isr. J. Agric. Res., 25, 1, p. 3-12.

Lomas, J. and J. Lewin, 1979. Yield prediction of a cotton crop by computer simulation. In: Agrometeorology by Seeman, Chirkov, Lomas, Primault (eds.), Springer – Verlag Berlin Heidelpergo N.Y., p. 1.264.

Ludlam, F.H. 1961. The hailstorm weather, 16, 5.

Ludlam, F.H. 1963. Severe local storms: a review. Amer. Meteorological. Society, Met. Monograph, 5, 27, p. 1-30.

Marshall, I.D. 1959. The influence of ambient temperature on the course of myxomatosis in rabbits. J. Hyg. Camb. 57, p. 484-97.

Massie, L.B. 1973. Modelling and simulation of southern corn leaf blight disease caused by race T. of Helmintho Sporium Maydis Nisik and Miyake. Ph.D. Thesis, Pennsylavania Sta. Univ., Univ. Park, Pennisylavania, p. 84.

Mathys, G. 1987. Novel approaches in crop protection. In: Proc. of seminar on "Agrometeorology and crop protection in the lowland humid and sub-humid tropics". July 7-11, 1986 at Cotonou, Benin, p. 29-36.

Mavi, H.S. 1986. Agrometeorology Ch. 9. and 11, Oxford & IBH Publishing Co., New Delhi.

McArdle et al. 1947. Mem. Med. Res. Counc. (London), Rep. RNP 47-391.

McKinion, J.M. and Lemon, H.E., 1985. Compute Electron Agric., 1, p. 31-40.

Menger, W. 1960. Der Kurverlauf bei asthama bronchialeim kindesalter wahrend verschiedener jahreszatem an der nordsec. In bioclimatology by W.P. Trong., p. 478. Permagon Press, London.

Mikhalevskii, 1953. Prediction of minimum temperature at surface. In: Meteorology and Agrometeorology by M.S. Kulik and V.V. Sinelshchikov, English translation from Russian language, p. 393, Meteorologiia Gidrologiia, Moscow.

Miller, D. 1992. Chained by regulations. Progressive Farmer Midwest, 107, p. 4-6.

Mills, W.D. 1944. N.Y. Agric. Exp. Stn. (Ithaca) Expt. Bull., 630: 4.

Minard, D., H.S. Belding and J.R. Kingston 1957. Pevention of heat casualities. J. Amer. Med. Ass., 165, p. 183-218.

Mitchell, K.J., Walshe, T.O. and Robertson, N.G. 1959. Weather conditions associated with outbreaks of facial eczema. N.Z.J. Agric. Res., 2, p. 584-604.

Molga, M. 1962. Agricultural Meteorology-outline of agrometeorologicl problems part I & II, Warsaw.

Moore, P.H. and R.V. Osgood 1985. Assessment of sugarcane crop damage and yield loss caused by high winds of hurricane. Agricultural and Forest Meteorology, 35, p. 267-89.

Mukherji, M.K. and S.H. Gage 1978. A model for estimating hatch and mortality of grass hopper egg population based on soil moisture and heat. Annals Entomolo. Soc. America, 71, 2, p. 183-89.

Nagarajan, K., J.W. Mishoe and W.L. Currey 1987. Development of an expert system for weed management in soybean. Amer. Soc. of Agricultural Engineering, Tech. paper no. 87 – 5024, St. Joseph, MI 96802.

Nagarajan, S. 1983. Plant disease epidemiology. Oxford and IBH. Publ. Co., New Delhi, p. 267.

Nagarajan, S. and H. Singh 1975. Indian Stem Rust Rules: An epidemiological concept on the spread of wheat stem rust. Plant Dis. Reptr., 5, p. 670-72.

Nagarajan, S. and H. Singh 1976. Preliminary studies on forecasting wheat stem rust appearance. Agric. Meteorol., 17, p. 281-89.

Nagarajan, S. and H. Singh 1990. Long distance dissipation of rust pathogens. Ann. Rev. Phytopathol., 28, p. 139-53.

Nagarajan, S. and Joshi, L.M. 1980. Investigation on predicting wheat rust appearance in central and Peninsular India. Indian Phyto. Pah., 33, p. 84-90.

Nix, H.A. 1987. Land evaluation for potential guayule rubber production in Australia. In: G.A. Stewart eds. The potential production of natural rubber from guayule in Australia. CSIRO Monograph, Canberra.

Nixon – Smith, W.F. 1972. Sheep weather alerts working paper No. 150, 40/145 of April, 1972 and No. 151, 40/3595 of April, 1972, Commonwealth Bureau of Meteorology, Melbourne, Victoria 3001.

Ollerenshaw, C.B. and Rowlands, W.T. 1959. A method of forecasting the incidence of fascioliasis in Anglesey. Vet. Res., 71, p. 591-98.

Palmer, W.C. 1965. Meteorological Drought. Weather Bureau Research paper No. 45, p. 58. Washington, D.C.

Palmer, W.C. 1968. Keeping track of crop moisture conditions, Nationwide: the new crop moisture index. Weatherwise, April, p. 156-61.

Payne, C.G. 1967. In: Environmental Control in Poultry Production. T.C. Carter (ed.) Oliver and Boyd, London, U.K. p. 40.

Perry, K.B. 1993. Future needs and directions in agricultural meteorology extension education. Agric. Meteorology (in review).

Penning de Vries, F.W.T., D.M. Jonsen, H.F.M. Ten Berge and A. Bakema, 1989. Simulation of ecophysiological processes of growth in several annual crops. PUDOC, Wageningen.

Phillips, W.F.R. 1897. Meteorological conditions for sunstrike. Trans. Amer. Climat. and Clin. Assoc., 13, p. 223-37.

Pisharoty, P.R. 1986. Meteorology for the Indian farmers. Indian Space Research Organisation, Bangalore, ISRO – PRL SP-20-86, p. 1-89.

Pitbaldo, R.E. 1992. The development and implementation of TOMCAST. Ontario Ministry of Agriculture and Food, Toronto, Canada.

Poole, K.A., K. and A.E. Causton 1985. AGVISER – Farm computer system. In: G.H. Jackson (ed.). Farm electronics and computing, pp. 375-76. Royal Agri. Soc. of England. Monograph Series No. 4, Stoneleigh, U.K.

Prasad, P., Awadesh Kumar and Mannu Ram 1994. Heavy rains and landslides in Pauri-Garhwal and Nainital districts. Mausam, 45, 2, p. 177-79.

Primault, B. 1958. Elements meeorologiques agissant sur l'apparition et I' extension de la fievre aphteuse schweiz. Arch. Tierheilk, 100, p. 383-99.

Rackliff, P.G. 1962. Application of an instability index to regional forecasting. Met. Mag., 91, p. 113.

Ramaswamy, G. 1988. Disease of crop plants in India. Prentice Hall of India, New Delhi.

Raulson, J.R., W.W. Wolf, P.D. Lingrena and N. Sparks 1982. Migration factor in Hiliothis management. Proc. Intern. Workshop on Hiliothis management, p. 75-86.

Reuter, H. 1951. Forecasting minimum temperatures. Telus, p. 141-47.

Rijks, D. 1992. WMO agricultural meteorology programme, Agric. Meteorol., 59, p. 319-24.

Rizk, D. 1988. Agrometeorology for sustainable and profitable land and water use in agriculture. WCP – 3688 report, WMO, Geneva, p. 1-16.

Robertson, G.W. 1968. A biometeorological time scale for a cereal crop involving day and night temperatures and photoperiod. Int. J. Biometeoro., 12, 3, p. 191-213.

Rosa, M., R. Ganesio, B. Gezzini, G. Maracchi, and S. Orlandini, 1993. PLASMO: A computer programme for grapevine downy mildew development forecasting. Comput. Electron Agric., 9, p. 205.

Rosen, D. 1979. La lutte integree contre les parasitesanimaux et vegetaux en Israel Entretiens ecologiques de Dijon, Cahiers trimestriels no. 3-4, p. 56-60.

Rosenthal, W.D., T.J. Gerik, B.P. Pendleton, G. Teetes, M. Makela and N. Stone 1989. Sorghum midge control using a plant insect model. ASAE paper No. 894043, Intern Seminar Quebec Municipal Convention Centre, Quebec, Canada.

Saxena, S.C. and S. Lal, 1986. Survival of Erwinia Chrysanthemi pv. Zea in soil. Indian. J. Mycol. Plant Pathol, 16, p. 37-41.

Seelay, M.W. 1994. The future of serving agriculture with weather/climate information and forecasting: some indications and observations. Agric. and For. Meteorol., 69, p. 47-59.

Seeley, M.W., J.M. Graham, and C.A. Schrader 1990. Importance and utilization of agricultural weather information in Minnesota: A Survey. J. Agron. Educ., 19, p. 86-91.

Shaffer M.J. and W.E. Larson 1987. NTRM – a soil crop simulation model for nitrogen, tillage, crop residue management, USDA - ARS, Conservation report no. 34-1.

Sharpley, A.N. and J.R., Williams 1970. EPIC – Erosion Productivity Impact Calculator model documentation. USDA Tech. Bull. No. 1768, p. 235.

Showalter, A.K. 1953. A stability index for thunderstorm forecasting. Bull. Amer. Met. Soc., 34, 6, p. 250-52.

Shrum, R.D. 1975. Progress report No. 347. Agril. Expt. Stn., Pennsylavania, State Univ., p. 81.

Siddaramiah A.L., S.A. Desai, and R.K. Hegde 1983. Studies on estimation of loss due to rust and tikka of groundnut, Mysore J. of Agric. Sci., 17, p. 365-67.

Sinha, P. 2005. Epidemiology and development of forecasting model for powdery mildew in jujube for Delhi. Indian Jour. or Agric. Sci., 75, 5, p. 272-76.

Smith, L.P. 1970. Weather and Animal diseases, Tech. Note. 13, WMO, Geneva, Switzerland, T., p. 152, 1970.

Smith, L.P. 1975. The modes of agricultural meteorology: Plants. In: "Methods in Agricultural Meteorology (Ed. L.P. Smith)", Elesvier Publ. Co. Amesterdam, New York, p. 67-96.

Smith, R.C. 1954. Trans Kansas Acad. Sci., 57, p. 397-433.

Soderberg, M.E. 1969. Minimum temperature forecasting during possible frost periods at agricultural weather stations in western Michigan, p. 8, WBTM CR-28, Weather Bureau, Kansas City, Missouri.

Solanki, V.A., B.K. Patel, and A.M. Shekh, 1995. Influence of weather on conidial concentration of mustard crop. Proc. Nat. Symp. on agriculture in relation to environment IARI, New Delhi.

Sonka, S.T., S.A. Changnon, and S.L. Hofing 1992. How agribusiness uses climate predictions: Implications for climate research and provision of predictions. Bull. Amer. Meteorol. Soc., 73, p. 1999-2008.

Stapper, M., R.C.G. Smith, R.E. Speed, and R. Colomb 1986. Framework of implementation and commercialization of SIRAGCROP. Tech. Paper No. 3, C.I.R., Griffth, NSW, Australia.

Steiner, H. 1979. Integrierte Produktion in Ackerbau, ein Entwicklungsprogrammn. Proc. Int. Symp. of IOBC/WPRS on Integrated Control in Agriculture and Forestry Vienna, p. 379-91.

Sterling, W.L., A.W. Hartstack and D.A. Dean 1992. The Texas Cotton insect model TEXCIM 5.0, M.P. – 1646, Texas Agril. Exp. Stn., College Station, Texas.

Sinha, P., N.V.K. Chakravarty, Prajneshu and A. Verma 1999. Production of mango powdery mildew. In: National seminar on dynamic crop simulation modelling for agroadvisory services, NCMWRF, DST.

Symmons, P. 1959. Bull. Entomol. Res., 50, p. 507-21.

Taylor, A.E.R. and R. Muller 1974. The effect of meteorological factors upon parasites. Symp. Brit. Soc. Parasitol., No. 12, Blackwell London, U.K.

Thomas, R.A. Davis, M.D., and Robert, J.T. Joy, Captain, M.C. 1960. Natural and artificial cold acclimatization in man. Proc. of Intern. Bioclim. Congress, Royal Soc. of Medicine, London, 4-10 Sept., Biometeorology, Pergumon Press, p. 286-303.

Tromp, S.W. 1963. Medical Biometeorology, Elsevier Amsterdam, the Netherlands.

Tromp, S.W. and J.J. Bouma (eds.) 1974. Progress in Human Biometeorology Part IB, Swets and Zeitlinger, Amsterdam, Netherlands, p. 512.

Tromp, S.W., H.D. Johnson and J.J. Bouma (eds.) 1976a. Progress in Animal Biometerology, I, Swets and Zeitlinger, Amsterdam, Netherland.

Tromp, S.W., H.D. Johnson and J.J. Bouma (eds.) 1976b. Progress in Animal Biometerology, II, Swets and Zeitlinger, Amsterdam, Netherland.

U.S. Deptt of agriculture, soil conservation service 1972. S.C.S. Hydrology Handbook, Hydrology, Section, 4, Ch. 4-10.

Valli, V.J. 1968. Weather and plant diseases forecasting. Agroclimatoogical mehods, Proc. of Reading Symp., UNESCO, p. 341-45.

van Keulen, H., F.W.T. Penning de Vries, F.M. Dress 1992. A summary model for crop growth. In: Penning de Vries and van Laar (eds.), Simulation of plant growth and crop production, p. 464.

Vetskevich, G.Z. 1961. Agrometeorology. Israel Programme of scientific translation, Jerusalem (ed.) AI Rudenko.

Waggoner, P.E. and J.G. Horsfall, 1969. EPIDEM – a simulator of plant disease written for a computer. Bull. Conn. Agric. Expt. Stn., No. 698, Conn., USA, p. 80.

Waggoner, P.E., J.G. Horsfall, R.J. Luken 1972. EPIMAY – A simulation of southern corn leaf blight. Bull. Conn. Agric. Exp. Stn. No. 729, Conn., USA, .p. 84.

Wallin, J.R. 1951. Phytopathol., 41, p. 37-38.

Wang, J.Y. 1967. Agricultural Meteorology. Agriculture Weather information service, San Jose, California, USA.

Wilkerson, G., J. Mishoe, J. Jones, J. Stimac, W. Boggess and D. Swaney 1983. SICM: soybean integrated crop management model : model description and user's guide, version 4.2, Agriculture Engineering Deptt., Res. Rep. AGE – 1. Agr. Engr. Deptt., Univ. of Florid, Gainesville, Fl. 32611.

Wilkerson, P.R. and Wilson, J.T. 1959. Survival of cattle ticks in central Queensland pastures. Aust. J. Agric. Res., 10, p. 129-43.

William, J.R., C.A. Jones and P.T. Dyke 1984. A modelling approach to determine the relationship between erosion and sol productivity. Trans. ASAE, 27, 1, p. 129-44.

WMO, 1978. Jordan meteorological society method for frost prediction. In WMO Tech note 157, p. 18.

Yaglou, C.P. 1926. The thermal index of atmospheric conditions and its application to sedentary and industrial life. J. Industr. Hyg., 8, p. 5-19.

Yaglou, C.P. and Minard, D. 1957. Control of heat causalities at military training centres. Arch. Indust. Health, 16, p. 302-16.

Yashida, S. 1981. Fundamentals of rice crop science. The international rice research institute, Manila, Phillopines.

Young, F.D. 1920. Forecasting minimum temperatures in Oregon and California. Mon. Wea. Rev., 16, Supplement, p. 53-60.

Zadoks, J.C. 1981. EPIPRE : a disease and pest management for winter wheat developed in the Nethelands. EPPO. Bull., 11, 3, p. 365-69.

Zadoks, J.C. 1984a. Analysing cost effectiveness of EPIPRE. Netherlands, EPPO Bull., 14, p. 401-07.

Zadoks, J.C. 1984b. EPIPRE – a computer based scheme for pest and disease control in wheat, p. 237-44. In: Cereal Production (ed.) E.J. Gallagher, Butterworths, London, p. 354.

Zadoks, J.C. 1984c. A quarter century of disease warning, 1950-1983. Plant disease, 60, p. 352-55.

Zadoks, J.C., F.H. Rijsdijk and R. Rabbinge 1984. EPIPRE: a system approach to supervised control of pests and disease of wheat in Netherlands. In: Pest and Pathogen control-strategic, tactical and policy models, Conway, G.R. (ed.) John Wiley & Sons, Chichester, p. 344.

10

Climate Change and its Impact on Agricultural Productivity

10.1 Introduction

Climate/weather is an important component of crop production causing high variability in yield potentials even in highly favourable and high technology agricultural ecosystems. During the last 50 years, human activities all over the world have exploited the natural resources to meet their needs inducing a disturbed ecosystem with dangerous signals of continuous global warming and climate change scenarios. It has now become a burning issue in every part of globe to assess the climate change behaviour using past records, and to predict future expected global changes in order to devise ways to mitigate the adverse effects on the society at large. These activities have induced a global climate change of increase of 1 to 6°C during the past 100 years resulting in extreme variability that 2002 and 2003 were declared as the warmest years (IPCC, 2001). Europe had a 1994 summer with a 6°C above the normal level. Now there is retreat of glaciers and warming of tundra's permafrost.

With development of fully coupled atmosphere-ocean-land interaction climate models, a significant advancement has been made in assessing the realistic climate scenarios, but it has not been possible to quantify climate processes over a timescale of a crop season.The resolution of both atmospheric and ocean models are inadequate for their coupling to quantify climatic variability. When the output of these climate models are used as inputs in any agricultural systems with such climatic interannual variability, these systems will become more vulnerable under the expected scenarios of climate change. Hoogenboom et al. (1995), Rosezweig and Iglesias (1998), and Singh and Ritchie (1993) with the use of simulation models quantified the impact of climate change on production of food crops with use of climate models (GCMs) scenarios of increasing temperature and varying precipitation levels. The impact on different crops varied from about 10 to 30 per cent crop yield reductions in different parts of the globe (Baethgen and Magrin, 1995; Singh and Padilla, 1995). Unfortunately, many of the impacts of global worming are irreversible. A common conclusion globally has been felt that the expected impact of climate change on crop production may be small, with negative impacts over low latitudes and positive impact over high latitudes. Therefore, it is worthwhile to have a brief review on climate change complexities in order to mitigate their adverse effects on agricultural production in the 21st century. In addition to a gradual change in climatic variables there is need to assess the risk in agricultural shift with global warming. Currently 95 per cent of our food is produced by 30 different kind of plant genetical varieties, but with declining stock of wild plant varieties will limit the production of new strains (Kellogg and Schware, 1986).

10.2 Forces of Global Environmental Change

Global, regional and national levels environmental problems are increasing due to man-made activities all over the globe. The situation has been so alarming that international agencies have focussed attention on a time frame preventive and corrective measures to protect the global environment with a strong commitment to scientific and technological advancement to meet the requirement/needs of human population. This would demand a strong emphasis on economic development with maintenance of the environmental quality. There are several factors influencing the global environmental change due to the economic development in the present day scenario. Coates (1991) listed following parameters inducing the global change as

(i) man-made activities all over the globe to sustain competitive environment at national, regional and global level,
(ii) economic development due to rapid industrialization in order to meet needs of human population,
(iii) population growth to create imbalance with the available resources,
(iv) rapid urbanization with reduction of productive land and shifting of human population to sustain livelihood,
(v) rapid industrialization with multinational involvement to exhaust natural resource without longterm impact planning,
(vi) Overexploitation of natural resources of landuse, soil health, water use with environmental degradation and loss of productivity potential,
(vii) occurrence of pests and disease epidemic with degraded environment,
(viii) impact of natural clamities of meteorological/climatological events such as ENSO, droughts, floods, cyclones, earth quakes, sonami and tidal waves, storms, typhoons etc. creating wide spread damage/losses,
(ix) complexity of induced environmental problems due to the above factors demanding a new scientific challenges to develop remedial measures, and
(x) non-coherence of public polices and attitudes at national, regional and global level to treat Earth as a single entity of God given Gift.

Keeping in view of severity of global environmental problems there is consensus on (World Commission on Environment and Development, 1987; Coates, 1991)

(i) high rate of emergence of green house effect due to rapid exhaustion of fossil fuel with industrialization,
(ii) deforestation to meet needs of human population in urbanization,
(iii) depletion of ozone layer,
(iv) nuclear warfare and radioactive waste matter destructions,
(v) overexploitation of natural resources for agricultural produce, energy production with nuclear power plants,
(vi) acid rain, industrial water and air pollution, and salinization of soil resource, and
(vii) collision of celestial bodies for environmental disastors.

Therefore, there is a growing demand for every nation to provide clean air, clean water, clean food material with preservation of environment and growing awareness to avoid toxics in any form. Growing awareness for use for nuclear energy for peaceful purposes in creating balance to meet the requirements of human population at national, regional and global level, as a single entity.

10.2.1 Environmental pollution

The above factors amounts to increase the environmental pollution problems. Therefore, environmental pollution is very complex with discharge of foreign (undesired) substances into the ecological system leading to measurable changes deteriorating the quality of natural environment as a whole which is prejudicial to living organisms on the earth's surface. Air, water and soil are natural sources of pollution. However, for global warming, we will restrict to air pollution. Air pollution is created with polluting compounds of the bi-products of sulphar, nitrogen, carbon compounds, hydrocarbons, ozone, metallic elements, dust, smoke and waste particles etc. Such substances not only affect the humans/animals (respiratory ailments, cardiovasucular diseases, cancers, allergies), but also plants (diseases affecting leaf tissues, chlorosis, chlorophyll loss, change in plant colour, reduced growth and grain yield). Pollutants penetrate into the stomata of leaves through the action of wind or rain or deposit of impurities on plant parts to cause phytotoxicity and reducing photosynthesis. Pollutants also influence the meteorological variables at the locations as winds carry solid particles in suspension over long distances. Clouds of ash or solid particles are very common over industrial areas, or due to volcanic activity. Forest fire also transmits large quantities of pollutants in form of smoke containing unburnt material and fumes of oily tree species. In addition pollen, microbes, organic fermentation, industrial fumes of ammonia, carbon dioxide, methane, sulphides, nitrogen oxides also pollutes the aeral environment severely. The acceptance level of these pollutants in the atmosphere for vegetations is given in Table 10.1 produced by U.S. Environmental Protection Agency. The liquid water in different forms over earth surface carries substances originating from the waste products of human activity which directly

Table 10.1
US Environmental Protection Agency (USEPA) ambient air quality standards.

Pollutant	Quality Standard
SO_2	80 $\mu g\ m^{-3}$ (0.03 ppm) annual (A.M.)
	365 $\mu g\ m^{-3}$ (0.14 ppm) 24 hr. maximum in new year
SPM	70 $\mu m\ g^{-3}$ annual (G.M.)
	260 $\mu m\ g^{-3}$ 24 hr. maximum once
CO	10 $\mu m\ g^{-3}$ (9 ppm) 8 hr. maximum once
	40 $\mu m\ g^{-3}$ (35 ppm) one hour maximum once a year
NO	160 $\mu m\ g^{-3}$ (0.08 ppm) one hour maximum once a year
	100 $\mu g\ m^{-3}$ (0.05 ppm as NO_2) annual (A.M.)
Hydrocarbon	160 $\mu g\ m^{-3}$ (0.24 ppm as CH_4) 3 hours maximum once a year.

(After: Lacasee and Treehow, 1976)

A.M. – airthmatic mean,
G.M. – Geometeric mean,
SPM – Solid particle matter

goes into the soil rendering it unproductive. The problem of water pollution have increased with industrial effluents, population increase and urbanization of big cities. These products are in the form of acids, industrial bases, hydrocarbons, cyanides, sulphides, detergents, pesticides, weedicides, which results in spread of infectious diseases, toxic and carcinogenic substances, reduction in quantative and qualitative of food production, and constraints for agricultural and industrial development. Biological effects resulting from the action of pathogenic bacteria, viruses and substances remove the free oxygen from water causing illness in animals and plants.

Pollution due to agricultural activities is characterized by the volume of residues produced, use of agrochemicals as fertilizers, herbicides and pesticides. Deposition of heavy metals- lead, mercury compounds and arsenic acid salts are persistent pollutants. With availability of land and water resources, farmers have been overexploiting the soil with 200 to 300 per cent of cropping intensity with advancement of agricultural technology with use of higher doses of fertilizer responsive hybrid varieties of rice, wheat, maize, sorghum, sugarcane, cotton etc. A widespread transformation of forest plant cover areas into agricultural farmlands resulted into the transformation of the complete vegetative cover ecology into a surface with higher surface evaporative and unsheltered topography of field crops with irrigation or rainfed ecology. The use of fertilizers is on increasing scale with higher concentrations leading to saturation of soil and pollution of ground water. The artificial fertilizers have become essential to meet the demands of intensive cropping system (Paddy/cotton – wheat or sugarcane). Therefore the artificial fertilizers having nitrogen, phosphorous or potassium base will result in pollution of soil and ground water if not applied on soil test and crop's requirement basis.

10.2.2 Causes of climate change

Inter-governmental Panel on Climate Change (IPCC) in 1996 referred climate change as a movement in climate system because of internal changes within the climate system or in the interaction of its components or because of changes in external forcing either by natural factors or anthropogenic activities. There are factors which cause the natural variation of climate systems such as

(i) changes in total solar irradiance over a long period of time.
(ii) variation in Earth's orbit around Sun as well as position of rotational axis with respect to the plane of the orbit.
(iii) water vapours and cloud variation influencing earth-atmosphere radiation balance, and
(iv) volcanic energy and aerosols.

These changes have a time scales above 10,000 years, but the other causes due to human activities which are more serious from climate change point of view in different parts of the globe such as

(i) the enhanced concentration of green house gases (GHGs),
(ii) changes in concentration of atmospheric aerosols,
(iii) variation in liquid water on earth's surface,
(iv) changes in global snow and ice cover,
(v) landuse surface changes due to large scale afforestation or deforestation influencing the energy balance components,
(vi) changes in climate system components such as ENSO and La Nena phenomena influcncing regional climate systems,
(vii) Changes in sea surface temperatures to induce the oceanic heat transport and sudden change in oceanic circulation in North Atlantic and Eastern Pacific regions, and
(viii) Anthropogenic causes due to combustion of fossil fuels, burning of forests and agricultural waste.

Among all these factors the rapid enhancement in the concentration of greenhouse gases is due to industrial revolution, agricultural mechanization, urbanization, widespread deforestation, combustion of fossil fuels and overexploitation of natural resources. The dominating GHG is CO_2, but variable concentration of water vapour acts as a passive companion to enhance the warming effect of other GHGs (Houghton et al. 1996).

Climate change strongly influence the agricultural production activities (Rosenzweig and Hillel, 1998) as they contribute in building up greenhouse gases particularly methane, nitrous oxide, water vapour, carbon dioxide and halocarbons. IPCC (2001) reported the enormous increase in the concentration of these gases over the last 200 years (Table 10.2), and their relative impact when concentration increases by one mole is greater than that of CO_2 by 10 to 100 times for CH_4, by 100 to 1000 times for N_2O, and more than 10,000 times for CFCs.

Global CO_2 emissions resulting from human activity (billion tons of carbon per year) is rising since 1850 (Fig. 10.1a) and contributing to atmospheric CO_2 concentrations (Fig. 10.1b) thereby reducing the efficiency with which the Earth's surface radiates the energy to space resulting in positive radiative forcing for warming the lower atmosphere and earth's surface (Fig 10.1c). Atmospheric CO_2 is utilized by plants for photosynthesis and is also dissolved in the ocean. About half of the anthropogenic CO_2 emissions is utilized by natural processes, but the net CO_2 in atmosphere have increased from 280 ppm from pre-industrial age to 370 ppm by the end of the 20th century (Houghton et al. 2001).

In order to sustain life of living organisms with present emission rate 0.75 Gtc/year of anthropogenic CO_2, it is desired to have immediate steps for equalization of emission and removal rate scenario. When emission is less than the net removal rate, scenario is maintained and it may take 2 to 3 centuries to stabilize atmospheric CO_2. In order to curtail net CO_2 scenario, the net removal rate must exceed the emission rate of anthropogenic gases so that

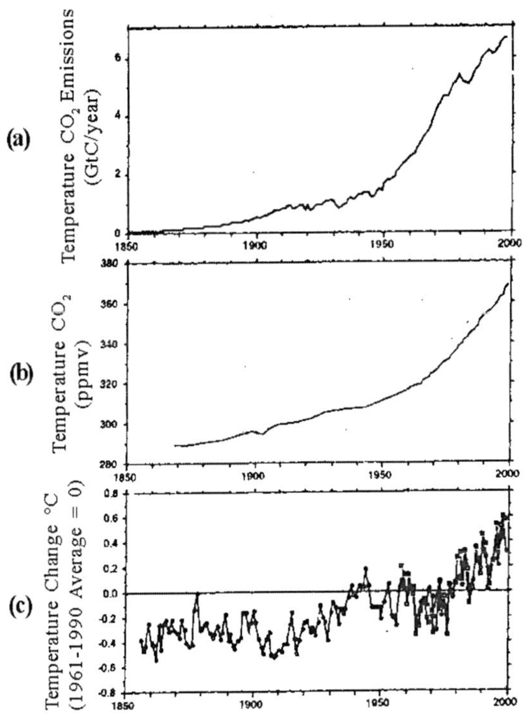

Fig. 10.1a: Global CO_2 emissions resulting from human activity (billion tons of carbon per year)
Fig. 10.1b: Atmospheric CO_2 concentrations parts per million.
Fig. 10.1c: Average global surface temperatures °C. The zero lines is set to the average for the period 1961-1990. [*After*: Houghton et al., 2001]

life of living organisms is saved from the adverse effects of global warming (Sterman and Sweeney 2007).

Budyko's expression for rise in temperature with the elevation of CO_2 levels in the atmosphere, climatic noise and anthropogenic factors expressed was of the form

$$\Delta T = \Delta T_c \left\{ \text{Ln}\left(\frac{C_2}{C_1}\right) \Big/ \text{Ln } 2 \right\} . \mu \qquad \ldots (10.1)$$

where, T_c is the change in temperature due to doubling of CO_2 concentration $\simeq 3 \pm 1.5$ °C, C_1 is 285 ± 5 ppm, in 1881, C_2 is 343 ± 5 ppm in 1984, μ is 0.7 ± 0.2 representing the thermal inertia of the climate system. Using these values ΔT will be 0.56 ± 0.17°C.

9.3 Global Climate Models

Climate is an integration of an enormous manifold of interacting physical, chemical and biological processes at all levels in the atmosphere and ocean surface locked with several problems in climate modelling such as:

Table 10.2
Changes in concentration of different gases in 20th century

Atmospheric greenhouse gas	Atmospheric residing time (yr)	Change in concentration during the period		Annual trend (%)	Increased terrestrial radiation after 1750 (Wm^{-2})	Anthropogenic Sources
		1000-1750	1751-2000			
Carbon dioxide (CO_2)	50-200	280 ppm	370 ppm	0.50	1.87	Fossil fuels, deforestation landuse and land erosion
Methane (CH_4)	10	700 ppb	1750 ppb	0.90	0.52	Rice cultivation, fossil fuels, biomass burning, industrial activity, by product of swamps, tundra and ruminant.
Nitrous oxide (N_2O)	150	270 ppb	316 ppb	0.25	0.12	Fertilizers, fossil fuels, biomass burning
Chlorofluoro-carbons (CCl_3F)	65	-	302 ppb	4.00	0.27	Refrigerants, solvents, foam production, propellants, destroying of stratospheric ozone.
(CCl_2F_2)	130	-	523 ppb	4.00	0.22	-do-
Tropospheric ozone (O_3)		15 ppb	35 ppb, varies in region to region	1.00	-	Nitrogen oxides and hydrocarbons in presence of sunlight in northern hemisphere.
Stratospheric ozone (O_3)		-	Decreased			
Total					3.0	
Increased terrestrial radiation by end of 21st centry			CO_2 equivalents		Increased terrestrial radiation after 200 (WM^{-2})	
Scenario 1 (present trend to contine)-A			1343 ppm		6.90	
Scenario 2 (Drastic reduction) – D			552 ppm		1.30	

(After IPCC, 2001)

(i) Complexity of each individual system,
(ii) Widely different timescales for the different climate subsystems,
(iii) Coupling problems of these systems, and
(iv) Numerical integration of these weather systems, of different time and space scales, which closely coupled together through different dynamic and thermodynamic processes. For a particular weather system it is necessary to specify temperature, wind velocity and humidity for a number of vertical levels from earth surface high into the stratosphere. Surface pressure and accurate boundary conditions over land and sea are also needed in order to carry out physical processes. Coupling of different climate subsystems is a complicated issue due to different characteristical time scales. The enormous computations are carried out on a regular grid basis covering the entire earth.

During the last 50 years, hydrodynamic models have been developed to depict the essential features of the large scale general circulation (IPCC, 1992). Manabe and Wetherald (1975) using a general circulation model studied the effects of changes in concentration of CO_2 and even for doubling of CO_2 concentration. Ocean was represented by a 50m deep surface layer. The promising results were presented with a general warming of troposphere attaining a maximum near the ground and a corresponding cooling in the stratosphere (Fig. 10.2). Stratospheric cooling due to proportionality increase in outgoing radiation in upper atmosphere. Maximum warming at high latitudes is strongly related to albedo feedbacks. Melting of ice and snow in warmer with doubled CO_2 climate will have less snow to reduce albedo.

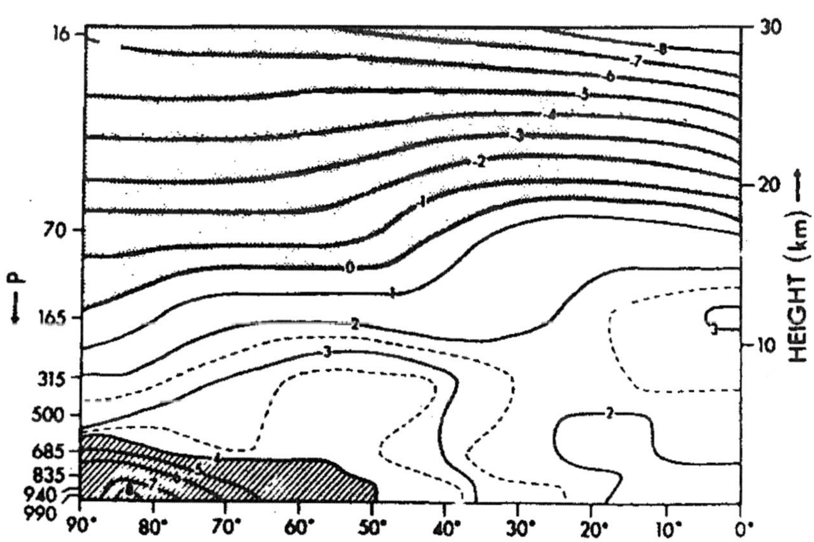

Fig. 10.2: Change in atmospheric temperature (°K) according to a model prediction when atmospheric CO_2 content are doubled. (After: Manabe and Wetherald, 1975).

Bengtsson (1994), using four different general circulation models, presented the cross-section of surface warming. Average warming is about 4°K with a maxima at high latitudes of northern hemisphere, whereas warming was less on southern hemisphere due to the effect of Antarctica as well as less land area (Fig. 10.3).

A vast amount of climate data of past 150 to 200 years is available to establish climate change trends of 0.3 to 0.6°C warming of earth's surface during the 20th century (Table 10.3). (Karl et al. 1996; Jones et al. 1999; IPCC, 2001; Rupakumar et al. 1992, 1994; Pant et al. 1999). There is simultaneous cooling of the stratosphere and rise of sea level by 10 to 25 cm. Peterson (2005) reported that a total of 25 climate change indices developed, based on daily temperature or precipitation values or some threshold values for particular locations, to study the interaction effects in different parts of the globe.

In the last two decades, more advanced coupled ocean-atmospheric models have been developed to account for oceanic circulation, deep ocean processes, feedback with clouds,

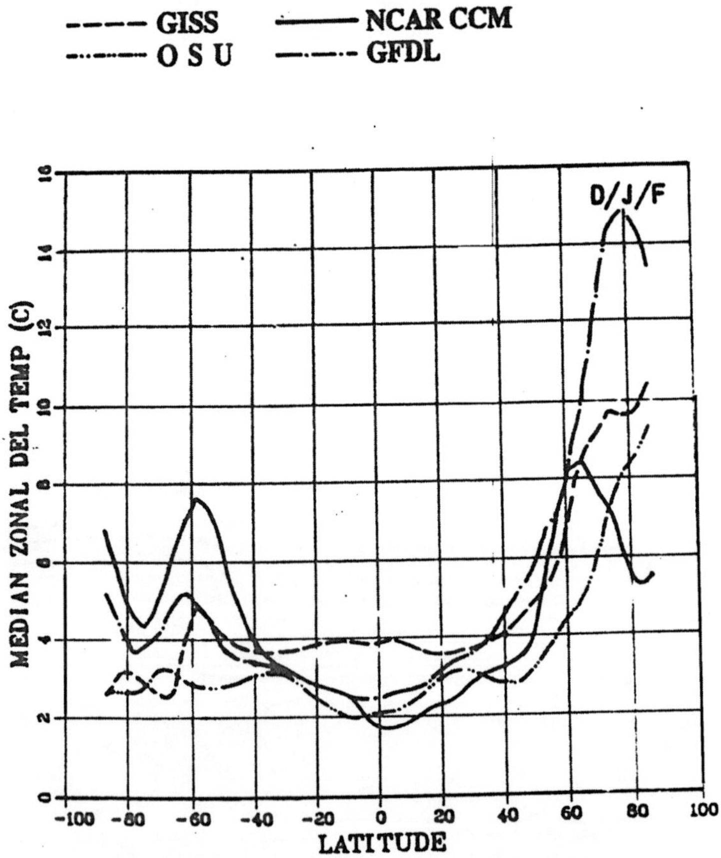

Fig. 10.3: Predicted average temprature change from doubling of CO_2 with four different general circulation models. (After: Bengtsson, 1994)

Table 10.3
Observed climate change during 20th century

Region	Temperature	Rainfall	References
All India	Mean temperature increase in 0.4°C/100 years	-	Hingane et al. (1985)
All India	Maximum temperature increase in 0.6°C/100 years, minimum temperature trend less	-	Rupakumar, et al. (1994)
Western Himalaya	Winter season increasing trend of 0.5°C/100 years, whereas shows decreasing trend in foothills	No trend	Pant et al. (1999)
Indo-Gangetic Plains	Surface temperature rising of 0.53°C/100 years during 1875-1958, whereas decreasing trend of 0.93°C/100 years during 1958-1997	Rising trend in western belt (170 mm/100 years) and decreasing trend in central belt (5 mm/100 years). A decreasing trend in eastern belt (50 mm/100 years) during 1900-1984 and increasing trend (480 mm/100 years) during 1984-1999. A westward shift in rainfall activities was noticed.	Singh and Sontakke (2002)
NE Peninsula, NE India, NW Peninsula, West Coast and Central Peninsula		NE Peninsula, NE India and NW Peninsula shows decreasing trend of -6 to 8% of normal Indian summer monsoon rainfall over 100 years, whereas increasing trend was noticed along the west coast and over central Peninsula of 10 to 12% of normal	Rupakumar et al. (1992)
Western and Eastern Himalaya	-	More snow over western Himalaya than eastern region during winter, whereas more rainfall over eastern Himalaya than western region during monsoon season	Kriplani et al. (1996)
Globe	0.3 to 0.6°C rise	-	IPCC, 2001
Global surface	0.3 to 0.6°C warming of earth's surface during 1860 to 1996 period	-	Jones et al. (1999)
USA	Diurnal temperature range decreased with higher increase of minimum temperature than maximum temperature		Karl et al. (1996)

sufficient resolutions and three dimensional global ocean circulations (Cubasch et al. 1991; Manabe et al. 1991; Hurrell, 1998; IPCC, 2001). These models have reported a global average surface temperature rise by 0.6°C during the 20th century. Model output estimated that average global surface temperature is likely to attain levels of 0.6 to 2.5°C in next 50 years and 1.4 to 5.8°C in next century by doubling the CO_2 concentration. The expected rise in temperature in higher latitudes will be likely much more than at equatorial regions (Houghton et al. 1996; IPCC, 2001). (Fig. 10.4).

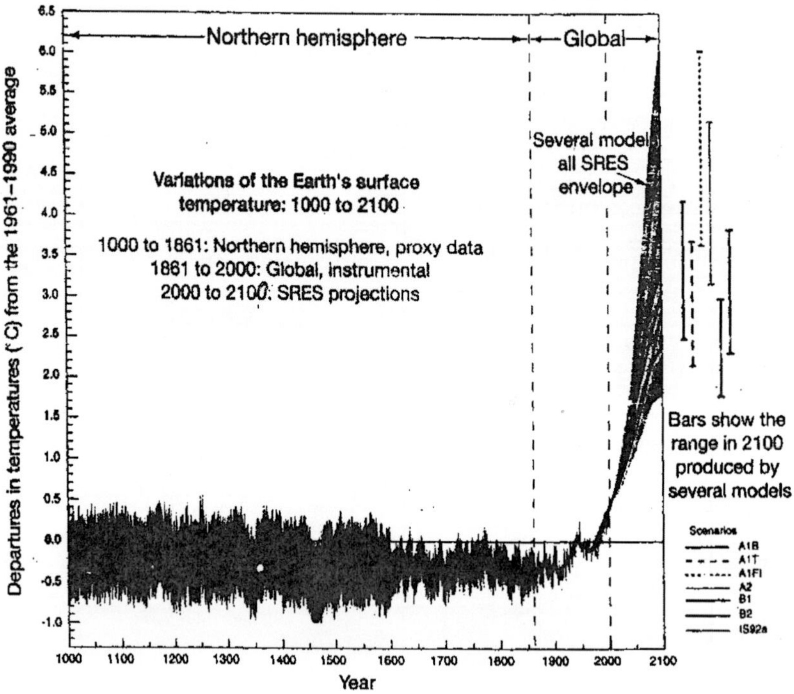

Fig. 10.4: Past 1000 years and future projections of surface air temperature. (After IPCC, 2001)

The global warming also has a strong impact on global precipitation patterns causing floods, droughts and extreme weather events. Global climate models, using the influence of rising concentration of GHGs, predicted an increase in rainfall amounts. Fischer et al. (1996) simulated temperature increase using three GCM models of the order of 4 to 5.2°C with doubled CO_2 level. Rosenweig (1990) using 5 GCM models estimated increase in global temperature of 3.9° (2.8 to 5.2°C) and global precipitation of 10.1 per cent (7.1 to 15.8%). The Goddard Institute of Space Studies (GISS), Geophysical Fluid Dynamics Laboratory (GFDL) and United Kingdom Meteorological Office (UKMO) models scenarios were of the order of 3.8°C to 5.2°C for doubled CO_2 concentration (IPCC, 1992). The GCM models have indicated a poleward shift in thermal regimes, more expansion of tropical agroecological zones, shift in temperate zones to high latitude, coastal lands to be covered under sea due to rise of sea levels (Fischer et al. 2002; Rosenwzeig and Parry, 1994) (Table 10.4).

De et al. (2000) showed that visibility over most of Indian cities showed a decreasing trend due to anthropogenic causes and rapid urbanization.

Table 10.4

Future projections on climate change over different parts of the globe.

	Temperature (°C)			Rainfall (mm)			
Region	**Summer**	**Winter**	**Annual**	**Summer**	**Winter**	**Annual**	**Reference**
All India	-	1 to 4°C with increased CO_2 concentration	-	−20% with increase in heavy rainfall	-	-	Bhaskaran, 1995
All India	-	-	2.33 to 4.78°C with doubling of CO_2 concentration	-	-	Increase in frequency of heavy rainfall events	Lonergan, 1998
Globe	-	-	0.6 to 2.5°C in next 50 years and 1.4 to 5.8°C by 2100 with doubled CO_2 levels	-	-	-	IPCC, 2001
All India	0.87 to 1.12°C by 2020	1.08 to 1.54?C by 2020	Increase in day and night temperature of 0.7 and 1.0°C by 2040 over 1980	1.8 to 5.1%	-2 to 4.7% by 2020	-	Lal et al. 2001; Lal et al. (1995)
	1.81 to 2.37°C by 2050	2.54 to 3.18°C by 2050	-	7.2 to 10.5% by 2050	-9.2 to 3.8% by 2050	-	
	2.91 to 4.62°C by 2080	4.14 to 6.31°C by 2080	-	10.1 to 15.2% by 2080	-24.8 to -4.5% by 2080	-	
Global	-	-	4 to 5.2°C increase with doubling of CO_2 levels	-	-	8 to 15 per cent increase	Fischer et al. (1996)
All India	-	-	2 to 4°C by 2050 for south of 25°N, >4°C by 2050 in north of 25°N; 1.3 to 1.7°C increase all over, but the increase in minimum temperature will be upto 4°C by 2050	6 to 13%	-	Decrease in rainy days over western and central part by more than 15 days, whereas in foothills of Himalaya, NE India rainy days increase by 5 to 10 days by 2050. Overall increase in rainy day intensity by 1 to 4 mm/day except for small pockets in NW India where intensity decreases by 1 mm/day	Rupa Kumar et al. 2003; Rupa Kuar and Ashrit, 2001.
Global	-	-	3.9°C (2.8 to 5.2°C) with doubling of CO_2 levels	-	-	10.1 per cent (7.1 to 15.8%) increase	Rosenweig (1990)

Climate variability and predictability project (CLIVAR) of the World Climate Research Programme (WCRP), on ocean-atmosphere interaction, provided input over the last 2 decades to meet challenges through stimulating efforts on climate change prediction, detection, and model integrations. Alexandrov and Hoogenboom (2000) compared the temperature and rainfall output of GCM models i.e., Planck Institute for Meteorology (ECHAM4), U.K. Hadley Centre for Climate Prediction and Research (Had CM2), Canadian Centre for Climate Modelling and Analysis (CGCM1), Australian Commonwealth Scientific and Industrial Research Organization (CSIRO – MK24) and Geophysical Fluid Dynamics Laboratory (GFDL – R15) for the period 2010 – 39, 2040 – 69, 2070 – 2099 referred as 2020, 2050 and 2080 with base period of 1961 – 90 (Table 10.5). Despite the increasing similarity of model sensitivities, the predicted climate changes were quite different from model to model (Busalacchi et al., 2005). Alexandrov and Hoogenboom (2000) predicted future changes which appears to be sensitive to parameterization, demonstrating the need to continue to refine them through the use of more discriminating diagnostics and analysis of the differences between the models and observations. Climate models have been used to simulate future climate change with use of different concentration of green house gases which influence the radiative forcing normally by doubling the CO_2 so that equilibrium is attained in relation to new forcing. First IPCC report (Hougton et al., 1990) adopted doubled green house gas equilibrium response of global surface temperature ranging from 1.5 to 4.5°C. In 1995 IPCC report (Hougton et al. 1996) with use of model sensitivities of 1.5, 2.5 and 3.5°C with emission scenarios, projected increase in global mean temperature of 0.9 to 3.5°C by 2100 (Kattenberg et al. 1996) based on the quantitative assumptions about population growth, economic growth and energy supplies in different parts of the world (Hougton et al. 1996).

10.4 Indian Scenario of Climate Change

Sarkar and Thapliyal (1988) reported climate change over India to indicate a slight warming trend in dry bulb temperature; but decadal study of Srivastava et al. (1992) presented a decreasing trend in temperature over most of northern part of the country (North of 23°N), whereas a rising trend of the order of 0.35°C during last 100 years. Rupakumar et al. (1994) found that the increase in mean temperature over India was mainly due to a rise in maximum temperature (Table 10.3), but Sinha Ray et al. (1997) study presented a increasing trend partly as a rise in minimum temperature due to urbanization. IPCC (1996) reported a slight warming trend of 0.4°C over 100 years where the changes were more clearly observed during winter as compared to small changes in monsoon season. Warming trend was observed during al the four seasons (Rupa Kumar et al. 2002) with -0.043°C and 0.027°C per decade warming during winter and summer seasons over 1901-90 period.

Srivastava et al. (1998) studied the rainfall and solar radiation trends over smaller spatial scales over India to show that a decreasing trend over hilly areas, increasing trend over

Table 10.5

Deviation in global climate model outputs in temperature and precipitation over base 1961-90 climate in Bulgaria

Model	Jan.	Feb.	Mar.	Apr.	May	June	July	Aug.	Sept.	Oct.	Nov.	Dec.
					Temperature °C							
ECHAM4	5.3	3.4	1.2	-0.9	-1.8	-1.3	-0.4	0.5	1.3	2.5	3.4	4.6
HadCM2	1.6	0.6	-0.7	-2.8	-3.1	-2.3	-0.3	0.5	-1.0	-0.6	0.0	0.6
CGCM1	5.1	3.7	1.7	-0.1	0	1.8	4.2	4.9	2.9	1.9	1.3	3.6
CSIRO-Mk2b	3.4	2.3	1.2	-1.0	-1.3	-0.9	-0.4	0.3	0.2	0.8	1.6	2.5
GFDL-R15	-0.4	-1.2	-3.1	-4.6	-4.7	-2.3	0.1	2.7	0.8	-0.8	-1.2	-1.6
					Temperature °C							
ECHAM4	88	56	89	12	-27	-40	-32	-55	-4	79	67	89
HadCM2	59	55	72	29	32	11	-25	-22	-2	18	52	34
CGCM1	19	3	21	-10	-32	-54	-70	-83	-65	-16	-25	5
CSIRO-Mk2b	58	51	47	+10	-8	-15	-4	-5	5	46	4	50
GFDL-R15	51	56	89	78	25	-42	-67	-78	-33	72	79	57

(After: Alxendrov and Hoogenboom, 2000)

drought prone areas in interior peninsula, and a decreasing trend in global radiation over Jodhpur, and increasing trend in diffuse radiation during winter was reported. Srivastava and Balakrishnan (1993) indicated an increase in sea level of about 8 cm during 1901 to 1950 with a decreasing trend in sea level pressure during the same period.

Sinha Ray et al. (1997) presented a significant decreasing trend in lower stratosphere temperature near equatorial Indian region, but during ENSO years it was warmer. In the upper stratosphere, a significant decreasing trend was observed during September to December, whereas decreasing trend was noticed in May and June in upper mesosphere. Rajeevan et al. (2000) showed an increasing trend in sea surface temperature (SST) and cloudiness over the equatorial Indian Ocean due to decreasing trend of outgoing long wave radiation indicating increasing convection and SST trend over northern Bay of Bengal and a decreasing trend in cloudiness over the same region due to activity of the monsoon trough and formation of monsoon depressions and storms.

De and Joshi (1999) studied interannual and interdecadal variability of tropical cyclones and reported that the maximum number of cyclonic storms occurred during 1921-30 and minimum during 1981-90 decades; and maximum number of cyclonic disturbances occurred during 1941-50 – the decade of fewest ENSO episodes. Srivastava et al. (2000) reported a significant decreasing trend in frequency of cyclonic storms and cyclonic disturbances with a decrease in intensification rate of cyclonic disturbances in recent years, however climatic extremes might become more frequent with climate change. Rao et al. (2000) found that extreme temperatures were increasing over southern India and falling over northern India. Bhujad et al. (2001) found that maximum temperatures above 40°C presented an increasing trend in pockets of northwest and pockets in south (Fig. 10.5a) India, whereas the frequency of temperatures less than 5°C showed an decreasing trend in northwest and central India (Fig. 10.5b). Widespread extended periods of extremely high temperature during summer months were expected to become more frequent with continued global warming. In some parts of south India summer heat wave conditions have become more frequent in recent years touching 48°C about 9 to 10°C above normal. A decreasing trend in frequency of rainfall (>7cm in 24 hours) during the winter season all over India, however, during the SW monsoon season most of the stations along west coast, Gujarat and Kutch, Karnataka, Rayalseema, Telangana and Islands of Aminidivid and Port Blair showed an increasing trend in heavy rainfall (Fig. 10.5c). Similarly a few pockets in Haryana, Punjab, Poona, Assam, and West Bengal showed an increasing trend. In the postmonsoon season a large number of stations over India showed a decreasing trend in frequency of heavy rainfall. Mall et al. (2006) reviewed the temperature increase and rainfall trends over regional parts of India including Himalayan and Indo-Gagentic regions based on the published work of scientists at Indian Institute of Tropical Meteorology, Pune.

The future projected climate change scenario for the Indian sub-continent as inferred by Lal et al. (2001), Mall et al. (2006); Rupa Kumar et al. (2003) using atmosphere-ocean GCMs models under the four SRES marker scenarios have been presented in Table 10.5.

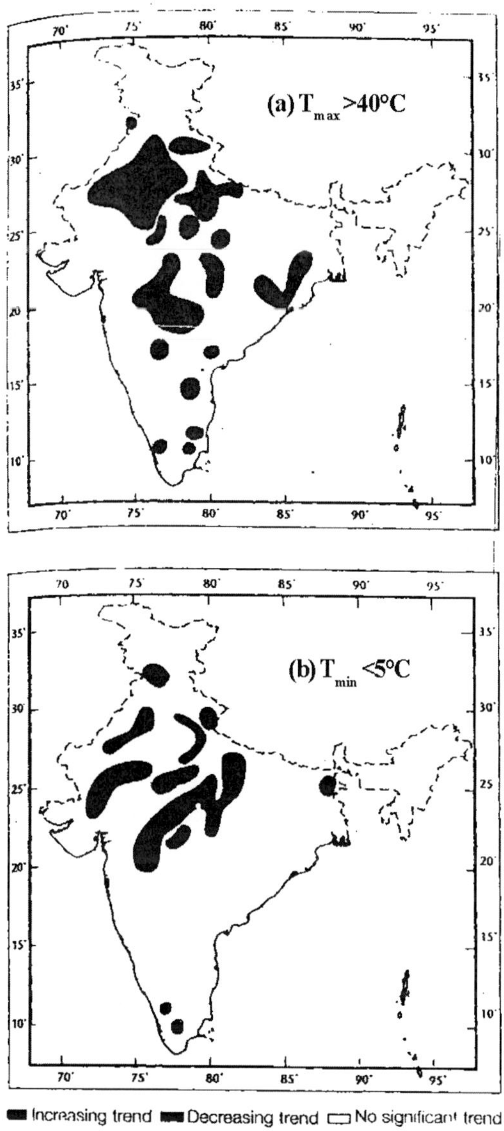

Fig. 10.5: (a) Trends in frequency of occurrence of temperature >40°C. (b) Trends in frequency of occurrence of temperature <5°C. (After: Bhujad et al., 2001)

These results present an annual increase in surface temperature due to global warming, with relatively higher warming over winter season than the summer monsoon. However the rainfall increase have been reported with intensive rainfalls all along the western coastal area, northeastern regions, and few pockets in northwest India, Andeman and Nicobar Islands.

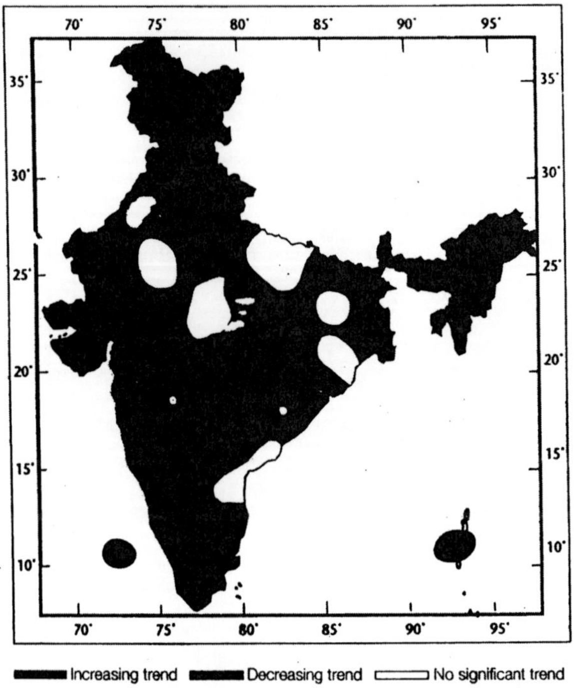

Fig. 10.5: (c) Trends in the frequency of rainfall >7 cm in 24 hours (1901-1990). (After: Bhujad et al., 2001)

10.5 Impact of Climate Change

10.5.1 Soil health

The decomposition of organic matter releases nutrients with reduction of soil fertility. Soil microbial activity are major contributor in decomposing organic matter as well as consumer of GHG because it provides energy and carbon source for growth and multiplication of microorganisms in the soil leading to soil humus formation, cycling of nutrients, soil tilth and structural changes. Soil microbial biomass (living part of organic matter)–C_m respond to climate changes, acting as an indictor of long term changes in soil organic matter contents - C_o, in the form of the microbial quotient i.e., ratio of C_m/C_o under equilibrium conditions is about 2.3 to 4 per cent (Powlson and Brookes, 1987; Dell Abate et al., 2005). Soils exhibiting a ratio higher or lower than this equilibrium value would be either accumulating or losing carbon or nitrogen contents. Therefore with the accumulation or losses of soil carbon contents we can predict soil fertility. With global warming higher temperature would result in rapid organic matter decomposition in irrigated soils. The rate of release of nutrients are accelerated, but residual organic matter accumulated is lower than relatively cooler soils.

With global warming, the tropical regions with small organic carbon reserves, will show a net efflux of CO_2 because rates of soil respiration increase exponentially with temperature. These days with intensive cropping system of rice-wheat rotation, even the well fertilized soils having higher organic carbon in the beginning (>80%) showeld a decling trend in organic carbon content, but reverse trend was noticed for organic carbon values lower than 0.8 per cent (Kalra et al. 2003). However, in poor tropical soils, the decline in organic carbon content was noticed with organic carbon value higher than 0.45 per cent. Rise of CO_2 concentrations will lower the pH to enhance nutrient availability and microbial activity. Therefore to maintain soil health, CO_2 and GHG emissions may be mitigated through implementation of strategies in landuse planning by increasing the soil organic matter through adoption of techniques such as flat bed planting, adoption of zero tillage practice and improving the microbial population of the region by growing of legumes to fix atmospheric nitrogen to overcome the nutrient limitations as well as the release of mass inoculums of organism will act as sink for the consumption of the prominent GHG in the region.

10.5.2 Rise in surface temperature and sea levels

Bengtsson (1994) computed increase in global mean surface temperature for scenario A (present trend to continue) and reported slow increase of only 0.5°C during first 40 years, but thereafter with a more rapid increase of about 0.35°C per decade (Fig. 10.6) with a total increase of 2.8°C in 100 years. However, for scenario D (drastic reduction), the temperature increase is significantly reduced and increases by 0.8°C in 100 years. The temperature increase over oceans is normally lower than over land due to its large heat capacity and ocean circulation. In Arctic region temperature increased by an annual mean of 12°C with the effect that North Pole is ice free in summer. As a consequence of global warming the sea level would rise with change in hydrological cycle, changes in mass of ice stored in glaciers, and thermal expansion of sea water. Fig. 10.7 shows the thermal expansion of ocean waters would increase by 16 cm for scenario A, whereas only 6 cm in scenario D. This would threaten valuable coastal agricultural lands and low lying Islands.

10.5.3 Vegetations

Claussen and Esch (1992) used a biome model (Prentice et al. 1992) to study the impact of climate change on the biosphere. Fourteen plant types were assigned climatic tolerances in terms of empirically determined values for temperature, moisture and sunshine hours for different seasons. The biome model predicted the plant types suitable in a given environment i.e., climatic variables. The biome model selected the potentially dominant vegetation types and compared with actual observations (Fig. 10.8). The agreement was very good, except in areas where agriculture or anthropogenic activities has obliterated the natural vegetation. Simultaneously biomes were calculated using the output of Humburg climate model (Fig. 10.9) and both maps showed a qualitative agreement between global distribution of biomes

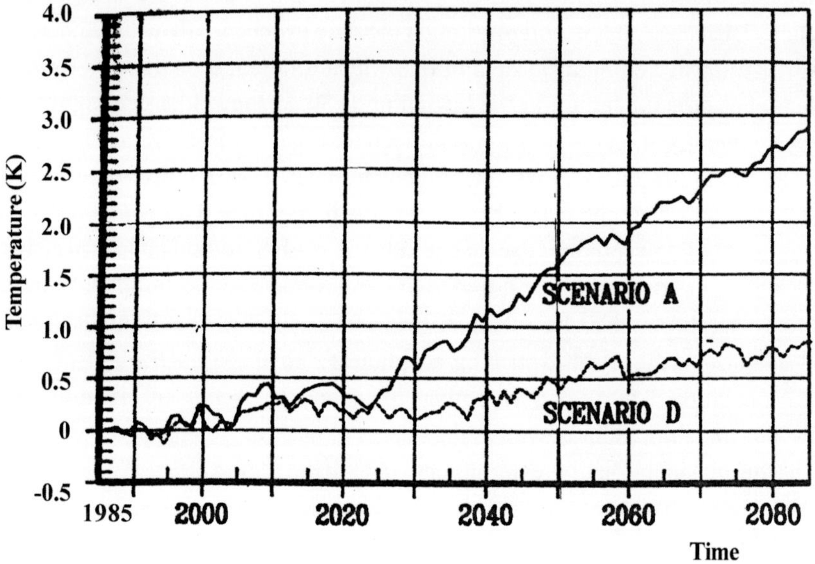

Fig. 10.6: Changes in the average global surface temperature during next 100 years by the coupled ocean-atmospheric Hamburg model. (After: Bengtsson, 1994)

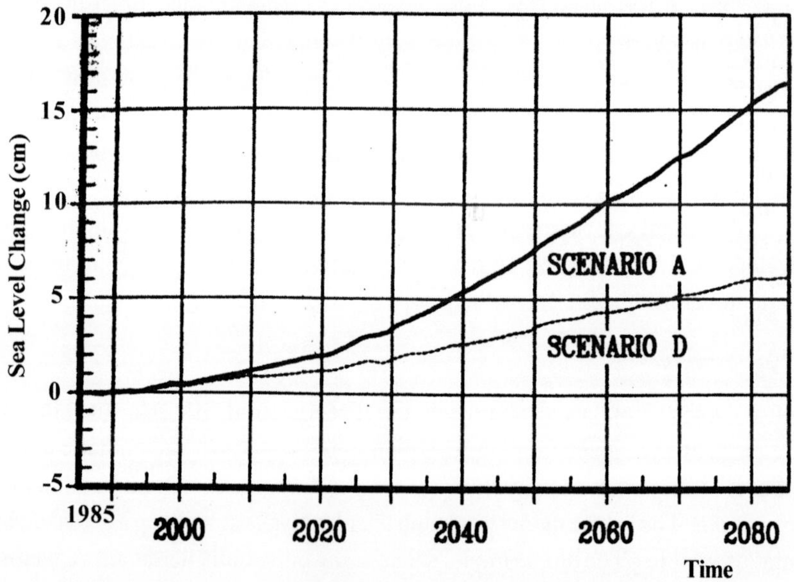

Fig. 10.7: Increase in the sea level height with rise of global surface temperature. (After: Bengtsson, 1994)

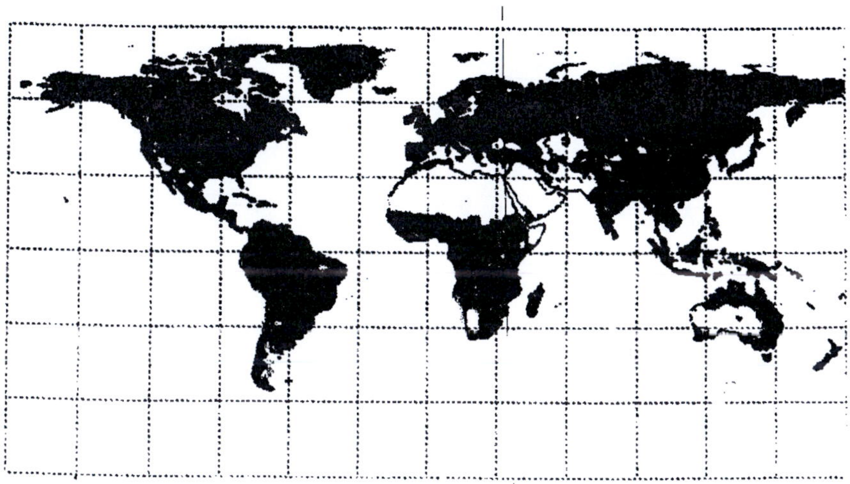

Fig. 10.8: Present biome distributions computed from observed climatology (IIASA data).
[*After*: Claussen and Esch, 1992]

from observed and from simulation methodology with the discrepancies that Australian desert was absent in simulation, but savanna covered the northern half Australia. Similarly, Kalahari Desert in South Africa was filled by xerophy woods. North America also had a warm bias.

Perlwitz (1992) adopted the same biome model (Prentice et al. 1992) for studying climate impact of CO_2 forcing for scenario A (1145 ppm) and presented the resulted biomass (Fig. 10.10) to highlight the potential natural vegetation in a state of equilibrium with he new climate conditions. Comparing with Humburg model Biomes (Fig. 10.9) it is reflected that there was marginally change in conditions for tropical rain forests, little change for Sahara as well as steppe and xerophytic woods south of it. Over Indian region, conditions were favourable for savanna to move into desert areas. In Europe, Siberia and North America, there was a northeast shift of climatic conditions. Alpine plants were migrating northwards. Changes in conditions for boreal biomes were strongest, as tundra, taiga and cold deciduous forests were reduced. It is felt that biological diversity would be reduced in some of the world's most fragile environment particularly tropical forests.

10.5.4 Crop responses

(i) Photosynthesis

During the last two decades CO_2 have increased by 1.5 μmol mol^{-1} per year on an average (Schimel et al. 1996) and it is necessary to quantify crop responses as how they influence growth and yield components. Elevated CO_2 increases the net photosynthesis rate (Pn) and decreases photorespiration (Pr) in C_3 plants. Ribisco enzyme is responsible for reaction of

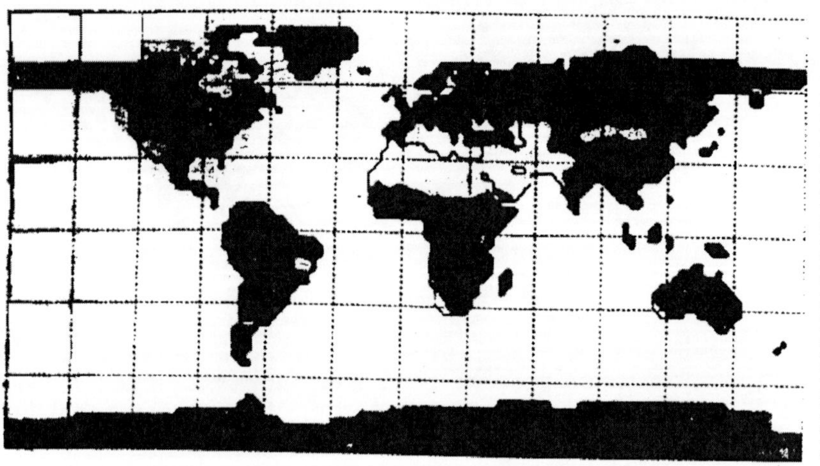

Fig. 10.9: Biome distributions computed from the Humburg climate model simulating present climate (ECHAM3-T42). [*After*: Claussen and Esch, 1992]

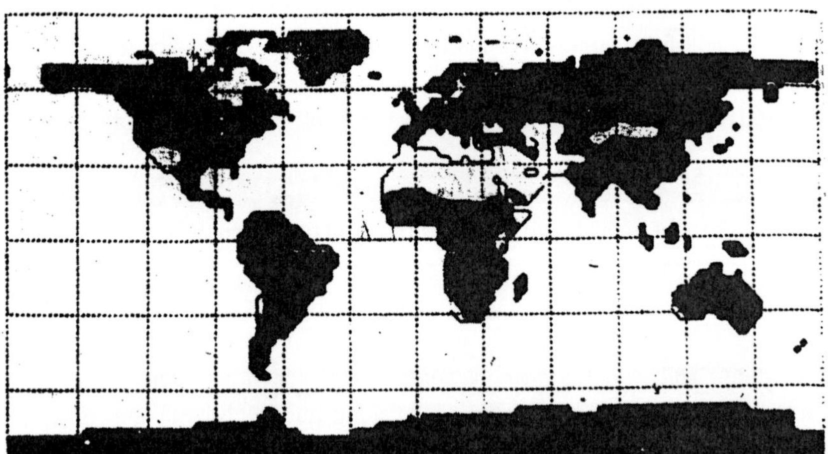

Fig. 10.10: Change of conditions favourable for biomes due to a climate change associated with an increase of CO_2 computed from a combined CHAM1 T21 and ECHAM3 T42 simulations. [*After*: Perlwitz, 1992]

CO_2 with ribulose bisphosphate (RuBP) as the present level of CO_2 reaction is not saturated and oxygen also reacts with RuBP producing phosphoglycolate to release CO_2 in photorespiration (Long, 1991). Therefore, increased or doubling CO_2 saturates Rubisco, reduces Pr and enhances Pn by 30 to 40 per cent provided radiation and soil water are not limited under optimum temperature (25 to 30°C) conditions. C_4 plants are already CO_2 saturated by present atmospheric levels and practically has no effect on C_4 photosynthesis assimilation rate; on the other hand photosynthesis in C_3 plants respond strongly to an increase in atmospheric CO_2 levels (even up to 1000 ppm) because of low affinity of Rubisco coupled with decrease of Pr due to increased CO_2/O_2 ratio (Berry and Downton, 1982). Amthor (2001) reported a higher ratio of increase in wheat yield in elevated CO_2 over ambient CO_2 levels (Fig. 10.11).

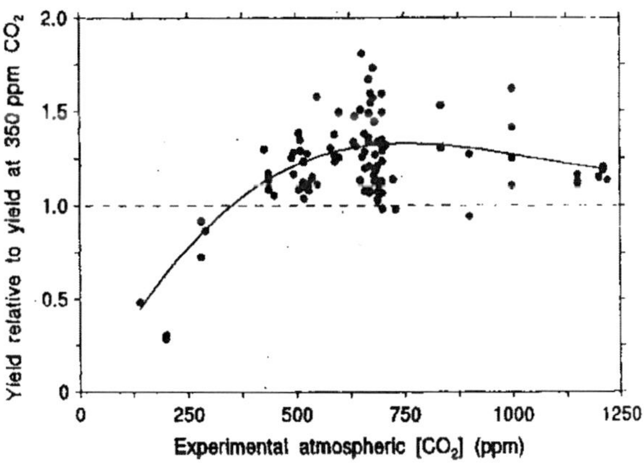

Fig. 10.11: Effect of CO_2 in the range 140-1220 ppm on wheat yield relative to yield at 350 ppm CO_2. (After: Amthor, 2001)

Elevated CO_2 also decreases the stomatal conductance (g_s) reducing evapotranspiration and increases internal CO_2 concentration to increase net photosynthesis to raise the photosynthetic capacity of plants grown in elevated CO_2 levels (Kang et al. 2002). In wheat g_s decreased by about 30-40 per cent by doubling CO_2 from present levels (Samarakoon et al. 1995) due to change in aperture since stomatal density (stomata/unit leaf area) and stomatal index (ratio of stomata/leaf epidermal cells) are unaffected in the season (Estiarte et al. 1994). Narrowing of stomata with increased CO_2 will help in protection from air pollutants. Leaf photosynthesis response in wheat is modified to a sigmoidal response with maxima over 20 to 32°C, but in maize response is linear up to 25°C and saturating around 32 to 35°C, and then sharply decreases upto 40°C (Stone, 2001). Sensitivity of Pn to CO_2 increased with temperature due to the characteristics of Rubisco (Lawlor and Keys, 1993). The optimum temperature increased by several degrees (5°C in wheat and 2°C in maize) with elevated CO_2 (Farquhar and Von Caemmerer, 1982; Long, 1991) due to temperature dependence of the kinetic parameters of Rubisco and relative solubility of CO_2 and O_2 increasing the relative photorespiration with temperature. Superiority of C_4 photosynthesis occurred above 25°C (Young and Long, 2000). There is decrease in respiration rate due to global environmental change when expressed per unit of accumulated dry matter (Drake et al. 1997; Gifford, 1995). Thus the influence of temperature on photosynthesis rate was greater at elevated CO_2 due to decrease of the ratio of photorespiration to photosynthesis at higher CO_2 (Berry and Bjorkman, 1980).

(ii) Crop development

Slightly faster development rates of leaf, ear, spike and tiller development has been reported in wheat grown at elevated CO_2 probably due to indirect effect of temperature increase with smaller g_s (Wheeler et al., 1996). The rate of development of wheat are sigmoidal than linear functions of temperature (Shaykewich, 1995). Therefore development rate increased above base temperatures of 1 to 5°C slowly and then rapidly to a maximum at 30°C, and then rapidly decreased thereafter. 1°C climate warming would reduce the crop duration by about 21 days (8%), respiration period by 8 days (6%), anthesis period reduced by 3 days and grain filling duration by 2 days (5%) in winter wheat (Mitchell et al. 1995). Therefore, slower developing varieties may be preferred to overcome the impact of warmer climate when soil water is not limiting.

(iii) Biomass and grain yield

Biomass results from the accumulation of carbon in plant products and therefore there is increase of net Pn upto 40 per cent with doubling CO_2 levels in C_3 plants. Response of wheat to extra carbon assimilation was flexible with the main stems least affected, but tiller production and survival and root growth were simulated most. Also leaf size, grain per ear and grain size also increased with elevated CO_2 in a similar manner to that of biomass (Mitchell et al. 1995; Batt et al. 1997). However, grain yield was more sensitive to biomass production during the reproductive phase of growth particularly around anthesis and vegetative growth phases (Mitchel et al. 1996), and an average increase in biomass and grain yield of 7 to 11 per cent per 100 μmol^{-1} enrichment was reported (Bender et al. 1999) under well fertilized and no water stress conditions. There is strong interaction of temperature and CO_2 for net photosynthesis in C_3 plants (Farquahar and von Caemmerer 1982), but is limited by the reduction of growing season with temperature rise (Long, 1991).

Warmer temperature reduced the duration of different development stages as there is less time to utilize the light, water and nutrient resources with decrease of cumulative biomass production. Later reproductive development stages will be severely affected than the earlier vegetative stages. Mitchell et al. (1995) reported that with a constant increase of 3.5°C during wheat growing season reduced the final winter wheat biomass by about 16 per cent and yield by 35 per cent. Grain yield was more severely affected than biomass due to the reduction of harvest index by warmer temperature (Mitchell et al. 1995; Batts et al. 1997) due to sensitivity of grain yield with the duration of grain filling phenophase and induction of partial sterility (Mitchell et al. 1993).

Effect of increasing CO_2 on transpiration are complex, but the influence on g_s is fairly consistent (Lawlor, 1998). At lower transpiration rate, there was less latent cooling which increased the water vapour pressure in the intercellular spaces of the leaves with decreased humidity gain in the boundary layer. Consequently, the gradient of vapour pressure increased to increase the transpiration. However, short term exposure to elevated CO_2 decreased transpiration. Increasing or doubling CO_2 decreased water use by about 20 per cent under wet conditions and slightly higher under dry conditions with improvement of water availability for longer period.

Despite the relatively small effect of CO_2 on ET, elevated CO_2 increased biomass substantially in wheat and thus increased WUE by 21 per cent, increased WUE by 9 to 19 per cent in sorghum (Conley et al. 2001), in sunflower by 26 per cent (Hui et al. 2001), in maize by 20 to 70 per cent (Mitchell, Mitchell and Lawlor, 2001), in cotton by 20 to 90 per cent (Kang et al., 2002); and in wheat by 10 to 80 per cent (Kang et al., 2002; Samarakoon et al., 1995). The relative increase in biomass due to doubling CO_2 was much greater under dry than wet conditions. Climate change will modify precipitation amount, evapotranspiration rates, run off, soil moisture storage and water availability to crops. It has been assessed that at mid-latitudes, evaporation would increase by 5 per cent for each °C rise in mean annual temperature. The increased evapotranspiration in plants will induce moisture stress. However, with CO_2 fertilization, the decrease of stomatal conductance will minimize the moisture stress on the other hand. High evapotranspiration rates in tropics and subtropics will induce moisture stress in crops accumulating into a agricultural drought reducing the grain yields.

Horie et al. (2000) reported that in rice WUE at elevated CO_2 (700 ppm) increased by 40 to 50 per cent over ambient CO_2 level between 24 to 26°C, but with warming dropped to 20 per cent at 30°C.

10.5.5 Landuse planning and seasonal changes

Global warming with rise of annual temperature from 1 to 6°C over equatorial to polar regions provided temperature gradient increase towards the higher latitudes. Therefore, over middle to higher latitudes there will be increase in length of potential growing season permitting early planting of crops in spring, earlier maturation and harvesting. Thus possibilities of cultivation towards poleward will be improved. However, in winter increased temperature in lower latitudes will enhance photorespiration, but with increasing photosynthesis rate the net biomass will be enhanced at slower rates as compared to the acceleration at higher latitude. Further the higher temperature will accelerate development of plants resulting in hastened maturity with reduction in crop yield. The climate and agroecology of crop zones would shift forcing the farmers to adapt to new conditions.

Aggarwal and Kalra (1994) used WTGROWS – a wheat growth simulator model for resource management in forecasting of wheat yield and climate variability. A decline in wheat yield was reported with the increase in January temperature i.e., tillering phenophase. The potential wheat yield in different agroecological zones varied due to difference in optimum date of sowing and wheat cultivar of the location. The yield variability was large in regions showing relatively higher yields indicting sensitivity towards interseasonal climatic variability. In India, the potential wheat yields were higher in NW region, but decreased towards eastward-southward i.e., with variations of seasonal temperature during the growing period with a decrease of 6.34 q/ha per °C rise in temperature (Table 10.6). Similarly rabi maize, barley, gram and mustard yield decreased by 1.52, 2.77, 0.32 q ha^{-1} per degree °C rise in seasonal temperature, (Kalra, et al. 2003).

Table 10.6

Grain yield (t ha^{-1}) of wheat in current weather and per cent change in response to climate change (425 ppm CO_2 and 2°C increase in mean temperature) in different regions of India.

Region	Potential Yield		Irrigated Yield		Rainfed Yield	
	Current	Percent Change	Current	Percent Change	Current	Percent Change
>27°N	6.66	-3.9	4.89	3.7	2.95	28.6
25.27°N	5.84	-1.5	4.78	-4.4	3.34	-7.2
23-25°N	5.86	-5.6	4.18	-10.8	1.17	-19.6
20-23°N	4.18	-18.4	2.29	-18.3	0.51	-11.8
<20°N	3.69	-17.3	2.43	-21.4	0.97	-23.9

(Aggarwal and Kalra, 1994).

The optimum date of sowing for wheat was spatially dependent due to prevailing seasonal climatic conditions. The optimum dates were earlier in the northwestern region (Punjab & Haryana) as compared to central and eastern regions. There is need to reduce the turnaround time to match the optimum dates of wheat sowing in northwestern region with adoption of zero tillage and flat bed planting. Temperature scenarios with global warming will shift the date of sowing later than the present optimum time to achieve optimum yeild of wheat in different agroenvironments to counteract the decrease of wheat yield with global warming or rise of temperature in subtropical areas. However, in mid and high latitude region rise of temperature will reduce mortality of winter plants.

Aggarwal and Kalra (1994) using the WTGROWS – wheat model demonstrated the shift of 2.5, 3.5 and 4.5 t ha^{-1} isoyield lines of irrigated wheat in India with 425 ppm of CO_2 concentration and a 2°C rise in temperature to northwards because with increase of temperature may cause heat stress. In temperate and mid-latitudes there will be reduction of winter chilling (Vernalization) period because many winter crops require a period of low temperature to either mitigate or accelerate the flowering process. Poor vernalization results in lower flowering bud initiation reducing the grain yield.

Similarly Gadgil et al. (1995, 1999a, 1999b) used PNUTGRO model to determine the sowing window for rainfed groundnut showing an optimum sowing window of 22 June – 17 August in monsoonal climate for minimizing the risk of crop failure, reduction of incidence of leaf minor and late leaf spot (Tikka) diseases when sowing was postponed after mid July.

C_3 crops in temperate and subtropical regions will also be benefitted from reduced weed infestation as 14 of world's 17 most troublesome terrestrial weed species are C_4 plants in C_3 crops. The difference in response to increased CO_2 may induce such weeds less competitive. In contrast, C_3 weeds in C_4 crops, particularly in tropical regions could become more of a problem depending on the relative response of crops vs weeds to the climate change scenario.

Increased CO_2 levels will promote plant growth with increase of carbon contents and decrease of nitrogen contents – thus implying a reduction of nutritional protein level for insect pests. With global warming longer growing season will enable the insects to complete more number of reproductive cycles. But with plant growth under elevated CO_2 level will modify the plant structure to sustain a higher level of pathogenic infection without reduction in yield. Since vigorous plants are able to resist infection from weak pathogens, such as facultative parasites, resulting in lower disease incidence and severity. However, larger plants provide more surface area for infection and disease severity. The elevated CO_2 may be beneficial for plant health and productivity by altering the morphology and physiology of plants to the detriment of pathogenic microbes. Temperature increase may extend the geographical range of some insect pests presently limited by low temperature.

10.5.6 Agriculture

Climate change and climate variability has a direct impact on crop productivity. The impact on global agricultural production is expected small, but regional vulnerability to food shortage may increase. Rozenzweig and Parry (1994) estimated the net effect of climate change on global production up to 5 per cent, but production may decrease in developing countries and increase in developed countries. Doubling CO_2 concentration may increase the photosynthetic rates by as much as 30 to 100 per cent in C_3 plants such as wheat, rice and soybean; whereas response in C_4 plants such as maize, sorghum, sugarcane, millets etc. may remain as such (IUCC, 1992). IPCC (2001) reported that, if no climate policy interventions are made in most tropical and subtropical areas, potential yields may decrease for projected increase in temperature, whereas in mid latitudes with warming effect due to increase of CO_2 levels may lead to positive impact responses by 2020, and generally negative impact responses with still greater warming beyond 2050 (Table 10.7). It is likely that the current imbalance of food production between cool to temperate regions and tropical to subtropical regions may become worse with climate change due to relatively opposite impacts of climate change in these regions. Even with the rise of sea surface temperature in equatorial regions, due to climate change, there would be variations in quantity of fish and sea foods availability.

Effect of temperature rise will enhance the biological, physical and chemical activity processes. Net photosynthesis rate is enhanced with reduction of photorespiration with increase of CO_2 levels. Evapotranspiration and water use efficiency are improved with decrease of stomata conductance with increase of CO_2 levels. These responses are more

Table 10.7
Impact of climate change on agriculture with no climate policy interventions

CO_2 Concentration	2025 (405-460 ppm)	2050 (445-640 ppm)	2100 (540-970 ppm)
Global mean temperature change from 1990	0.4 to 1.1°C	0.8 to 2.6°C	1.4 to 5.8°C
Global mean sea level rise from 1990	3 to 14 cm	5 to 32 cm	9-88 cm
Average crop yields	Cereal yields may increase in mid and high latitude regions, whereas in tropical and subtropical regions decreasing trend	Cereal yield decreases in tropical and subtropical regions, but mixed effect in mid latitude regions	Reduction in cereal yields in mid latitude regions with warming of few °C
Extreme low and high temperature events	Reduction in frost damage for winter crops, but increased heat stress damage in summer crops and livestock population	Effects of changes in extreme temperatures is increased	Effects of changes in extreme temperatures is amplified

(After: IPCC, 2001)

pronounced in C_3 plants as compared to C_4 plants whereby increasing the productivity in C_3 plants (Rosenzweig and Hillel, 1995; Porter, 1993). Crop yields of C_3 plants respond favourably to increased CO_2 levels leading to an increase in yield by 15 to 22 per cent, whereas in C_4 plant yield increase is limited within 5 per cent (Agarwal and Sinha, 1993; Saseendran et al. 2000; Reilly et al. 1996).

With use of different crop models, it has been inferred that there is negative impact in equatorial and subtropical regions, whereas improvements in mid latitude to poleward belt (Reilly et al. 1996). By doubling of CO_2 levels and using GCM scenarios, crop simulation models in rice, wheat, soybean, maize, millets, etc., have been widely used in different parts of the globe to highlight their impact on crop yields. In maize yield the impact varied from –61 to increase per cent (Latin America), –30 to increase per cent (Europe), – 55 to + 62 per cent (North America and Canada), –15 to + 6 per cent (Africa), –65 to –10 per cent (South Asia). In wheat yield impact will be –50 to +5 per cent (Latin America), –19 to 41 per cent (former Soviet Union), increase/decrease in Europe, 100 to 234 per cent in North America and Canada, –60 to + 67 per cent (South Asia), –41 to + 65 per cent (Japan, South Korea, Australia and New Zealand). Impact on soybean yield varied from –1 to + 40 per cent (Latin America) –91 to 38 per cent (North America and Canada). Rice yields varied from –22 to 28 per cent in South Asia, –78 to –28 per cent in China and Taiwan, –45 to 30 per cent variations in Japan and South Korea due to climate change scenarios with the doubling of CO_2 levels.

In India an increase of rice crop yield was reported (Aggarwal and Mall, 2002; Rathore et al. 2001; Mohandas et al. 1995; Achanta, 1993), whereas decreasing trend was noticed by (Sasendran et al. 1999; Lal et al. 1998; Sinha and Swminathan, 1991). Similarly in wheat decrease in yield was reported over all India basis (Sinha and Swaminathan, 1991; Aggarwal and Sinha, 1993; Gangandhar Rao and Sinha, 1994; Aggarwal, 2000; Aggarwal, 2003), and decrease/increase trend on scenario basis (Lal et al. 1998; Aggarwa and Kalra, 1994). Similarly decrease in yields have been reported in maize (Hundal and Kaur, 1997); chickpea (Mandal, 1998); pigeonpea (Mandal, 1998), groundnut (Hundal and Kaur, 1996), sorghum (Chatterjee, 1998); whereas increase in brassica yield (mustard) was reported in North India (Upreaty et al. 1996). Therefore, it is essential to understand the location specific crop ecosystem responses which are influenced during the crop growing season so as to influence the variation in crop yield due to rise of CO_2 levels or greenhouse gases concentration and global warming. Therefore to sustain agricultural productivity, there is strong need for adaptations and adjustments with the climatic scenario of the agroecology zone such as

(i) use of varieties that are matched with the regional growing season.
(ii) in high latitudes regions short growing season is the primary limiting factor for crop production and therefore global warming would allow the use of more productive medium/longer growing season varieties already adapted in lower latitude locations.
(iii) in regions, where mid and late summer temperature and water stress become more severe to interfere with the plant's reproductive cycle, the shorter duration/seasonal varieties may be adapted.
(iv) In tropical and subtropical zone, there is need for development or introduction of more drought and heat resistant strains of major crop species in response to global warming trends.
(v) The uneconomic crops should be replaced with economic crops in the zone with sensitivity analysis adoption with changing climatic scenario.
(vi) There is strong need for maintaining soil health with maintenance of nutrition balance in the zone with adoption of moisture conservation through zero tillage and use of mulches. Monsoon season rainfall be conserved to adapt fertilizer responsive varieties to improve yield potential.

10.6 Impact of Air Pollutants on Vegetation

Climatic and edaphic factors influence the susceptibility of vegetations to the pollutants exposure. Climate factors such as relative humidity, air temperature, light intensity and photoperiod strongly influence the vegetations to pollutants (UNEPA, 1986; Heck et al. 1988). Similarly edaphic factors such as soil moisture, soil nutrients and soil oxygen concentration also influence vegetations to pollutants (UNEPA, 1986). It was noticed that those factors which control stomatal conductance also control the plant response to pollutants

(Tingey and Anderson, 1991). The site of action of pollutants varied as a function of pollutant and the plant organs such as heavy metals and acid precipitation affected the roots, whereas SO_2, N_2O, NO_2, and O_3 affected the shoot directly. The basic assumption is that the pollutants enter the leaf through stomata and interact with the mesophyll cells and/or guard cells. The effect of air pollutants can occur from molecular to organism level disturbances induced by air pollutants in the form of leaf injury, decreased growth and yield, changes in crop quality, decreased reproduction, inducing foliar lesions and decreasing growth, development of necrosis of the causative pollutants. Air pollutants which are deposited on the soil (heavy metals) affect the functioning of roots and interfere with soil resource capture by the plant. The reduced resource utilization (i.e., photosynthesis, nutrient uptake, soil water uptake) will affect plant growth through biomass distribution in different plant organs. When air pollution stress occurs along with moisture stress, the growth reduces sharply reducing economic yield. Therefore, plant acts as a indicator for recognition of air pollutants for their monitoring and surveillance. For phytotoxic air pollutants, normally wild and cultivated plants in polluted areas are used as indicators (Guderian and van Haut, 1970). Observations on external injury such as necrosis or change in growth habit, chlorosis, discolouration, injury patterns on leaf. The standardized grass culture for surveillance of pollutants in areas with SO_2, HF and HCl used is grass L multiflorum for 2 weeks at site (Scholl, 1972).

10.6.1 Adverse effects

(i) SO_2

Crops can be grouped into three resistance groups according to their tolerances to SO_2. Fodder plants are not sensitive to SO_2; wheat, leafy vegetables, beans and rose flowers are moderately sensitive; and roots and cabbage as the least sensitive. Weeds are sensitive to air pollutants as are commercial plants.

Injury is caused by absorption of high concentration of SO_2 resulting in rapid deposition of sulphite which is toxic to the metabolic processes taking place in the mesophyll cells. Sulphite is oxidized to sulphate. Chronic injury occurs when sulphate accumulates beyond a threshold value than the plant cells can tolerate. Estimates show that sulphate is about 30 times less toxic than sulphite (Jacobson and Hill, 1970). Acute injury occurs between veins on broad leaves in the form of lesions. The metabolic processes are completely disrupted in the dead areas of leaf. Rarely young leaves display necrotic markings, whereas fully expanded leaves are most sensitive to SO_2 injury. Injury on monocotyledonous leaves can occur at the tips and in lengthwise areas between main veins (Heck et al. 1977). Chronic injury appears as a yellowing or chlorosis of the leaf. A silvering will occur under the surface of broad leaves. In conifers species, chronic injury occurs on older needles appearing as yellowish green colour, then changes to reddish brown, initiating at the tips before developing further. There is reduction in growth and yields even without any visible marking. SO_2 pollutes tomato and cotton with chronic injury during flowering and fruit formation, and even leaf destruction (Thomas and Hendriction, 1957).

Fumigation experiments support that injury to vegetation will rarely occur if 8 hr average concentration of SO_2 does not exceed 0.25 ppm SO_2 (Mudd, 1975). Other potentially damaging concentrations to vegetation are 0.35 ppm for 4 hr, 0.55 ppm for 2 hr and 0.95 ppm for 1 hr. An average annual SO_2 concentration of 0.03 ppm indicates that chronic or acute vegetation injury occurred during the growing season from higher potentially damaging concentrations (Linzon, 1969). Barley, cotton, beans, potato, oats, wheat, sunflower are slightly sensitive to SO_2 pollutants than alfafa; whereas vegetable crops, castor bean, onion, corn, citrus, corn stalks, apple buds are resistant to SO_2 pollutants, (O'Gara, 1956).

Environmental factors which are conducive to optimum plant growth are usually the same factors which instigate SO_2 injury. These are sunlight, moderate temperature, higher relative humidity, wind and adequate soil moisture. In addition, time of day and season, and plant factors such as genotype, nutrition, stage of growth, tissue maturation determine the sensitivity of particular species to SO_2 injury. Vegetation is most susceptible to SO_2 during the active growth periods. It the environmental factors and growth stages of the plants are not conducive to injury, the plants will escape injury even in the presence of unfavourable concentration of SO_2. It has been reported that potato plant is sensitive to SO_2 both during day and night because stomates do not close at night (Barrett and Benedict, 1970).

(ii) Oxidants

Ozone and Pan (peroxyacetyl nitrate) are the main phytotoxicants. Ozone injury result in form of lesions i.e., pigmentation, chlorosis, bleaching and necrosis due to thickening and pigmentation of cell walls. Mosophyll cells are most susceptible to O_3 injury with the brown, black, red or purple lesions primarily occurring on the upper surface of leaves. Chlorotic lesions occur in the upper leaf surfaces due to injury to palisade cells. Bleaching is due to collapse of the palisade cells and sometimes the upper epidemical cells. Necrosis on leaves takes the form of leaf tip injury. Ozone injury is usually intervenal although lesions frequently concentrate along the sides of main veins. The young leaves are resistant, but with expansion becomes susceptible at middle and basal portions, and with complete maturation the leaves become resistant again to O_3. The sub-cellular effects of O_3 involve the formulation of granulation in chloroplast stroma and then there is disruption of the cells with cellular contents in centre of the cell (Hill et al. 1970) resulting in decrease of photosynthesis and changing of cell membrane permeability.

Sensitive varieties of alfalfa, spinach, clover, oats, radish, sweet corn, beans have been injured by 10 to 12 pphm for 2 hr (Manning and Feder, 1980). Various environmental factors predispose plants to O_3 injury. These include poor nutrition, low light, high relative humidity, and moderately high temperatures. Stomatal closer due to moisture stress can protect plants from O_3 injury. Low carbohydrate content within the plant increases the sensitivity and can protect plants from O_3 injury by increasing sugar levels.

Legumes such as soybeans and clover were more susceptible to O_3 than grains such as corn or wheat (UNEPA, 1986; Heagle et al. 1988). Susceptibility to pollutants are also under genetic control (Scholz et al. 1989). A number of studies have been reported in USA that ambient O_3 concentration can be sufficiently high to impair the yields and quality of a number of crops including citrus, grape, tobacco, cotton, tomato, bean, soybean, peanut, onion, and potato (USEPA, 1986). Crop losses in wheat and soybean have been reported using ambient air quality data (U.S. Deptt. of Commerce, 1987).

(iii) Nitrogen oxides

NO and NO_2 occur in atmosphere in significant concentrations, but rarely exceeds 1 ppm. Concentration of over 25 ppm failed to injure several plant species (Lacasse and Treshow, 1976). NO_2 produces plant injury symptoms similar to those of SO_2. White or tan coloured lesions are produced between the main veins of leaf and grey discolouration on poor leaf surfaces (Taylor and Maclean, 1970).

Environment factors significantly affect the dosages on NO_2 required to cause plant injury. Under full sunlight about 6 ppm of NO_2 for 2 hrs injuries sensitive species such as tomato and cucumer. Under low light, sensitive plant species may be injured when exposed to 2.5 to 3.0 ppm of NO_2 for 2 hr. Low concentration of NO_2 (Less than 0.5 ppm) inhibited the growth of tomato plants under continuous exposure for 10 to 22 days (Spierings, 1971).

(iv) Flourides

Flourides occur in the atmosphere as HF, SiF_4 and H_2 SiF_6. It enters the leaf through stomates and are translocated through the margins of broad leaves and to the tips of monocotyledons and coniferous needles. The injury takes the form of chlorosis, before necrosis occurs. Fluoride injury start as a grey or light green water soaked discolouration of leaf tissue which changed to a tan or radish brown colour (Spierings, 1967). On corn, sorghum and grasses the symptoms first appear as scattered chlorotic fleck at the tips and upper margins of middle aged leaves and becomes necrotic. Growth of leaves during exposure can result in a crinkled appearance. The marginal chlorosis may become necrotic. Coniferous needles display a light brown to radish brown tip necrosis. Needles are most sensitive when they are elongating and fully expanded in the season (Threshow and Pack, 1970).

Flower petals are rarely injured by fluorides. Soft suture of peach is characterized by a premature reddening and softening of the flesh along the suture lines. Atmospheric fluoride exceeding 44 mg over a 30 days period has produced suture of peaches.

Fluroides inhibit photosynthesis with the impairment initially being greater than degree of visible leaf injury, with continued fumingation, the decrease in the rate of photosynthesis paralleled the increase in leaf tissue necrosis as fluorides are required in the glycolytic pathway of plant respiration. In pine needles phloem and xylem parenchyma cells are enlarged greatly so that mesophyll cells exhibited granulation, vocuolation, and finally total collapse. The resin duct epithelial cells were enlarged greatly so that the duct canals were frequently occluded. (Chang, 1975).

Sensitive plant species are affected by about 0.4 to 1μg/H F/m³ for periods ranging from one to several weeks and that more resistant species required about 10 times as much as fluorides to be injured. Certain sensitive gladiolus varieties can display leaf injury with only 20 ppm fluorides in their leaves, whereas bitternut chicory can concentrate upto 1000 ppm fluoride in its leaves without showing visible injury. Fluoride contaminated fodder are sensitive to effect cattles, sheep, horse, swine and poultry in decreasing sensitivity. Animal feed should not be in excess of 40 ppm fluoride on an annual average based on monthly sampling in order to avoid the adverse effects of the disease. Animal flourosis is quite common in cattle, sheep and chickens due to protoplasmic poisonic nature of fluorine (Largent, 1952; Phillips, 1952).

Plant grown under, optimum moisture conditions are sensitive to fluorides and more sensitive when K, P or Mg were deficient. Tomato plant grown under deficient Ca have increased injury. Deficient and super optimal levels of N reduced fluorides toxicity in tomato, whereas optimum levels of N enhanced fluoride toxicity in tomato.

(v) Particulate matter

These are cement-kiln dust, magnesium lime dust, carbon soot, H_2SO_4 aerosols, fluorides, lead compounds, Ni compounds, salt spray having size of 1 mg to 100 mg effecting vegetation. Cement dust in combination with dew and light rain formed a thick crust on upper surface of citrus tree, and interfered with the penetration of light required in photosynthesis with a resultant decrease of starch formation (Anon 1977). Reduced yield are obtained when the sides of trees faced a cement plant which prevented pollen germination on dust covered stigmas of flowers. Cement kiln dust deposition on bean leaves @0.6 to 3.8 gm m^{-2} for 8 hrs for a period of 2 to 3 days reduced CO_2 exchange by 30 per cent (Manning, 1971). Cement dust may cause chlorosis and death of leaf tissue by the combination of a thick crust deposit and allkaline toxicity produced in wet weather. Inhaling of dusty air results in deposition of silica particles in lungs along the arterial trunks (Clelland et al. 1952).

Soot smoke suppressed the growth of pine trees in the vicinity. Injuries were related to the acidity of the soot particles and the tarry coating with clogged stomates interferred the normal exchange of gases. (Lerman and Darley, 1975).

(vi) Smog

Normally smog is a fog made heavier and darker by smoke of area. However, if smog is generated in industrial areas during foggy, anticyclonic weather condition, coal smoke with enough sulphar dioxide vapours reduces the visibility and causes heavy loss of human and animal. London smog occurred in 1952 with similar conditions led to the respiratory tract symptoms (Douglas and Stewart, 1953). In rye grass there was reduction of tillers and leaves by 54 and 44 per cent and dry weight by 33 per cent (Bleadale, 1952). Also during such foggy temperature inversion anticyclonic weather conditions in industrial area, smog is generated with mixture of ozone and peroxidized organic compounds formed by the photochemical reactions between oxides of nitrogen and organic compounds such as gasoline vapours or partially burned fuel. Such conditions were generated in Log Angeles in 1951 to create eye, nose and throat irritation and damaging effects on plants (Littman, Ford and Endow, 1956). Green vegetable leaves develop lesions of silvery or bronze colour in lower epidermis cells (Middleton et al., 1956).

10.6.2 Beneficial effects

Some atmospheric contaminants contain elements which are essential nutrients for normal growth and development of vegetations. Sulphur, being essential component of proteins and lipids as well as some vitamins if present in atmosphere as SO_2. Good yielding soils need 20 to 50 kg S/ha annually, while unfertilized forests needs about 15 kg S/ha/year for new growth. Natural sources could supply about 10 kg S/ha/year if evenly distributed over land area, while anthropogenic sources from combustion processes and smelting could supply another 7 kg S/ha/year on global basis. Rate of absorption of atmospheric SO_2 concentration of 40 mg/m^3 would provide 20 kg S/ha/year. Assuming that all of the sulphar from rain was lost in drainage, this value of 40 mg/m^3 was considered to be sufficient for a large number of crops. Based on fumigation results of a number of crops with 400 mg/m^3 SO_2 or less increases the yield. However, the duration of the exposure should be controlled to avoid damage. A concentration of 400 mg/m^3 was 10 times higher than the amount needed to provide adequate sulphar for fertilization so that there is in fact an adequate margin between the concentration that are beneficial and those that are damaging. Therefore where the soil are not sufficiently alkaline, an increase in acidity may occur which in turn will result in leaching of cations and decrease of pH resulting in biological damage.

Plants grown under sulphar or nitrogen deficiency were less susceptible to damage by high concentrations of atmospheric SO_2 than in the normal case. These can tolerate higher ambient levels of SO_2 and results in higher yields under such conditions. Similarly crops grown under certain deficient conditions are less susceptible to O_3 damage and indeed O_3 may even have a beneficial effect which is manifested by increased root development. Air pollutants other than SO_3 which could contribute to increased crop yields are CO_2, NO, NO_2, NH_3, and certain trace metals.

CO_2 role in photosynthesis is well known and green house operators have been using increased CO_2 level as a fertilizer for many years. Nitrogen oxides from pollution sources other than applied fertilizers are insignificant. Nitrogen in the form of NH_3 is released into atmosphere in considerable quantities from sewage treatment plants and large cattle feed operations. NH_3 can be absorbed by natural water bodies as well as by vegetation.

Due to large differences in resistant species offers possibility of reducing injury due to air pollutants. Table below shows the risk that exists for various agricultural landuse types from the chronic effects of SO_2, HF and HCl based on reduction of plant productivity. However to reduce action of pollutants, fertilization improves the negative effects, but directly influence the injury in above ground organs. Optimum balanced fertilization reduces the uptake of pollutants. Table 10.8 shows the threshold permissible limits of pollutants levels for their safer use.

Crops	**Degree of risk in**		
	SO_2	**HF**	**HCl**
Agricultural field crops	Very slight	Slight to medium	Slight to medium
Fodder plants	Slight	Very severe	Slight to medium
Vegetables	Slight to medium	Slight to severe	Slight to medium
Pastures	Slight to medium	Very severe	Medium
Fruits	Slight to severe	Very severe	Very severe
Forests vegetation	Medium to severe	Medium to severe	Medium to severe
Ornamental plants	Very slight	Slight to severe	Slight

(After: Guderian, 1977).

Table 10.8

Threshold permissible pollution levels standards

Pollutants	**Concentration over short and long term basis**	**Reference**
SO_2	0.15 mg SO_2/m^3 air for 24 hr average	USSR, 1949, 1951
	0.50 mg SO_2/m^3 air for any short term value	Anonymous, 1979, 1982
SO_2	0.14 mg SO_2/m^3 air for long term value	TA Luft, 1974
	0.40 mg SO_2/m^3 air for short term value	
	Primary standards	
	80 µg SO_2/m^3 air (0.03 ppm) yearly average	JAPCA, 1971
	365 µg SO_2/m^3 air (0.14 ppm) maximum 24 h average	Lacasse and
	exceeded once in a year	Treshow, 1976
	Secondary standard	
	60 µg SO_2/m^3 air (0.02 ppm) yearly average	
	260 µg SO_2/m^3 air (0.10 ppm) maximum 24 hr average	
	exceeded once in a year	
	1300 µg SO_2/m^3 air (0.5 ppm) maximum 3 hr average	
	exceeded once in a year	
$SO_2 + NO_2$	$\geq$50 ppb SO_2 + 50 ppb NO_2 in air on long term average	ECE, 1986
HF	2 µg HF/m^3 air of long term average	
	4 µg HF/m^3 air for short term average	TA Luft, 1974
Cl_2	0.03 mg/m^3 air on 24 hr average	USSR (1949, 1951)
	0.03 mg/m^3 air at any time	Anonymous (1979, 82)

Fluoride	0.5 μg F/m³ air (0.65 ppb) during growing period	Weinstein, 1971
	0.78 μg F/m³ air (1 ppb) during 30 days average	
	1.55 μg F/m³ air (2 ppb) during 7 days average	
	2.70 μg F/m³ air (3.5 ppb) during 24 hours	
	3.50 μg F/m³ air (4.5 ppb) during 12 hr. period	
SPM	75 μg/m³ yearly geometric mean	USEPA, 1976
	260 μg/m³ once maxima in 24 hours	
O_3	≥80 ppb O_3 in air on long term average	ECE, 1986
HCl	0.10 mg HCl/m³ on long term average	TA Luft, 1974
	0.20 mg HCl/m³ on short term average	
C_2H_4	2.0 mg/m³ ethene in air on long term average	Anonymous, 1979, 82
	3.0 mg/m³ ethene in air on short term average	
NH_3	0.1 mg/m³ air on short term average	Anonymous, 1979, 1982
	0.3 mg/m³ air on short term average	
NO_2	0.04 mg/m³ air on long term average	Anonymous, 1979, 1982
	0.1 mg/m³ air on short term average	
Soot, dust	0.05 to 0.15 mg/m³ air on long term average	Anonymous, 1979, 1982
	0.15 to 0.5 mg/m³ air on short term average	USSR, 1949, 1951
CO	1 mg/m³ air on short term average	Anonymous, 1979, 1982
	3 mg/m³ air on short term average	
H_2S	0.008 to 0.01 mg/m³ air short term average	Anonymous 1979, 1982
	0.015 to .03 mg/m³ air on short term average	USSR, 1949, 1951
$SO_2 + NO_2 + O_3$	>40 ppb SO_2 + 40 ppb NO_2 + 40 ppb O_3 in air on long term average	ECE, 1986
CO	10 μg/m³ (9 ppm) 24 hour maximum once	USEPA, 1976
	40 μg/m³ (35 ppm) 8 hour maximum once	Lacasse and Treshow, 1976
	160 μg/m³ (0.08 ppm) one hour maximum once a year	
NO	100 μg/m³ (0.05 pp NO_2) yearly average	Lacasse and Treshow, 1976
Hydrocarbon	160 μg/m³ (0.24 ppm as CH_4) 3 hour maximum once a year	Lacasse and Treshow, 1976
	2 mg/m³ as C_2H_4 for long term average	Anonymous 1979, 1982
	3 mg/m³ as C_2H_4 for short term average	Anonymous 1979, 1982
	0.1 mg/m³ as C_2H_6 for long term average	Anonymous 1979, 1982
	0.3 mg/m³ as C_2H_6 for short term average	Anonymous 1979, 1982

1 ppm = 2.67 SO_2 = 0.83 HF = 1.51 HCl = 1.91 NO_2 = 1.99 O_3 = 0.71 NH_3 = 1.16 C_2H_4 mg/m³

10.7 Simulation Models with Climate Change Scenarios

Aggarwal et al. (2002) indicated that the impact of higher temperature for irrigated wheat in north India resulted in decrease of wheat yield by 17 per cent with rise of 2°C temperature (Fig. 10.12). However, this decrease was compensated to some extent by increasing levels of CO_2 due to its positive influence on crop growth. Clifford et al. (1993) reported that in groundnut and sorghum yield per plant was higher in groundnut (30.1%) as compared to

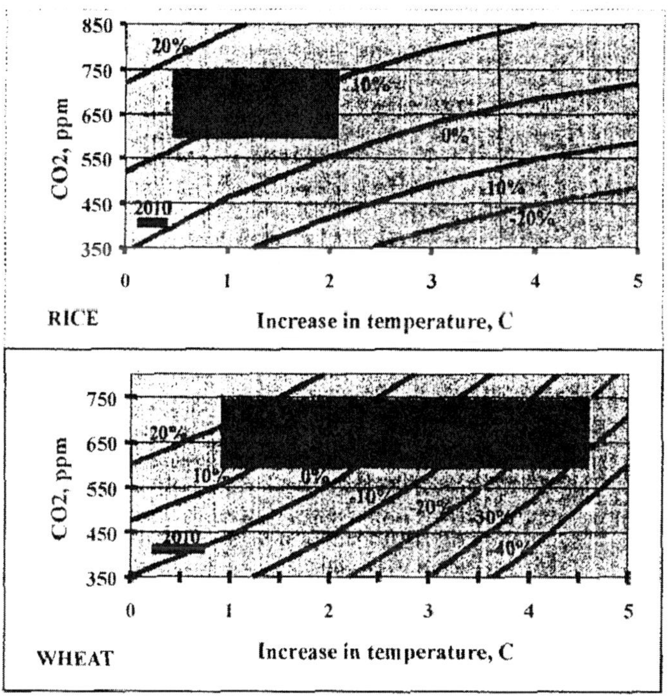

Fig. 10.12: Rising carbon dioxide-temperature interaction effect on productivity of rice and wheat crops in New Delhi environment (*After*: Kalra et al., 2003).

sorghum (6.2%) with increase of CO_2 from 350 ppm to 700 ppm. Similarly Kalra et al. (2003) showed the effect of climate change on rice and wheat yields with different levels of management. Simulation crop models were modified to include the effects of CO_2 concentration as the interaction of CO_2 with water use, crop growth and yield were modified (Penning de Veries, 1989; Rosenzweig, 1990; Wolf, 1993; Curry et al., 1990; Singh and Ritchie, 1993; Easterling et al., 1992, 1993; Boote et al. 1992; Sinclir and Rawlins, 1993) in order to study the impact of climate change. Table 10.9 shows impact of doubled CO_2 levels in different modified crop models to highlight the alteration in response functions.

Dhakhwa et al. (1997) assessed the impact of global warming and CO_2 fertilization using CERES and EPIC models. The climate change scenarios included changes in temperature, precipitation and solar radiation from two GCMs scenario – GISS and GFDL. When GCM predicted an increase of 4°C temperature, the first hypothesis assumed a symmetrical temperature i.e., maximum and minimum temperature each assumed to increase by 4°C with equal day/night warming (EW), second hypothesis assumed a differential day/night warming (DW) where daily nightime temperature increase (warming) was three times larger than the increase in daytime temperature i.e., the change in mean diurnal temperature range was equal to change in daily mean temperature (Karl et al. 1993). Table 10.10 shows simulated maize yield and dry matter under different scenarios confirming that crop models can be modified to reflect changes in biomass, yield and WUE in response to various combinations of direct and indirect effects of climate change.

CERES – Maize was modified to account for the beneficial effects of increased CO_2 concentration in crop growth and transpiration. Doubling of CO_2 levels induced in crop growth increase by 34 ± 6 per cent for C_3 plants and about 14 ± 11 per cent for C_4 plants in controlled environment (Porter, 1992). The increase of CO_2 concentration resulted in partially closure of stomates and reduced transpiration (Rosenberg et al. 1990). With doubled CO_2 levels in atmosphere, the average transpiration reduction for C_3 and C_4 crops was assumed in the range 30 to 51 per cent (Kimball and Idso, 1983), and stomatal resistance increased by 67 ± 14 per cent with a doubling CO_2 from 330 to 660 ppm (Rosenberg et al. 1990). With decrease of stomatal conductance due to elevated CO_2 in atmosphere, WUE also increased (Eamus, 1991). Therefore, in CERES crop models a 10 per cent improvement in photosynthesis and a 30 per cent reduction in transpiration variables were used to account for direct effects due to CO_2 fertilization.

Similarly in EPIC model, RUE was assumed to increase by 10 per cent and decrease of transpiration by 30 per cent (Kimball, 1983; Sinclair and Rawlins, 1993) for maize with doubling of CO_2 concentration.

Considering the direct effects of CO_2, the simulated maize yield increased by 14 to 18 per cent. However, the impact of higher temperature on simulated yield and biomass decreased under both GCM scenarios due to the reduction of growth cycle particularly the grain filling period. The biomass, yield and WUE declined for the warmer, drier climate offered by UKMO as compared to GFDL. In both scenarios for DW hypothesis an increase in yield was noticed as compared to EW hypothesis. EPIC model was more sensitive to night time minima as compared to CERES as interannual variability of yield and biomass was less. Olszyk et al. (1999) simulated rice yield using ORYZAI process model using climate change scenarios to compare with the experimental yields for 1995 and 1996 seasons (dry season) and 1994 (Wet season). Under current climatic conditions of ambient CO_2 and temperature, rice simulated yield was 14 per cent lower than experimental yield in 1996; but, was higher by 17 and 37 per cent during 1995 and 1994 seasons. Biomass was also underestimated, whereas simulated methane were same of experimental values. Improvements are needed in ORYZAI model for the effects of global climate change (Table 10.11) for grain yield and biomass estimation. Elevated CO_2 increased both simulated as well as experimental yields of grain and biomass, but elevated temperature decreased them; whereas both variables increased the grain yield and biomass of rice in between of their independent values.

Alexandrov and Hoogenboom (2000) studied the impact of climate change scenarios of GCM models (Table 10.5) on crop yields in Bulgaria. CERES v 3.5 (Ritchi et al 1998) with DSSAT v. 3.5 (Tsuji et al., 1994) was used for crop yield simulation (Table 10.12). A decrease in winter wheat and maize yields for 2020, 2050 and 2080 was reported probably due to

Table 10.9

Impact of doubled CO_2 levels in modification of response functions in different crop simulation models

Crop model	Region	Climate Change	Modified response of functions in the model	Impact on crop yield	Reference
CERES Wheat CERES Maize	Southern Great Plains of USA	Doubled CO_2 levels	Light use efficiency increased by 25 per cent in wheat and 10 per cent in maize. Potential plant transpiration was modified by Penman-Monteith equation for evapotranspiration to include the effects on stomatal conductance (g_s) decrease. Rogers *et al.* (1983) relation of g_s decrease with increased CO_2 levels was introduced. GCM models – GISS (Hansen *et al.*, 1983) and GFDL (Manabe and Wetherald, 1987) were used.	Simulated yield decreased for both crops due to shorter life cycle of crop under elevated temperature. Water stress increased due to higher evapo-transpiration. Simulated yield decreased for irrigation crop were less than rainfed crops. Direct effect of CO_2 partially offset the negative effect of temperature increase, but not entirely.	Rosenzweig, 1990
SOYGROW	Southeast USA	Doubled CO_2	Two climate scenarios GISS and GFDL with doubled CO_2 levels	Soybean yield increased at 12 locations and decreed at 7 locations using GISS scenario in rainfed, but with GFDL scenario yield declined at 17 locations with irrigation yield increased	Wilkerson *et al.*, 1983
CERES- Maize	Southeast USA	Doubled CO_2 levels	Two climate scenarios with doubled CO_2 levels	Rainfed maize yield decreased more than soybean yield. The beneficial effects of doubled CO_2 were less for maize than soybean. Irrigation requirement considerably increased and WUE reduced for maize in both scenarios	Curry *et al.*, 1990
WOFOST winter wheat	Europe	Doubled CO_2 levels	Effect of increased CO_2 levels on photosynthesis, transpiration and specific leaf area were modified.	Yield increased under water limitations with direct effects on net photosynthesis	Wolf, 1993

AFRCWHEAT2	Europe	Doubled CO_2	Included direct effect on photosynthesis rate, photosynthetic efficiency at low light levels and stomatal resistance.	Simulated wheat yield increased under climate change scenarios, but yield variability increased also.	Semenov *et al.*, 1993.
EPIC model for wheat, maize, sorghum, and soybean	MINK Region of USA (Missouri IOWA-Nabraska-Kansas)	Doubled CO_2 levels	Used weather data of 1931-40 which was warmer by 1.1 to 1.6°C over 1951-80 period with lower precipitation by 21-59 mm in the dust boil MINK region. EPIC model was modified with direct effect of CO_2 on net photosynthesis rate (Peart *et al.*, 1989). In EPIC, vapour pressure deficits (vpd) affects the photosynthetic light use efficiency. The changes in vpd and CO_2 concentration influence the stomatal conductance to reduce transpiration. With doubling of CO_2 about 40 per cent reduction in conductance (Morison and Gifford, 1983).	Simulated yield of dryland maize, sorghum, soybean declined by 25 per cent with warming without CO_2 effects, but with increased CO_2 levels compensated these losses partially.	Easterling *et al.*, 1992, 1993.

Table 10.10

Simulated maize yield attributes under different scenarios using modified CERES and EPIC models.

Maize	Maize Model	Yield (kg/ha)	Biomass (kg/ha)	PET (mm)	ET (mm)	WUE (kg/ha/mm)	HI	Per cent departure from the baseline					
								Yield	Biomass	PET	WUE	ET	HI
Base	CERES	6159	14594	591	462	13.3	0.42						
	EPIC	6095	17101	1383	562	18.29	0.47						
Direct 2xCO_2	CERES	7272	16686	590	424	17.2	0.44	18	14	0	29	-8	3
	EPIC	6966	19312	1360	521	23.39	0.48	14	13		28		1
Indirect GFDL-EW	CERES	4577	13324	602	447	10.2	0.34	-26	-9	2	-23	-3	-19
	EPIC	4604	13379	1525	585	15.7	0.48	-24	-22		-17		-4
GFDL-DW	CERES	4652	13636	558	427	10.9	0.34	-24	-7	-5	-18	-8	-19
	EPIC	5210	14860	1395	549	18.29	0.45	-15	-13		0		-4
UKMO-EW	CERES	3969	12179	679	474	8.4	0.33	-36	-17	15	-37	3	-23
	EPIC	3750	10870	1643	610	13.1	0.45	-38	-36		-28		-3
UKMO-DW	CERES	4081	13266	560	424	9.6	0.31	-34	-9	-5	-28	-8	-15
	EPIC	4630	13070	1436	553	17.69	0.47	-24	- 24		-3		0
Combined													
GFDL-EW	CERES	5627	15601	601	426	13.2	0.36	-9	7	2	-1	-8	-15
	EPIC	5430	15300	1498	553	19.68	0.47	-11	-11		8		0
UKMO-EW	CERES	5009	14580	677	458	10.9	0.33	-19	2	15	-18	-1	-22
	EPIC	4400	12410	1616	578	16.97	0.47	-28	-27		-7		0
UKMO-DW	CERES	4922	15499	559	403	12.2	0.32	-20	6	-5	-8	-13	-25
	EPIC	5210	14550	1429	528	21.85	0.47	-15	-15		19		1
GFDL-DW	CERES	5639	15775	557	403	14.0	0.36	-9	8	-6	5	-13	-15
	EPIC	5990	16760	1384	517	23.18	0.47	-2	-2		27		0

(After : Dhakhwa *et al.*, 1997)

Table 10.11
Comparison of simulated and experimental yield and biomass of IR 72 rice in different seasons

Seasons	Parameter (g/m^2)	1994		1995		1996	Experimental
		Simulated	Experimental	Simulated	Experimental	Simulated	
Amb. CO_2, ambT	Grain yield	575	366(419)	910	683(781)	858	874(999)
	Biomass above ground	1020	1002(1146)	1560	1414(1617)	1570	1622(1855)
	Biomass below ground	173	214(245)	265	389(444)	267	270(308)
	Total biomass	1193	1216(1391)	1825	1803(2061)	1837	1892(2163)
Elv. CO_2 AmbT	% change in grain yield	34	29	35	24	40	15
	% change in above ground biomass	34	21	33	27	27	17
	% change in below ground biomass	34	61	33	30	27	79
	% change in total biomass	34	28	33	28	27	27
Amb CO_2, ElvT	% change in grain yield	-15	-26	-13	-16	3	-13
	% change in above ground biomass	-7	0	-7	1	-18	-4
	% change in below ground biomass	-7	-25	-7	-31	-18	22
	% change in total biomass	-7	-4	-7	-6	-18	-1
Elv CO_2, ElvT	% change in grain yield	3	-14	33	9	43	-6
	% change in above ground biomass	21	11	22	31	13	7
	% change in below ground biomass	21	78	22	25	13	21
	% change in total biomass	21	23	22	30	13	9

Amb CO_2 ~ 340 ppm, Elv CO_2 (Experimental measured) = 370 ppm, Amb T using climatic data IRRI, Elev T = Amb T + 4°C, Elv CO_2 = Amb CO_2 + 300 ppm

(After: Olszyk et al., 1999).

Table 10.12
Impact of climate change on maize and wheat yields using different GCM models

GCM scenarios	Percentage departure of grain yield over the current climate					
	Maize			Winter Wheat		
	2020	2050	2080	2020	2060	2080
ECHAM4	-5.5	-11.6	-15	14	27	30.4
HadCM2	9	-1.3	-17	10.5	30	26
CGCM1	-6	-10.2	-19	18	26	32
CSIRO-Mk2b	-7	-9.6	-11	16	27	36
GFDL-R15	-12	-19.0	NA	13	24	NA
CO_2 level (ppm)	447	554	697	447	554	697

shortened vernalization period of winter wheat and a short growing season duration for both crops. However, when direct impact of CO_2 was considered, the winter wheat yield increased. Therefore there is scope for adaptation and possible changes in sowing date and selection of hybrids in order to mitigate the adverse impact of warming on maize yields. Changes in cropping system, irrigation and agricultural landuse may provide additional options for adaptation in agriculture. Alexandrov and Hoogenboom (2000) tried different sowing dates of spring crops in Bulgaria to shift under GCM climate change scenarios in order to reduce the yield loss caused by an increase in temperature. The selection of earlier sowing date for maize is appropriate response to offset the negative effect of a potential increase in temperature (Fig. 10.13). The change of sowing date will allow for the crop to develop

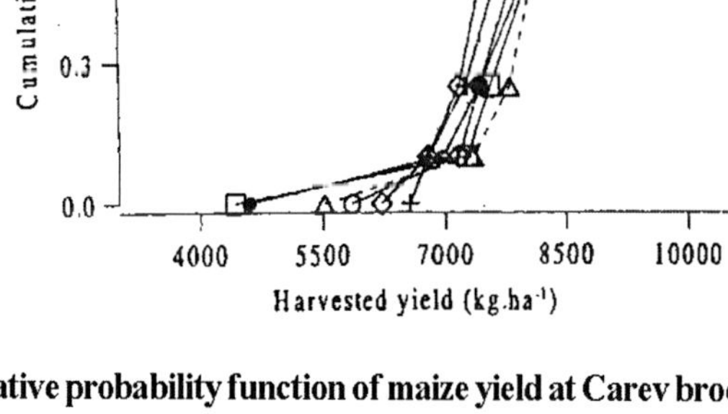

Fig. 10.13: Cumulative probability function of maize yield at Carev brod under the ECHAM4 climate change scenario for the 2080s and different sowing dates (CO_2 = 697 ppm). (After: Alexandrov and Hoogenboom, 2000)

during the lower temperature periods, which would decrease the development rates and increase the growth duration particularly the grain filling period. The sowing date of maize for northeast Bulgaria should occur at least 2 weeks earlier in 2080 under ECHAM4 scenario, relative to the current climate conditions and would also help in adjustments of cropping system in remaining part of the year. Another benefit of earlier shifting of maize sowings would also help us in using hybrids of medium duration growing season which are beneficial for maize productivity over long and short duration maize cultivars under the HadCM2 transient climate change scenarios for the 2020, 2050 and 2080 (Alexandrov and Hoogenboom, 2000) (Fig. 10.14).

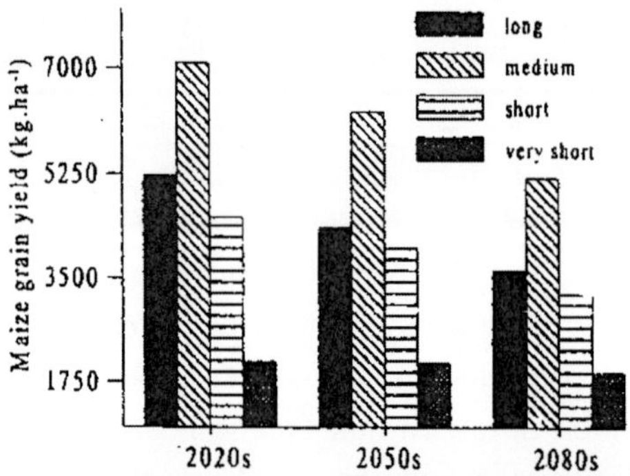

Fig. 10.14: Simulated maize yield in Ognjanovo under the HadCM2 transient climate scenarios for the 2020s, 2050s, and 2080s and hybrids with different growing duration. (After: Alexandrov and Hoogenboom, 2000)

Maayar et al. (1997) studied the combined effect of climate change and CO_2 fertilization on a variety of C_3 and C_4 crops in Quebec region using the outputs of the Canadian Climate Centre (CCC) general circulation model coupled with the food and agriculture organization (FAO) crop model.

Fig. 10.15 shows the relative yield changes owing to doubling of CO_2 concentrations without CO_2 fertilization in different C_3 and C_4 crops, whereas Fig. 10.16 shows the relative yield changes owing to the combined effects of climate change and CO_2 fertilization in variety of field crops including pulses, oilseeds and vegetables. Without considering CO_2 fertilization effect, C_4 cereals (corn and sorghum) were most favoured with climate change in Quebec region by the future climatic conditions; whereas C_3 cereals and legumes will have a decreasing trend in the region. With incorporation of CO_2 fertilization effect, C_4 cereals appeared not very sensitive to increased CO_2 levels, but C_3 crops like barley, phaseolus beans and onions were sensitive to increased CO_2 levels. Even the possibility of cultivation of sugar beet in southern Quebec was feasible.

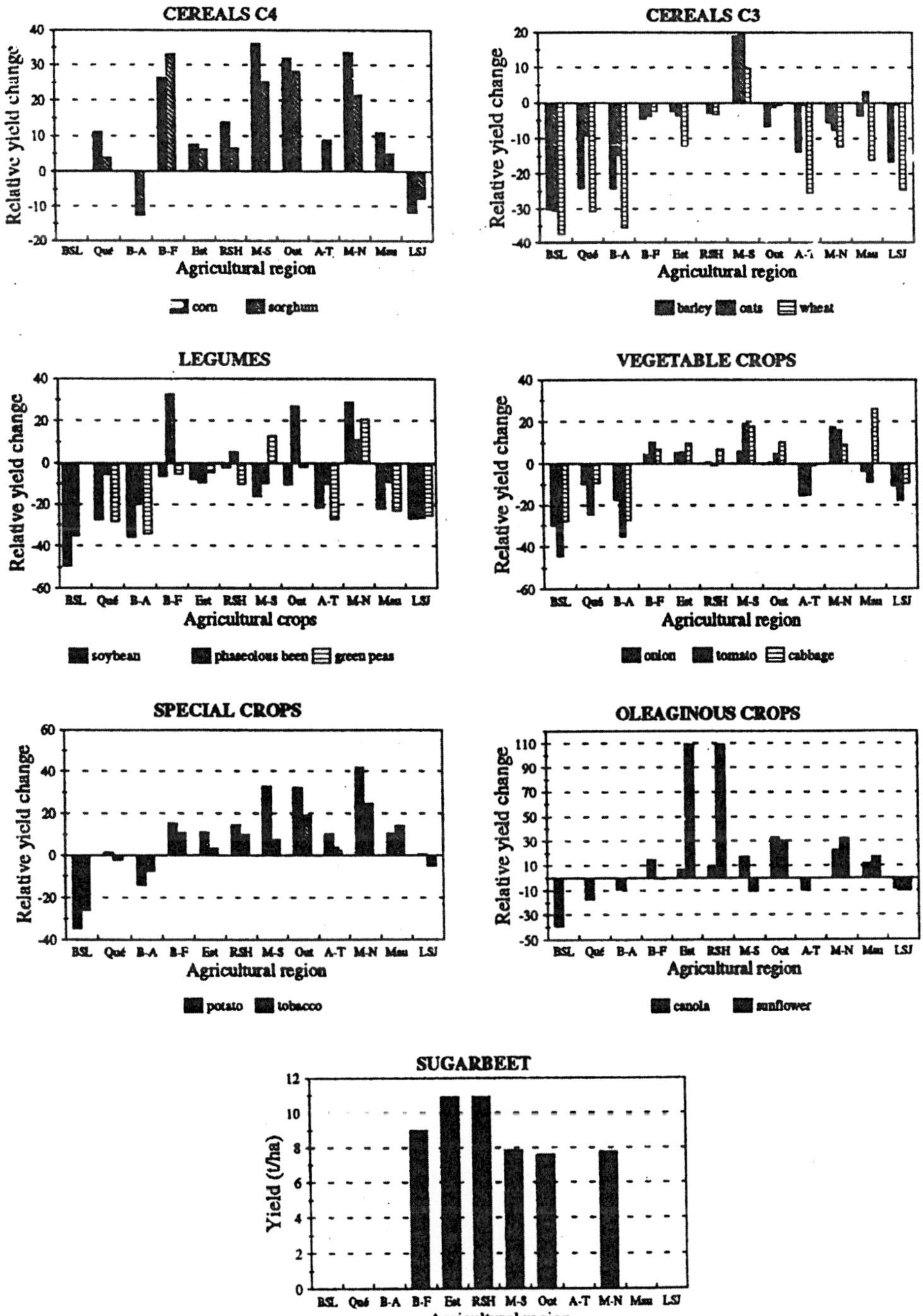

Fig. 10.15: Relative yield changes owing to 2 × CO_2 climate scenario without CO_2 fertilization. (After: Maayar et al., 1997).

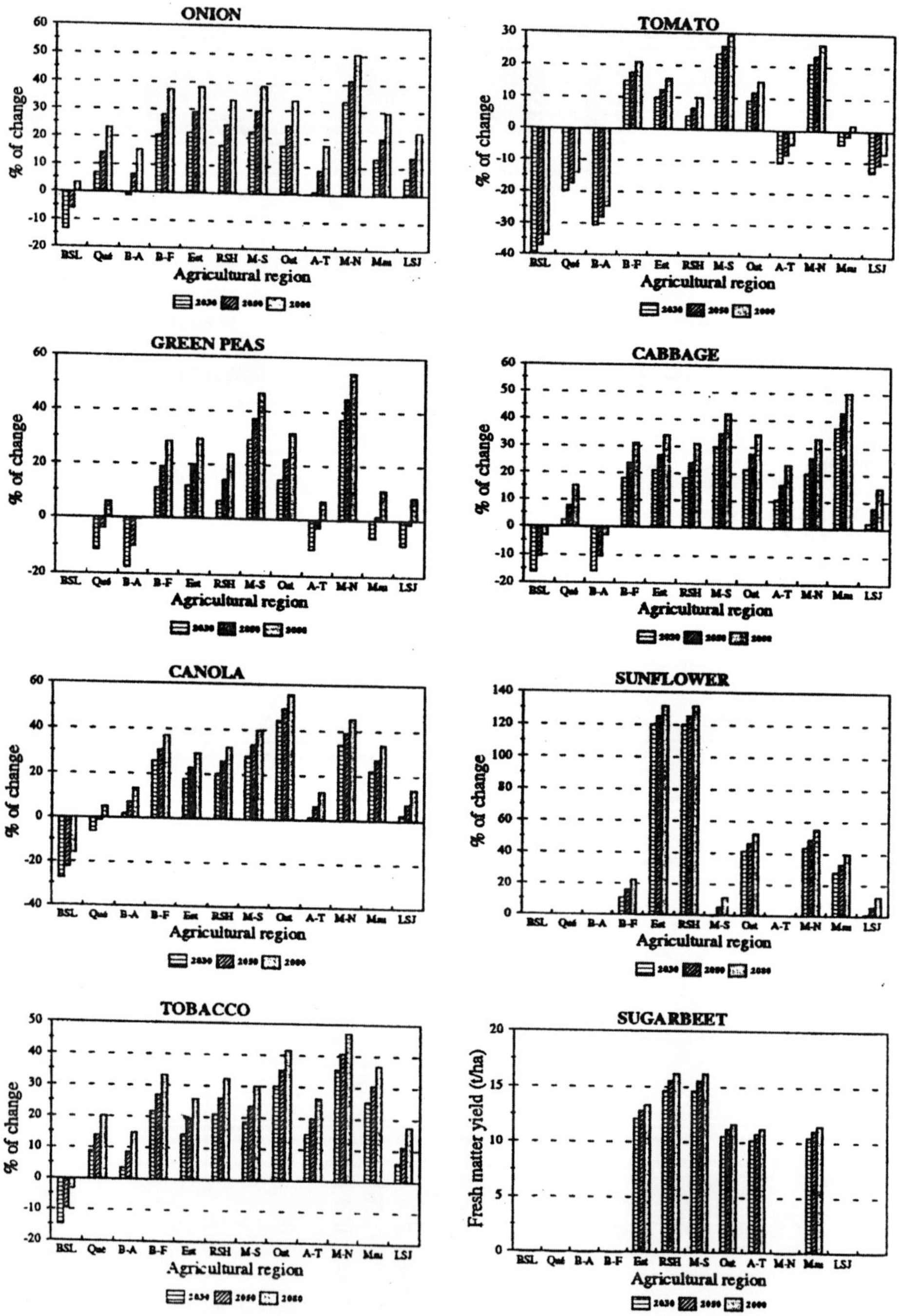

Fig. 10.16: Relative yield changes owing to cimbined effect of climate change and CO_2 fertilization. (After: Maayar et al., 1997).

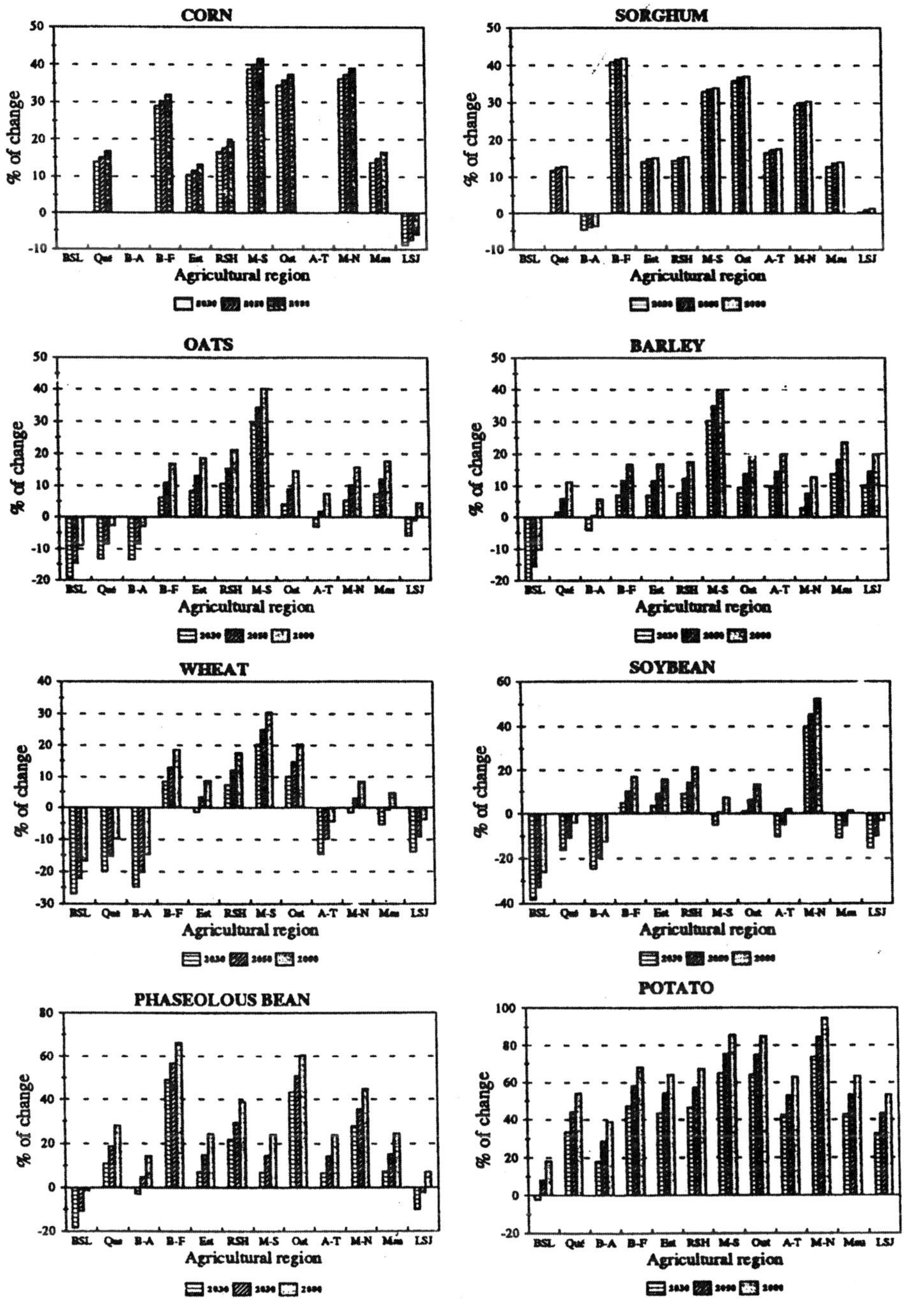

Fig. 10.16: (Continued)

10.8 Impact of Climate Change vs. Technological Advancement on Crop Productivity

Advances in agricultural technology may affect the response of crop yields to a future climate change. Mckenney et al. (1992) analyzed the impact of climate change on crop yields using EPIC model with the contribution of technological advancements with time by the year 2030. Following scenarios were generated:

(i) C-Future baseline crop productivity in 2030 in the absence of climate change-control (1951-80) data.

(ii) D_1, D_2 – Climate analog data of 1931 – 40 with current (330 ppm) (D_1) and elevated atmosphere CO_2 concentration (D_2) superimposed on future baseline (C). The period 1931-40 was warmer by 1.1 to 1.6°C over 1951-80 period i.e., a analog condition of green house warming.

(iii) D_3 – analog climate with use of currently available and future technologies which might be developed in adaptation to warmer and drier conditions (decreased stomatal conductance and increased irrigation efficiency). Task $D_{3,1}$ with current CO_2 and task $D_{3,2}$ with elevated CO_2.

The selected future technological advancement by 2030, not due to climate change, but to improve agricultural productivity with the passage of time considered in term of:

(i) Photosynthetic efficiency (LUE) improvement – two levels of 10 and 20 per cent variation.

(ii) Harvest index (HI) improvement varied by 10 and 20 per cent.

(iii) Pest control (PSTF) including insects, weeds and disease – varied by 10 and 20 per cent.

(iv) Rate and timing of leaf area development (DLAI) – earlier leaf development would increase leaf area duration providing greater photosynthetic capacity and deeper plant roots over the growing season. Two levels of DLAI to provide 5 and 10 per cent earlier in the growing season.

(v) Harvest efficiency (HE) – to avoid field losses during harvest (management).

(vi) Stomatal conductance (g_s) – decrease will increase drought tolerance, reduction of evapotranspiration, increase of WUE. g_s decreased by 0.001 and 0.002 m/sec from base value of 0.007 for crops.

(vii) Irrigation efficiency (EFI) – with control of time and amount of irrigation water (management).

A sensitivity analysis was conducted using EPIC model to select appropriate future technology under controlled climate (Table 10.13) as well as climate change conditions (Table 10.14) to highlight their impacts on maize and wheat yields in MINK region (Missouri – Iowa – Nebraska – Kansas).

Table 10.13

Impact on maize and wheat yields under controlled climate

Future Technology	Percent change in yield of					
	Rainfed Maize			Wheat		
	C	D_1	D_2	C	D_1	D_2
LUE + 10%	9	10	11	10	10	10
LUE + 20%	19	21	21	20	20	20
HI + 10%	10	10	10	9	9	9
HI + 20%	19	20	19	18	18	18
PSIF + 10%	11	11	11	12	12	12
PSIF + 20%	21	21	21	24	24	25
DLAI + 5%	8	5	5	3	1	1
DLAI + 10%	17	11	12	6	3	3
1/rs + 0.001	3	6	5	6	8	8
1/rs + 0.002	8	12	12	15	18	17

(After: Mckenney et al. 1992)

These is practically no interaction with climate or CO_2 for HI and PSIF for these crops. Maize shows a slight increase in sensitivity to LUE under the analog climate, but no impact in wheat. With added CO_2, corn shows small increase in sensitivity to LUE and also more response to earlier leaf development in control climate over the analog climate (poor). A decrease of stomatal conductance have a greater impact on yield under analog climate as WUE is increased, and even has a greater impact under rainfed conditions. Fig. 10.17 presents under task A and B, the potential impacts of climate change imposed on crops, however with time some of the technological changes may affect the ways in which both crops and farmers respond to mitigate the adverse effects. EPIC model was used in task C and D with incorporation of existing trends and responses by manipulating parameters of future technologies. Imposition of analog climate with and without CO_2 added (Task D_1 and D_2) caused yield to decline by almost same percentage as current technology scenarios (Task B_1 and B_2). With addition of CO_2 all crops (when adapted and adjusted) yielded above future control (Task C) (Fig. 10.18). With new future technologies (Task A) yield would increase for all crops by 72 per cent, whereas for wheat 88 per cent. The pattern of yield response to analog climate (D_1 and D_2) with future technologies is similar to that which occurs with current technologies (B_1 and B_2).

10.9 Future Guidelines

Climate variables are so difficult to compute that even the best model will be far from perfect in assessing what the future will be in tropical/subtropical belts. Global climate change, in terms of increased CO_2 concentrations, temperature rise and different precipitation

Table 10.14

Per cent change in yield under various analog climate scenarios with current (Task B vs A) and future technology (Task D vs C) for rainfed and irrigated conditions.

Crops	Task A yield (t/ha)	% change in relative yield to A				Task C yield (t/ha)	% change in relative yield to C			
		B_1	B_2	$B_{3.1}$	$B_{3,2}$		D_1	D_2	$D_{3.1}$	$D_{3,2}$
Dryland corn	6.4	-28	-19	-24	-15	10.9	-28	-19	-17	-7
Dryland soybean	2.5	-27	-16	-18	-5	4.1	-25	-14	-12	1
Dryland sorghum	3.3	-21	-10	-8	+5	5.7	-21	-11	-4	9
Dryland wheat	3.1	+1	+13	8	21	5.6	-7	8	5	20
Dryland Alfafa	6.9	-5	+12	4	33	12.1	-7	10	4	23
Irrigated corn	9.9	-7	-1	2	7	17.4	-9	-4	3	10
Irrigated sorghum	6.0	-11	-1	4	15	10.3	-10	0	7	18
Irrigated wheat	4.5	+8	+11	3	5	8.7	-1	9	6	18

Task A – Control (1951-80 climate; Task B_1 – Analog climate (1931-40), 330 ppm CO_2; Task B_2 – Analog climate, 450 ppm CO_2; Task $B_{3.1}$ – Analog climate, 350 ppm CO_2 with adjustments; Task $B_{3,2}$ – Analog climate, 450 ppm CO_2 with adjustments; Task C – Future technology with (1951-80) climate (Future Control); Task D_1 -- Future technology with analog climate (1931-40) 330 pm CO_2; Task D_2 – Future technology with analog climate, 450 ppm CO_2; Task $D_{3.1}$ – Future technology, 350 ppm CO_2 with adjustments with analog climate; Task $D_{3.2}$ – Future technology, 450 ppm CO_2 with adjustments with analog climate.

(After: Mckenney et al. 1992)

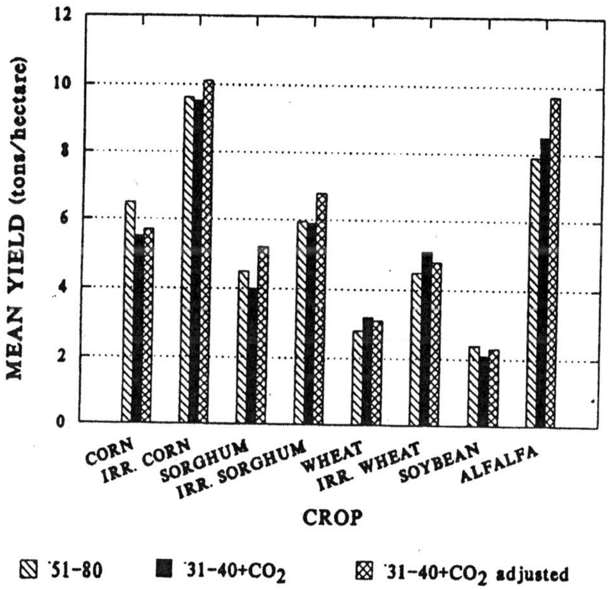

Fig. 10.17: Mean yield under the control climate (1951-80) (A), the analog climate (1931-40) with 450 ppm CO_2 and no adjustment (B_2) and the analog climate with 450 ppm CO_2 and adjustment (B3.2). (After: Easterling et al., 1992)

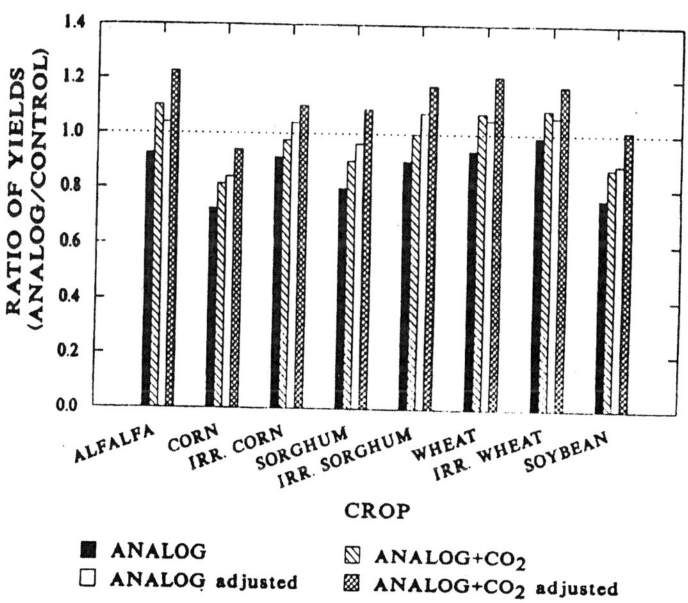

Fig. 10.18: Mean crop yield under analog climate scenarios with future technology (Tasks D_{1-3}) relative to yields under the control climate with future technology (Task C). (After: McKenney et al., 1992).

patterns, will further aggravate this vulnerability to climatic variability and increase in frequency of their impacts in term of abnormal weather events in form of droughts, cyclones and floods etc. In spite of all these variabilities, global climate models have been developed to simulate future scenarios with difference in simulated results in different agroecological zones. Climate change – as observed through rising CO_2 concentration and temperature precipitation trends, is posing a great threat to agricultural productivity in different agroecological zones of the globe. Climatic response to increased CO_2 and temperature levels in C_3 and C_4 crops needs to be assessed over spatial and temporal resolution scales for agricultural adaptations and proper management of agricultural resources. Crop growth models have been widely adopted for different crops in different regions under different GCMs scenarios to assess their impacts on grain yield and their attributes. Thermal and drought tolerant genotypes will be suitable for climate change scenarios. The impact of temperature rise will decline the yield levels, but with rise of CO_2 level will enhance the net productivity with better soil water utilization due to decrease of stomatal conductivity; and therefore, partial compromising results are available for different crops all over the globe. Technological advancement will counteract the negative impacts on productivity due to global warming with better farming management practices.

For sustainable agriculture due to global warming there is a general need to emphasize on

(i) reduction of the use of chemical fertilizer and energy intensity of agriculture,

(ii) adoption of field management techniques to reduce impact of direct sunlight and wind on crops in tropical/subtropical belts particularly,

(iii) adoption of water conservation techniques,

(iv) conservation and maintenance of existing diversity in different ecosystems,

(v) risk covering cropping plans be adopted during the abnormal weather situations occurred during monsoon, extreme events like drought, floods, extreme weather conditions,

(vi) use of microclimatic modification approaches to alter radiation energy and water balance conditions for improving productivity,

(vii) large scale afforestation and more use of available carbon dioxide than emission rate (improving efficiency to reverse warming trend) and more use of conventional and renewable energy sources, and

(viii) reduction of amount of waste generated with increasing use of waste material's recycling.

Exercise 9

Q.1. Use past 100 years of meteorological data on temperature and rainfall to develop climate change trends in your agroclimatic zones taking 1960-90 mean value as a reference level. Assess the rate of change of temperature during the last 15 years.

Q.2. Develop the rate of change of CO_2 concentration levels in the main urban cities of your agroclimatic zone and link it with the rate of change of temperature.

Q.3. Using any global climate change model assess the future scenarios of temperature change w.r.to. increase of present CO_2 levels at your location to 25, 50, 75 and 100 percent increase in next 10, 20, 50 and 100 years.

Q.4. Use crop simulation models of major crops of your agroclimatic zone with use of input scenarios of climate change at your location i.e., present trends established through question no. 1 for rainfall and maximum-minimum temperature, and study the impact behaviour on crop yield, yield attributes, growing season, potential productivity.

Q.5. Use crop simulated models for major crops of your agroclimatic zone with use input scenarios of future climate change trends established through question no. 3 at your location.

References

Achanta, A.N., 1993. An assessment of the potential impacts of global warming on Indian rice production. In: Achanta, A.N. (ed.) The Climate Change Agenda: The Indian perspective, TERI, New Delhi.

Aggarwal, P.K. 2000. Application of systems simulation for understanding and increasing yield potential of wheat and rice. Ph.D. Thesis, Wageningen University, The Netherlands, p. 176.

Aggarwal, P.K. 2003. Impact of climate change on Indian agriculture. J. Plant Biol., 30, p. 189-98.

Aggarwal, P.K. and N. Kalra, 1994. Analysing the limitations set by climate factors, genotype, water and nitrogen availability on productivity of wheat II. Climatically potential yields and optimal management strategies. Field Crops Res., 38, p. 93-103.

Aggarwal, P.K. and R.K. Mall, 2002. Climate change and rice yields in diverse agroenvironments of India II. Effect of uncertainties in scenarios and crop models on impact assessment. Climate Change, 52, p. 331-43.

Aggarwal, P.K. and S.K. Sinha, 1993. Effect of probable increase in carbon dioxide and temperature on wheat yields in India. J. Agril. Meteorol., 48, p. 811-14.

Aggarwal, P.K., S. Nagarajan, M.E. Shibu and Y.S. Ramakrishna, 2002. Impact of climate change scenarios on Indian agriculture, South Asia Expert Workshop on Adaptation to Climate Change for Agricultural Productivity. 1-3 May, 2002, New Delhi.

Alexandrov, V.A. and G. Hoogenoom, 2000. The impact of climate variability and change on crop yield in Bulgaria. Agril. for Meteorol., 104, p. 315-27.

Anon, 1977. Particulates and air pollution: An Annotated Bibliography. US & PA Office of Air Programme, Pub. EPA-450/1-77-001.

Anonymous, 1979, 1982. Durchfuhrungs bestimanung zur 5. Durchfuhrngsverordnung zum landeskulturgesetz – Reinhultung der Luft – Gesetzblutt der DDR, Teil I, Nr. 31, Vom. 28.6.1979, Nr. 21, Vom 15.3.1982.

Amthor, G.S., 2001. Effect of atmospheric CO_2 concentration on wheat yield. Review of results from experiments using various approaches to control CO_2 concentration. Field Crops Research, 37, p. 1-34.

Baethgen, W.E. and G.O. Margin, 1995. Assessment impact of climate change in winter crop production in Uruguary and Argentina using crop simulation models. In: Climate change and Agriculture: Analysis of potential international impacts. C. Rosenzweig et al. (eds.) American Soc. Agron. Special Pub. 59, Madison, W.I., 207-228.

Barrett, T.W. and H.M. Benedict, 1970. Sulphur dioxide. In: Recognition of air pollution injury to vegetation: A pictorial atlas (Jacobson, J.S. and A.C. Hills (eds.), Air Pollution Control Association, Pittsburgh PA, pp. C1-C17.

Batt, G.R., J.I.R. Morrison, R.H. Ellis, P. Hadley and T.R. Wheeler, 1997. Effects of CO_2 and temperature on growth and yield of crops of winter wheat over several seasons. European Jour. Agronomy, 7, p. 43-52.

Bender, J., U. Hertsein and C. Black, 1999. Growth and yield responses of spring wheat to increasing carbon dioxide, ozone and physiological stresses: A statistical analysis of ESPACE-wheat results. European Jour. Agron., 10, p. 185-95.

Bengtsson, L. 1994. Climate change: Climate of the 21st Century, Agril. and for Meteorology, 72, p. 3-29.

Berry, J. and O. Bjorkman 1980. Photosynthetic response and adaptation to temperature in higher plants. Annual review of Plant Physiology, 31, p. 491-543.

Berry, J.A. and J.S. Downton, 1982. Environmental regulation of photosynthesis. In: Development, Carbon Metabolism and Plant Productivity, Vol. II, (ed.) Govingee, N.Y. Academic Press, p. 263-393.

Bhaskaran, B., J.F.B. Mitchell, J.R. Lavery and M. Lal, 1995. Climate response of Indian sub-continent to doubled CO_2 concentrations. Int. Jour. Climatol., 15, p. 873-92.

Bhjuad, G.S., S.R. Kshirasagar, K.C. Sinha Ray, 2001. Trends in frequency of surface temperature above and below certain threshold values over various stations in India. Proc. Tropmet, 2000, National Symp. on Tropical Meteorology, p. 523-27.

Bleadale, J.K.A. 1952. Phytophathology, 42, p. 558; Atmospheric pollution and plant growth, Nature, 169, p. 376-77.

Boote, K.J., J.W., Jones and P. Singh, 1992. Modelling growth and yield of groundnut. In: Nigam, S.N. (ed.) Groundnut – a global perspective. Proc. Int. Workshop. ICRISAT, Hyderabad, India, p. 331-43.

Busalacchi, A., D. Legler, H. Cattle, T. Palmer and J. Gould, 2005. Climate change and variability – what can we predict? WMO bulletin, 54, p. 51-57.

Chang, C.W. 1975. Fluorides. In: Responses of plants to air pollution, Academic Press, N.Y. Mudd. J.B. and Kozlowski, T.T. (eds.), pp. 57-58.

Chatterjee, A., 1998. Simulating the impact of increase in CO_2 and temperature on growth and yield of maize and sorghum. M.Sc. Thesis, Division of Environmental Sciences, IARI, Delhi.

Crawford, M. 1962. Diseases of livestock as influenced by weather. In: Biometeorology, W.W. Tromp (ed.) Pergamon Press, London, U.K., p. 162-173.

Claussen, M. and M. Esch, 1992. Biomass computed from simulated climatologies. Max-Planck-Inst. Meteor. Report, 89, Humburg.

Clifford, S.C., I.M. Stronch, A.D. Mohamed, S.N. Azam Ali, and N.M.J. Crout, 1993. The effect of elevated atmospheric CO_2 and water stress on lght interception, organic matter production and yield in stands of groundnut. Exptal. Bot., 44, p. 1763-70.

Coates, J.F. 1991. Factors shaping and shaped by the environment: 1990-2010. Futures Research Quarterly, 7, 3, p. 5-55.

Conley, M.M., B.A. Kimball, T.J. Brooks, P.J. Pinter, D.J. Hunsaker, G.W. Wall, N.R. Adam, R.L. LaMorte, A.D. Matthias, T.L., Thompson, 2001. CO_2 enrichment increases water use efficiency in sorghum. New Phytologist, 151, 407-12.

Cubaseh, U., K. Hasselmann, H. Hock, E. Maier-Reimer, U. Mikolajewicz, B.D. Santer and R. Sausen, 1991. Time-dependent greenhouse warming computations with a coupled ocean-atmosphere model. Max-Planck Inst. Meteor. Report 67, Humberg.

Curry, R.B., R.M. Peart, J.W. Jones, K.J. Boote and L.H. Allen, 1990. Simulation as a tool for analysing crop response to climate change. Tans. Am. Soc. Agric. Eng. 33, p. 981-90.

De, U.S. and K.S. Joshi, 1999. Interannual and interdecadal variability of tropical cyclones over the Indian seas. The Decan Geographer, 37, 1, p. 5-21.

De, U.S., G.S. Prakasa Rao and A.K. Jaswal, 2000. A study of variability over Indian cities Mausam.

Dhakhwa, G.B., C.E., Campbell, S.K. LeDue and E.J. Cooter, 1997. Maize growth: assessing the effects of global warming and CO_2 fertilization with crop models. Agril. for Meteorol., 87, p. 253-72.

Dell Abate, M.T., L. Pompili, A., Benedetti, and C. Dazzi, 2005. Soil microbial activity in toposequence under Mediterranean climate. Biol. Fert. Soils, 22, p. 293-98.

Douglas C.K.M. and K.H. Stewart, 1953. Meteorol. Mag., 82, p. 67.

Drake, B.G, M.A. Gonzalez-Maler and S.P. Long, 1997. More efficient plants: A consequence of rising atmospheric CO_2? Annual Review of Plant Physiol. and Molecular Biology, 48, p. 609-39.

Eamus, D. 1991. The interaction of rising CO_2 and temperature with water use efficiency. Plant Cell and Environment 14, p. 843-52.

Easterling, W.E., N.J. Rosenberg, M.S. McKenney and C.A. Jones, 1992. A introduction to the methodology: the region of study, and a historic analog of climate change. Agric. for Meteorol, 59, 3-15.

ECE, 1986. UN Economic and Social Council, 1986.

Esterling, W.E., N.J. Rosenberg, M.S. McKenney, L.A. Katz and K.M. Lemon, 1993. Agricultural impacts and responses to climate changes in MINK region. Clim. Change 24, p. 23-61.

Estiarte, M., J. Penuelas, B.A. Kimball, S.B. Idso, R.L. LaMorte, P.J. Pinter, G.W. Wall and R.L. Garcia, 1994. Elevated CO_2 effects on stomatal density of wheat and sour orange trees. Jour. Exptl. Botany, 45, p. 1665-68.

Farquhar, G.D. and S. von Caemmerer, 1982. Modelling of photosynthetic response to environmental conditions. In: Lange, Nobel, Osmond and Ziegler (eds.) Physiological Plant Ecology, 12B, Springer, Verlag, Berlin, p. 549-87.

Fischer, G., K. Frohberg, M.L. Parry and C. Rosenweig, 1996. The potential effects of climate change on world food production and security. In: Global Climate Change and Agricultural Production (eds. F. Bazzaz and D. Sonhroek). FAO and John Wiley & Sons, Rome, p. 354.

Fischer, G., M. Shah and Harrij van Velthuizen, 2002. Climate change and agricultural vulnerability. A special report by Inter. Institute for Applied System Analysis under UNICA No. 1113, World Summit on Sustainable Development, Johannesburg, 2002.

Gadgil, S., 1995. Climate change and agriculture – an Indian perspective. Current Science, 69, p. 649-59.

Gadgil, S., Y.P. Abrol and P.R. Seshagiri Rao, 1999a. On growth and fluctuation of Indian foodgrain production. Current Science, 76m p. 548-56.

Gadgil, S., P.R. Seshagiri Rao and S. Sridhar, 1999b. Modelling impact of climate variability on rainfed groundnut. Current Science, 76, p. 557-69.

Gangadhar Rao, D. and S.K. Sinha, 1994. Impact of climate change on simulated wheat production in India. In: C. Rosenweig and I. Iglesias (eds.) Implication of Climate Change for Intern. Agriculture: Crop Modelling Study, USEPA 230-B-94-003. USEPA, Washington, DC, p. 1-17.

Gifford, B.M., 1995. Whole plant respiration and photosynthesis of wheat under increased CO_2 concentration and temperature: long term vs short term distinctions for modelling. Global Change Biology, 1, p. 385-96.

Gore, P.G. and K.C. Sinha Ray, 2001. Drought and aridity over districts of Gujarat. Jour. Agrometeorology, Anand.

Guderian, R. 1977. Air Pollution. Ecological Studies 22, Springer Verlag Berlin Heidelbergy, New York, p. 1-123.

Guderian, R. and H. von Hunt, 1970. Nachweis Von Schwefeldioxide Wirkungen an Pflanzen. Staub Reinh. Luft, 30, p. 17-26.

Hansen, J., G. Russell, D. Rind, P. Stone, A. Lacis, S. Lebedeff, R. Ruedy and L. Travis, 1983. Efficient three dimensional globe models for climate studies. Models I & II Mon. Wealth. Rev. III, 4, p. 609-35.

Heagle, A.S., L.W. Kress, P.T. Temple, R.J. Kohut, J.E. Miller, and H.E. Heggestad 1988. Assessment of crop loss from air pollutants, W.H. Heck, O.C. Taylor and D.T. Tingey (eds.) Eslsevier London, p. 141-79.

Heck, W.W., A.S. Heagle, and E.B. Cowling 1977. Air pollution: Impact on plant. In: Proc. 32nd Meeting Soil Conservation Soc. of Am., p. 193-203.

Heck, W.W., O.C. Taylor, and D.T. Tingey 1988. Assessment of crop loss from air pollutants, Elsevier, London.

Higane, L.S., K. Rupa Kumar and B.V. Ramana Murty, 1985. Long term trends of surface air temperature in India. J. Climatol. 5, p. 521-26.

Hill, A.C., H.E. Heggestad, and S.N. Linzon, 1970. Ozone. In: Recognition of Air Pollution Injury to Vegetation: A Atlas (Jacobson, J.S. and Hill A.C. (eds.), Air Pollution Control Association, Pittsburgh, PA, p. B1-B22.

Hoogenboom, G., G.Y. Tsuji, J.W. Jones, U. Singh, D.C. Godwin, N.B. Pickering and R.B. Curry, 1995. Decision support system to study climate change impacts on crop production. In: Climate change and Agriculture: Analysis of Potential international impacts. C. Rozenzweig et al. (eds.) ASA Epec. Publ. 59, ASA, Madison, W.I., 51-75.

Horie, T., J.T. Baker, H. Nakagawa, T. Matsui and H.Y. Kim, 2000. Crop ecosystem responses to climatic change: rice. In: Climate change and global crop productivity (eds.) K.K. Reddy and H.F. Hodges, Wallingford, U.K., CABI, Publishing, p. 81-106.

Houghton, J., Y. Ding, D. Griggs, M. Nouger, P., Van der Linde, X. Dai, K. Maxkell and C. Johnson (eds.) Climate Change 2001. The scientific basis. Cambridge Univ. Press, Cambridge, U.K.

Houghton, J.T., G.J. Jenkins and J.J. Ephrannus (eds.) 1990. Climate Change. The IPCC Scientific Assessment, WMO, Cambridge Univ. Press, Cambridge.

Houghton, J.T., L.G. Meira Fillio, B.A. Callander, N. Harris, A. Kattenberg and K. Maskell, 1996. Climate Change, 1995. The Science of Climate Change. Cambridge Univ. Press, Cambridge.

Hui, D., Y. Luo, W. Cheng, J.S. Coleman, D.W. Johnson and D.A. Suns, 2001. Canopy radiation and efficiencies an affected by elevated CO_2 global change. Biology, 7, p. 75-91.

Hundal, S.S. and P. Kaur, 1996. Climate change and its impact on crop productivity in Punjab, India. In: Climate Variability and Agriculture (eds.) Abrol, Y.P., Gadgil, G. and Pant, G.B. Narosa Publ. House, New Delhi, p. 410.

Hundal, S.S. and P. Kaur, 1997. Application of CERES – Wheat model to yield prediction in irrigated plains of the Indian Punjab. Jour. Agril. Science, 129, p. 13-18.

Hurrell, J.W., 1998. Relationships among recent atmospheric circulation changes, global warming and satellite temperatures. Science Progress, 81, p. 205-24.

IPCC, 1990. Climate change, the IPCC Scientific Assessment. J. Houghton, G.J. Jenkins and J.J. Ephraums (eds.) Cambridge Univ. Press, Cambridge, p. 364.

IPCC, 1990. Intergovernmental Panel on Climate Change: Introduction. In: Houghton, J., Jenkins, G.J., Ephraums, J.J. (eds.). Climate Change: The IPCC Scientific Assessment. Cambridge Univ. Press, Cambridge.

IPCC, 1992. Climate change, the supplementary report of IPCC Scientific Assessment. J. Houghton, B.A. Callendar and S.K. Varnay (eds.) Cambridge Univ. Press, Cambridge, p. 198.

IPCC, 1996. Climate change, 1995. Impacts, adaptations and mitigation of climate change: Scientific technical analysis (eds.) R.J. Watson, M.C. Zinyowera and R.H. Moss. Contribution of Working Group to second assessment report of IPCC. Cambridge Univ. Press, Cambridge.

IPCC, 2001. Climate Change 2001. The Scientific basic (eds.), Houghton, J., Y. Ding, Griggs, M., Nouger; X., Dai; K., Mexkell; C. Johnson. Combridge University Press, Cambridge, U.K.

IUCC, 1992. Information unit on climate change. Fact sheet No. 101, Impacts of climate change. UNEP, Polaisdes Nations, CH 1211, Geneva.

Jacobson, J.S. and Hill, A.C. 1970. Recognition of air pollution injury to vegetation. A Pictorial Atlas. Air Pollution Control Assoc. Pittsburgh, PA.

JAPCA, 1971. Environmental protection agency sets national air quality standards, JAPCA 21, p. 352-53.

Jones, P.D., D.E. Parker, T.J. Osborn and K.R. Briffa, 1999. Global and hemispheric temperature anomalies-land and marine instrumental records. In: Trends: a Compendium of Data on Global Change, CO_2 Information Analysis Centre, Oak Ridge National Laboratory, Oak, Ridge, Tennessee.

Kalra, Naveen, P.K. Aggarwal, S. Chander, H. Pathak, A. Choudhary, Mukesh Sehgal, H.K. Rai, U.A. Soni, Anil Sharma, Monica Jolley, U.K. Singh, Owes Ahemed and M.Z. Hussain, 2003. Impacts of climate change on agriculture. In: Climate change and Indian: Vulnerability Assessment and Adaptation. Univ. Press (India) Pvt. Ltd., Hyderabad, p. 193-226.

Kang, S., F. Zhang, X. Hu and J. Zhang 2002. Benefits of CO_2 enrichment on crop plants are modified by soil water status. Plant and Soil, 238, p. 69-77.

Karl, T.R., P.D. Jones, R.W. Knight, G. Kukla, N. Plummer, V. Razuvayev, K.P. Gallo, J. Lindseay, R.J. Charlston and T.C. Peterson, 1993. Asymmetric trends of daily maximum and temperatures. Bull. Am. Meteor. Soc., 74, 6, p. 17-23.

Karl, T.R., R.W. Knight, D.R. Easterling and R.G. Quayle, 1996. Indices of climatic change for the USA. Bulletin of the American Meteorological Society, 77, p. 279-292.

Kattenberg, A., et al., 1996. Climate models – projection of future climate. In: Houghton, J.T., Meira, L.G. Ellho, B.A. Callander, N. Harris, A. Kattenberg and K. Maskell (eds.) Climate change, 1995. The scientific of climate change. Cambridge Univ. Press, Cambridge.

Kellogg William and Schware Robert, 1981. Climate change and society, p. 68-69. Boulder, Co. Westview Press.

Kimball, B.A. and S.B. Idso, 1983. Increasing atmospheric CO_2 effects on crop yield, water use and climate. Agric. Water Manag. 7, p. 55-72.

Kimball, B.A., 1983. Carbon dioxide and agricultural yield: an assemblage and analysis of 770 prior observations. Agron. J., 75, p. 779-85.

Kriplani, R.H., S.R. Inamdar and N.A. Sontakke, 1996. Rainfall variability over Bangladesh and Nepal: Comparison and connection with features over India. Inter. Jour. Climatology, 16, p. 689-703.

Kupakumar, K., K. Krishna Kumar, G.B. Pant and G. Srinivasan, 2002. Climate change – The Indian Scenario. In: Background paper prepared by FICCI, Intern. Conf. on Science & Technology capacity building for climate change, Oct. 20-22, New Delhi, p. 5-17.

Lacasse, N.L. and M. Treshow (eds.) 1976. Diagnosing vegetation injury caused by air pollution, U.S. Environmental Protection Agency Handbook.

Lal, M., P.H. Whettori, A.B. Pittodi and B. Chakraborty, 1998. The greenhouse gas induced climate change over the Indian sub-continent as projected by GCM model experiments. Terrestrial Atmosphere and Oceanic Sciences, TA09 (4), p. 663-69.

Lal, M., R. Vos Cubasch and J. Waszkewitz, 1995. Effect of transiet increase in greenhouse gases and sulphate aerosols on monsoon climate. Current Science, 69, 9, p. 752-63.

Lal, M., T. Nozawa, S. Emori, H. Harasawa, K. Kimoto Takahashi, A. Abe-Quichi, T. Nakjima, T. Takemura and A. Numaguti, 2001. Future climate change: Implications for Indian summer monsoon and its variability. Current Science, 81, p. 1196-1207.

Largent, E.J. 1952. The effects of air borne fluorides in livestock. In: McCabe L.C. (ed.) Air Pollution. Proc. of United States Tech. Conf. on . . . , New York, p. 64.

Lawlor, D.W. and A.J. Keys, 1993. Understanding photosynthetic adaptation to changing climate. In: Fowden, Mansfield and Stoddart (eds.). Plant Adaptation to Environmental Strees, Chapman and Hall, London, p. 85-106.

Lawlor, D.W., 1998. Plant response to global change: temperature and drought stress. In: Kok and Stulen (eds.). Responses of Plant Metabolism to Air Pollution and Global Change. Backbuys Publisher, Leiden, Netherland, p. 193-207.

Lerman, S.L. and E.F. Darley, 1975. Particulates. In: Responses of plants to air pollution (Mudd. J.B. and Kozlowski, T.T. (eds.), Academic Press, N.Y.

Linzon, S.N. 1969. Symptomatology of sulphur dioxide injury on vegetation. In Handbook of Effects Assessment. Vegetation Damage (Lacasse, N.L. and W.J. Moroz (eds.). Penn. State Univ., pp. VIII-1 to VIII-13.

Littman, E.E., H.W. Ford, and N. Endow, 1956. Industr. Engng. Chem., 48, p. 1492.

Lonergan, S., 1998. Climate warming and India. In: Measuring impacts of climate change on Indian Agriculture (ed.) A. Dinar et al., Washington, D.C. World Bank.

Long, S.P. 1991. Modification of the responses of photosynthetic productivity to rising temperature by atmospheric CO_2 concentrations: has its importance being underestimated? Plant, Cell and Environment, 14, p. 729-39.

Maayar, Mustapha El, Bhawan Singh, Pierre Andre, Christopher, R.B., Thouez, Jean - Pierre 1997. The effects of climate change and CO_2 fertilization on agriculture in Quebec. Agril. and For. Meteorol., 85, p. 193-208.

Mall, R.K. Ranjit Singh, Akhilesh Gupta, G. Sriniavasan and L.S. Rathore, 2006. Impact of climate change in Indian agriculture: A review. Climate Change, 78, p. 445-78.

Manabe, S. and R.J. Wetherald, 1975. The effect of doubling CO_2 concentration on the climate of a general circulation model. Jour. Atmospheric Sciences, 32, p. 3-15.

Manade, S. and R.T. Wetherald, 1975. The effect of doubling the CO_2 concentration on the climate of a general circulation model. J. Atm. Sci., 32, p. 3-15.

Manabe, S. and R.T. Wetherald, 1987. Large scale changes of soil wetness induced by an increase in atmospheric CO_2. J. Atm. Sci., 44, p. 1211-35.

Manabe, S., R.J., Stouffer, M.J. Spelman and K. Bryan, 1991. Transient responses of a coupled ocean-atmosphere model to graduate changes of atmospheric CO_2 Part I. Annual Mean Responses. J. Climate, 4, p. 785-818.

Mandal, N., 1998. Simulating the impact of climate variability and climate change on growth and yield of chick pea and pigeon pea crops. M.Sc. Thesis, Division of Environmental Sciences, IARI, Delhi.

Manning, W.J. 1971. Effects of limeston dust on leaf condition, foliar disease incidence, and leaf surface microflora of native plants. Environ. Pollut. 2, p. 67-74.

Manning, W.J. and Feder, W.A. 1980. Biomonitoring Air Pollutants with Plants. Applied Science Publishers Ltd., London, p. 11-137.

Mckenney, M.S., W.E. Easterling and N.J. Rosenberg, 1992. Simulation of crop productivity and response to climate change in the year 2030. The role of future technologies, adjustments and adaptations. Agril. for Meteorol., 59, p. 103-27.

Middleton, J.T., Crafts, A.S. Brewer, R.F. and Taylor, O.C. 1956. California Agriculture, 10, 6, p. 9.

Middleton, J.T., D.F. Darley, and R.F. Brewer 1957. Damage to vegetation from polluted atmosphere. Philadelphia. Proc. Amer. Petrol Inst., Sect III, Refining, p. 85.

Middleton, J.T., Emik, L.O., Taylor, O.C., 1965. Air quality criteria and standards for agriculture, JAPCA, 15, p. 476-80.

Mitchell, R.A.C., C.L. Gibbard, V.J. Mitchell and D.W. Lawlor, 1996. Effects of shading in different development phases on biomass and grain yield of winter wheat at ambient and elevated CO_2. Plant, Cell and Environment, 19, p. 615-21.

Mitchell, R.A.C., D.W. Lawlor, V.J. Mitchell, C.L. Gibband, E.M. White and J.R. Porter, 1995. Effects of elevated CO_2 concentration and increased temperature on winter wheat: test of ARCWHEAT simulation model. Plant, Cell and Environment 18, p. 736-48.

Mitchell, R.A.C., V.J. Mitchell, and D.W. Lawlor 2001. Response of wheat canopy CO_2 and water gen-exchange to soil water content under ambient and elevated CO_2. Global Change Biology, 7, 599-611.

Mitchell, R.A.C., V.J. Mitchell, S.P. Driscole, J. Franklin and D.W. Lawlor, 1993. Effects of increased CO_2 concentration and temperature on growth and yield of winter wheat at 2 levels of nitrogen application. Plant Cell and Environment, 16, p. 521-29.

Mohandas, S., A.A., Kareem, T.B. Ranganahan and S. Jeyaraman, 1995. Rice production in India under current and future climates. In: Mathewas R.B., Kropff, M.J., Bachelet, D. and Laar Van, H. (eds.) Modeling the impacts of climate change on rice production in Asia. CAB Intern., U.K., p. 165-181.

Mooley, D.A. and B. Parthasarthy, 1984. Fluctuations of all India summer monsoon rainfall during 1871-1978. Climate Change, 6, p. 287-301.

Mudd, J.B. 1975. Sulphur dioxide and peroxyacyl nitrates. In: Responses of plants to air pollution, Mudd. J.B. and Kozlowski, T.T. (eds.), Academic Press, N.Y., p. 9-22., 97-116.

O'Gara, P.J. 1956. In Middleton, J.T., Crafts A.S., Brewer, R.F. and Taylor, O.C. Calif. Agric., 10, 6, p. 9.

Olszyk, D.M., H.G.S. Centeno, L.H. Ziska, J.S. Kern and R.B. Mathews, 1999. Global climate change, rice productivity and methane emissions: Comparison of simulated and experimental results. Agril. for Meteorol., 97, p. 87-101.

Ottman, M.J., B.A. Kimball, P.J. Pinter, G.W. Wall, R.L. Vanderlip, S.W. Leavitt, R.L., LaMorte, A.D. Matthias, and T.J. Brooks 2001. Elevated CO_2 increases sorghum biomass under drought conditions. New Phytologist, 50, p. 261-73.

Pant, G.B. and L.S. Higane, 1988. Climate change in and around the Rajasthan desert during the 20th Century, J. Climatol., 8, p. 391-401.

Pant, G.B., K. Rupa Kumar and H.P. Borgaonkar, 1999. Climate and its long term variability over the Western Himalaya during the past two centuries. The Himalayan Environment (eds.) S.K. Dash and J. Bahadur, New age Intern. Pvt. Ltd., New Delhi, p. 172-84.

Peart, R.M., J.W. Jones, R.B. Curry, K. Boote and L.H. Allen, 1982. Impact of climate change on crop yield in the SE USA. In: Smith, J.B., Tirpak, D.A. (eds.). The potential effects of global climate change on the United States. Report MEPA-230-05-89-050. USEPA, Washington, p. 1-54.

Penning de Vries, F.W.T., D.M., Jansen, H.F.M. Ten Berge and A. Bakema, 1989. Simulation of ecophsiological processes of growth in several annual crops. Simulation Monographs, PUDOC, Wageningen, Netherlands and IARI, Philippines.

Perlwitz, J., 1992. Preliminary results of global SST anomaly experiment with a T42 GCM Annals Geophysicae Abstracts of the VII General Assembly of the European Geophysical Society in Edinburgh, April 6-10.

Peterson, T.C. 2005. Climate change indices. WMO Bulletin, 54, p. 82-86.

Phillips, P.H. 1952. Flourosis in livestock. In: McCabe, L.C. (ed.) Air pollution. Proc. of US Tech. Conf. on . . . , New York, p. 123.

Porter, H., 1992. Inter-specific variations in the growth response of plants to an elevated ambient CO_2 concentration. Vegetation, 104-105, p. 77-79.

Porter, J.R., 1993. AFRCWHEAT2: a model of growth and development of wheat incorporating response to water and nitrogen. Eur. J. Agron.

Powlson, D.S. and P.C. Brookes, 1987. Measurement of soil microbial biomass provides early indicators of changes in total soil organic matter due to straw incorporation. Soil Biology and Biochemistry, 19, p. 159-64.

Prentice, I.C., S.P. Harrison, R. Leemans, R.A. Monserud and A.M. Solomon, 1992. A global biome model based on plant physiology and dominance, soil properties and climate. J. Biogeogr., 19, p. 117-34.

Rajeevan, M., U.S. De and R.K. Prasad, 2000. Decadal variations of sea surface temperatures, cloudiness and monsoon depressions in the north Indian Ocean. Current Science, 79, 3, p. 283-85.

Rao, G.S., Prakash, A.K. Jaswal and U.S. De, 2000. Extreme temperature events over Indian sub-continents. Proc. Tropmet., 2000, p. 533-39.

Rathore, L.S., K.K. Singh, S.A. Saseendran and A.K. Baxla, 2001. Modelling the impacts of climate change on rice production in India. Mausam, 52, 1.

Reilly, J., W.E. Baethgen, F.E. Chege, S.C. Van de Geijn, Lin Erda, A. Iglesias, G. Kenny, D. Patterson, J. Rogasik, R. Ritter, C. Rosenzweig, W. Sombroek and J. Westbrook, 1996. Agriculture in a changing climate: Impacts and adaptation. In: Changing climate 1995. Impacts, adaptation and mitigation of climate change: Scientific Technical Analysis, Report of Working Group II of the Inter-government Panel on Climate Change, Cambridge Univ. Press, London, 427-67.

Reynolds, J.F. and Acock, B., 1985. Modeling approaches for evaluating vegetation response of CO_2 concentration. In: Strain B.R., Cure, J.D. (eds.). Direct Effects of Increasing CO_2 concentration on vegetation. DOE/ER-0283, USDE, CO_2 Research Division, Washington.

Richie, J.T., U. Singh, D. Godwin and L.A. Hunt, 1989. A user's guide to CERES maize-V2.10. International Fertilizer Development Centre, Muscle Shoals, Albama.

Ritchie, J., U. Singh, D. Godwin and W. Bowen, 1998. Cereal growth, development and yield. In: G. Tsuiji, G. Hoogenboom, Thornton, P. (eds.) Understanding options for agricultural production, Kluwer Academic Publishers, Dordrecht, p. 79-98.

Ritchie, J.T. and S. Otter, 1985. Description and performance of CERES wheat: a user oriented wheat yield model. Publ. No. ARS-38, USDA, ARS, Raleigh, NC.

Rogers, H.H., G.E. Bingham, J.D. Cure, J.M. Smith and K.A. Surano, 1983. Response of selected plant species to elevated CO_2 in the field. J. Environ. Qual., 12,, p. 569-74.

Rosenberg, N.J., B.A. Kimball, P. Martin and C.F. Cooper, 1990. From climate and CO_2 enrichment to evapotranspiration. In: Waggoner, P.E. (ed.) Climate Change and Water Resources, Ch. 7, John Wiley, N.Y., p. 151-75.

Rosenzweig, C. and A. Iglesias, 1998. The use of crop models for international climate change impact assessment. In: Understanding options for agricultural production. G.Y. Tsuji, G. Hoogenboom and P.K. Thornton (Eds.). Kluwer Acad. Publishers, Dortrecht, The Netherlands, p. 267-292.

Rosenzweig, C. and D. Hillel, 1995. Potential impacts of climate change on Agriculture and food supply consequences, 1, 2.

Rosenzweig, C. and D. Hillel, 1998. Climate change and the global harvest. Oxford Univ. Press, Oxford, U.K., p. 324.

Rosenzweig, C. and Parry, M.L., 1994. Potential impacts of climate change on world food supply. Nature 367, p. 133-38.

Rosenzweig, C., 1990. Crop response to climate change in the Southern Great Plains: a simulation study. Prof. Geog., 42, 1, p. 20-37.

Rudroff, B.F.T., C.L. Mulchi, C.S.T. Daughtry, and E.H. Lee 1996. Growth, radiation and canopy reflectance of wheat and corn under elevated ozone and carbon dioxide atmosphere. Remote Sensing of Environment, 55, p. 163-73.

Rupakumar, K. and R.G. Ashrit, 2001. Regional aspects of global climate change simulations: Validation and assessment of climate response over Indian monsoon region to transient increase of greenhouse gases and sulphate aerosols. Mausam, 52, 1, p. 229-44.

Rupakumar, K., G.B. Pant, B. Parthasarthy and N.A. Sontakko, 1992. Spatial and subseasonal patterns of long term trends of Indian summer monsoon rainfall. Int. J. Climatology, 12, p. 257-68.

Rupakumar, K., K. Krishnakumar, G.B. Pant and G. Srinivasan 2002. Climate change – The Indian Scenario. In: Background paper prepared by FICCI. Intern. Conf. on Science and Technology Capacity Building for Climate Change. Oct. 20-22, New Delhi, p. 5-17.

Rupakumar, K., K. Krishnakumar, V. Prasanna, K. Ramala, N.R. Deshpande, S. Patwardhn and G.B. Pant, 2003. Future climate scenario. In: Climate change and India: Vulnerability, assessment and adaptation. Univ. Press (India) Pvt. Ltd. Hyderabad.

Rupakumar, K., Krishankumar and G.B. Pant, 1994. Diurnl asymmetry of surface temperature trends over India. Geophy. Res. Lett., 21, 8, p. 77-80.

Samarakoon, A.B., Muller, W.J. and Gifford, R.M., 1995. Transpiration and leaf area under elevated CO_2 effects of soil water status and genotypes in wheat. Australian Jour. Physiology, 23, p. 33-44.

Sarkar, R.P. and V. Thapliyal, 1988. Climate change and variability. Mausam, 39, p. 127-38.

Saseendran, S.A., K.K. Singh, L.S. Rathore, S.V. Singh and S.K. Sinha, 2000. Effects of climate change on rice production in the tropical humid climate of Kerala, India. Climate Change, 44, 4, p. 459-514.

Saseendran, S.A., K.K. Singh, L.S., Rathore, S.V. Singh and S.K. Sinha, 1999. Effects of climate change on rice production in the tropical humid climate of Kerala, India. Climate Change, 123, p. 1-20.

Schimel et al. 1996. Radiative forcing of climate change. In: Houghton, J.T., Meira, Filho, L.G., Callornder,

B.A., Harris, N., Kattenberg, A. and Maskell, K. (eds.) Climate Change, 1995. The Science of Climate Change, Cambridge Univ. Press, Cambridge, p. 65-131.

Schlesinger, M.E. and J.F.B. Mitchell, 1987. Climate simulation of the equilibrium climate response to increased carbon dioxide. Reviews of Geophysics, 25, p. 760-98.

Scholl, G. 1972. Ein biologisches Verfahren zum Nachweis Von Fluorverbindungen in Immissionen. Wien: Osterr. Agrarverlag, Mitt. Forstt. Bundes Versuchsanstalt. Wein: 97/101, p. 255-270.

Scholz, F., H.R. Gregorius, and D. Rudin (eds.) 1989. Genetic effects of air pollutants in forest tree plantations, Proceedings of the joint meeting of the IUFRO working parties. Genetic aspects of air pollution, population and ecological genetics, and biochemical genetics, Spring – Verlag, Berlin.

Seimenov, M.A., J.R. Porter and R. Delecolle, 1993. Simulation the effects of climate change on growth and development of wheat in UK and France. In: Kenny, G.L., Harrison, P.A., Parry, M.L. (eds.). The effects of climate change on agricultural and horticultural potential in Europe. Res. Rep. No. 2, Environmental Change Unit, Univ. of Oxford, Oxford, U.K.

Shaykewich, C.F., 1995. An appraisal of cereal crop phenology modelling. Canad. Jour. Plant Science, 75, p. 329-31.

Sinclair, T.R. and S.L. Rawlins, 1993. Inter-seasonal variation in soybean and maize yield under global environmental change. Agron. J., 85, 406-09.

Singh, N. and N.A. Sontakko, 2002. On climate fluctuations and environmental changes of the Indo-Gangetic plains, India. Climatic Change, 52, p. 287-313.

Singh, R.S., P. Narian and K.D. Sharma, 2001. Climate changes in Luni river basin of arid western Rajasthan. Vayu Mandal, 31, 1-4, p. 103-06.

Singh, U. and H. Padilla, 1995. Simulating rice response to climate change. In: Climate change and agriculture: Analysis of potential international impacts. C. Rosenzweig et al., eds. ASA Special Publ. 59, ASA, Madison, WI, 99-121.

Singh, U. and J.T. Ritchie, 1993. Simulating the impact of climate change on crop growth and nutrient dynamics using the CERES – Rice model. J. Agric. Meteorol., 48, 5, 819-22.

Singh, U., D.C. Godwin and R.J. Morrison, 1990. Modeling in impacts of climate change on agriculture production in the South Pacific. In: Global warming related effects in agriculture and human health and comfort in the South Pacific. Philip., Hughes and Glenn McGregor (ed.), University of Papua New Guinea, Port Moresby, p. 24-40.

Sinha Ray, K.C. and U.S. De, 2003. Climate change in India as evidenced from instrumental records. WMO Bulletin, 52, 1, p. 53-59.

SinhaRay, K.C., R.K. Mukhopadhyay and S.K. Chowdhury, 1997. Trends in minimum and maximum temperatures and sea level pressure over India. Proc. of Tropmet, 1997, New Delhi.

Sinha, Ray, K.C., S.R. Khirsagar and G.S. Bhujad, 2001. Trends in sea surface temperature over Indian seas. Proc. Topmet. 2000. National Symp. on Tropical Meteorology, p. 114-117.

Sinha, S.K. and M.S. Swaminathan, 1991. Deforestation, climate change and sustainable nutrition security. Climate Change, 16, p. 33-45.

Spierings, F. 1967. Method for detemining the susceptibility of trees on air pollution by artificial fumigation. Atmos. Environ., 1, p. 205-10.

Spierings, F. 1971. Influence of fumigation with NO_2 on growth and yield of tomato plants. Neth. J. Plant Pathol., 77, p. 194-200.

Srivastava, H.N., B.N. Dewan, S.K. Dikshit, G.S. Prakash Rao, S.S. Singh and K.R. Rao, 1992. Decadal trends in climate over India. Mausam, 28, p. 1-2, p. 41-45.

Srivastava, A.K., K.C., Sinha Ray and U.S. De, 2000. Trends on frequency of cyclonic disturbances and their intensification over Indian seas. Mausam, 51, 2, p. 113-18.

Srivastava, H.N. and R.C. Sinha Ray, 1994. Climate change scenario: Global and national perspective. Vayu Mandal, 24, 3-4, p. 41-50.

Srivastava, H.N., S.K. Dikshit, R.C. Sinha Ray and R.K. Mukhopadhyay 1998. Trends in rainfall and radiation over India. Vaya Mandal, 28, 1-2, p. 41-45.

Srivastava, H.N. and T.K. Balkrishnan, 1993. Synthesis of meteorological and tide guage observations on a decadal scale near the Indian Coast. TOGA Notes, 15-17.

Sterman, J.D. and L.B. Sweeney, 2007. Understanding public complacency about climate change: adults mental models of climate change violate conservation of matter. Climate Change, 82, p. 225-251.

Stone, P. 2001. The effect of heat stress on cereal yield and quality. In: Crop Responses and Adaptation to Temperature Stress (ed.) A.S. Basra, Binghamton N.Y., Food Products Press, p. 243-91.

TA Luft, 1974. Erste Allgemeine Verwaltungsvorschrift zum Bundes – Immissionsschutzezetz. Technische Anleitung zur Reinhaltung der luft vom 28.8.1974 (Gm Bl. S. 426).

Taylor, O.C. and D.C. Maclean, 1970. Nitrogen oxides and the peroxyacyl nitrates. In: Recognition of air pollution injury to vegetation: A Pictorial Atlas, Jacobson, J.S. and Hill, A.C. (eds.). Air Pollution Contr. Assoc., Pittsburgh, PA, E1-E5.

Thapliyal, V. and S.M. Kulshretha, 1991. Climate changes and trends over India. Mausam, 42, p. 333-38.

Thomas, M.D. and R.H. Hendriction 1957. Effect of air pollution on plant. In: Magill, P.L. Holden, F.R. and Ackley, C. (eds.) Air Pollution Handbook, NY, Sec. 9, p. 1.

Tingey, D.T. and C.P. Anderson 1991. Ecological Genetics and Air Pollution (G.E. Taylor, Jr., L.F. Pitelka, and M.T. Clegg, eds.) Springer – Verlag, N.Y., p. 209-35.

Treshow, M. and M.R. Pack, 1970. Fluorides. In: Recognition of air pollution injury to vegetation: A Pictorial Atlas (Jacobson, J.S. and Hill, A.C.) (eds.), Air Poll. Contr. Assoc. Pittsburgh, PA, pp. D1-D17.

Tsuji, G., G. Uchara, S. Balas (eds.) 1994. DSSAT Version 3, Vol. 1-3, IBSNAT, Univ. of Hawali, Honolulu.

U.S. Deptt. of Commerce, 1987. Census of Agriculture, Bureau of the Census, Crops Harvested Maps, Vol. 2, Subject series, Part I.

USEPA, 1976. Diagnosing vegetation injury caused by air pollution, USEPA Handbook.

USSR, 1949 & 1951. G'giena i Sanitariya, 1949. No. 8, p. 62. In: Int. Dig. Helth. Leg. 1951, 2, p. 454.

UNEPA, 1986. Air quality criteria for ozone and other photochemical oxidants, EPA-600/8-84-020. Environmental Criteria and Assessment Office, Research Triangle Park, N.C.

Upreaty, D.C., N.V.K. Chakravarty, R.K. Katiyal and Y.P. Abrol., 1996. Climate variability and Brassica. Climate variability and Agriculture (eds.) Abrol., Y.P., Sulochana, Gadgil, Pant, G.B., p. 264-80. Narosa Publ. House, New Delhi, India.

USEPA, 1988. Review the national ambient air quality standards for oxides of nitrogen, EPA – 600/8-82-026, 1982; for particulate watter and sulphar oxides EPA- books-82-029, 2982; for ozone, 1988; Environmental Criteria and Assessment Office, Research Triangle Park, N.C.

Wall, G.W., F., Wechsung, G., Wechsang, B.A. Kimball, , T. Kartshall, P. Pinter and R.L. LaMorte, 1996. Effect of free air CO_2 environment and two soil nitrogen regime on varietal and horizontal distribution of root length density, surface area and density of wheat. Annual Res. Report, ASDA-ARS, Arizona, Arizona, p. 56-59.

Weinstein, L.H. 1971. Boyce Thomson institute for plant research, Jonkers, New York, Schreiben Vom. 15.11971 an den Verein Deutschr Ingenieure, Dusseldorf.

Weir, A.H., P.L. Bragg, J.R. Porter and J.H. Rayner, 1984. A winter wheat crop simulation model without water on nutrient limitation. J. Agric. Sci., 102, p. 371-82.

Wheeler, T.R., G.R. Batt, R.H., Ellis, P. Hadley and J.I.L. Morison, 1996. Growth and yield of winter wheat crops in response to CO_2 and temperature. Jour. Agril. Sciences, Cambridge, 127, p. 37-48.

Wilkerson, G.G., J.W. Jones, K.J. Boote and K.J. Ingrim, 1983. Modelling soybean growth for management. Trans. ASAE, 26, p. 6373.

Wolf, J., 1993. Effect of climate change on wheat and maize production potential in the EC. In: Kenny, O.J., Harrison, P.A., Parry, M.L. (eds.). The effect of climate change on the agricultural and horticultural potential in Europe. Res. Rep. 2. Environment Change, Univ. of Oxford, Oxford, UK.

World Commission on Environment and Development, 1987. Our common future. New York, Oxford University Press.

Young, K.J. and S.P. Long, 2000. Crop ecosystem responses to climate change: Maize and sorghum. In: Climate Change and Global Crop Productivity (eds.) K.R. Reddy and H.F. Hodges, Wallingford, U.K.: CABI Publishing, p. 102-32.

Appendix I

Mean January and July temperatures at Hisar (Lat. 29°10'N, Longitude 75°76'E) for the period 1966-2005.

Year	January (°C)			July (°C)		
	T_{Max}	T_{Min}	T_{Mean}	T_{Max}	T_{Min}	T_{Mean}
1966	21.5	4.5	13.0	40.8	28.4	34.6
1967	20.8	7.2	14.0	40.5	28.0	34.2
1968	20.4	4.8	12.6	40.8	27.4	34.1
1969	20.8	6.2	13.5	40.2	27.5	33.8
1970	20.4	5.0	12.7	37.1	26.9	32.0
1971	19.8	1.0	10.4	35.5	24.3	29.9
1972	21.3	4.5	12.9	37.6	26.8	32.2
1973	20.2	3.6	11.9	37.7	27.1	32.4
1974	20.9	2.4	11.9	33.6	24.2	30.4
1975	18.5	4.7	11.6	35.8	25.3	30.5
1976	21.2	5.9	13.5	36.6	26.4	31.5
1977	19.9	3.9	11.9	34.2	25.8	30.0
1978	20.5	2.9	11.7	33.7	26.9	30.3
1979	20.6	6.3	13.4	37.3	26.3	31.8
1980	20.2	3.8	12.0	35.4	26.7	30.4
1981	19.9	7.2	13.5	34.2	26.7	30.4
1982	20.0	5.7	12.8	37.7	26.7	32.2
1983	19.5	4.7	12.1	35.4	26.1	30.7
1984	19.6	2.6	11.1	35.9	26.1	30.7
1985	19.8	5.5	12.6	35.1	26.1	30.6
1986	19.8	3.8	11.8	34.8	24.9	29.8
1987	20.6	5.5	13.0	40.4	28.4	34.4
1988	22.2	5.2	13.7	34.6	26.8	30.7
1989	19.2	4.1	11.6	37.2	25.9	31.5
1990	22.2	6.1	14.1	34.2	25.8	30.0
1991	19.7	4.0	11.8	39.3	27.6	33.4
1992	20.9	6.3	13.6	34.6	24.7	29.6
1993	19.7	5.6	12.6	35.3	25.4	30.3
1994	20.6	6.3	13.4	35.2	26.2	30.7
1995	18.6	5.4	12.0	37.4	27.2	32.3
1996	20.1	3.5	11.8	36.8	25.8	31.3
1997	20.1	1.9	11.0	37.2	25.4	31.3
1998	19.0	3.8	11.4	36.1	26.5	31.3
1999	17.2	7.7	12.4	37.5	26.9	32.2
2000	18.4	5.4	11.9	35.6	25.2	30.4
2001	18.2	3.0	10.6	34.8	24.1	29.4
2002	19.8	4.5	12.2	39.8	28.5	34.1
2003	16.4	3.7	10.0	33.8	25.7	29.7
2004	17.4	6.4	11.9	39.2	27.4	33.3
2005	18.3	4.0	11.1	34.6	26.1	30.3
N	40	40	40	40	40	40
$\sum x^2$	15830.58	967.92	6058.79	53938.33	27826.83	39752.54
$\sum x$	794.2	188.6	490.7	1466.5	1054.1	1259.6
$\overline{x}$	19.85	4.715	12.2675	36.66	26.35	31.49
σ_x	1.242	1.480	.9890	2.078	1.1029	1.481
σ_{N-1}	1.258	1.499	1.0016	2.105	1.11699	1.4998

Appendix II
Probability of a Random Value of z = (X – μ)/σ Being Greater than the Value tabulated in the Margins

Z	.00	.01	.02	.03	.04	.05	.06	.07	.08	.09
.0	.5000	.4920	.4920	.4880	.4840	.4801	.4761	.4721	.4681	.4641
.1	.4602	.4562	.4522	.4583	.4443	.4404	.4364	.4325	.4286	.4247
.2	.4207	.4168	.4129	.4090	.4052	.4013	.3974	.3936	.3897	.3859
.3	.3821	.3783	.3745	.3707	.3639	.3632	.3594	.3557	.3520	.3483
.4	.3446	.3409	.3372	.3336	.3300	.3264	.3228	.3192	.3156	.3121
.5	.3085	.3050	.3015	.2981	.2946	.2912	.2877	.2843	.2810	.2776
.6	.2743	.2709	.2676	.2643	.2611	.2578	.2546	.2414	.2483	.2451
.7	.2420	.2389	.2358	.2327	.2296	.2266	.2236	.2206	.2177	.2148
.8	.2119	.2090	.2061	.2033	.2005	.1977	.1949	.1922	.1894	.1867
.9	.1841	.1814	.1788	.1762	.1736	.1711	.1685	.1660	.1635	.1611
1.0	.1587	.1562	.1539	.1515	.1492	.1469	.1446	.1423	.1401	.1339
1.1	.1357	.1335	.1314	.1292	.1271	.1251	.1230	.1210	.1190	.1170
1.2	.1151	.1131	.1112	.1093	.1075	.1056	.1038	.1020	.1003	.0985
1.3	.0968	.0951	.0934	.0918	.0901	.0885	.0869	.0853	.0838	.0823
1.4	.0808	.0793	.0778	.0764	.0749	.0735	.0721	.0708	.0694	.0981
1.5	.0668	.0655	.0643	.0630	.0618	.0606	.0594	.0582	.0571	.0559
1.6	.0548	.0537	.0526	.0516	.0505	.0495	.0485	.0475	.0465	.0455
1.7	.446	.436	.0427	.0418	.0409	.0101	.0392	.0384	.0375	.0367
1.8	.0359	.0351	.0344	.0336	.0329	.0322	.0314	.0307	.0301	.0294
1.9	.0287	.0281	.0274	.0268	.0262	.0256	.0250	.0244	.0239	.0233
2.0	.0228	.0222	.0217	.0212	.0207	.0202	.0197	.0192	.0188	.0183
2.1	.0179	.0174	.0170	.0166	.0162	.0158	.0154	.0150	.0146	.0143
2.2	.0139	.0136	.0132	.0129	.0125	.0122	.0119	.0116	.0113	.0110
2.3	.0107	.104	.0102	.0099	.0096	.0094	.0091	.0089	.0087	.0084
2.4	.0082	.0080	.0078	.0075	.0073	.0071	.0069	.0068	.0066	.0064
2.5	.0062	.0060	.0059	.0057	.0055	.0064	.0052	.0051	.0049	.0048
2.6	.0047	.0045	.0044	.0043	.0041	.0040	.0039	.0038	.0037	.0036
2.7	.0035	.0034	.0033	.0032	.0031	.0030	.0029	.0028	.0027	.0026
2.8	.0026	.0025	.0024	.0023	.0023	.0022	.0021	.0021	.0020	.0019
2.9	.0019	.0018	.0018	.0017	.0016	.0016	.0015	.0015	.0014	.0014
3.0	.0013	.0013	.0013	.0012	.0012	.0011	.0011	.0011	.0010	.0010
3.1	.0010	.0009	.0009	.0009	.0008	.0008	.0008	.0008	.0007	.0007
3.2	.0007	.0007	.0006	.0006	.0006	.0006	.0006	.0005	.0005	.0005
3.3	.0005	.0005	.0005	.0004	.0004	.0004	.0004	.0004	.0004	.0003
3.4	.0003	.0003	.0003	.0003	.0003	.0003	.0003	.0003	.0003	.0002
3.6	.0002	.0002	.0001	.0001	.0001	.0001	.0001	.0001	.0001	.0001
3.9	.0000									

Standard table available in a book on statistical principles.

Appendix III
Area Under the Standard Normal Curve

Z	.00	.01	02	.03	.04	.05	.06	.07	.08	.09
0.0	.00000	.00399	.00798	.01197	.01595	.01994	.02392	.02790	.03188	.03586
0.1	.03983	.04383	.04776	.05172	.05567	.05962	.06356	.06749	.07142	.07535
0.2	.07926	.08317	.08706	.09095	.09483	.09871	.10257	.10642	.11026	.11409
0.3	.11791	.12172	.12552	.12930	.13307	.13683	.14058	.14431	.14803	.1517
0.4	.15542	.15910	.16276	.16640	.17003	.17364	.17724	.18082	.18439	.18793
0.5	.19146	.19497	.19847	.20194	.20540	.20884	.21226	.21566	.21904	.22240
0.6	.22575	.22907	.23237	.23565	.23891	.24215	.24537	.24857	.25175	.25490
0.7	.25804	.26115	.26424	.26730	.27035	.27337	.27637	.27935	.27230	.28524
0.8	.28814	.29103	.29389	.29673	.29955	.30234	.30511	.30785	.31057	.31372
0.9	.31594	.31859	.31121	.32381	.32639	.32894	.33147	.33398	.33646	.33891
1.0	.34134	.34375	.34614	.34849	.35083	.35314	.35543	.35769	.35993	.36214
1.1	.36433	.36650	.36864	.37076	.37286	.37493	.37698	.37900	.38100	.38298
1.2	.38493	.38686	.38877	.39065	.38251	.94435	.39617	.39796	.39973	.40147
1.3	.40320	.40490	.40658	.40824	.40988	.41198	.41309	.41466	.41621	.41774
1.4	.41921	.42073	.42220	.42364	.42507	.42647	.42785	.42922	.43056	.43189
1.5	.43319	.43448	.43574	.43699	.43822	.43943	.44062	.44179	.44295	.44408
1.6	.44520	.44630	.44738	.44845	.44950	.45053	.45154	.45254	.45352	.45449
1.7	.45543	.45637	.45728	.45818	.45907	.45994	.46080	.46164	.43246	.46327
1.8	.46407	.46485	.46562	.46638	.46712	.46784	.46856	.46926	.46995	.47062
1.9	.47128	.47193	.47257	.47320	.47381	.47441	.47500	.47558	.47615	.47670
2.0	.47725	.47784	.47831	.47882	.47932	.47982	.48030	.48077	.48124	.48169
2.1	.48214	.48257	.48300	.48341	.48382	.48422	.48461	.48500	.48537	.48574
2.2	.48610	.48645	.48679	.48713	.48745	.4778	.48809	.48840	.48870	.48899
2.3	.48928	.48956	.48983	.49010	.49036	.49061	.49086	.49111	.49134	.49158
2.4	.49180	.49202	.49224	.49245	.49266	.49286	.49305	.49324	.49343	.49361
2.5	.49379	.49396	.49413	.49430	.49446	.49461	.49477	.49492	.49506	.49520
2.6	.49534	.49847	.49560	.49573	.49585	.49598	.49609	.49621	.49632	.49643
2.7	.49653	.49664	.49674	.49683	.49693	.49702	.49711	.49720	.49728	.49736
2.8	.49744	.49752	.49760	.49767	.49774	.49781	.49788	.49795	.49801	.49807
2.9	.49813	.19819	.49825	.49831	.49836	.49841	.49846	.49851	.49856	.49861
3.0	.49865	.49869	.49874	.49878	.49882	.49886	.49889	.49893	.49896	.49900
3.1	.49903	.49906	.49910	.49913	.49916	.49918	.49921	.49924	.49926	.49929
3.2	.49931	.49934	.49936	.49938	.49940	.49942	.49944	.49946	.49948	.49950
3.3	.49952	.49953	.49955	.49957	.49958	.49960	.49961	.49962	.49964	.49965
3.4	.49966	.49968	.49969	.49970	.49971	.49972	.49973	.49974	.49975	.49976
3.5	.49977	.49978	.49978	.49979	.49980	.49981	.49981	.49982	.49983	.49983
3.6	.49984	.49985	.49985	.49986	.49986	.49987	.49987	.49988	.49988	.49989
3.7	.49989	.49990	.49990	.49991	.49991	.49992	.49992	.49992	.49992	.49992
3.8	.49993	.49993	.49993	.49994	.49994	.49994	.49994	.49995	.49995	.49995
3.9	.49995	.49995	.49996	.49996	.49996	.49996	.49996	.49996	.49997	.49997
4.0	. 49997	-	-	-	-	-	-	-	-	-
4.5	.499997	-	-	-	-	-	-	-	-	-
5.0	.4999997	-	-	-	-	-	-	-	-	-

Standard table available in a book on statistical principles.

Appendix IV

Relative Height of Ordinate of the Normal Curve (Y_z) for different Value of Standard Deviations $\frac{(x_1 - \overline{x})}{S.D.}$ from the Mean

$\frac{(x_1 - \overline{x})}{S.D.}$	0.00	0.01	0.02	0.03	0.04	0.05	0.06	0.07	0.08	0.09
0.0	.3989	.3989	.3989	.3988	.3986	.3984	.3982	.3980	.3977	.3973
0.1	.3970	.3965	.3961	.3956	.3950	.3945	.3939	.3932	.3925	.3918
0.2	.3910	.3902	.3894	.3885	.3876	.3867	.3757	.3847	.3836	.3825
0.3	.3814	.3802	.3790	.3778	.3765	.3752	.3739	.3726	.3712	.3697
0.4	.3683	.3668	.3653	.3637	.3621	.3605	.3589	.3572	.3555	.3538
0.5	.3521	.3503	.3485	.3467	.3448	.3429	.3410	.3391	.3372	.3352
0.6	.3332	.3312	.3292	.3271	.3251	.3230	.3209	.3187	.3166	.3144
0.7	.3122	.3100	.3078	.3056	.3034	.3011	.2989	.2966	.2943	.2920
0.8	.2897	.2874	.2850	.2827	.2803	.2780	.2856	.2732	.2809	.2685
0.9	.2661	.2637	.2613	.2589	.2563	.2541	.2516	.2492	.2468	.2444
1.0	.2420	.2396	.2371	.2347	.2323	.2299	.2275	.2251	.2226	.2202
1.1	.2178	.2155	.2131	.2107	.2083	.2059	.2036	.2012	.1989	.1965
1.2	.1942	.1919	.1895	.1872	.1849	.1827	.1804	.1781	.1858	.1736
1.3	.1714	.1992	.1669	.1647	.1626	.1604	.1582	.1561	.1540	.1518
1.4	.1497	.1476	.1456	.1435	.1415	.1394	.1374	.1354	.1334	.1315
1.5	.1295	.1275	.1257	.1238	.1219	.1200	.1182	.1163	.1145	.1127
1.6	.1109	.1092	.1074	.1057	.1040	.1023	.1006	.0989	.0973	.0957
1.7	.0941	.0925	.0909	.0893	.0878	.0863	.0848	.0833	.0818	.0804
1.8	.0790	.0775	.0761	.0748	.0734	.0721	.0707	.0694	.0681	.0669
1.9	.0656	.0644	.0632	.0620	.0608	.0596	.0584	.0573	.0562	.0551
2.0	.0540	.0529	.0519	.0508	.0498	.0488	.0478	0.468	.0459	.0449
2.1	.0440	.0431	.0422	.0413	.0404	.0396	.0387	.0379	.0371	.0363
2.2	.0355	.0347	.0339	.0332	.0325	.0317	.0310	.0303	.0296	.0290
2.3	.0283	.0277	.0270	.0264	.0258	.0252	.0246	.0241	.0235	.0229
2.4	.0224	.0219	.0213	.0208	.0203	.0198	.0194	0.189	.0184	.0180
2.5	.0175	.0171	.0167	.0162	.0158	.0154	.0151	.0147	.0143	.0139
2.6	.0136	.0132	.0129	.0126	.0122	.0119	.0116	.0113	.0010	.0107
2.7	.0104	.0101	.009	.0096	.0094	.0091	.0088	.0086	.0084	.0084
2.8	.0079	.0077	.0075	.0073	.0071	.0069	.0067	.0065	.0063	.0061
2.9	.0060	.0058	.0056	.0054	.0053	.0051	.0050	.0048	.0047	.0046
3.0	.0044	.0043	.0042	.0040	.0039	.0038	.0037	.0036	.0035	.0034
3.1	.0033	.0032	.0031	.0030	.0029	.0028	.0027	.0026	.0025	.0025
3.2	.0024	.0023	.0022	.0022	.0021	.0020	.0020	.0019	.0018	.0018
3.3	.0017	.0017	.0016	.0016	.0015	.0015	.0014	.0014	.0013	.0013
3.4	.0012	.0012	.0012	.0011	.0011	.0010	.0010	.0010	.0009	.0009
3.5	.0009	.0008	.0008	.0008	.0008	.0007	.0007	.0007	.0007	.0006
3.6	.0006	.0006	.0006	.0006	.0005	.0005	.0005	.0005	.0005	.0004
3.7	.0004	.0004	.0004	.0004	.0004	.0004	.0003	.0003	.0003	.0003
3.8	.0003	.0003	.0003	.0003	.0002	.0002	.0002	.0002	.0002	.0002
3.9	.0002	.0002	.0002	.0002	.0002	.0002	.0002	.0002	.0001	.0001
4.0	.0001									

Standard table available in a book on statistical principles.

Appendix V

Specified percentiles of gamma distribution function $\left[P(z) = \int_0^z \frac{e^{-z} z^{\hat{\alpha}-1} dz}{\Gamma(\hat{\alpha})}\right]$

$\hat{\alpha}$ / P(z)	0.3	0.4	0.5	0.6	0.7	0.8	0.9	1.0	1.1	1.2
0.05	0.3211×10^{-4}	0.4147×10^{-3}	0.1966×10^{-2}	0.5645×10^{-2}	0.1216×10^{-1}	0.2190×10^{-1}	0.3496×10^{-1}	0.5129×10^{-1}	0.7175×10^{-1}	0.9315×10^{-1}
0.10	0.3237×10^{-4}	0.2349×10^{-2}	0.7895×10^{-2}	0.1806×10^{-1}	0.3315×10^{-1}	0.5298×10^{-1}	0.7720×10^{-1}	0.1054	0.1371	0.1719
0.15	0.1252×10^{-2}	0.6492×10^{-2}	0.1788×10^{-1}	0.3589×10^{-1}	0.6009×10^{-1}	0.8974×10^{-1}	0.1241	0.1625	0.2044	0.2494
0.20	0.3270×10^{-2}	0.1339×10^{-1}	0.3209×10^{-1}	0.5880×10^{-1}	0.9234×10^{-1}	0.1315	0.1754	0.2231	0.2742	0.3280
0.25	0.6900×10^{-2}	0.2356×10^{-1}	0.5070×10^{-1}	0.8677×10^{-1}	0.1298	0.1783	0.2312	0.2877	0.3471	0.4090
0.30	0.1273×10^{-1}	0.3754×10^{-1}	0.7424×10^{-1}	0.1200	0.1725	0.2302	0.2919	0.3567	0.4241	0.4936
0.35	0.2142×10^{-1}	0.5591×10^{-1}	0.1029	0.1588	0.2210	0.2877	0.3579	0.4308	0.5058	0.5826
0.40	0.3374×10^{-1}	0.7936×10^{-1}	0.1375	0.2038	0.2757	0.3514	0.4300	0.5108	0.5933	0.6771
0.45	0.5061×10^{-1}	0.1087	0.1787	0.2558	0.3375	0.4223	0.5092	0.5578	0.6876	0.7784
0.50	0.7313×10^{-1}	0.1451	0.2275	0.3157	0.4074	0.5014	0.5967	0.6531	0.7903	0.8879
0.55	0.1027	0.1898	0.2853	0.3851	0.4871	0.5903	0.6942	0.7985	0.9030	1.0076
0.60	0.1413	0.2448	0.3542	0.4659	0.5785	0.6913	0.8039	0.9163	1.0283	1.1400
0.65	0.1913	0.3126	0.4367	0.5610	0.6847	0.8074	0.9291	1.0498	1.1697	1.2887
0.70	0.2566	0.3973	0.5371	0.6747	0.8101	0.9432	1.0744	1.2040	1.3320	1.4587
0.75	0.3429	0.5048	0.6617	0.8137	0.9615	1.1059	1.2473	1.3863	1.5231	1.6581
0.80	0.4601	0.6456	0.8212	0.9890	1.1506	1.3074	1.4601	1.6094	1.7560	1.9001
0.85	0.6266	0.8391	1.0361	1.2220	1.3994	1.5702	1.7356	1.8971	2.0548	2.2094
0.90	0.8848	1.1298	1.3528	1.5605	1.7571	1.9453	2.1267	2.3026	2.4740	2.6415
0.95	1.3723	1.6620	1.9207	2.1590	2.3826	2.5951	2.7990	2.9957	3.1867	3.3727

$\hat{g}$ / P(z)	1.3	1.4	1.5	1.6	1.7	1.8	1.9	2.0	2.1	2.2	2.3	2.4
0.05	0.1183	0.1459	0.1759	0.2018	0.2422	0.2783	0.3160	0.3554	0.3962	0.4385	0.4820	0.5268
0.10	02096	0.2497	0.2922	0.3367	0.3831	0.4311	0.4808	0.5318	0.5842	0.6778	0.6925	0.7484
0.15	0.2970	0.3465	0.3989	0.4527	0.5082	0.5652	0.6236	0.6832	0.7440	0.8058	0.8686	0.9323
0.20	0.3842	0.4425	0.5026	0.5643	0.6275	0.6990	0.7577	0.8244	0.8921	0.9607	1.0301	0.1003
0.25	0.4730	0.5389	0.6063	0.6750	0.7451	0.8162	0.8883	0.9613	1.0351	1.1097	1.1849	1.2600
0.30	0.5649	0.6377	0.7118	0.7871	0.8634	0.9406	1.0186	1.0973	1.1766	1.2568	1.3373	1.4184
0.35	0.6608	0.7403	0.8208	0.9022	0.9844	1.0674	1.1509	1.2350	1.3197	1.4048	1.4903	1.5763
0.40	0.7521	0.8479	0.9346	1.0219	1.1098	1.1982	1.2871	1.3764	1.4661	1.5561	1.6464	1.7369
0.45	0.8699	0.9621	1.0547	1.1478	1.2413	1.3351	1.4292	1.5235	1.6180	1.7128	1.8077	1.9027
0.50	0.9860	1.0844	1.1830	1.2818	1.3808	1.4799	1.5791	1.6783	1.7778	1.8771	1.9766	1.0762
0.55	1.1123	1.2169	1.3215	1.4260	1.5305	1.6349	1.7393	1.8436	1.9478	2.0519	2.1560	2.2600
0.60	1.2514	1.3624	1.4731	1.5835	1.6936	1.8034	1.9130	2.0223	2.1314	2.2403	2.3491	2.4576
0.65	1.4070	1.5246	1.6416	1.7580	1.8739	1.9893	2.1043	2.2188	2.3330	2.4469	2.5604	2.6736
0.70	1.5873	1.7068	1.8324	1.9552	2.0772	2.1985	2.3191	2.4392	2.5588	2.6778	2.7964	2.9145
0.75	1.7915	1.9235	2.0542	2.1837	2.3123	2.4399	2.5666	2.6926	2.8179	2.9425	3.0665	3.1899
0.80	2.0421	2.1823	2.3208	2.4579	2.5937	2.7283	2.8618	2.9943	3.1259	3.2567	3.3867	3.5160
0.85	2.3613	2.5109	2.6585	2.8043	2.9484	3.0911	3.2324	3.3724	3.5114	3.6493	3.7863	3.9224
0.90	2.8056	2.9670	3.1257	3.2822	3.4366	3.5893	3.7403	3.8897	4.0378	4.1846	4.3302	4.4747
0.95	3.5544	3.7325	3.9074	4.0793	4.24987	4.4156	4.5808	4.7439	4.9052	5.0648	5.2230	5.3798

$\hat{\varphi}$ / P(z)	2.5	2.6	2.7	2.8	2.9	3.0	3.1	3.2	3.3	3.4	3.5	3.6
0.05	0.5727	0.6198	0.6678	0.7169	0.7668	0.8177	0.8694	0.9218	0.9751	1.0290	1.0837	1.1390
0.10	0.8052	0.8629	0.9215	0.9809	1.0411	1.1021	1.1637	1.2260	1.2889	1.3525	1.4166	1.4812
0.15	0.9969	1.0623	1.1283	1.1951	1.2626	1.3306	1.3993	1.4685	1.5382	1.6084	1.6791	1.7503
0.20	1.1713	1.2429	1.3151	1.3878	1.4612	1.5350	1.6094	1.6842	1.7594	1.8351	1.9112	1.9876
0.25	1.3373	1.4143	1.4910	1.5699	1.6464	1.7273	1.8066	1.8863	1.9663	2.0467	2.1274	2.2084
0.30	1.5000	1.5819	1.6643	1.7471	1.8303	1.9138	1.9976	2.0817	2.1630	2.2507	2.3357	2.4208
0.35	1.6626	1.7492	1.8361	1.9234	2.0109	2.0986	2.1866	2.2749	2.3616	2.4520	2.5408	2.6299
0.40	1.8277	1.9188	2.0101	2.1016	2.1932	2.2851	2.3771	2.4693	2.5616	2.6540	2.7466	2.8393
0.45	1.9980	2.0933	2.1888	2.2844	2.3801	2.4759	2.5718	2.6678	2.7639	2.8600	2.9563	3.0525
0.50	2.1757	2.2753	2.3750	2.4746	2.5743	2.6741	2.7738	2.8736	2.9733	3.0731	3.1729	3.2727
0.55	2.3639	2.4677	2.5715	2.6753	2.7790	2.8826	2.9862	3.0897	3.1932	3.2966	3.4000	3.5033
0.60	2.5659	2.6741	2.7822	2.8900	2.9978	3.1054	3.3129	3.3202	3.4275	3.5346	3.6416	3.7485
0.65	2.7865	2.8992	3.0166	3.1237	3.2357	3.3474	3.4589	3.5702	3.6813	3.7923	3.9031	4.0137
0.70	0.0322	3.1496	3.2666	3.3832	3.4995	3.6156	3.7313	3.8468	3.9620	4.0770	4.1917	4.3062
0.75	3.3128	3.4352	3.5572	3.6787	3.7998	3.9204	4.0407	4.1607	4.2808	4.3696	4.5186	4.6373
0.80	3.4646	3.7726	3.9000	4.0269	4.1532	4.2790	4.4043	4.5293	4.6538	4.7779	4.9016	5.0250
0.85	4.0576	4.1921	4.3258	4.4588	4.5913	4.7231	4.8543	4.9849	5.1151	5.2448	5.3739	5.5027
0.90	4.6182	4.7607	4.9023	5.0431	5.1831	5.3223	5.4608	5.5987	5.7359	5.8725	6.0085	6.1440
0.95	5.5352	5.6895	5.8426	5.9947	6.1457	6.2958	6.4450	6.5933	6.7408	6.8876	7.0336	7.1789

$\hat{\gamma}$ / P(z)	3.7	3.8	3.9	4.0	4.1	4.2	4.3	4.4	4.5
0.05	1.1949	1.2515	1.3086	1.3663	1.4246	1.4833	1.5426	1.6023	1.6626
0.10	1.5454	1.6120	1.6782	1.7448	1.8118	1.8793	1.9472	2.0154	2.0841
0.15	1.8219	1.8939	1.9663	2.0391	2.1123	2.1858	2.2596	2.3338	2.4083
0.20	2.0644	2.1415	2.2190	2.2968	2.3749	2.4533	2.5319	2.6108	2.6900
0.25	2.2898	2.3713	2.4532	2.5353	2.6177	2.7003	2.7831	2.8662	2.9494
0.30	2.5062	2.5918	2.6777	2.7637	2.8499	2.9364	3.0230	3.1097	3.1967
0.35	2.7191	2.8084	2.8980	2.9876	3.0775	3.1674	3.2575	3.3478	3.4381
0.40	2.9332	3.0251	3.1182	3.2113	3.3046	3.3979	3.4914	3.5849	3.6785
0.45	3.1489	3.2453	3.3418	3.4383	3.5349	3.6315	3.7282	3.8249	3.9217
0.50	3.3725	3.4724	3.5722	3.6721	3.7719	3.8718	3.9717	4.0715	4.1714
0.55	3.6066	3.7099	3.8131	3.9163	4.0194	4.1225	4.2256	4.3286	4.4316
0.60	3.8553	3.9621	4.0687	4.1753	4.2817	4.3881	4.4944	4.6007	4.7068
0.65	4.1241	4.2345	4.3446	4.4547	4.5646	4.6744	4.7840	4.8936	5.0030
0.70	4.4205	4.5326	4.6485	4.7622	4.8758	4.9891	5.1023	5.2153	5.3282
0.75	4.7557	4.8739	4.9918	5.1094	5.2269	5.3441	5.4610	5.5778	5.6944
0.80	5.1480	5.2707	5.3930	5.5150	5.6368	5.7583	5.8795	6.0004	6.1211
0.85	5.6310	5.7589	5.8864	6.0135	6.1403	6.2667	6.3928	6.5186	6.6440
0.90	6.2789	6.4134	6.5473	6.6808	6.8138	6.9464	7.0786	7.2104	7.3418
0.95	7.3235	7.4675	7.6109	7.7537	7.8959	8.0375	8.1787	8.3193	8.4595

Source: Mooley, 1974.

Appendix VI
Critical Value of Rank Correlation Coefficient

N	Significance level (one-tailed test)	
	.05	**.01**
4	1.000	
5	.900	1.000
6	.820	.843
7	.714	.893
8	.643	.833
9	.600	.783
10	564	.716
12	.506	.712
14	.456	.645
16	.425	.601
18	.399	.564
20	.377	.534
22	.359	.508
24	.343	.485
26	.329	.465
28	.317	.448
30	.306	.432

Standard table available in a book on statistical procedures.

Appendix-VII

Soil Moisture Retention Table – 150 mm

Soil moisture retained after different amount of potential removal of water from the soil have occurred. Water holding capacity of soil is 150 mm

Σ(P - PE)	0	1	2	3	4	5	6	7	8	9
	Water Retained in Soil									
0	150	149	148	147	146	145	144	143	142	141
10	140	139	138	137	136	135	134	133	132	131
20	131	130	129	128	127	127	126	125	124	123
30	122	122	121	120	119	116	117	116	115	114
40	114	113	113	112	111	111	110	109	108	107
50	107	106	106	105	104	103	103	102	101	100
60	100	99	98	97	97	97	96	95	94	93
70	93	92	92	91	90	90	89	89	88	87
80	87	86	86	85	84	84	84	83	83	82
90	82	81	81	80	79	79	78	77	77	76
100	76	76	78	75	74	74	73	72	72	71
110	71	71	70	70	69	69	68	68	67	67
120	66	66	66	65	65	64	64	63	63	62
130	62	62	61	61	60	60	60	59	59	58
140	58	58	57	58	56	56	55	55	54	54
150	54	53	53	53	52	52	52	52	51	51
160	51	51	50	50	50	49	49	48	48	47
170	47	47	47	46	46	46	45	45	45	44
180	44	44	44	43	43	43	42	42	42	41
190	41	41	41	40	40	40	40	39	39	39
200	39	38	38	38	37	37	37	37	36	36
210	36	36	35	35	35	34	34	34	34	34
220	34	34	33	33	33	33	33	32	32	32
230	31	31	31	31	31	31	30	30	30	30
240	30	29	29	29	29	28	28	28	28	28
250	28	27	27	27	27	27	26	26	26	26
260	26	26	25	25	25	25	25	24	24	24
270	24	24	24	23	23	23	23	23	23	23
280	22	22	22	22	22	22	22	22	21	21
280	21	21	21	20	20	20	20	20	20	20
300	20	19	19	19	19	19	19	19	18	18
310	18	18	18	18	18	18	18	17	17	17
320	17	17	17	17	17	17	17	16	16	16
330	16	16	16	16	16	16	16	15	15	15
340	15	15	15	15	15	15	14	14	14	14
350	14	14	14	14	14	14	14	13	13	13
360	13	13	13	13	13	13	13	12	12	12
370	12	12	12	12	12	12	12	12	11	11
380	11	11	11	11	11	11	11	11	11	11
390	11	11	11	10	10	10	10	10	10	10
400	10	10	10	10	10	10	10	10	9	9
410	9	9	9	9	9	9	9	9	9	9
420	9	9	9	8	8	8	8	8	8	8

Σ(P - PE)	0	1	2	3	4	5	6	7	8	9
430	8	8	8	8	8	8	8	8	8	8
440	8	8	8	7	7	7	7	7	7	7
450	7	7	7	7	7	7	7	7	7	7
460	7	7	7	7	6	6	6	6	6	6
470	6	6	6	6	6	6	6	6	6	6
480	6	6	6	6	6	6	6	6	5	5
490	5	5	5	5	5	5	5	5	5	5
500	5	5	5	5	5	5	5	5	5	5
510	5	5	5	5	5	5	5	5	4	4
520	4	4	4	4	4	4	4	4	4	4
530	4	4	4	4	4	4	4	4	4	4
540	4	4	4	4	4	4	4	4	4	4
550	4	4	4	4	4	4	4	3	3	3
560	3	3	3	3	3	3	3	3	3	3
570	3	3	3	3	3	3	3	3	3	3
580	3	3	3	3	3	3	3	3	3	3
590	3	3	3	3	3	3	3	3	3	3
600	3	3	3	3	3	2	2	2	2	2
610	2	2	2	2	2	2	2	2	2	2
620	2	2	2	2	2	2	2	2	2	2
630	2	2	2	2	2	2	2	2	2	2
640	2	2	2	2	2	2	2	2	2	2
650	2	2	2	2	2	2	2	2	2	2
660	2	2	2	2	2	2	2	2	2	2
670	2	2	2	2	2	2	2	2	2	2
680	2	2	1	1	1	1	1	1	1	1
690	1	1	1	1	1	1	1	1	1	1
700	1	1	1	1	1	1	1	1	1	1
710	1	1	1	1	1	1	1	1	1	1
720	1	1	1	1	1	1	1	1	1	1
730	1	1	1	1	1	1	1	1	1	1
740	1	1	1	1	1	1	1	1	1	1

****

	0	5		0	5		0	5
750	1	1	790	1	1	830	1	1
760	1	1	800	1	1	840	1	1
770	1	1	810	1	1			
780	1	1	820	1	1			

For any other W.H.C. value use Eq. 3.2 (*After*: Thornthwaite and Mather, 1955)

Appendix VIII
Radiation balance parameters at different temperature levels

Temperature (°C)	Blackbody Emittance (Wm^{-2})	Radiative Resistance (sm^{-1})	Clear Sky Emissivity (∈)	$4\sigma T^3$ ($Wm^{-2}\,{}^{\circ}k^{-1}$)	Saturated Vapour Pressure (mb)	Saturated Vapour Density ($g\,m^{-3}$)	Slope Δ(T) ($mb^{\circ}C^{-1}$)	$gm^{-3}\,{}^{\circ}C^{-1}$	$\frac{\Delta}{\Delta+\gamma}$
1	2	3	4	5	6	7	8	9	10
-5	293	274	0.70	4.4	4.21	3.41	0.32	0.25	0.32
-4	298	271	0.70	4.4	4.55	3.66	0.34	0.26	0.33
-3	302	268	0.71	4.5	4.90	3.93	0.37	0.28	0.35
-2	306	265	0.71	4.5	5.28	4.22	0.39	0.30	0.36
-1	311	262	0.72	4.6	5.68	4.52	0.42	0.32	0.38
0	316	260	0.72	4.6	6.11	4.85	0.45	0.33	0.39
1	320	257	0.73	4.7	6.57	5.19	0.48	0.36	0.41
2	325	254	0.73	4.7	7.05	5.56	0.51	0.38	0.42
3	330	251	0.74	4.8	7.58	5.95	0.54	0.40	0.44
4	335	249	0.74	4.8	8.13	6.36	0.57	0.42	0.45
5	339	246	0.75	4.9	8.72	6.80	0.61	0.45	0.47
6	344	243	0.75	5.0	9.35	7.26	0.65	0.48	0.48
7	349	241	0.76	5.0	10.01	7.75	0.69	0.51	0.50
8	354	238	0.76	5.1	10.72	8.27	0.73	0.53	0.51
9	359	236	0.77	5.1	11.47	8.82	0.78	0.57	0.53
10	364	233	0.77	5.2	12.27	9.40	0.83	0.60	0.54
11	370	231	0.78	5.2	13.12	10.02	0.88	0.63	0.55
12	375	228	0.78	5.3	14.02	10.66	0.93	0.67	0.57
13	380	226	0.79	5.3	14.97	11.35	0.98	0.70	0.58
14	385	223	0.79	5.4	15.98	12.07	1.04	0.74	0.60
15	391	221	0.80	5.4	17.04	12.83	1.10	0.78	0.61
16	396	219	0.80	5.5	18.17	13.64	1.17	0.83	0.62
17	402	217	0.81	5.6	19.37	14.49	1.23	0.87	0.63
18	407	214	0.81	5.6	20.63	15.38	1.30	0.92	0.65
19	413	212	0.82	5.7	21.96	16.32	1.37	0.96	0.66
20	419	210	0.82	5.7	23.37	17.31	1.45	1.01	0.67
21	424	208	0.83	5.8	24.86	18.35	1.53	1.07	0.68

1	2	3	4	5	6	7	8	9	10
22	430	206	0.83	5.8	26.43	19.44	1.62	1.12	0.69
23	436	204	0.84	5.9	28.09	20.59	1.70	1.18	0.70
24	442	202	0.84	6.0	29.83	21.08	1.79	1.24	0.72
25	448	200	0.85	6.0	31.67	23.06	1.89	1.30	0.73
26	454	198	0.85	6.1	33.61	24.39	1.99	1.36	0.74
27	460	196	0.86	6.2	35.65	25.79	2.10	1.43	0.75
28	466	194	0.86	6.2	37.80	27.25	2.21	1.50	0.75
29	473	192	0.87	6.3	40.06	28.79	2.32	1.57	0.76
30	479	190	0.87	6.3	42.43	30.40	2.44	1.65	0.77
31	485	188	0.88	6.4	44.93	32.08	2.57	1.72	0.78
32	492	186	0.88	6.5	47.55	33.85	2.69	1.81	0.79
33	498	184	0.89	6.6	50.31	35.69	2.83	1.89	0.80
34	505	183	0.89	6.6	53.20	37.63	2.97	1.98	0.80
35	511	181	0.90	6.7	56.24	39.65	3.12	2.07	0.81
36	518	179	0.90	6.7	59.42	41.76	3.27	2.16	0.82
37	525	177	0.91	6.8	62.76	43.97	3.43	2.26	0.83
38	531	176	0.91	6.9	66.26	46.28	3.57	2.36	0.83
39	538	174	0.92	6.9	69.93	48.68	3.76	2.46	0.84
40	545	172	0.92	7.0	73.78	51.20	3.94	2.57	0.84
41	552	171	0.93	7.1	77.80	53.82	4.13	2.68	0.85
42	559	169	0.93	7.1	82.02	56.56	4.32	2.80	0.86
43	566	167	0.94	7.2	86.42	59.42	4.52	2.92	0.86
44	574	166	0.94	7.3	91.03	62.39	4.73	3.04	0.87
45	581	164	0.95	7.3	95.86	65.50	4.94	3.17	0.87

(*After*: Adapted from Monteith, 1973)

Appendix IX

Total daily solar radiation at the top of the atmosphere (Wm^{-2}) (solar constant 1354 Wm^{-2})

	Longitude of Sun															
Longitude	0	22½°	45°	67½°	90°	112½°	135°	157½°	180°	202½°	225°	247½°	270°	292½°	315°	337½°
	Approximate Date															
	Mar. 21	Apr. 13	May 6	May 29	Jun. 22	Jul. 15	Aug. 8	Aug. 31	Sep. 23	Oct. 16	Nov. 8	Nov. 30	Dec. 22	Jan. 13	Feb. 4	Feb. 26
90°	-	205.0	374.2	484.2	522.0	481.8	370.8	202.6	-	-	-	-	-	-	-	-
80°	75.1	205.0	368.4	476.9	513.8	475.0	365.5	202.6	74.1	3.4	-	-	-	-	-	3.4
70°	148.8	254.5	363.0	455.1	490.5	452.7	359.6	251.6	146.9	62.5	11.6	-	-	-	11.6	63.5
60°	216.7	307.8	392.1	452.7	474.5	450.3	388.2	304.9	214.2	132.3	70.8	34.9	23.8	35.4	40.8	133.5
50°	278.7	354.8	420.2	464.3	479.4	462.4	416.4	351.4	275.3	200.7	138.6	98.9	85.3	99.4	140.1	203.1
40°	332.5	391.1	441.1	471.1	480.3	468.7	436.7	386.8	328.1	264.2	207.9	168.7	153.6	169.6	210.4	268.0
30°	375.6	419.3	450.3	468.7	472.6	465.3	446.4	414.9	370.8	321.4	273.4	238.5	225.9	239.4	275.3	324.7
20°	407.6	433.3	447.4	453.2	453.2	450.8	444.0	428.5	402.8	368.4	332.0	303.9	293.2	305.4	334.9	372.7
10°	427.5	439.8	432.8	427.0	423.1	425.1	429.4	429.9	422.2	404.7	382.4	362.5	355.3	364.5	385.3	409.6
0°	433.8	423.1	405.7	389.7	382.9	387.8	402.3	418.3	429.0	429.4	421.7	412.5	408.6	414.4	425.5	434.3
–10°	427.5	399.4	368.4	342.7	333.0	341.2	365.0	394.5	422.2	441.1	449.3	451.2	452.2	453.7	453.7	446.4
–20°	407.6	363.5	319.9	287.4	274.8	286.0	317.0	359.2	402.8	439.6	464.8	478.9	484.2	481.3	469.2	445.0
–30°	375.6	317.0	263.2	225.4	211.3	224.4	260.8	313.1	370.8	425.1	467.2	494.4	504.6	496.8	471.6	430.5
–40°	332.5	260.8	200.2	153.5	144.0	159.0	198.2	258.3	328.1	397.0	457.6	497.8	513.3	500.2	461.9	401.3
–50°	278.7	197.8	133.8	93.5	80.0	93.1	132.8	195.8	275.3	360.1	436.7	491.5	511.8	493.4	440.6	364.5
–60°	216.7	130.4	67.9	33.0	22.8	33.0	67.4	128.9	214.2	312.1	407.1	478.4	507.0	480.0	410.5	316.0
–70°	148.8	61.6	11.1	-	-	-	11.1	61.1	146.9	257.8	377.1	481.3	524.0	483.7	380.5	261.2
–80°	75.1	3.4	-	-	-	-	-	3.4	74.1	207.9	382.9	504.6	548.7	507.0	385.8	210.4
–90°	-	-	-	-	-	-	-	-	-	207.9	388.2	511.8	556.9	514.8	392.1	210.4

Appendix X

Relative humidity: Non-aspirated psychrometer (natural aspiration of 1m s^{-1})

Dry bulb	Wet bulb depression (°C)																										
(°C)	0·2	0·4	0·6	0·8	1·0	1·2	1·4	1·6	1·8	2·0	2·5	3·0	3·5	4·0	4·5	5·0	6·0	7·0	8·0	9·0	10	11	12	13	14	15	16
0	96	92	88	84	80	76	73	69	65	61	52	43	33	24	16	7											
1	96	92	88	85	81	78	75	71	67	64	55	46	37	29	20	12											
2	96	93	89	85	82	78	75	71	67	64	58	49	41	33	24	16	1										
3	96	93	89	86	83	79	76	72	69	66	57	49	44	36	28	21	6										
4	97	93	90	87	83	80	77	74	70	67	59	51	44	36	32	25	10										
5	97	94	90	87	84	81	78	75	72	69	61	53	46	39	31	24	15	2									
6	97	94	91	88	85	82	79	76	73	70	63	55	48	41	34	28	14	6									
7	97	94	91	88	85	82	80	77	74	71	64	57	50	44	37	31	18	5									
8	97	94	92	89	86	83	80	78	75	72	66	59	52	46	40	33	21	9									
9	97	95	92	89	86	84	81	79	76	73	67	61	54	48	42	36	24	13	2								
10	97	95	92	90	87	84	82	79	77	74	68	62	56	50	44	38	27	16	5								
11	97	95	92	90	87	85	83	80	78	75	69	63	58	52	46	41	30	19	9								
12	98	95	93	90	88	86	83	81	78	76	70	65	59	54	48	43	32	22	12	3							
13	98	95	93	91	88	86	84	81	79	77	71	66	61	55	50	45	35	26	17	6							
14	98	95	93	91	89	86	84	82	80	78	72	67	62	57	52	47	37	28	18	10	1						
15	98	96	93	91	89	87	85	83	81	78	73	68	63	58	53	49	39	30	21	13	4						
16	98	96	94	91	89	87	85	83	81	79	74	69	64	60	55	50	41	32	24	16	7						
17	98	96	94	92	90	88	86	84	82	80	75	70	65	61	56	52	43	34	26	18	10	3					
18	98	96	94	92	90	88	86	84	82	80	76	71	66	62	58	53	45	36	28	21	13	6					
19	98	96	94	92	90	88	86	85	83	81	76	72	67	63	59	55	46	38	31	23	16	9					
20	98	96	94	92	91	89	87	85	83	81	77	73	68	64	60	56	48	40	33	25	18	11	5				
21	98	96	94	93	91	89	87	85	84	82	78	73	69	65	61	57	49	42	34	27	21	14	8	1			
22	98	96	95	93	91	89	88	86	84	82	78	74	70	66	62	58	51	43	36	29	23	16	10	4			
23	98	96	95	93	91	90	88	86	84	83	79	75	71	67	63	59	52	45	38	31	25	19	13	7	1		
24	98	97	95	93	91	90	88	86	85	83	79	75	71	68	64	60	53	46	39	33	27	21	15	9	4		
25	98	97	95	93	92	90	88	87	85	84	80	76	72	68	65	61	54	47	41	35	29	23	17	11	6	1	
26	98	97	95	93	92	90	89	87	86	84	80	76	73	69	66	62	55	49	42	36	30	25	19	14	8	3	
27	98	97	95	94	92	90	89	87	86	84	81	77	73	70	66	63	56	50	44	38	32	26	21	16	11	6	1
28	98	97	95	94	92	91	89	88	86	85	81	77	74	70	67	64	57	51	45	39	33	28	23	18	13	8	3
29	98	97	95	94	92	91	89	88	86	85	81	78	74	71	68	65	58	52	46	40	35	30	24	19	15	10	5
30	98	97	95	94	93	91	90	88	87	85	82	78	75	72	68	65	59	53	47	42	36	31	26	21	16	12	8
31	99	97	96	94	93	91	90	88	87	86	82	79	75	72	69	66	60	54	48	43	38	32	28	23	18	14	9
32	99	97	96	94	93	91	90	89	87	86	83	79	76	73	70	67	61	55	49	44	39	34	29	24	20	16	11
33	99	97	96	94	93	91	90	89	87	86	83	80	76	73	70	67	61	56	50	45	40	35	30	26	21	17	13
34	99	97	96	94	93	92	90	89	87	86	83	80	77	74	71	68	62	56	51	46	41	36	32	27	23	19	15
35	99	98	97	95	93	92	90	90	88	87	83	80	77	74	71	68	63	57	52	47	42	37	33	29	24	20	16
36	99	98	97	95	93	92	90	90	88	87	84	81	78	75	72	69	63	58	53	48	43	39	34	30	26	22	18
37	99	98	97	95	93	92	90	90	88	87	84	81	78	75	72	69	64	59	54	49	44	40	35	31	27	23	19
38	99	98	97	95	93	92	90	90	88	87	84	81	78	75	73	70	65	59	54	50	45	41	36	32	28	24	21
39	99	98	97	95	94	93	91	91	89	87	84	81	78	75	73	70	65	60	55	50	46	41	37	33	29	26	22

$e = e_s(w) - 0.799(T_d - T_w)$ for normal conditions

$e = e_s(w) - 0.720\,(T_d - T_w)$ for ice formation on wet bulb. $RH = \dfrac{e}{e_s(T_d)} \times 100$

Appendix XI

Relative humidity: Non-aspirated psychrometer with aspiration rate of 3m s^{-1}.

Dry bulb (°C)	Wet bulb depression (°C) 0·2	0·4	0·6	0·8	1·0	1·2	1·4	1·6	1·8	2·0	2·5	3·0	3·5	4·0	4·5	5·0	6·0	7·0	8·0	9·0	10	11	12	13	14	15	16
0	96	93	89	86	82	79	75	72	69	65	57	49	41	33	25	17	2										
1	97	93	90	86	83	81	77	74	71	68	60	52	44	36	29	21	7										
2	97	93	90	87	84	80	77	74	71	68	62	54	47	40	32	25	11										
3	97	94	91	88	84	81	78	75	72	69	62	54	50	43	36	29	16	3									
4	97	94	91	88	85	82	79	76	73	70	63	56	49	42	39	33	20	7									
5	97	94	91	88	86	83	80	77	74	72	65	58	51	45	38	32	25	12									
6	97	94	92	89	86	83	81	78	75	73	66	60	53	47	41	35	23	16	5								
7	97	95	92	89	87	84	81	79	76	74	67	61	55	49	43	37	26	14	9								
8	97	95	92	90	87	85	82	80	77	75	69	63	57	51	45	40	29	18	7	3							
9	97	95	93	90	88	85	83	80	78	76	70	64	58	53	47	42	31	21	11	1							
10	98	95	93	90	88	86	83	81	79	77	71	65	60	54	49	44	34	24	14	5							
11	98	95	93	91	88	86	84	82	80	77	72	66	61	56	51	46	36	26	17	8							
12	98	95	93	91	89	87	84	82	80	78	73	68	63	57	53	48	38	29	20	11	3						
13	98	96	93	91	89	87	85	83	81	79	74	69	64	59	54	49	40	31	23	14	6						
14	98	96	94	92	90	87	85	83	81	79	74	70	65	60	56	51	42	33	25	17	9	2					
15	98	96	94	92	90	88	86	84	82	80	75	71	66	61	57	52	44	36	27	20	12	5					
16	98	96	94	92	90	88	86	84	82	81	76	71	67	62	58	54	46	37	30	22	15	8	1				
17	98	96	94	92	90	89	87	85	83	81	77	72	68	64	59	55	47	39	32	24	17	10	4				
18	98	96	94	92	91	89	87	85	83	82	77	73	69	65	60	56	49	41	34	27	20	13	7				
19	98	96	94	93	91	89	87	86	84	82	78	74	70	65	61	58	50	43	35	29	22	15	9	3			
20	98	96	95	93	91	89	88	86	84	83	78	74	70	66	62	59	51	44	37	30	24	18	12	6			
21	98	96	95	93	91	90	88	86	85	83	79	75	71	67	63	60	52	46	39	32	26	20	14	8	3		
22	98	97	95	93	92	90	88	87	85	83	79	76	72	68	64	61	54	47	40	34	28	22	16	11	5		
23	98	97	95	93	92	90	89	87	85	84	80	76	72	69	65	62	55	48	42	36	30	24	18	13	8	3	
24	98	97	95	94	92	90	89	87	86	84	80	77	73	69	66	62	56	49	43	37	31	26	20	15	10	5	
25	98	97	95	94	92	91	89	87	86	84	81	77	74	70	67	63	57	50	44	38	33	27	22	17	12	7	3
26	98	97	95	94	92	91	89	88	86	85	81	78	74	71	67	64	58	51	46	40	34	29	24	19	14	9	5
27	98	97	95	94	92	91	89	88	87	85	82	78	75	71	68	65	59	52	47	41	36	30	25	21	16	11	7
28	98	97	95	94	93	91	90	88	87	85	82	79	75	72	69	65	59	53	48	42	37	32	27	22	18	13	9
29	99	97	96	94	93	91	90	88	87	86	82	79	76	72	69	66	60	54	49	43	38	33	28	24	19	15	11
30	99	97	96	94	93	91	90	89	87	86	83	79	76	73	70	67	61	55	50	44	39	34	30	25	21	17	13
31	99	97	96	94	93	92	90	89	88	86	83	80	77	73	70	67	62	56	51	45	41	36	31	27	22	18	14
32	99	97	96	94	93	92	90	89	88	86	83	80	77	74	71	68	62	57	52	46	42	37	32	28	24	20	16
33	99	97	96	94	93	92	90	89	88	87	83	80	77	74	71	69	63	58	52	47	43	38	34	29	25	21	17
34	99	98	97	95	93	92	90	89	88	87	84	81	78	75	72	69	64	58	53	48	44	39	35	30	26	23	19
35	99	98	97	95	93	92	90	89	88	87	84	81	78	75	72	70	64	59	54	49	44	40	36	32	28	24	20
36	99	98	97	95	94	92	90	89	88	87	84	81	78	76	73	70	65	60	55	50	45	41	37	33	29	25	21
37	99	98	97	95	94	93	91	90	88	87	85	82	79	76	73	70	65	60	55	51	46	42	38	34	30	26	23
38	99	98	97	95	94	93	91	90	88	88	85	82	79	76	74	71	66	61	56	51	47	43	39	35	31	27	24
39	99	98	97	95	94	93	91	90	88	88	85	82	79	77	74	71	66	61	57	52	48	44	40	36	32	28	25

$e = e_s(w) - 0.666(T_d - T_w)$ for normal conditions

$e = e_s(w) - 0.594\,(T_d - T_w)$ for ice formation on wet bulb, $RH = \dfrac{e}{e_s(T_d)} \times 100$

Appendix XII

Relative humidity with dew point equipment

Relative Humidity: Dew-point and Air Temperature

Dew-point depression (°C)	Air temperature (°C)						
	0	5	10	15	20	25	30
0·5	96·5	96·6	96·7	96·8	96·9	97·0	97·1
1·0	93·1	93·3	93·5	93·8	94·0	94·2	94·3
1·5	89·8	90·1	90·5	90·8	91·1	91·4	91·6
2·0	86·6	87·0	87·5	87·9	88·3	88·6	89·0
2·5	83·5	84·0	84·6	85·1	85·5	86·0	86·4
3·0	80·5	81·1	81·7	82·3	82·9	83·4	83·9
3·5	77·6	78·3	79·0	79·7	80·3	80·9	81·4
4·0	74·8	75·6	76·4	77·1	77·8	78·4	79·1
4·5	72·1	73·0	73·8	74·6	75·3	76·0	76·7
5·0	69·5	70·4	71·3	72·1	72·9	73·7	74·5
5·5	67·0	67·9	68·9	69·8	70·6	71·5	72·3
6·0	64·5	65·5	66·5	67·5	68·4	69·3	70·1
6·5	62·1	63·2	64·3	65·3	66·2	67·1	68·0
7·0	59·8	61·0	62·1	63·1	64·1	65·1	66·0
7·5	57·6	58·8	59·9	61·0	62·0	63·1	64·0
8·0	55·5	56·7	57·8	59·0	60·0	61·1	62·1
8·5	53·4	54·7	55·8	57·0	58·1	59·2	60·2
9·0	51·4	52·7	53·9	55·1	56·2	57·3	58·4
9·5	49·5	50·8	52·0	53·2	54·4	55·5	56·6
10	47·6	48·9	50·2	51·4	52·6	53·8	54·9
11		45·4	46·7	48·0	49·2	50·4	51·6
12		42·1	43·5	44·8	46·0	47·3	48·5
13		39·1	40·4	41·7	43·0	44·3	45·5
14		36·2	37·6	38·9	40·2	41·5	42·7
15		33·5	34·9	36·2	37·5	38·8	40·1
16			32·4	33·7	35·0	36·3	37·6
17			30·0	31·4	32·7	33·9	35·2
18			27·9	29·2	30·4	31·7	33·0
19			25·8	27·1	28·4	29·6	30·9
20			23·9	25·2	26·4	27·7	28·9
21				23·4	24·6	25·8	27·0
22				21·7	22·9	24·1	25·3
23				20·1	21·3	22·4	23·6
24				18·6	19·8	20·9	22·1
25				17·2	18·4	19·5	20·6

$$RH = \frac{e_s(T_{dew})}{e_s(T_d)} \times 100$$

Appendix XIII

Vapour pressure (in millibars) from relative humidity and temperature

Vapour Pressure (in millibars) from Relative Humidity and Temperature

Temp.	Relative humidity (%)																				
(°C)	100	96	92	88	84	80	76	72	68	64	60	56	52	48	44	40	36	32	28	24	20
0	6·1	5·9	5·6	5·4	5·1	4·9	4·6	4·4	4·2	3·9	3·7	3·4	3·2	2·9	2·7	2·4	2·2	2·0	1·7	1·5	1·2
1	6·6	6·3	6·0	5·8	5·5	5·3	5·0	4·7	4·5	4·2	3·9	3·7	3·4	3·2	2·9	2·6	2·4	2·1	1·8	1·6	1·3
2	7·1	6·8	6·5	6·2	5·9	5·6	5·4	5·1	4·8	4·5	4·2	3·9	3·7	3·4	3·1	2·8	2·5	2·3	2·0	1·7	1·4
3	7·6	7·3	7·0	6·7	6·4	6·1	5·8	5·5	5·2	4·9	4·5	4·2	3·9	3·6	3·3	3·0	2·7	2·4	2·1	1·8	1·5
4	8·1	7·8	7·5	7·2	6·8	6·5	6·2	5·9	5·5	5·2	4·9	4·6	4·2	3·9	3·6	3·3	2·9	2·6	2·3	2·0	1·6
5	8·7	8·4	8·0	7·7	7·3	7·0	6·6	6·3	5·9	5·6	5·2	4·9	4·5	4·2	3·8	3·5	3·1	2·8	2·4	2·1	1·7
6	9·4	9·0	8·6	8·2	7·9	7·5	7·1	6·7	6·4	6·0	5·6	5·2	4·9	4·5	4·1	3·7	3·4	3·0	2·6	2·2	1·9
7	10·0	9·6	9·2	8·8	8·4	8·0	7·6	7·2	6·8	6·4	6·0	5·6	5·2	4·8	4·4	4·0	3·6	3·2	2·8	2·4	2·0
8	10·7	10·3	9·9	9·4	9·0	8·6	8·1	7·7	7·3	6·9	6·4	6·0	5·6	5·1	4·7	4·3	3·9	3·4	3·0	2·6	2·1
9	11·5	11·0	10·6	10·1	9·6	9·2	8·7	8·3	7·8	7·3	6·9	6·4	6·0	5·5	5·0	4·6	4·1	3·7	3·2	2·8	2·3
10	12·3	11·8	11·3	10·8	10·3	9·8	9·3	8·8	8·3	7·9	7·4	6·9	6·4	5·9	5·4	4·9	4·4	3·9	3·4	2·9	2·5
11	13·1	12·6	12·1	11·5	11·0	10·5	10·0	9·4	8·9	8·4	7·9	7·3	6·8	6·3	5·8	5·2	4·7	4·2	3·7	3·1	2·6
12	14·0	13·5	12·9	12·3	11·8	11·2	10·7	10·1	9·5	9·0	8·4	7·9	7·3	6·7	6·2	5·6	5·0	4·5	3·9	3·4	2·8
13	15·0	14·4	13·8	13·2	12·6	12·0	11·4	10·8	10·2	9·6	9·0	8·4	7·8	7·2	6·6	6·0	5·4	4·8	4·2	3·6	3·0
14	16·0	15·3	14·7	14·1	13·4	12·8	12·1	11·5	10·9	10·2	9·6	8·9	8·3	7·7	7·0	6·4	5·8	5·1	4·5	3·8	3·2
15	17·0	16·4	15·7	15·0	14·3	13·6	13·0	12·3	11·6	10·9	10·2	9·5	8·9	8·2	7·5	6·8	6·1	5·5	4·8	4·1	3·4
16	18·2	17·4	16·7	16·0	15·3	14·5	13·8	13·1	12·4	11·6	10·9	10·2	9·4	8·7	8·0	7·3	6·5	5·8	5·1	4·4	3·6
17	19·4	18·6	17·8	17·0	16·3	15·5	14·7	13·9	13·2	12·4	11·6	10·8	10·1	9·3	8·5	7·7	7·0	6·2	5·4	4·6	3·9
18	20·6	19·8	19·0	18·2	17·3	16·5	15·7	14·9	14·0	13·2	12·4	11·6	10·7	9·9	9·1	8·3	7·4	6·6	5·8	5·0	4·1
19	22·0	21·1	20·2	19·3	18·4	17·6	16·7	15·8	14·9	14·1	13·2	12·3	11·4	10·5	9·7	8·8	7·9	7·0	6·1	5·3	4·4
20	23·4	22·4	21·5	20·6	19·6	18·7	17·8	16·8	15·9	15·0	14·0	13·1	12·2	11·2	10·3	9·3	8·4	7·5	6·5	5·6	4·7
21	24·9	23·9	22·9	21·9	20·9	19·9	18·9	17·9	16·9	15·9	14·9	13·9	12·9	11·9	10·9	9·9	8·9	8·0	7·0	6·0	5·0
22	26·4	25·4	24·3	23·3	22·2	21·1	20·1	19·0	18·0	16·9	15·9	14·8	13·7	12·7	11·6	10·6	9·5	8·5	7·4	6·3	5·3
23	28·1	27·0	25·8	24·7	23·6	22·5	21·3	20·2	19·1	18·0	16·9	15·7	14·6	13·5	12·4	11·2	10·1	9·0	7·9	6·7	5·6
24	29·8	28·6	27·4	26·3	25·1	23·9	22·7	21·5	20·3	19·9	17·9	16·7	15·5	14·3	13·1	11·9	10·7	9·5	8·4	7·2	6·0
25	31·7	30·4	29·1	27·9	26·6	25·3	24·1	22·8	21·5	20·3	19·0	17·7	16·5	15·2	13·9	12·7	11·4	10·1	8·9	7·6	6·3
26	33·6	32·3	30·9	29·6	28·2	26·9	25·5	24·2	22·9	21·5	20·2	18·8	17·5	16·1	14·8	13·4	12·1	10·8	9·4	8·1	6·7
27	35·7	34·2	32·8	31·4	29·9	28·5	27·1	25·7	24·2	22·8	21·4	20·0	18·5	17·1	15·7	14·3	12·8	11·4	10·0	8·6	7·1
28	37·8	36·3	34·8	33·3	31·8	30·2	28·7	27·2	25·7	24·2	22·7	21·2	19·7	18·1	16·6	15·1	13·6	12·1	10·6	9·1	7·6
29	40·1	38·5	36·9	35·3	33·7	32·0	30·4	28·8	27·2	25·6	24·0	22·4	20·8	19·2	17·6	16·0	14·4	12·8	11·2	9·6	8·0
30	42·4	40·7	39·0	37·3	35·6	33·9	32·2	30·5	28·9	27·2	25·5	23·8	22·1	20·4	18·7	17·0	15·3	13·6	11·9	10·2	8·5
31	44·9	43·1	41·3	39·5	37·7	35·9	34·1	32·3	30·6	28·8	27·0	25·2	23·4	21·6	19·8	18·0	16·2	14·4	12·6	10·8	9·0
32	47·6	45·6	43·7	41·8	39·9	38·0	36·1	34·2	32·3	30·4	28·5	26·6	24·7	22·8	20·9	19·0	17·1	15·2	13·3	11·4	9·5
33	50·3	48·3	46·3	44·3	42·3	40·2	38·2	36·2	34·2	32·2	30·2	28·2	26·2	24·1	22·1	20·1	18·1	16·1	14·1	12·1	10·1
34	53·2	51·1	48·9	46·8	44·7	42·6	40·4	38·3	36·2	34·0	31·9	29·8	27·7	25·5	23·4	21·3	19·2	17·0	14·9	12·8	10·6
35	56·2	54·0	51·7	49·5	47·2	45·0	42·7	40·5	38·2	36·0	33·7	31·5	29·2	27·0	24·7	22·5	20·2	18·0	15·7	13·5	11·2
36	59·4	57·0	54·7	52·3	49·9	47·5	45·2	42·8	40·4	38·0	35·7	33·3	30·9	28·5	26·1	23·8	21·4	19·0	16·6	14·3	11·9
37	62·8	60·2	57·7	55·2	52·7	50·2	47·7	45·2	42·7	40·2	37·7	35·1	32·6	30·1	27·6	25·1	22·6	20·1	17·6	15·1	12·6
38	66·3	63·6	61·0	58·3	55·7	53·0	50·4	47·7	45·1	42·4	39·8	37·1	34·5	31·8	29·2	26·5	23·9	21·2	18·6	15·9	13·3
39	69·9	67·1	64·3	61·5	58·7	55·9	53·1	50·3	47·6	44·8	42·0	39·2	36·4	33·6	30·8	28·0	25·2	22·4	19·6	16·8	14·0

$$e(\text{mb}) = \frac{e_s(T_d)}{100} \times \text{RH}$$

Appendix XIV

Vapour pressure dificition mb. from RH and temperature

Temp °C	RH (%) 100	96	92	88	84	80	76	72	68	64	60	56	52	48	44	40	36	32	28	24	20
0	0·0	0·2	0·5	0·7	1·0	1·2	1·5	1·7	2·0	2·2	2·4	2·7	2·9	3·2	3·4	3·7	3·9	4·2	4·4	4·6	4·9
1	0·0	0·3	0·5	0·8	1·1	1·3	1·6	1·8	2·1	2·4	2·6	2·9	3·2	3·4	3·7	3·9	4·2	4·5	4·7	5·0	5·3
2	0·0	0·3	0·6	0·8	1·1	1·4	1·7	2·0	2·3	2·5	2·8	3·1	3·4	3·7	3·9	4·2	4·5	4·8	5·1	5·4	5·6
3	0·0	0·3	0·6	0·9	1·2	1·5	1·8	2·1	2·4	2·7	3·0	3·3	3·6	3·9	4·2	4·5	4·9	5·2	5·5	5·8	6·1
4	0·0	0·3	0·7	1·0	1·3	1·6	2·0	2·3	2·6	2·9	3·3	3·6	3·9	4·2	4·6	4·9	5·2	5·5	5·9	6·2	6·5
5	0·0	0·3	0·7	1·0	1·4	1·7	2·1	2·4	2·8	3·1	3·5	3·8	4·2	4·5	4·9	5·2	5·6	5·9	6·3	6·6	7·0
6	0·0	0·4	0·7	1·1	1·5	1·9	2·2	2·6	3·0	3·4	3·7	4·1	4·5	4·9	5·2	5·6	6·0	6·4	6·7	7·1	7·5
7	0·0	0·4	0·8	1·2	1·6	2·0	2·4	2·8	3·2	3·6	4·0	4·4	4·8	5·2	5·6	6·0	6·4	6·8	7·2	7·6	8·0
8	0·0	0·4	0·9	1·3	1·7	2·1	2·6	3·0	3·4	3·9	4·3	4·7	5·1	5·6	6·0	6·4	6·9	7·3	7·7	8·1	8·6
9	0·0	0·5	0·9	1·4	1·8	2·3	2·8	3·2	3·7	4·1	4·6	5·0	5·5	6·0	6·4	6·9	7·3	7·8	8·3	8·7	9·2
10	0·0	0·5	1·0	1·5	2·0	2·5	2·9	3·4	3·9	4·4	4·9	5·4	5·9	6·4	6·9	7·4	7·9	8·3	8·8	9·3	9·8
11	0·0	0·5	1·0	1·6	2·1	2·6	3·1	3·7	4·2	4·7	5·2	5·8	6·3	6·8	7·3	7·9	8·4	8·9	9·4	10·0	10·5
12	0·0	0·6	1·1	1·7	2·2	2·8	3·4	3·9	4·5	5·0	5·6	6·2	6·7	7·3	7·9	8·4	9·0	9·5	10·1	10·7	11·2
13	0·0	0·6	1·2	1·8	2·4	3·0	3·6	4·2	4·8	5·4	6·0	6·6	7·2	7·8	8·4	9·0	9·6	10·2	10·8	11·4	12·0
14	0·0	0·6	1·3	1·9	2·6	3·2	3·8	4·5	5·1	5·8	6·4	7·0	7·7	8·3	8·9	9·6	10·2	10·9	11·5	12·1	12·8
15	0·0	0·7	1·4	2·0	2·7	3·4	4·1	4·8	5·5	6·1	6·8	7·5	8·2	8·9	9·5	10·2	10·9	11·6	12·3	13·0	13·6
16	0·0	0·7	1·5	2·2	2·9	3·6	4·4	5·1	5·8	6·5	7·3	8·0	8·7	9·4	10·2	10·9	11·6	12·4	13·1	13·8	14·5
17	0·0	0·8	1·5	2·3	3·1	3·9	4·6	5·4	6·2	7·0	7·7	8·5	9·3	10·1	10·8	11·6	12·4	13·2	13·9	14·7	15·5
18	0·0	0·8	1·7	2·5	3·3	4·1	5·0	5·8	6·6	7·4	8·3	9·1	9·9	10·7	11·6	12·4	13·2	14·0	14·9	15·7	16·5
19	0·0	0·9	1·8	2·6	3·5	4·4	5·3	6·1	7·0	7·9	8·8	9·7	10·5	11·4	12·3	13·2	14·1	14·9	15·8	16·7	17·6
20	0·0	0·9	1·9	2·8	3·7	4·7	5·6	6·5	7·5	8·4	9·3	10·3	11·2	12·2	13·1	14·0	15·0	15·9	16·8	17·8	18·7
21	0·0	1·0	2·0	3·0	4·0	5·0	6·0	7·0	8·0	8·9	9·9	10·9	11·9	12·9	13·9	14·9	15·9	16·9	17·9	18·9	19·9
22	0·0	1·1	2·1	3·2	4·2	5·3	6·3	7·4	8·5	9·5	10·6	11·6	12·7	13·7	14·8	15·9	16·9	18·0	19·0	20·1	21·1
23	0·0	1·1	2·2	3·4	4·5	5·6	6·7	7·9	9·0	10·1	11·2	12·4	13·5	14·6	15·7	16·9	18·0	19·1	20·2	21·3	22·5
24	0·0	1·2	2·4	3·6	4·8	6·0	7·2	8·4	9·5	10·7	11·9	13·1	14·3	15·5	16·7	17·9	19·1	20·2	21·5	22·7	23·9
25	0·0	1·3	2·5	3·8	5·1	6·3	7·6	8·9	10·1	11·4	12·7	13·9	15·2	16·5	17·7	19·0	20·3	21·5	22·8	24·1	25·3
26	0·0	1·3	2·7	4·0	5·4	6·7	8·1	9·4	10·8	12·1	13·4	14·8	16·1	17·5	18·8	20·2	21·5	22·9	24·2	25·5	26·9
27	0·0	1·4	2·9	4·3	5·7	7·1	8·6	10·0	11·4	12·8	14·3	15·7	17·1	18·5	20·0	21·4	22·8	24·2	25·7	27·1	28·5
28	0·0	1·5	3·0	4·5	6·0	7·6	9·1	10·6	12·1	13·6	15·1	16·6	18·1	19·7	21·2	22·7	24·2	25·7	27·2	28·7	30·2
29	0·0	1·6	3·2	4·8	6·4	8·0	9·6	11·2	12·8	14·4	16·0	17·6	19·2	20·8	22·4	24·0	25·6	27·2	28·8	30·4	32·0
30	0·0	1·7	3·4	5·1	6·8	8·5	10·2	11·9	13·6	15·3	17·0	18·7	20·4	22·1	23·8	25·5	27·2	28·9	30·5	32·2	33·9
31	0·0	1·8	3·6	5·4	7·2	9·0	10·8	12·6	14·4	16·2	18·0	19·8	21·6	23·4	25·2	27·0	28·8	30·6	32·3	34·1	35·9
32	0·0	1·9	3·8	5·7	7·6	9·5	11·4	13·3	15·2	17·1	19·0	20·9	22·8	24·7	26·6	28·5	30·4	32·3	34·2	36·1	38·0
33	0·0	2·0	4·0	6·0	8·0	10·1	12·1	14·1	16·1	18·1	20·1	22·1	24·1	26·2	28·2	30·2	32·2	34·2	36·2	38·2	40·2
34	0·0	2·1	4·3	6·4	8·5	10·6	12·8	14·9	17·0	19·2	21·3	23·4	25·5	27·7	29·8	31·9	34·0	36·2	38·3	40·4	42·6
35	0·0	2·2	4·5	6·7	9·0	11·2	13·5	15·7	18·0	20·2	22·5	24·7	27·0	29·2	31·5	33·7	36·0	38·2	40·5	42·7	45·0
36	0·0	2·4	4·8	7·1	9·5	11·9	14·3	16·6	19·0	21·4	23·8	26·1	28·5	30·9	33·3	35·7	38·0	40·4	42·8	45·2	47·5
37	0·0	2·5	5·0	7·5	10·0	12·6	15·1	17·6	20·1	22·6	25·1	27·6	30·1	32·6	35·1	37·7	40·2	42·7	45·2	47·7	50·2
38	0·0	2·7	5·3	8·0	10·6	13·3	15·9	18·6	21·2	23·9	26·5	29·2	31·8	34·5	37·1	39·8	42·4	45·1	47·7	50·4	53·0
39	0·0	2·8	5·6	8·4	11·2	14·0	16·8	19·6	22·4	25·2	28·0	30·8	33·6	36·4	39·2	42·0	44·8	47·6	50·3	53·1	55·9

v.p.d. $e_s(T_d) - e$ in mb.

Appendix XV

Absolute humidity from relative humidity and temperature

	RH (%)																				
Tem. °C	100	96	92	88	84	80	76	72	68	64	60	56	52	48	44	40	36	32	28	24	20
0	4.8	4.7	4.5	4.3	4.1	3.9	3.7	3.5	3.3	3.1	2.9	2.7	2.5	2.3	2.1	1.9	1.7	1.6	1.4	1.2	1.0
1	5.2	5.0	4.8	4.6	4.4	4.2	3.9	3.7	3.5	3.3	3.1	2.9	2.7	2.5	2.3	2.1	1.9	1.7	1.5	1.3	1.0
2	5.6	5.3	5.1	4.9	4.7	4.4	4.2	4.0	3.8	3.6	3.3	3.2	2.9	2.7	2.5	2.3	2.1	1.8	1.6	1.4	1.1
3	6.0	5.7	5.5	5.2	5.0	4.8	4.5	4.3	4.0	3.8	3.6	3.3	3.1	2.9	2.6	2.4	2.1	1.9	1.7	1.4	1.2
4	6·4	6·1	5·9	5·6	5·3	5·1	4·8	4·6	4·3	4·1	3·8	3·6	3·3	3·1	2·8	2·5	2·3	2·0	1·8	1·5	1·3
5	6·8	6·5	6·3	6·0	5·7	5·4	5·2	4·9	4·6	4·4	4·1	3·8	3·5	3·3	3·0	2·7	2·4	2·2	1·9	1·6	1·4
6	7·3	7·0	6·7	6·4	6·1	5·8	5·5	5·2	4·9	4·6	4·4	4·1	3·8	3·5	3·2	2·9	2·6	2·3	2·0	1·7	1·5
7	7·7	7·4	7·1	6·8	6·5	6·2	5·9	5·6	5·3	5·0	4·6	4·3	4·0	3·7	3·4	3·1	2·8	2·5	2·2	1·9	1·5
8	8·3	7·9	7·6	7·3	6·9	6·6	6·3	6·0	5·6	5·3	5·0	4·6	4·3	4·0	3·6	3·3	3·0	2·6	2·3	2·0	1·7
9	8·8	8·5	8·1	7·8	7·4	7·1	6·7	6·3	6·0	5·6	5·3	4·9	4·6	4·2	3·9	3·5	3·2	2·8	2·5	2·1	1·8
10	9·4	9·0	8·6	8·3	7·9	7·5	7·1	6·8	6·4	6·0	5·6	5·3	4·9	4·5	4·1	3·8	3·4	3·0	2·6	2·3	1·9
11	10·0	9·6	9·2	8·8	8·4	8·0	7·6	7·2	6·8	6·4	6·0	5·6	5·2	4·8	4·4	4·0	3·6	3·2	2·8	2·4	2·0
12	10·7	10·2	9·8	9·4	9·0	8·5	8·1	7·7	7·2	6·8	6·4	6·0	5·5	5·1	4·7	4·3	3·8	3·4	3·0	2·6	2·1
13	11·3	10·9	10·4	10·0	9·5	9·1	8·6	8·2	7·7	7·3	6·8	6·4	5·9	5·4	5·0	4·5	4·1	3·6	3·2	2·7	2·3
14	12·1	11·6	11·1	10·6	10·1	9·7	9·2	8·7	8·2	7·7	7·2	6·8	6·3	5·8	5·3	4·8	4·3	3·9	3·4	2·9	2·4
15	12·8	12·3	11·8	11·3	10·8	10·3	9·7	9·2	8·7	8·2	7·7	7·2	6·7	6·2	5·6	5·1	4·6	4·1	3·6	3·1	2·6
16	13·6	13·1	12·5	12·0	11·4	10·9	10·4	9·8	9·3	8·7	8·2	7·6	7·1	6·5	6·0	5·4	4·9	4·4	3·8	3·3	2·7
17	14·5	13·9	13·3	12·7	12·2	11·6	11·0	10·4	9·8	9·3	8·7	8·1	7·5	6·9	6·4	5·8	5·2	4·6	4·1	3·5	2·9
18	15·4	14·7	14·1	13·5	12·9	12·3	11·7	11·1	10·4	9·8	9·2	8·6	8·0	7·4	6·8	6·1	5·5	4·9	4·3	3·7	3·1
19	16·3	15·6	15·0	14·3	13·7	13·0	12·4	11·7	11·1	10·4	9·8	9·1	8·5	7·8	7·2	6·5	5·9	5·2	4·6	3·9	3·3
20	17·3	16·6	15·9	15·2	14·5	13·8	13·1	12·4	11·8	11·1	10·4	9·7	9·0	8·3	7·6	6·9	6·2	5·5	4·8	4·1	3·5
21	18·3	17·6	16·9	16·1	15·4	14·7	13·9	13·2	12·5	11·7	11·0	10·3	9·5	8·8	8·1	7·3	6·6	5·9	5·1	4·4	3·7
22	19·4	18·6	17·9	17·1	16·3	15·5	14·8	14·0	13·2	12·4	11·6	10·9	10·1	9·3	8·5	7·8	7·0	6·2	5·4	4·7	3·9
23	20·6	19·7	18·9	18·1	17·3	16·5	15·6	14·8	14·0	13·2	12·3	11·5	10·7	9·9	9·0	8·2	7·4	6·6	5·8	4·9	4·1
24	21·8	20·9	20·0	19·2	18·3	17·4	16·5	15·7	14·8	13·9	13·1	12·2	11·3	10·4	9·6	8·7	7·8	7·0	6·1	5·2	4·4
25	23·0	22·1	21·2	20·3	19·3	18·4	17·5	16·6	15·7	14·7	13·8	12·9	12·0	11·1	10·1	9·2	8·3	7·4	6·4	5·5	4·6
26	24·4	23·4	22·4	21·4	20·5	19·5	18·5	17·5	16·6	15·6	14·6	13·6	12·7	11·7	10·7	9·7	8·8	7·8	6·8	5·8	4·9
27	25·8	24·7	23·7	22·7	21·6	20·6	19·6	18·5	17·5	16·5	15·5	14·4	13·4	12·4	11·3	10·3	9·3	8·2	7·2	6·2	5·2
28	27·2	26·1	25·0	23·9	22·9	21·8	20·7	19·6	18·5	17·4	16·3	15·2	14·2	13·1	12·0	10·9	9·8	8·7	7·6	6·5	5·4
29	28·7	27·6	26·4	25·3	24·1	23·0	21·8	20·7	19·5	18·4	17·2	16·1	14·9	13·8	12·6	11·5	10·3	9·2	8·0	6·9	5·7
30	30·3	29·1	27·9	26·7	25·5	24·3	23·1	21·8	20·6	19·4	18·2	17·0	15·8	14·6	13·4	12·1	10·9	9·7	8·5	7·3	6·1
31	32·0	30·7	29·5	28·2	26·9	25·6	24·3	23·1	21·8	20·5	19·2	17·9	16·7	15·4	14·1	12·8	11·5	10·2	9·0	7·7	6·4
32	33·8	32·4	31·1	29·7	28·4	27·0	25·7	24·3	23·0	21·6	20·3	18·9	17·6	16·2	14·9	13·5	12·2	10·8	9·5	8·1	6·8
33	35·6	34·2	32·8	31·4	29·9	28·5	27·1	25·7	24·2	22·8	21·4	20·0	18·5	17·1	15·7	14·3	12·8	11·4	10·0	8·6	7·1
34	37·6	36·0	34·5	33·0	31·5	30·0	28·5	27·0	25·5	24·0	22·5	21·0	19·5	18·0	16·5	15·0	13·5	12·0	10·5	9·0	7·5
35	39·6	38·0	36·4	34·8	33·2	31·7	30·1	28·5	26·9	25·3	23·7	22·2	20·6	19·0	17·4	15·8	14·2	12·7	11·1	9·5	7·9
36	41·7	40·0	38·3	36·7	35·0	33·3	31·7	30·0	28·3	26·7	25·0	23·3	21·7	20·0	18·3	16·7	15·0	13·3	11·7	10·0	8·3
37	43·9	42·1	40·4	38·6	36·9	35·1	33·3	31·6	29·8	28·1	26·3	24·6	22·8	21·1	19·3	17·5	15·8	14·0	12·3	10·5	8·8
38	46·2	44·3	42·5	40·6	38·8	36·9	35·1	33·2	31·4	29·5	27·7	25·9	24·0	22·2	20·3	18·5	16·6	14·8	12·9	11·1	9·2
39	48·6	46·6	44·7	42·7	40·8	38·9	36·9	35·0	33·0	31·1	29·1	27·2	25·3	23·3	21·4	19·4	17·5	15·5	13·6	11·7	9·7

$$\text{Absolute Humidity (gm}^{-3}) = \frac{216.7e}{T + 273}, \text{ T in °C.}$$

Index